THE MARTIAN ANOMALIES

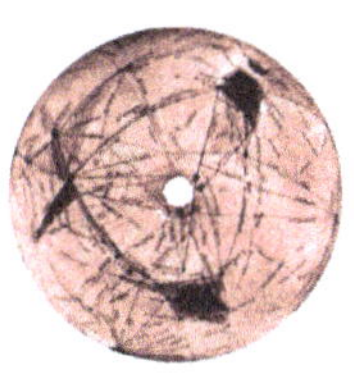

THE LOCHLAINN SEABROOK COLLECTION

AMERICAN CIVIL WAR
Abraham Lincoln Was a Liberal, Jefferson Davis Was a Conservative: The Missing Key to Understanding the American Civil War
Confederacy 101: Amazing Facts You Never Knew About America's Oldest Political Tradition
Confederate Blood and Treasure: An Interview With Lochlainn Seabrook
Everything You Were Taught About African-Americans and the Civil War is Wrong, Ask a Southerner!
Everything You Were Taught About the Civil War is Wrong, Ask a Southerner!
Give This Book to a Yankee! A Southern Guide to the Civil War For Northerners
Heroes of the Southern Confederacy: The Illustrated Book of Confederate Officials, Soldiers, and Civilians
Lincoln's War: The Real Cause, the Real Winner, the Real Loser
The Great Yankee Coverup: What the North Doesn't Want You to Know About Lincoln's War!
The Ultimate Civil War Quiz Book: How Much Do You Really Know About America's Most Misunderstood Conflict?
Women in Gray: A Tribute to the Ladies Who Supported the Southern Confederacy

CONFEDERATE MONUMENTS
Confederate Monuments: Why Every American Should Honor Confederate Soldiers and Their Memorials

CONFEDERATE FLAG
Confederate Flag Facts: What Every American Should Know About Dixie's Southern Cross
What the Confederate Flag Means to Me: Americans Speak Out in Defense of Southern Honor, Heritage, and History

SECESSION
All We Ask Is To Be Let Alone: The Southern Secession Fact Book

SLAVERY
Everything You Were Taught About American Slavery is Wrong, Ask a Southerner!
Slavery 101: Amazing Facts You Never Knew About America's "Peculiar Institution"

CHILDREN
Honest Jeff and Dishonest Abe: A Southern Children's Guide to the Civil War
Saddle, Sword, and Gun: A Biography of Nathan Bedford Forrest For Teens

NATHAN BEDFORD FORREST
A Rebel Born: A Defense of Nathan Bedford Forrest - Confederate General, American Legend (winner of the 2011 Jefferson Davis Historical Gold Medal)
A Rebel Born: The Screenplay (film about N. B. Forrest)
Forrest! 99 Reasons to Love Nathan Bedford Forrest
Give 'Em Hell Boys! The Complete Military Correspondence of Nathan Bedford Forrest
I Rode With Forrest! Confederate Soldiers Who Served With the World's Greatest Cavalry Leader
Nathan Bedford Forrest and African-Americans: Yankee Myth, Confederate Fact
Nathan Bedford Forrest and the Battle of Fort Pillow: Yankee Myth, Confederate Fact
Nathan Bedford Forrest and the Ku Klux Klan: Yankee Myth, Confederate Fact
Nathan Bedford Forrest: Southern Hero, American Patriot - Honoring a Confederate Icon and the Old South
Saddle, Sword, and Gun: A Biography of Nathan Bedford Forrest For Teens
The God of War: Nathan Bedford Forrest As He Was Seen By His Contemporaries
The Quotable Nathan Bedford Forrest: Selections From the Writings and Speeches of the Confederacy's Most Brilliant Cavalryman

QUOTABLE SERIES
The Alexander H. Stephens Reader: Excerpts From the Works of a Confederate Founding Father
The Quotable Alexander H. Stephens: Selections From the Writings and Speeches of the Confederacy's First Vice President
The Quotable Jefferson Davis: Selections From the Writings and Speeches of the Confederacy's First President
The Quotable Nathan Bedford Forrest: Selections From the Writings and Speeches of the Confederacy's Most Brilliant Cavalryman
The Quotable Robert E. Lee: Selections From the Writings and Speeches of the South's Most Beloved Civil War General
The Quotable Stonewall Jackson: Selections From the Writings and Speeches of the South's Most Famous General
The Unquotable Abraham Lincoln: The President's Quotes They Don't Want You To Know!

CIVIL WAR BATTLES
Encyclopedia of the Battle of Franklin - A Comprehensive Guide to the Conflict that Changed the Civil War
Nathan Bedford Forrest and the Battle of Fort Pillow: Yankee Myth, Confederate Fact
The Battle of Franklin: Recollections of Confederate and Union Soldiers
The Battle of Nashville: Recollections of Confederate and Union Soldiers
The Battle of Spring Hill: Recollections of Confederate and Union Soldiers

CONSTITUTIONAL HISTORY

America's Three Constitutions: Complete Texts of the Articles of Confederation, Constitution of the United States of America, and Constitution of the Confederate States of America

The Articles of Confederation Explained: A Clause-by-Clause Study of America's First Constitution

The Constitution of the Confederate States of America Explained: A Clause-by-Clause Study of the South's Magna Carta

VICTORIAN CONFEDERATE LITERATURE

Rise Up and Call Them Blessed: Victorian Tributes to the Confederate Soldier, 1861-1901

Support Your Local Confederate: Wit and Humor in the Southern Confederacy

The God of War: Nathan Bedford Forrest As He Was Seen By His Contemporaries

The Old Rebel: Robert E. Lee As He Was Seen By His Contemporaries

Victorian Confederate Poetry: The Southern Cause in Verse, 1861-1901

ABRAHAM LINCOLN

Abraham Lincoln: The Southern View - Demythologizing America's Sixteenth President

Lincolnology: The Real Abraham Lincoln Revealed in His Own Words - A Study of Lincoln's Suppressed, Misinterpreted, and Forgotten Writings and Speeches

Lincoln's War: The Real Cause, the Real Winner, the Real Loser

The Great Impersonator! 99 Reasons to Dislike Abraham Lincoln

The Unholy Crusade: Lincoln's Legacy of Destruction in the American South

The Unquotable Abraham Lincoln: The President's Quotes They Don't Want You To Know!

NATURAL HISTORY

North America's Amazing Mammals: An Encyclopedia for the Whole Family

The Concise Book of Owls: A Guide to Nature's Most Mysterious Birds

The Concise Book of Tigers: A Guide to Nature's Most Remarkable Cats

PARANORMAL

Carnton Plantation Ghost Stories: True Tales of the Unexplained from Tennessee's Most Haunted Civil War House!

UFOs and Aliens: The Complete Guidebook

FAMILY HISTORIES

The Blakeneys: An Etymological, Ethnological, and Genealogical Study - Uncovering the Mysterious Origins of the Blakeney Family and Name

The Caudills: An Etymological, Ethnological, and Genealogical Study - Exploring the Name and National Origins of a European-American Family

The McGavocks of Carnton Plantation: A Southern History - Celebrating One of Dixie's Most Noble Confederate Families and Their Tennessee Home

MIND, BODY, SPIRIT

Autobiography of a Non-Yogi: A Scientist's Journey From Hinduism to Christianity (Dr. Amitava Dasgupta, with Lochlainn Seabrook)

Britannia Rules: Goddess-Worship in Ancient Anglo-Celtic Society - An Academic Look at the United Kingdom's Matricentric Spiritual Past

Christ Is All and In All: Rediscovering Your Divine Nature and the Kingdom Within

Christmas Before Christianity: How the Birthday of the "Sun" Became the Birthday of the "Son"

Jesus and the Gospel of Q: Christ's Pre-Christian Teachings As Recorded in the New Testament

Jesus and the Law of Attraction: The Bible-Based Guide to Creating Perfect Health, Wealth, and Happiness Following Christ's Simple Formula

Seabrook's Bible Dictionary of Traditional and Mystical Christian Doctrines

Sea Raven Press Blank Page Journal: For Reflections, Notes, and Sketches

The Bible and the Law of Attraction: 99 Teachings of Jesus, the Apostles, and the Prophets

The Book of Kelle: An Introduction to Goddess-Worship and the Great Celtic Mother-Goddess Kelle, Original Blessed Lady of Ireland

The Goddess Dictionary of Words and Phrases: Introducing a New Core Vocabulary for the Women's Spirituality Movement

The Martian Anomalies: A Photographic Search for Intelligent Life on Mars

Vintage Southern Cookbook: Delicious Dishes From Dixie

WOMEN

Aphrodite's Trade: The Hidden History of Prostitution Unveiled

Princess Diana: Modern Day Moon-Goddess - A Psychoanalytical and Mythological Look at Diana Spencer's Life, Marriage, and Death (with Dr. Jane Goldberg)

Women in Gray: A Tribute to the Ladies Who Supported the Southern Confederacy

REPRINTS

A Short History of the Confederate States of America (author Jefferson Davis; editor Lochlainn Seabrook)

Prison Life of Jefferson Davis (author John J. Craven; editor Lochlainn Seabrook)

Life of Beethoven (author Ludwig Nohl; editor Lochlainn Seabrook)

The New Revelation (author Arthur Conan Doyle; editor Lochlainn Seabrook)

Lochlainn Seabrook does not author books for fame and fortune, but for the love of writing and sharing his knowledge.

SeaRavenPress.com

Warning:
SEA RAVEN PRESS
BOOKS WILL EXPAND
YOUR MIND!

The MARTIAN ANOMALIES

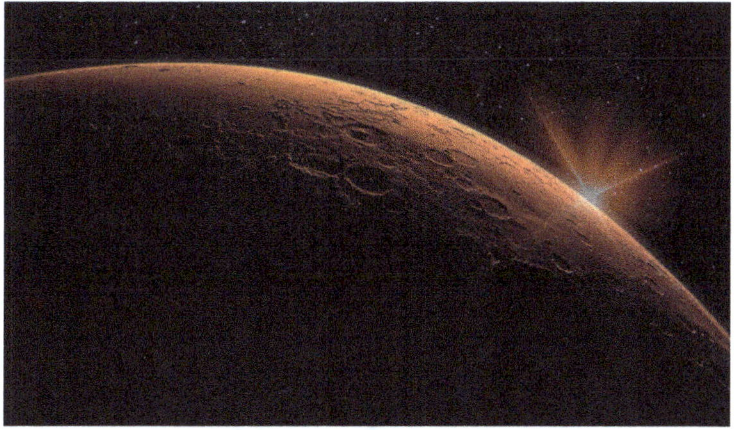

A Photographic Search for Intelligent Life on Mars

LOCHLAINN SEABROOK

JEFFERSON DAVIS HISTORICAL GOLD MEDAL WINNER

Diligently Researched and Generously Illustrated by the Author for the Elucidation of the Reader

2022

Sea Raven Press, Nashville, Tennessee, USA

THE MARTIAN ANOMALIES

Published by
Sea Raven Press, Cassidy Ravensdale, President
Nashville, Tennessee, USA
SeaRavenPress.com • searavenpress@gmail.com

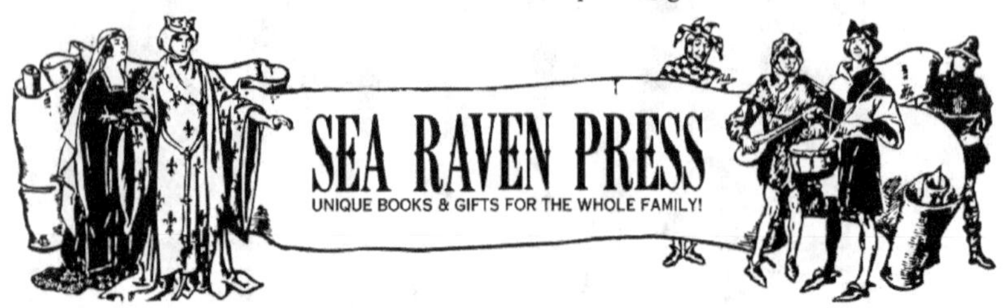

PRINTING HISTORY
1st SRP paperback edition, 1st printing, February 2022 • ISBN: 978-1-955351-14-0
1st SRP hardcover edition, 1st printing, February 2022 • ISBN: 978-1-955351-15-7

ISBN: 978-1-955351-14-0 (paperback)
Library of Congress Control Number: 2022932063

The Martian Anomalies: A Photographic Search for Intelligent Life on Mars, by Lochlainn Seabrook. Includes an introduction, illustrations, notes, appendices, and a bibliography.

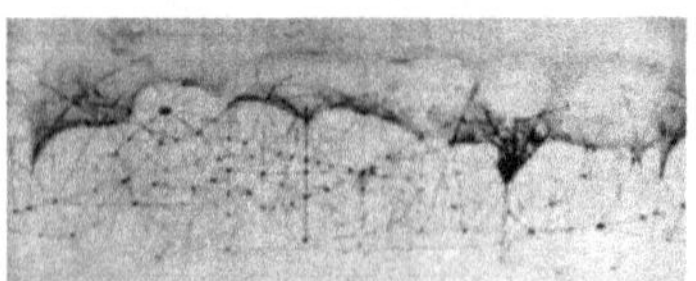

ARTWORK
Front and back cover design and art, book design, layout, font selection, and interior art by Lochlainn Seabrook
All images, image captions, graphic design, and graphic art copyright © Lochlainn Seabrook
All images selected, placed, manipulated, cleaned, colored, tinted, and/or created by Lochlainn Seabrook
Cover image: "Sunrise Over Mars," vectorized NASA photo art

All persons who approve of the authority and principles of Lochlainn Seabrook's literary work, and realize its benefits as a means of reeducating the world about vital facts left out of our science and history books, are hereby requested to avidly recommend his books to others and to vigorously cooperate in extending their reach, scope, and influence around the globe.

The astronomical, archaeological, and anthropological views documented in this book are those of the publisher.

PROUDLY WRITTEN, PUBLISHED IN NASHVILLE, TENNESSEE, UNITED STATES OF AMERICA

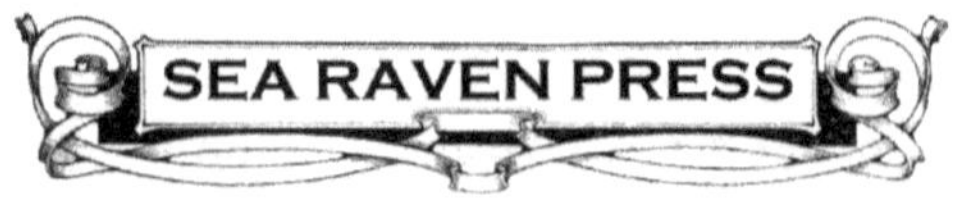

Dedication

To the unbiased, imaginative, tolerant, creative, and receptive individuals who explore our Universe with an open and inquisitive mind. You are true scientists.

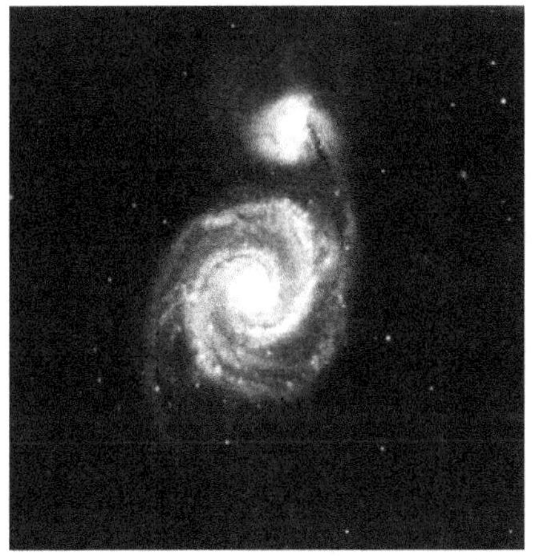

Epigraph

"There is every reason to believe that Mars and other planets are inhabited. Why should the Earth be the only planet supporting human life? It is not singular in any other respect. But if intelligent creatures do exist, as we may assume they do elsewhere in the universe, I should not expect them to communicate with the Earth by wireless [radio]. Light rays, the direction of which can be controlled much more easily, would more probably be the first method attempted."

Albert Einstein

JANUARY 31, 1920

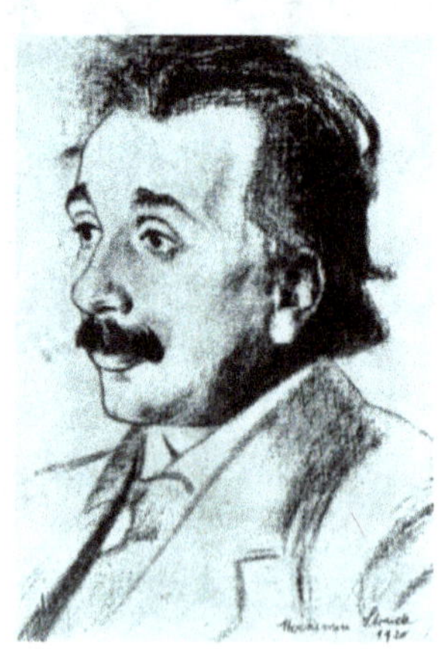

CONTENTS

SECTION ONE
Images From the Surface of Mars

SECTION TWO

Images From Above the Surface of Mars

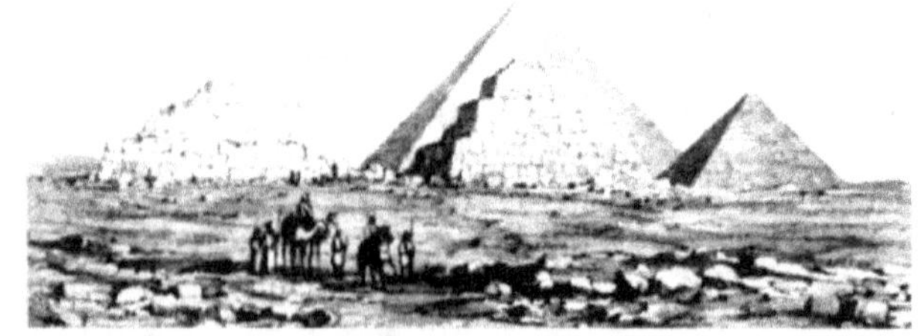

NOTES TO THE READER

☛ This book is not meant to be an all-inclusive look at NASA's massive library of Mars photographs. At the time of this writing (Thanksgiving Day 2021) NASA's Mars Exploration Program contains a collection of 444,265 raw images taken by the Curiosity rover alone (this number grows hourly). Hundreds of thousands of additional photos have been shot by other rovers, as well as other spacecraft, before and since, totaling in the many millions. *The Martian Anomalies* is only intended to serve as an introduction to this vast and complex subject.

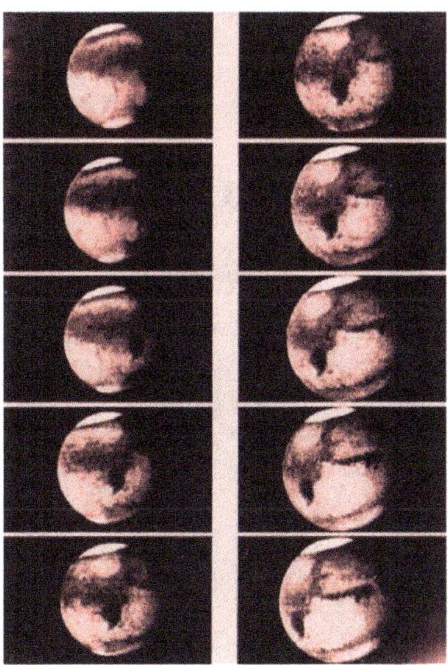

Mars as seen from Earth through a telescope shows dark markings and polar caps, all which vary with the changing Martian seasons. Image credit: NASA.

☛ To eliminate possible confusion with the creators of the images used in this book (i.e., NASA and its affiliates), I am the "author" referred to throughout the text.

☛ I have eschewed the fascinating issue of Martian measurements, considering it too convoluted and problematic for a work of this nature. Having said this, please bear in mind that while much of what can be seen in the enclosed photos appears to be quite grand in scale, most of the cliffs and "mountains" shown in Section One, for instance, are typically no more than 10-20 feet in height, while most of the "boulders" and other seemingly "large" objects seen are seldom more than a few feet or even a few inches in width. In other words, in most cases the manner in which the photos were taken and the lack of any reference points (such as well-known features like people and cars), make the objects in these particular Martian photos appear to be much larger than they actually are.

☛ In keeping with NASA's detailed media usage guidelines, credit is gratefully given to the many NASA-created photographs and illustrations that I used in this book. My thanks also to Bert Ulrich (at NASA's Office of Communications) for his time.

☛ With a few minor edited exceptions, the technical oriented text in captions accompanying photos marked "original image" is primarily by NASA—and is always set in quote marks to identify them as such. (I refer to these specific images as "parent" photos or "full frame" images.)

☛ Although it can be, this book was not necessarily designed to be read from front to back. You may open it anywhere and begin your journey.

☛ For best results in using this book, I not only recommend using a magnifying instrument of some kind, but also reading with an open but critical mind. If one wants to get at the inner meaning of sacred scripture, for example, one must learn to ignore orthodox literal interpretation and read below the surface, using both common sense and intuition (2 Corinthians 3:6). The same holds true with the study of anomalous phenomena.

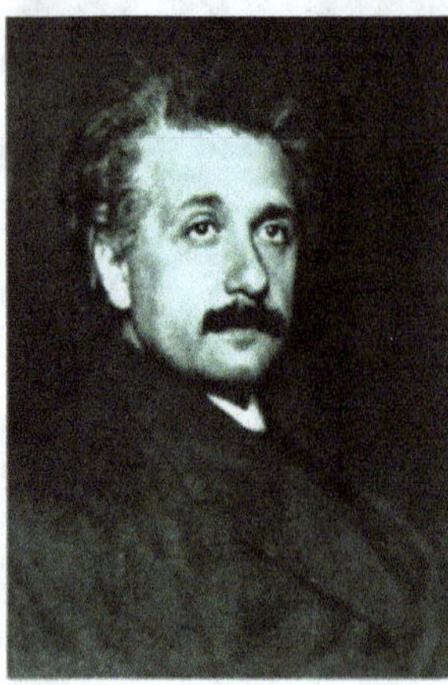

Dr. Albert Einstein believed that creative vision is more important than book learning, a scientific approach with which the author wholeheartedly concurs.

The eye must be conditioned to see beyond what we have been told is true and what we have been told is false. Look through and past the obvious (dust, rock, weathering, etc.) and into the subtle dimensions of the enclosed images. Think of the immense span of time that has passed since these probable ancient artifacts were first created and abandoned. We are talking on the order of hundreds of thousands if not millions of years.

While we must always be cautious of pareidolia (the human tendency to perceive a specific, often meaningful image in a random or ambiguous visual pattern), open-minded observational practice, combined with insight and imagination, will bring about a heightened level of awareness, and with it, scientific enlightenment.

Yes, establishment science ridicules such ideas. Yet, like many other scientists, Einstein embraced them, saying:

> "I believe in intuitions and inspirations. I sometimes feel that I am right. I do not know that I am . . . [but] I would have been surprised if I had been wrong. I am enough of the artist to draw freely upon my imagination. Imagination is more important than knowledge. Knowledge is limited. Imagination encircles the world."[1]

In dealing with the subject of *The Martian Anomalies*, I side with the great German American physicist.

L.S.

---

1. From a 1929 interview with the *Post*.

HOW TO USE THIS BOOK

TO AID THE READER IN plunging into this often labyrinthian and always controversial topic, I have divided my collection of NASA Mars photographs into two segments. Section One: images from the surface of Mars; and Section Two: images from above the surface of Mars. Each section in turn is comprised of individual chapters, with, generally speaking, one parent photo (or sometimes region) per chapter.

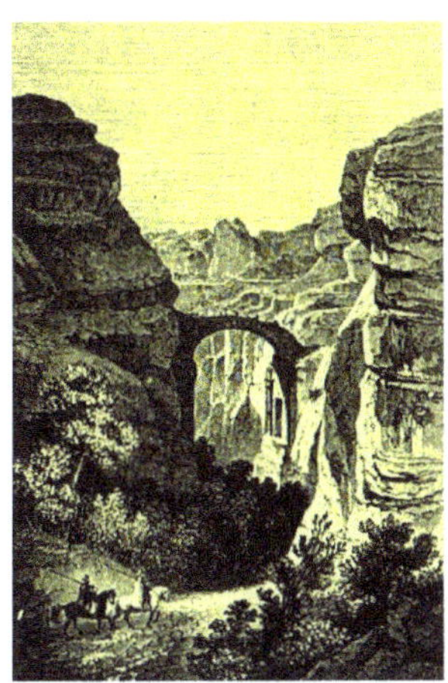

Our planet hosts the ruins of many lost and obscure ancient civilizations. Mars may as well—if only we are willing to look. For instance, recently an unknown civilization was identified in the mountainous Sierra Nevada de Santa Marta region of Colombia. Created by a mysterious people known as the Tairona, their most famous city, La Ciudad Perdida, housed as many as 10,000 people—yet it was not discovered until the 1970s. Lost civilizations like this one continue to be found on Earth. What might Mars be hiding beneath its red dust?

Each chapter begins with the original, full frame, untouched, unprocessed parent NASA image that is the focus of that particular chapter. Beneath this image I provide pertinent details about the image, usually including its name or title (all of which I have created, since NASA seems to have no ordered photographic cataloging or numbering system), the name of the Mars craft or vehicle that took the image, which of its onboard cameras was used, the Earth date the photo was taken, the Mars date the photo was taken (e.g., "Sol 729"), and the time it was taken according to UTC. This specific caption ends with the image credit.

Next is my analysis image of that chapter's parent photo, which I have brightened (or darkened), tinted, colored, sharpened, and sometimes cropped, to enhance viewability. This particular image contains numbered (and at times unnumbered) circles and ovals around the areas and objects I wish to highlight as possible or likely evidence of a past (or current) advanced race on Mars. This section also contains my comments on the parent image.

Following these initial two photos are the individual extracted images that I have cut from the original, large, parent photograph highlighted in that chapter, along with my observations on what I believe may be visible in these smaller photos.

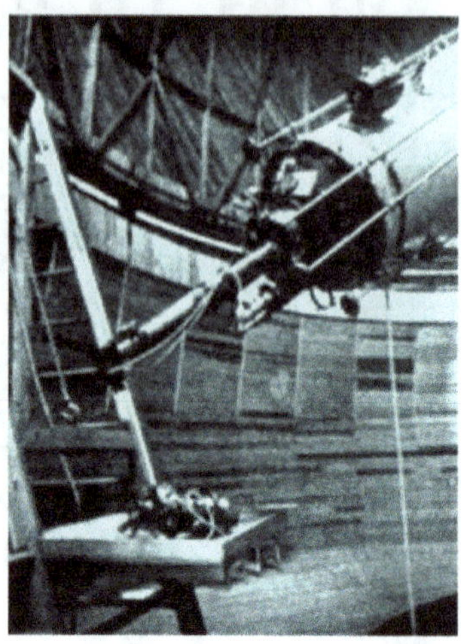

Photographic apparatus at the Lowell Observatory, Flagstaff, Arizona, circa early 1900s. Percival Lowell used it to photograph what he called "canals on Mars."

Despite the expensive state-of-the-art cameras NASA mounts on its Mars exploratory vehicles, a number of the photos that have come down to us (the public) are anything but high resolution. In fact, some are low to medium resolution, with the result that many of the objects in them are impossible to delineate let alone identify. In these cases I have been forced to shrink the photo before inserting it into my book to prevent it from pixelating and blurring beyond recognition. Still, the pixelation of thousands of NASA's Mars photos is a serious issue, one for which there is no solution at this time.

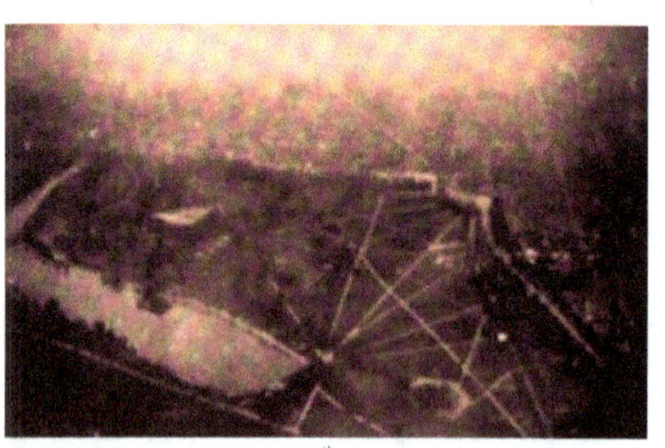

Around the turn of the 20th Century, Lowell took this photo of Hyde Park and the Serpentine in London, UK, via a "free balloon" from a height of 2,200 feet. It shows "the artificial markings of Earth [as] seen from space," which the forward thinking astronomer maintained were comparable to the "artificial markings" on Mars.

For these reasons, and others, I recommend using a magnifying glass while perusing these pages. This will prove beneficial with the higher resolution images as well, as some contain an enormous amount of detail and information in a small space that can be easily missed with the naked eye.

ABBREVIATIONS
Used in This Book

- **AOIMSG**: Airborne Object Identification and Management Synchronization Group.
- **AU**: Astronomical unit.
- **ASU**: Arizona State University.
- **CALTECH or CIT**: California Institute of Technology.
- **CIA**: Central Intelligence Agency.
- **DIA**: Defense Intelligence Agency.
- **DNI**: Director of National Intelligence.
- **DOD**: Department of Defense.
- **DSN**: Deep Space Network.
- **JPL**: Jet Propulsion Laboratory.
- **LTP**: Lunar Transient Phenomenon.
- **MSSS**: Malin Space Science Systems.
- **NASA**: National Aeronautics and Space Administration.
- **NSA**: National Security Agency.
- **SRI**: Stanford Research Institute.
- **SUA**: Special Use Airspace.
- **TLP**: Transient Lunar Phenomenon.
- **UAP**: Unidentified aerial phenomenon/phenomena (i.e., UFOs).
- **UArizona**: University of Arizona.
- **USGS**: United States Geological Survey.
- **UTC**: An abbreviation for *Coordinated Universal Time*, defined by NASA as: "The world-wide scientific standard of timekeeping. It is based upon carefully maintained atomic clocks and is highly stable. Its rate does not change by more than about 100 picoseconds per day. The addition or subtraction of leap seconds, as necessary, at two opportunities every year adjusts UTC for irregularities in Earth's rotation. UTC is used by astronomers, navigators, the Deep Space Network (DSN), and other scientific disciplines. Its reference point is Greenwich, England: when it is midnight there on Earth's prime meridian, it is midnight (00:00:00.000000)."[2]

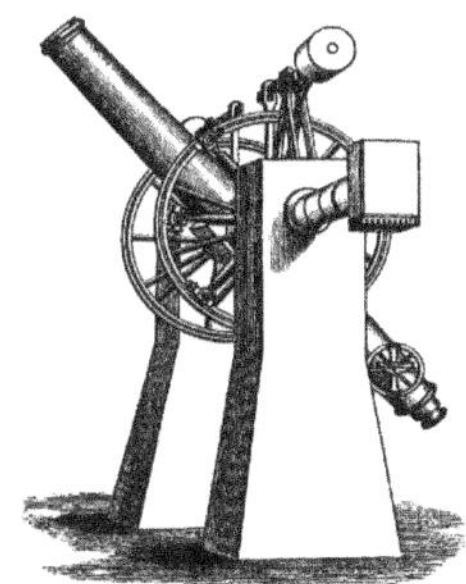

A telescope from 1871.

Note also that while UTC and GMT (Greenwich Mean Time) indicate the same time, the former is a time standard while the latter is a time zone. Other important differences between GMT and UTC: GMT observes Daylight Saving Time while UTC does not, and GMT uses pendulum clocks while (as noted in NASA's explanation above) UTC uses atomic clocks, which provide a more scientifically accurate measurement of time. Nonetheless, as global timekeeping systems, both track the same time and are used for essentially the same purpose: universal time coordination.[3]

---

2. Website: https://solarsystem.nasa.gov/basics/chapter2-3/
3. For more about time on Mars, see Website: www.giss.nasa.gov/tools/mars24/help/notes.html

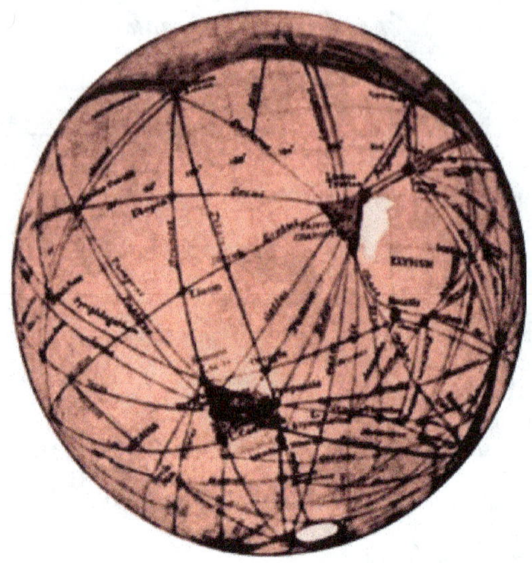

Above: A 1910 drawing of Mars based on observations using low resolution telescopes. Contrary to conventional scientific wisdom, the photos in this book reveal a planet that appears to be far from the "cold lifeless cinder" that many modern establishment scientists claim it to be.

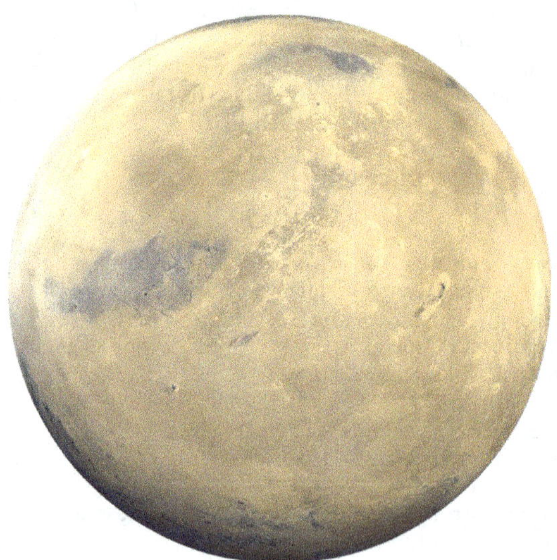

Above: A high resolution photo of Mars' Cerberus hemisphere, taken by the Viking 1 Orbiter on February 11, 1980, from 1,245 miles overhead. A number of Martian anomalies—all suggesting the possibility of advanced ancient life—are so obvious they can be seen even from this great height. Image credit: NASA.

INTRODUCTION

by Lochlainn Seabrook

"The real conspiracy theory may turn out to be the conventional idea that Mars has never been home to large, complex, intelligent life forms." — LOCHLAINN SEABROOK

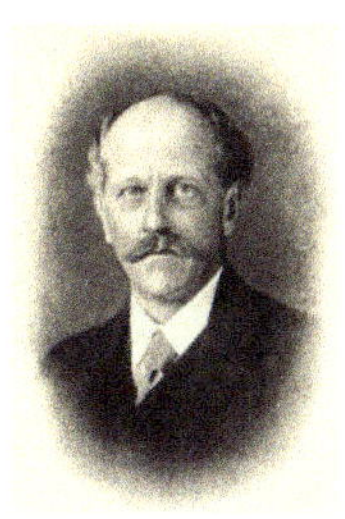
Percival Lowell.

IN THE LATE 1800s AMERICAN astronomer Percival Lowell attributed the "canals" on Mars (first "discovered" by Italian astronomer Giovanni Schiaparelli in 1877) to intelligent life. Though upon closer inspection the "canals" later turned out to be nonexistent, Lowell's hypothesis that advanced beings once thrived on the Red Planet continues enthusiastically and unabated into the present day. Why? Is it justified? I believe that this book may help provide the answer.

One of the chief differences between Victorian Era astronomy and 21st-Century astronomy is that today we have high resolution cameras in both the skies above Mars and on its surface. The latter are mounted to mobile NASA robots, sophisticated exploratory vehicles known as rovers, of which, so far, five have been sent to Mars: Sojourner (landed on Mars July 1997), Spirit (landed on Mars January 2004), Opportunity (landed on Mars January 2004), Curiosity (landed on Mars August 2012), and Perseverance (landed on Mars February 2021).

The express mission of the more recent Mars rovers has been to "search for signs of ancient water," as well as past conditions "that could have been suitable for sustaining microbial life." Revealingly, the twin rovers Spirit and Opportunity found bedrock showing "extensive altercation by water,"[4] while the Mars Reconnaissance Orbiter recently found salt minerals left behind by surface water on the Red Planet as recently as 2 billion years ago.[5]

---

4. *NASA Facts: Mars Exploration Rover*. Website: https://d2pn8kiwq2w21t.cloudfront.net/documents/mars03rovers.pdf

5. "NASA's MRO Finds Water Flowed on Mars Longer Than Previously Thought": https://mars.nasa.gov/news/9119/nasas-mro-finds-water-flowed-on-mars-longer-than-previously-thought/?site=msl

But their stereoscopic cameras caught something else, something quite unexpected—to mainstream scientists, at least. Peering mysteriously out of the red dust in numerous Mars photos appear to be what many consider the archaeological, anthropological, and perhaps fossilological remains of a remote but advanced society, a human or human-like civilization that once occupied the Red Planet. In a word, Martians. If such a race did truly exist in the distant past, I for one am intrigued, for as a historian I am deeply interested in the rise and fall of any and all intelligent, pre 21st-Century peoples, whether they happened to inhabit our planet or one far outside our solar system.

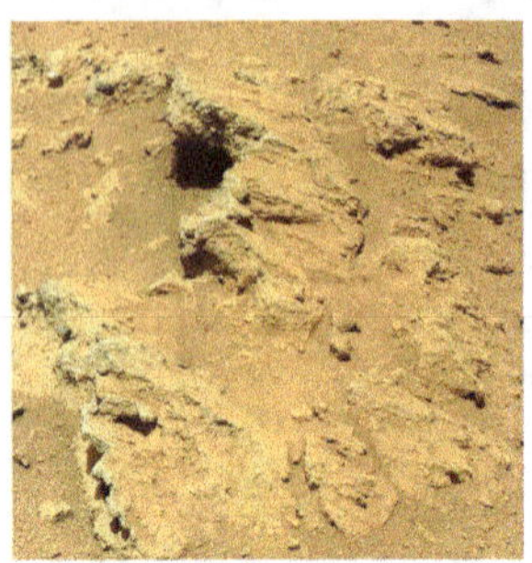

Evidence of an ancient stream on Mars, taken by Curiosity on Sol 39 (September 14, 2012). Note rounded rocks, gravel, rock outcrop, and sedimentary conglomerate. Image credit: NASA/JPL-Caltech/MSSS.

If a society of technological beings did live on Mars for a time, who were they and, more importantly, what happened to them?

There are many hypotheses to choose from, but one of the more realistic theories goes something like this.

Mars was once a habitable world with a livable atmosphere, surface water, and thriving flora and fauna. (Even many of today's most hardheaded scientists believe that Mars was once fairly warm and wet, with volcanoes and flowing near neutral-pH water, as well as a thicker carbon-dioxide atmosphere than it has now, one capable of sustaining lakes and salty seas—and by inference, life.)[6] As on Earth, the inhabitants ranged from microbial to large complex life forms, the latter perhaps actual human beings, but more probably a humanoid-like creature; possibly a thin, gray, large-headed extraterrestrial, similar to those so often reported by alien abductees and made popular in such films as *Fire in the Sky* and *Close Encounters of the Third Kind*.

Whoever these original Martians were, over time they developed a highly specialized civilization on the Red Planet,

6. For more, see Website: https://mars.nasa.gov/mer/mission/overview/

complete with cities, suburbs, buildings, houses, streets, parks, sidewalks, monuments, and factories, along with all of the infrastructure needed to run an advanced technological society.

Then, at some point, perhaps between 200,000 and 1 million years ago, a great shift of some kind occurred. Perhaps it was due to drastic climate change or a calamitous astronomical event; maybe it was a massive rebellion in the wake of an overly tyrannical big government; or maybe it was simply overpopulation and the accompanying problems of increased pollution, growing crime, and dwindling land, food supplies, and water resources. While we do not know the cause, we have clues that the aftermath seems to have led to the eruption of a Mars-wide conflict, possibly a nuclear war. I will call this presumed struggle "*The Great Martian Apocalypse,*" for it would have decimated entire towns and cities, leaving atomic bomb-like carnage and wreckage in its wake.

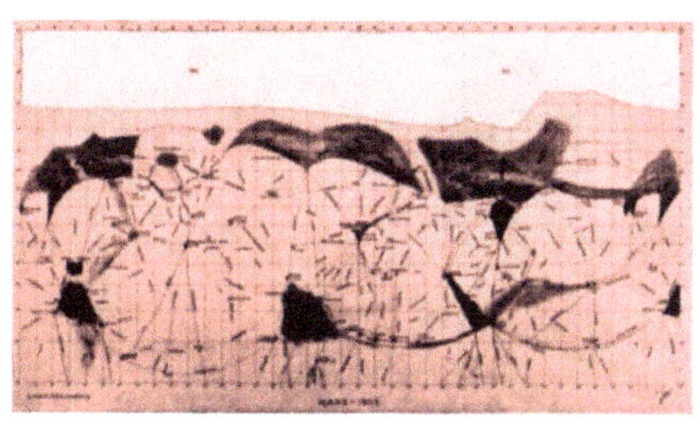

Lowell's 1905 map of Mars.

Whatever its nature, this cataclysmic occurrence seems to have also poisoned or even completely destroyed the planet's air, soil, and water, resulting in the near annihilation of all Martian life. With its protective magnetic field now gone, solar wind and radiation began to dominate, leading to the thinning of its once protective atmosphere. The ground became dry, parched, and desert-like, seas, lakes, and rivers evaporated (though some water probably became locked away underground), and the weather turned bitterly cold, with an average temperature of -80°F (-62°C).[7]

Despite this epoch catastrophe, a few individual Martians

7. Note that equatorial temperatures on Mars can attain 70°F. Overall, "temperatures on the surface generally reach 32°F (0°C) in early afternoon with some areas reaching 80°F (27°C). At night the temperature plunges to a frigid -190°F (-123°C)." See National Aeronautics and Space Administration, p. 10.

survived the now lethal climate and treacherous dust storms by hiding inside special shelters. From here they quickly deployed reconnaissance ships in search of a safe place to escape to. The probes returned with the news that the only other nearby livable planet was Earth. And so the remaining population boarded its dilapidated spacecraft and fled to Gaia, the Blue Planet.

At the time, 200,000 to 1 million years ago, Earth was already inhabited by various now extinct hominids—such as *Homo erectus*, Neanderthal, and the Denisovans. Over time, as the theory goes, the Martian survivors blended in and mixed with these similar but distinct groups, eventually picking up their DNA along the way, but ultimately outliving them, becoming the ancestors of we ourselves: *Homo sapiens*. In connection with this, it is interesting to note that our species arose quite suddenly, and some believe quite inexplicably, in the fossil record some 200,000 years ago.

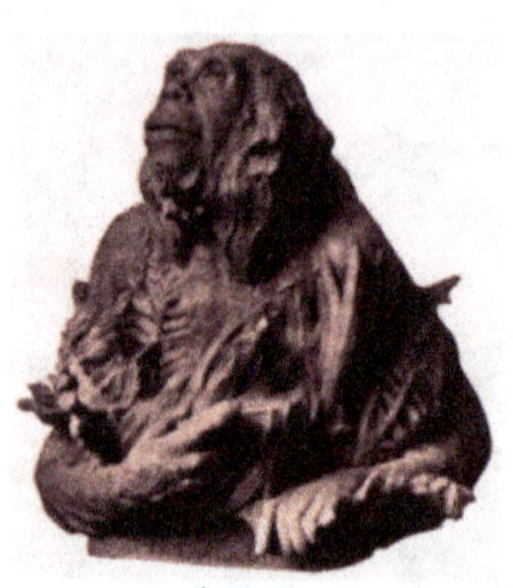

An early 20th-Century model of Pithecanthropus or Java Man, today classified as *Homo erectus*.

If this story has any truth to it, then all modern humans would be descendants of Martians.[8] Either that, or ancient humans here on Earth were far more advanced than mainstream science believes (or will admit), and a group traveled to Mars and colonized it millennia ago, returning to our planet following The Great Martian Apocalypse.

8. If modern humans evolved on Earth rather than on Mars (as is the current conventional view), then, according to the latest objective (as opposed to politically correct) evidence, our hominid ancestors arose in Europe, *not* Africa. See, for example, the following Websites:
https://bigthink.com/life/evolution-europe/
https://www.newscientist.com/article/mg24232263-300-did-the-ancestor-of-all-humans-evolve-in-europe-not-africa/
https://www.utoronto.ca/news/human-ancestors-originated-europe-not-africa-u-t-part-international-team-studying-pre-human
https://www.telegraph.co.uk/science/2017/05/22/europe-birthplace-mankind-not-africa-scientists-find/
https://www.scientificamerican.com/article/is-the-out-of-africa-theory-out/#:~:text=All%20the%20ancestors%20of%20contemporary,early%20settlers%20hailed%20from%20Asia
https://sitn.hms.harvard.edu/flash/2018/oldest-human-fossil-found-outside-africa-sheds-light-behavior-early-homo-sapiens/
https://www.natureasia.com/en/nmiddleeast/article/10.1038/nmiddleeast.2018.15

Does all of this sound far-fetched? Some may think it reads like bad science fiction. But at one time the U.S. government did not think so. In fact, it invested considerable time, energy, and taxpayer dollars in an effort to further explore the idea.

As just one example: In 1978 the Defense Intelligence Agency (DIA) and the Stanford Research Institute (SRI) founded a program later known as the "Stargate Project," whose goal was to explore the possibility of using mental telepathy or extrasensory perception (ESP) to secretly spy on hostile countries and collect vital intelligence that would otherwise be difficult or impossible to acquire.

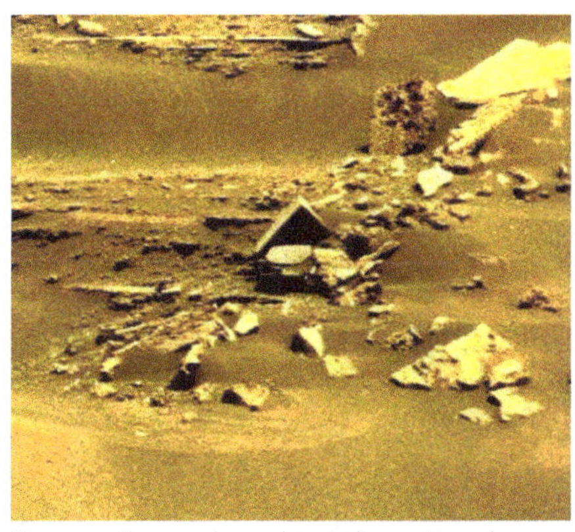

Dozens of anomalous objects lie scattered over a Martian hillside. All are explained away by conventional scientists as "rocks." NASA's Mars rover Curiosity acquired this image using its Mast Camera on Sol 1448 (2016-09-01, 21:47:36 UTC). Image credit: NASA/JPL-Caltech/MSSS.

One of the groups' early experiments, one that specifically concerns the topic of this book, is described in a declassified CIA document entitled "Mars Exploration." In it we learn that on May 22, 1984, the Stargate Project brought in a psychic (known in the paper as "subject") for a remote viewing session. Remote reviewing is defined as the psychic ability to perceive and gather information and imagery from and about remote geographical targets.

The psychic was not given any information other than a sealed envelope containing a 3 X 5 card with the following information:

- The planet Mars.
- Time of interest approximately 1 million years B.C.

The psychic was allowed to hold the envelope during the "interview," but not open it.

According to the transcript, at the start of the experiment (shortly after 10:00 AM) the psychic was told by the CIA operative, the interviewer (a scientist known as a "monitor"), to focus on the contents of the envelope as well as a set of "geographic coordinates." These were verbally given to him as:

- 40.89 degrees north
- 9.55 degrees west

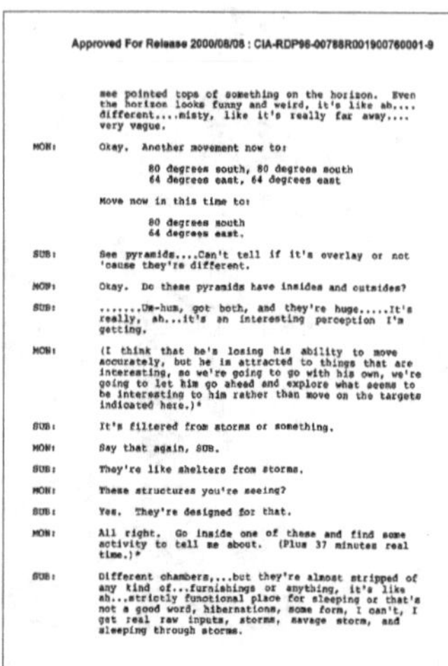

Approved For Release 2000/08/08 : CIA-RDP96-00788R001900760001-9

see pointed tops of something on the horizon. Even the horizon looks funny and weird, it's like ah.... different....misty, like it's really far away.... very vague.

MON: Okay. Another movement now to:

80 degrees south, 80 degrees south
64 degrees east, 64 degrees east

Move now in this time to:

80 degrees south
64 degrees east.

SUB: See pyramids....Can't tell if it's overlay or not 'cause they're different.

MON: Okay. Do these pyramids have insides and outsides?

SUB: ....Um-hum, got both, and they're huge.....It's really, ah...it's an interesting perception I'm getting.

MON: (I think that he's losing his ability to move accurately, but he is attracted to things that are interesting, so we're going to go with his own, we're going to let him go ahead and explore what seems to be interesting to him rather than move on the targets indicated here.)*

SUB: It's filtered from storms or something.

MON: Say that again, SUB.

SUB: They're like shelters from storms.

MON: These structures you're seeing?

SUB: Yes. They're designed for that.

MON: All right. Go inside one of these and find some activity to tell me about. (Plus 37 minutes real time.)*

SUB: Different chambers,...but they're almost stripped of any kind of...furnishings or anything, it's like ah...strictly functional place for sleeping or that's not a good word, hibernations, some form, I can't, I get real raw inputs, storms, savage storm, and sleeping through storms.

5

Page 5 from the declassified U.S. government document called "Mars Exploration," featuring an explosive 1984 ESP session with a remote viewer. The documentary film *Third Eye Spies*, directed by Lance Mungia, reveals that the CIA, the NSA, and the DIA, have all utilized remote viewing and that "the evidence for ESP is overwhelming."

Not only did the psychic have no idea what was in the envelope, he also could not have known that these coordinates were on the surface of Mars; nor could he have known that they pinpointed the exact location of the Cydonia region on Mars: the area containing the famous "Face on Mars," numerous pyramid like structures, and what appear to be the ruins of a large ancient city.

Starting off the interview, the CIA operative asks the psychic what he sees (telepathically). He answers that he can see a "a yellowish or okra colored pyramid or pyramid form . . . it's very high, it's kind of sitting in a large depressed area." When asked what he sees at the time "indicated in the envelope" (i.e., 1 million B.C.), he replies: "I'm tracking severe clouds, more like a dust storm, ah . . . it's a geologic problem . . . I'm looking at [the] after effect of a major geologic problem."

The CIA then asks the psychic to "go back to the time before the geologic problem." The psychic says: ". . . um, total difference, it's ah . . . [I] see ah . . . large flat surfaces, very ah . . . smooth . . . angles, walls, they're really large though, I mean they're megalithic . . ."

The CIA interviewer then directs the psychic to "look around . . . before the geologic activity . . . in and around this area [i.e., the Cydonia region on Mars] and see if you can find any activity." The psychic replies: "I'm seeing ah . . . It's like a perception of a shadow of people, very tall . . . thin, it's only a shadow. It's as if they were there and they're not . . . not there anymore."

The CIA operative then asks the psychic to "go back to a period of time when they are there," but "don't try to put things together, just report the raw data." Psychic: "I just keep seeing very large people. They appear thin and tall, but they're very large. Ah . . . wearing some kind of strange clothes."

Next the CIA scientist asks the psychic to stay in the same time period, but "move from your physical location in space to another physical location," namely:

- 46.45 degrees north
- 353.22 degrees east

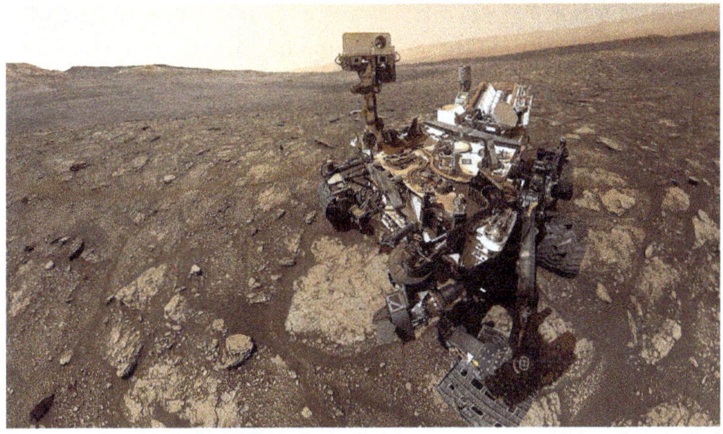

On November 12, 2020, NASA's Curiosity Mars rover snapped this self portrait at a location nicknamed "Mary Anning" after a 19th century English paleontologist. What are we *not* seeing in this photo? Image credit: NASA/JPL-Caltech/MSSS.

Unbeknownst to the psychic, these coordinates point to an area on Mars with deep craters and two, long, nearly perfectly straight lines running to the Esk crater (45.6 degrees north, 7.1 degrees west).

The psychic states: "[I'm] deep inside of a cavern, not a cavern, more like a canyon. Um, I'm looking up, up the sides of a steep wall that seem to go on forever. And there's like ah . . . a structure with a . . . it's like the wall of the canyon itself has been carved. Again I'm getting very large structures, no . . . ah . . . no intricacies, [just] huge sections of smooth stone."

CIA scientist: "Do the structures have insides and outsides?"

Psychic: "Yes, they're very, it's like a rabbit warren, corners of rooms, they're really huge . . . [my] perception is that the ceiling is very high, walls very wide."

The CIA scientist then asks the psychic to "move from this point in time" to another location nearby, at these coordinates:

- 45.86 degrees north
- 354.1 degrees east

Psychic: "They have a ah . . . appears to be the end of a very large road and there's a . . . marker thing that's very large, keep getting [a] Washington Monument overlay, it's like an obelisk."

CIA scientist: ". . . From this point then, let us move to another point"—in this case to the following coordinates:

- 35.26 degrees north
- 213.24 degrees east

Psychic: ". . . It's like I'm in the middle of a . . . huge circular basin . . . of the range mountains by [sic] almost all the way around . . . very ragged mountains, very tall. Basin's very, very, very large . . . [I] see a right angle corner to something but that's all, I don't see anything else."

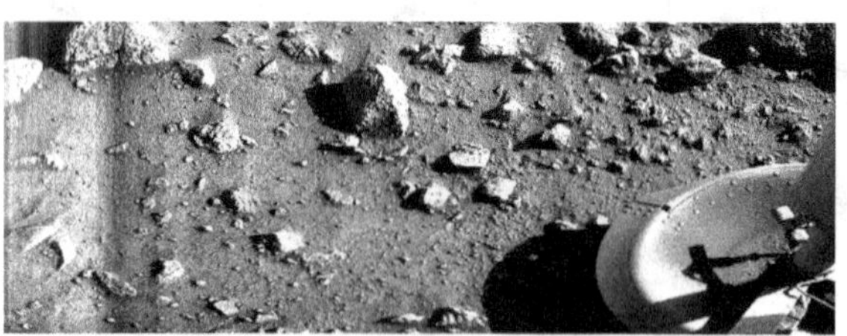

According to NASA this is "the first photograph ever taken on the surface of the planet Mars." As we will see, however, many of their own photos cast doubt upon this statement. This image was snapped by Viking 1 only minutes after the spacecraft landed early on July 20, 1976. Image credit: NASA/JPL.

CIA scientist: "Okay. Then let's move into a little different place, very close." The coordinates given were:

- 34.6 degrees north
- 213.09 degrees east

Psychic: "The cluster of squares up and down. . . . They're almost flush with the ground and it's like they're connected. . . something very white or reflects light."

CIA scientist: "Now move over to [the following coordinates]:"

- 34.57 degrees north
- 212.22 degrees east

Psychic: "It's like I can just perceive ah . . . ah . . . like a radiating pattern of some kind. It's like some really . . . ah . . . strange intersecting kind of roads that are dug into valleys . . . [Their shapes are] like real neat channels cut, they're very deep . . ."

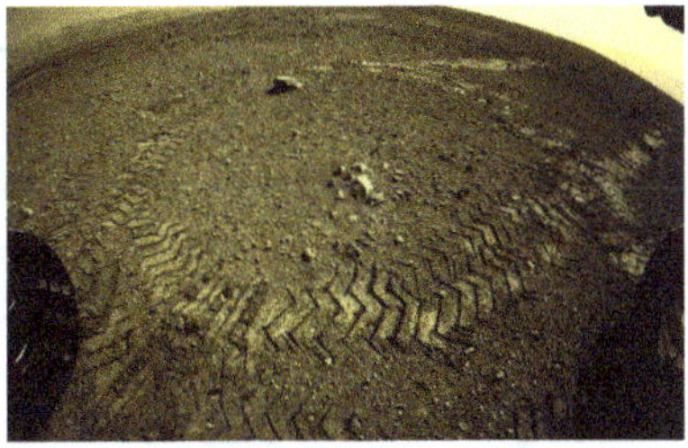

This image shows the tracks left by NASA's Curiosity rover on Aug. 22, 2012, as it completed its first test drive on Mars. Image credit: NASA/JPL-Caltech.

CIA scientist: ". . . it's very important that you maintain your focus. I have a movement exercise again for you and this is some considerable distance away, so holding the focus in time, remember the focus in time that you had before and moving now to [this location]":

- 15 degrees north
- 198 degrees east

Psychic: "[I] see the . . . um . . . intersecting . . . ah . . . whatever these are, are aqueduct type things . . . these . . . rounded bottom carved channels, like road beds. [I] see pointed tops of something on the horizon. . ."

CIA scientist: "Okay. Another movement now to [this spot]:"

- 80 degrees south
- 64 degrees east

Psychic: "[I] see pyramids . . . can't tell if it's overlay or not 'cause they're different."

CIA scientist: "Okay. Do these pyramids have insides and outsides?"

Psychic: ". . . um-hum, [they] got both, and they're huge . . ."

At this point the psychic seems to be "losing his ability to move accurately." Without the psychic being able to hear them, the CIA research team decides to let the psychic "go ahead and explore what seems to be interesting to him rather than move on the targets indicated here."

The psychic continues talking about the giant pyramids at the current coordinates: "[They're] filtered [hidden?] from storms or something . . . they're like shelters from storms."

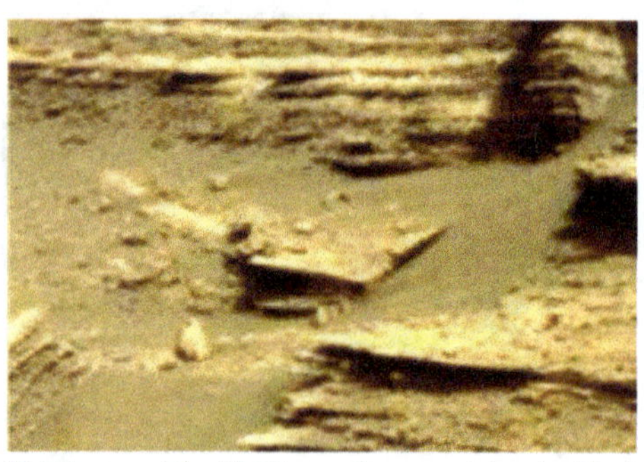

This image of the Martian surface was taken by Mast Camera (Mastcam) onboard NASA's Mars rover Curiosity on Sol 1448 (2016-09-01, 21:48:48 UTC). Though there are numerous objects with blatantly artificial characteristics at this location, mainstream science, harnessed by its blind adherence to the non-extraterrestrial theory, chooses to ignore them. Image credit: NASA/JPL-Caltech/MSSS.

CIA scientist: "These structures you're seeing?"

Psychic: "Yes. They're designed for that."

CIA scientist: "All right. Go inside one of these and find some activity to tell me about."

Psychic: "Different chambers . . . but they're almost stripped of any kind of . . . furnishings or anything, it's like ah . . . [a] strictly functional place for sleeping or that's not a good word, hibernations, some form, I can't, I can't get real raw inputs, storms, savage storms, and sleeping through storms."

CIA scientist: "Tell me about the ones who sleep through the storms."

Psychic: ". . . Ah . . . very . . . tall again, very large . . . people, but they're thin, they look thin because of their height and they dress like in, oh hell, it's like a real light silk, but it's not flowing type clothing, it's like cut to fit."

CIA scientist: "Move close to one of them and ask them to tell you about themselves."

Psychic: "They're ancient people. They're ah . . . they're dying, it's past their time or age. . . . They're very philosophic about it. They're looking for ah . . . a way to survive and they just can't. . . . they can't seem to find their way out . . . so they're hanging on while they look or wait for something to return or something coming with the answer . . ."

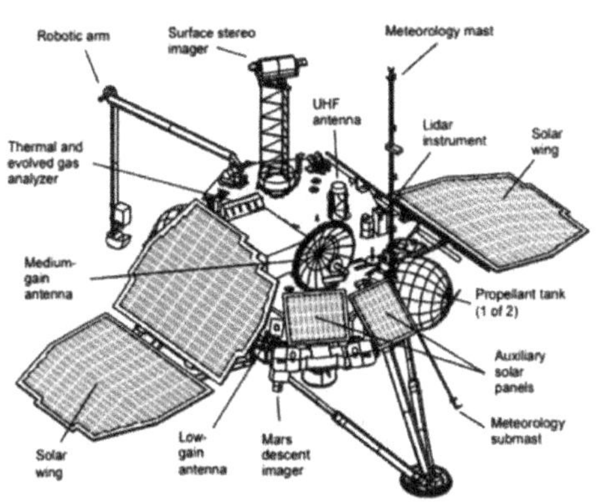

Diagram of the Mars Polar Lander, whose mission ran from January 3, 1999, to December 3, 1999. The spacecraft puzzlingly vanished on the latter date and has never been found. Conventional astronomers assume that it "crashed." But conspiracy theorists have a different view: The Mars Polar Lander was seized by an advanced race of extraterrestrial beings that did not appreciate us violating their air space. Image credit: NASA.

CIA scientist: "What is it they're waiting for?"

Psychic: ". . . There ah . . . evidently was a . . . a group or a party of them that went to find ah . . . [a] new place to live. It's like I'm getting all kinds of overwhelming input of the . . . corruption of their environment. It's failing very rapidly and this group went somewhere, like a long way to find another place to live."

CIA scientist: "What was the cause of the atmospheric disturbance or the environment disturbance?"

Psychic: ". . . Oh, I get a globe . . . a . . . it's like a globe that goes through a comet's tail or . . . it's through a river or something, but it's all very cosmic. It's like space pictures."

CIA scientist: "All right, now before you leave this individual, ask him if there is any way that you, ask him if he knows who you are and is there any way you can help him in his present predicament?"

Psychic: "All I get is that they must just wait. [He] doesn't know who I am. [I] think he perceives I'm a hallucination or something."

CIA scientist: "Okay, when the others left, these people are waiting, when the others left, how did they go?"

Psychic: ". . . [I] get an impression of ah . . . don't know what the hell it is. It looks like the inside of a larger boat. Very rounded walls and shiny metal."

CIA scientist: "Go along with them on their journey and find out where it is they go."

Psychic: "[I get the] impression of a really crazy place [Earth during the Pleistocene age?] with volcanos and gas pockets and strange plants, very volatile place, it's very much like going from the frying pan into the fire. Difference is [that at the new place, Earth?] there seems to be a lot of vegetation where the other place [Mars] did not have it. . . ."

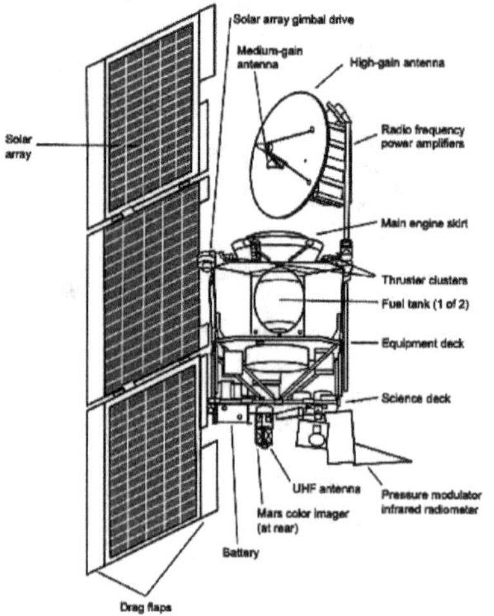

The Mars Climate Orbiter, another NASA spacecraft intended to survey the Red Planet, but which mysteriously disappeared without a trace on September 23, 1999. Was it destroyed by a technologically sophisticated Martian civilization that views us as interlopers? Image credit: NASA.

CIA scientist: "All right it's time to come back now to the sound of my voice into present time to right now the 22nd of May 1984, the sound of my voice. Move now back to the room, back to the sound of my voice, back further to the sound of my voice on the 22nd of May 1984." End of interview.[9]

Clearly, the psychic had no idea that the subject of the interview was the planet Mars. Thus he was either a charlatan with a superb imagination or he was a genuine clairvoyant who was able to tune into the last days of a dying Martian civilization and the escape of its surviving members to Pleistocene Earth. Since the Stargate Project was run by hardheaded, scientifically minded members of the U.S. Army and (at one time) funded by such groups as the CIA, I think we can be sure that the psychics used in these experiments were trusted,

---

9. Source, Website: https://www.cia.gov/readingroom/docs/CIA-RDP96-00788R001900760001-9.pdf

experienced, and credible individuals. Otherwise they would not have been called on in the first place.[10]

And what of ESP itself, the technique used in the above government experiment? Skeptics label it an "alleged" ability, and continue to discredit, belittle, and slander it. The educated and well-read, however, are fully aware that the effectiveness and validity of ESP was scientifically proven long ago, in the last century, as a matter of fact, by such well respected people as military veteran, psychologist, and Duke University professor J. B. Rhine. As early as the 1930s, Rhine was conducting controlled, repeatable experiments in precognition, psychokinesis, clairvoyance, and telepathy—all which conclusively determined the scientific reality of psychic powers.[11]

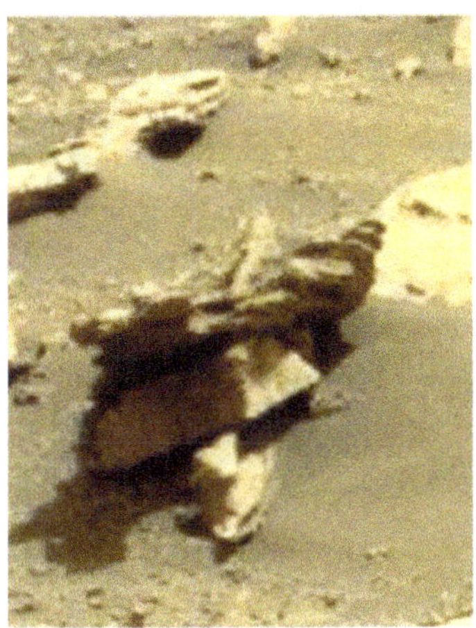

Do you see anything seemingly artificial in this photo? I see plenty, and so do others. Conventional scientists continue to maintain that *everything* on the surface of Mars is a product of natural geological forces and processes, and that anyone claiming anything different are victims of pareidolia. But is the matter really this black and white? Image taken by Mast Camera (Mastcam) onboard NASA's Mars rover Curiosity on Sol 1448 (2016-09-01, 21:48:48 UTC). Image credit: NASA/JPL-Caltech/MSSS.

In 1944, Dr. H. S. Burr, professor of anatomy at the Yale School of Medicine, experimented with electricity and discovered that a mysterious electrical field surrounds every living thing, from plants to humans.[12] Today, this electrical force field is known as an "aura," and is believed to be a form of radiation that links all creatures together as one.[13] Could our aura—a boundless, timeless, invisible vibratory medium—be part of what makes ESP, as well as

10. The U.S. government was not the first to travel to Mars telepathically. A Victorian psychic exploration of the Red Planet can be found in the 1905 book *Journeys to the Planet Mars*, which found, among other things, that Martians refer to their home planet as "Ento" and to themselves as "Entoans." See Weiss, passim.

11. See Rhine, passim.

12. H. S. Burr, "Electrical Correlates of Growth in Corn Roots," *Yale Journal of Biology and Medicine*, July 1942, Vol. 14, pp. 582-588.

13. For more on the science behind the paranormal abilities of the mind, see Talbot.

phenomena like faith and prayer, possible? I believe so, for I have seen and experienced their effects firsthand.[14]

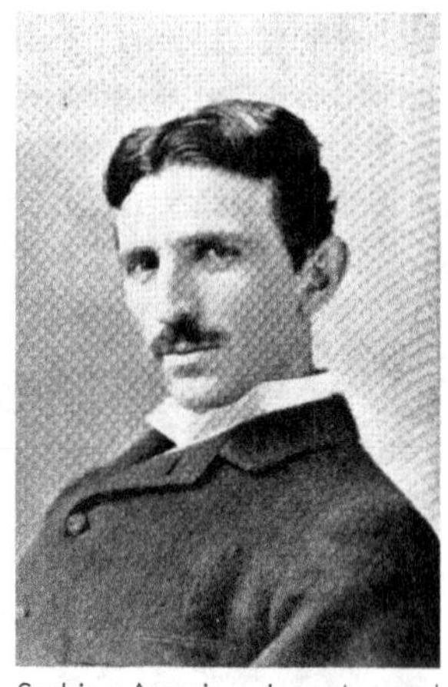

Serbian American inventor and engineer Nikola Tesla, one of our most brilliant scientists, once commented: "The day science begins to study non-physical phenomena, it will make more progress in one decade than in all the previous centuries of its existence." Sadly for humanity, many of Tesla's modern day counterparts ridicule this idea.

Many individuals of note would agree with me, including most of the great early electrical scientists, men such as Thomas Edison, Nikola Tesla, Guglielmo Marconi, and Charles P. Steinmetz, all who were deeply interested in mental telepathy.[15] There have been many other highly esteemed individuals throughout history who were open about their belief in the powers of the mind.[16] The Nobel Prize winning surgeon and botanist Alexis Carrel, for example, publicly declared that telepathy should be viewed and studied by scientists just like any other physiological phenomenon.[17] Many members of the CIA, the NSA, and the DIA would no doubt support this view, since all three organizations have used psychics at one time or another (and, according to some, probably still do).[18]

Despite the evidence for both remote viewing and a probable ancient Martian race, those who embrace and promote these ideas are often considered "crackpots" or worse. Yet, something rational

---

14. For more on the power of thought (as well as the "magic" powers behind thoughts, beliefs, feelings, and words) from a spiritual, mystical, and Christian perspective, see my books: *Jesus and the Law of Attraction*; *Christ Is All and In All*; *The Bible and the Law of Attraction*; and *Seabrook's Bible Dictionary of Traditional and Mystical Christian Doctrines*.

15. See Bristol, p. 135.

16. For example: Jesus, Buddha, Paracelsus, Hermes Trismegistus, Arthur Conan Doyle, Ernest Rutherford, Hippolyte Baraduc, Arthur Eddington, Dana Sleeth, Gustav Geley, Sigmund Freud, Ella Wheeler Wilcox, Hanbury Hankin, Ralph Waldo Emerson, Albert Pike, Phillips Thomas, Paul C. Ferrell, Theodore C. Foote, Louis Bromfield, William Seabrook, Walter D. Scott, Napoleon, Erna F. Grabe, Ann Lee, Theos Bernard, René Fauvel, Thomas Jay Hudson, Annie Besant, Alexander the Great, Ben H. Lampman, Morton Prince, Gene Sarazen, William James, Knute Rockne, Marcus Bach, Douglas MacArthur, William Shakespeare, Mary Baker Eddy, Francis Bacon, Alexander Cannon, Frederick Kalz, Albert Heim, Marie Corelli, Robert Gault, Ashley C. Dixon, Maurice Bucke, Walt Whitman, George Russell, and Herbert D. Seibert.

17. Bristol, p. 135. See also pp. 129-135.

18. For more, see the 2019 documentary expose *Third Eye Spies*. Website: www.thirdeyespies.com

is needed to explain the many strange objects in the photos that I have included in this book. As a writer and researcher I have been studying these odd shapes and seemingly out-of-place forms for many years. To call them "nothing more than rocks," as many do, is an insult to the intelligence of every thinking person.

Most of our orthodox scientists are among them, of course, and routinely scoff at the mere mention of theories like this. Indeed, the official scientific stance on this topic is that no intelligent life, including either modern humans or archaic hominids (pre-human, human-like beings), have ever lived or even walked on Mars.

There is an old saying, however: The camera never lies. Does this include the high resolution stereoscopic cameras fitted to the Mars rovers? If it does, then it is time to reexamine their findings in a detached and unemotional manner and discuss them in an open and impartial forum. This, indeed, is one of the purposes of my book: to encourage closer scrutiny and study of NASA's Mars images, while raising awareness of this subject, both in the lay community and the scientific community.

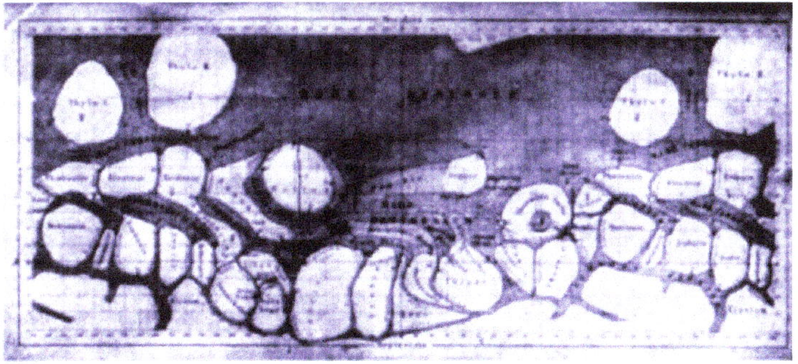

Based on his observations, Italian astronomer Giovanni Schiaparelli drew up this map of Mars in 1877. It highlights the "canals," or alleged artificial water channels, upon which Lowell later based his theory that a superior life form once existed on the Red Planet. Though the *canali* turned out to be a product of pareidolia (in this case, a result of late 19th- and early 20th-Century low resolution telescopes), 21st-Century high resolution photography is chronicling a wealth of highly compelling evidence that Lowell may have been right—as this very book demonstrates.

Why is this important?

The many startling NASA images I have chronicled here confer upon us a singular proposal: a highly evolved culture seems to have once populated the Red Planet. Though at this time we do not

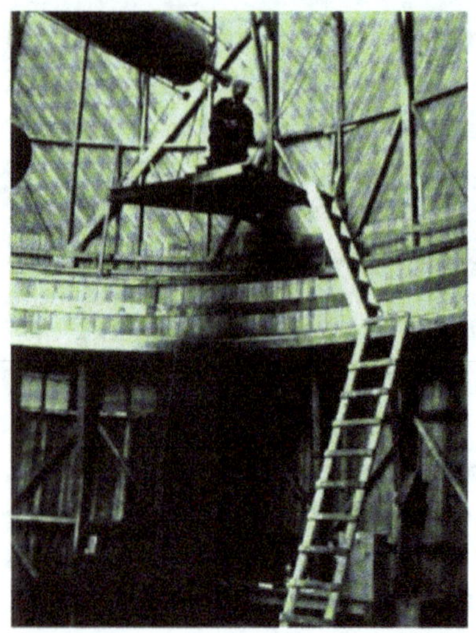

Lowell "at the telescope."

know where they originated, who they were, how they lived, when they lived there, or what happened to them, the Mars photos reach out to us from 140 million miles away in a most convincing, profound, personal, and emotional way by presenting us with countless signs, articles, and items that they left behind—most of them quite "manmade" in appearance.

Among these are confusing debris fields scattered over with thousands of strangely familiar items (but perhaps constructed of unknown materials), the result of what seems to have been, as noted, a violent climatological or astronomical episode, or possibly nuclear war. Various sections of the Martian surface are unevenly covered with, for example, objects that have every appearance of artificiality, including:

- The building and house foundations of once thriving villas, towns, and cities.
- The remains of stone, cement, and brick walls, cellars, stairs and staircases, windows, doors, walkways, roofing, crown molding, porticoes, and masonry from various types of structures.
- Construction support beams, pillars, and metal reenforcements.
- Machinery of every description, from aerial and marine craft to land vehicles, from military weapons and factory equipment to farming and domestic implements.
- Statuary, some seemingly strictly ornamental; others perhaps religious, cultic, or ritualistic in nature; some animal, some humanoid; some realistic, some mythological; some small, some colossal.

- Infrastructure, including what seem to be vestiges of roads, bridges, sidewalks, subways, airports, telecommunications, power grids, pipelines, sewage systems, levees, tunnels, channels, water lines, dams, parks, seaports, and riverports.
- Magnificent megalithic monuments, such as pyramids and ornate shrines, along with intricate artwork (carvings, inscriptions, paintings, murals, reliefs), palaces, altars, obelisks, halls, pavilions, corridors, terraces, temples and temple complexes, pylons, basins, decorative floors, towers, walls, courts, columns, sanctuaries, colonnades, chapels, storage chambers, tombs, cemeteries, and mortuary architecture.

There are objects on Mars covered with letter like impressions, many of them similar in appearance to ancient Egyptian hieroglyphs, like this 4,500 year old inscription of King Sneferu found in the Mines of Sinai.

Most astounding are those objects that are emblazoned with various forms of writing, some of it jumbled and unrecognizable, some of it appearing to be carved or stamped into rock, metal, and other materials, using, sometimes very clear, Latin (i.e., Roman) lettering and numbers—the script upon which the modern English alphabet was formed. Amidst the Martian debris I have also spotted what appear to be letters from the Hebrew alphabet, as well as hints of ancient Egyptian like hieroglyphs—all three writing systems which are related, I might add.

Naturally most of the remnants of this alleged Martian race are covered in multiple layers of dirt and dust, some inevitably taking on a rock-like appearance, along with the often drab brownish, yellowish, and reddish hues of the Martian soil. This has rendered these particular items quite nondescript—which is what one would expect of damaged and destroyed items that have been sitting on the surface of a harsh, sandy, barren, cold, wind-blown planet for thousands, perhaps tens or hundreds of thousands, or even millions, of years.

What are we actually seeing in these spectacular sometimes disturbing NASA images?

Let us begin by acknowledging that, whether on Earth or Mars, geology is a master at creating highly unusual looking objects and formations. Here on our own planet we have, for instance:

- Yellowstone National Park (Wyoming).
- The Devil's Postpile (California).
- The Valley of the Moon (Argentina).
- The Stone Forest (China).

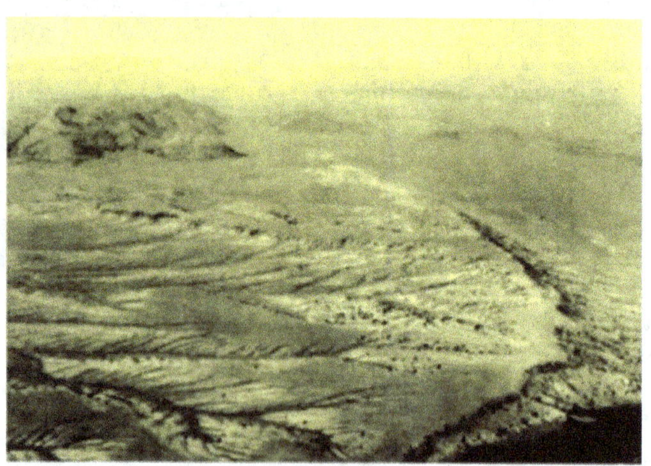

Moonstone rock formations, central Wyoming, USA.

- The Pancake Rocks (New Zealand).
- The Richat Structure (Mauritania).
- Turnip Rock (Michigan).
- Mima Mounds (Washington).
- The Cave of Crystals (Naica, Mexico).
- The Reed Flute Cave (China).
- Goblin Valley State Park (Utah).
- Pamukkale (Turkey).

- Yamal Craters (Russia).
- The Moeraki Boulders (New Zealand).
- The Giant's Causeway (Northern Ireland).
- Hoodoos (Utah).
- The Chocolate Hills (Philippines).
- The Fairy Chimneys (Turkey).
- Kummakivi (Finland).
- The Devil's Slide (Utah).
- Mesas (southwestern USA).
- Elephant Rock (Nevada).
- Arches National Park (Utah).

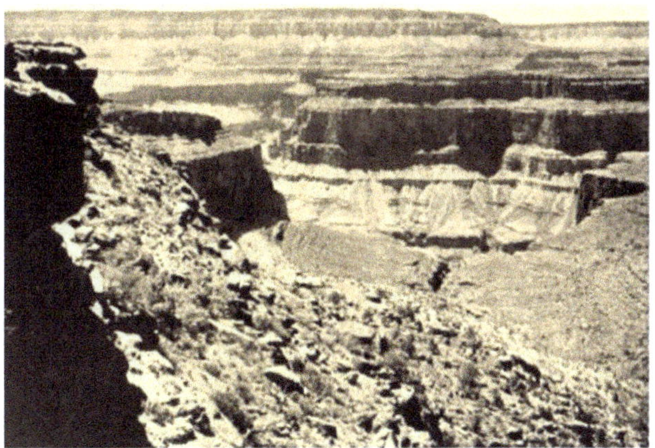

The Grand Canyon, Arizona, USA.

- Danxia Landforms (China).
- Babele (Romania).
- Nastapoka Arc (Canada).
- Fingal's Cave (Scotland).
- The Wave (Arizona).
- The Grand Canyon (Arizona).
- Fairy Circles (Namibia).
- Batholiths (California).
- Skaftafell (Iceland).
- Salar de Uyuni (Bolivia).
- Skull Rock (California).
- Gibson Steps (Australia).
- The Great Unconformity (Arizona).

- The Devil's Tower (Wyoming).
- The White Desert (Egypt).

Like many of my readers, I have visited a number of these locations and stood before them filled with awe and wonder.

I also recognize that specific geological processes can and do create odd and even extremely bizarre features and formations all over the Earth's surface. Among these there are:

- Concretion.
- Alluvium.
- Mudcracks.
- Cirque.
- Hogbacks.
- Rift valleys.
- Lamination.

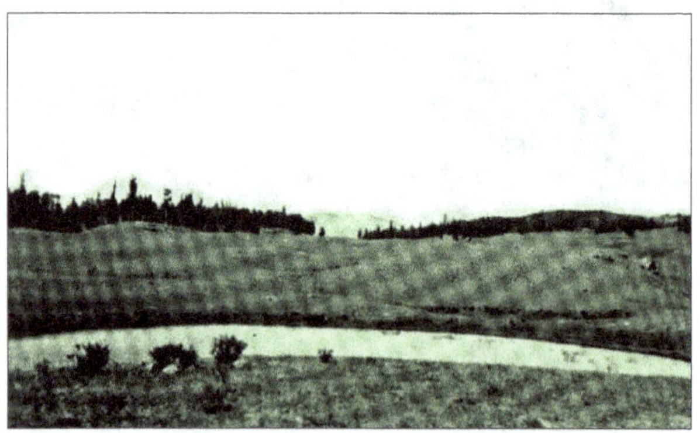

An example of a moraine, this one located near Alamosa Canyon, in the vicinity of the headwaters of La Jara Creek, Colorado, USA.

- Slickenside.
- Abrasions.
- Veins.
- Cinder cones.
- Liesegang rings.
- Fault scarps.
- Amygdules.
- Phenocrysts.

• Flood basalt.
• Ripples.
• Desert pavement.
• Volcanic plugs.
• Cavernous weathering.
• Fiamme.
• Kame.
• Moraines.
• Boudinage.
• Exfoliating rocks.
• Frost heaves.

Yet, despite their sometimes artificial resemblance to either human-made or even alien-like objects, none of these natural earthly geological processes and features could ever be confused with artificial structures made by intelligent beings. All of them are the obvious result of the forces of physical geology, especially to the trained eye.

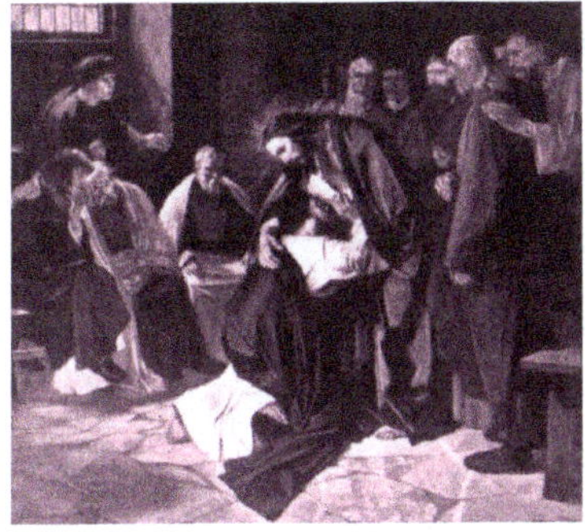
The Apostle Thomas could not accept the reality of the resurrected Jesus until he experienced it firsthand, prompting this terse comment from the Master: "Thomas, because thou hast seen me, thou hast believed: blessed are they that have not seen, and yet have believed."

The same cannot be said, however, for many of the objects in the Mars photos in this book, a preponderance which—as my more dispassionate and impartial readers will quickly conclude—while defying positive identification, still bear numerous characteristics of unnaturalness. Nonetheless, hardcore doubting Thomases persist in refusing to accept even the slightest possibility that intelligent life once existed on Mars, or that, in fact, advanced beings may still visit or even live there.

Why is there so much disagreement over the nature of the NASA Martian images? To better understand the answer to this question, let us look more closely at the claimants.

On the one side we have the skeptics, nonbelievers, and

conservative mainstream scientists. On the other we have the believers, the flexible-minded, and independent scientists.

The first group, the nonbelievers, steadfastly maintain that every object on the surface of Mars—i.e., apparently anything larger than a grain of sand—is a rock (or is composed of rock-like material), a *natural* product of geological processes, no matter what its age, shape, size, color, or outline. Believers, however, hold that many of the objects photographed by the cameras on NASA's Mars exploratory vehicles are plainly *artificial*; i.e., they have been conceived and produced by a superlative intelligence.

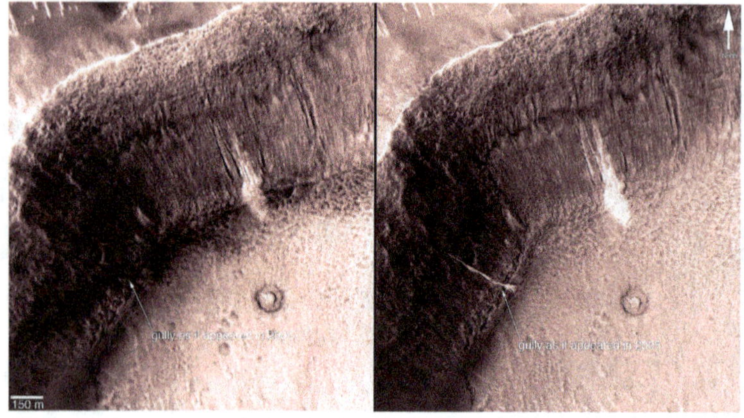

NASA may have proven that water is still flowing on Mars. But at the time the following commentary was written (December 6, 2006), it was still unwilling to commit either way. The photo on the left was taken of a gully in 2001. The photo on the right was taken of the same gully in 2005. The 2005 photo shows the appearance of a "new light-toned deposit" of some kind, suggesting that water may have flowed in this particular gully between 2001 and 2005. This is *not* proof, but, according to NASA, only "very tantalizing evidence that this may have occurred." NASA's description: "This set of images shows a comparison of the gully site as it appeared on Dec. 22, 2001 (left), with a mosaic of two images acquired after the change occurred (the two images are from Aug. 26, 2005, and Sept. 25, 2005). Sunlight illuminates each scene from the northwest (top left). The 150-meter scale bar represents 164 yards." Image credit: NASA/JPL/Malin Space Science Systems.

In that *authentic science is meant to be an open-ended search for truth, whatever that search turns up*, are skeptics just neophobes behaving non-scientifically, acting out of fear, prejudice, spite, and short-sightedness, thereby missing a golden opportunity to forward human knowledge? Or are believers overly confident, illogical, gullible, and simple-minded, thus impeding the advancement of our understanding of both our solar system and our species by adding unnecessary noise to the discussion?

Are skeptics trapped in their own parochial echo chamber, too intellectually cavalier, myopic, and rigid to see the obvious; or are believers prone to pareidolia, foolishly misinterpreting mere dirt and stone for intelligent design? Are skeptics partisan bigots, locked in a confirmation bias loop that prevents them from seeing reality, or do believers lack the ability to think critically?

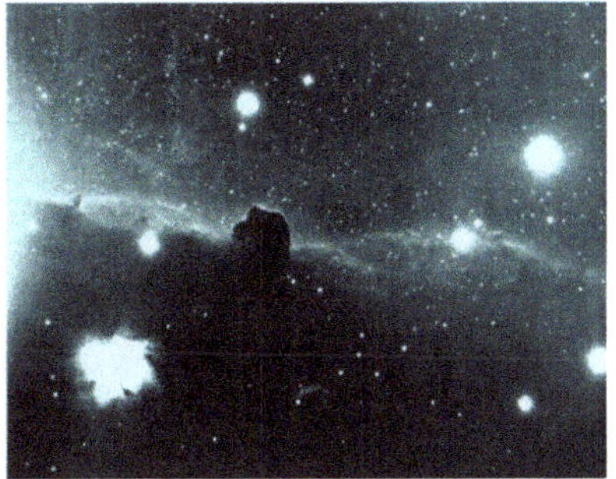

The Dark Horse Nebula in the constellation Orion. Reasonable people are interested in and often accept the extraterrestrial hypothesis, which states that advanced non-human life forms must exist, not only in our solar system, but in our galaxy and beyond. Using a revised version of the Drake Equation, in the summer of 2020 British astrophysicists Tom Westby and Christopher Conselice calculated that there are probably (at least) 36 intelligent civilizations in the Milky Way that possess the technology to communicate with us. Popular astronomer Carl Sagan believed there could be anywhere from a few to many millions in our galaxy.

Where, you might be asking yourself at this point, do I the author stand on this topic?

After years of carefully scrutinizing the NASA Mars photos in countless magazines and books, on all manner of electronic devices and computer monitors, and using a variety of magnifying glasses, lighting, and filters, I have come to the following conclusion, the only one that does not offend my sense of logic: *The NASA Mars images strongly suggest that an advanced race once lived on the Red Planet*.

Having said this, I am the first to acknowledge that until honest and forthright astrogeologists, astrobiologists, and astroarchaeologists stand on the surface of Mars and examine the planet's anomalies in situ—film them, handle them, test them, and study them in detail—we cannot know for sure. In the meantime, however, I am quite confident that most of my readers will discover, as I have, that the enclosed NASA images are not only highly compelling, they are extremely persuasive that *many of the Martian anomalies are very likely the sophisticated products of an ancient and ingenious extraterrestrial life form*, one perhaps far more technologically advanced than we are.

Unlike many mainstream scientists, in keeping with the Drake Equation (which postulates that there are a number of other advanced civilizations in our galaxy, the Milky Way), I assert that there are, without question, large complex life forms—i.e., intelligent technological species—that exist outside Earth and even outside both our solar system and our galaxy. This would include the planet Mars, and I encourage my readers to remain open to this idea, and to view the enclosed NASA photographs as not just possible, but credible even realistic circumstantial evidence.

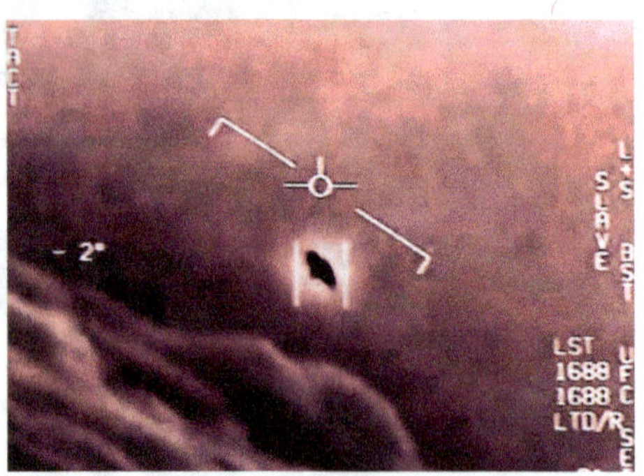

This frame from an authenticated video released by the DOD, shows footage captured during an encounter between Navy pilots and a UFO, or what the U.S. military now calls UAP: unidentified aerial phenomena. Though government officials have not publicly stated that the mysterious craft are piloted by extraterrestrials, thus far they have not denied the possibility. Are some of them based on Mars? Image credit: U.S. Navy.

Thus, my perspective is quite simple: I fall midway between the skeptics and the believers. To me, both views are non-scientific, for both are based on assumptions, preconceived ideas about what is possible and what is not possible—all of this without a shred of hard evidence either way. Having said this, after perusing the hundreds of objects I have selected for inclusion in this book, level headed individuals must admit, at the very least, that something odd once transpired (and may still be transpiring) on Mars, and that, at the very most, whatever it was or is, deserves serious scientific attention. Let us heed the words of Aristotle, who once remarked: “The more you know, the more you realize you don’t know.”

I did not write this book to promote my personal views. Rather, it is my effort to present a photographic perspective on this contentious scientific phenomena: the anomalous, and still to be definitively identified, objects found scattered over the surface of the Martian landscape as imaged by NASA's cutting-edge photographic equipment. These photos will permit those with "eyes to see" to delve more deeply into this fascinating areological enigma than ever before.

At this point the question logically arises: If the anomalies on Mars are indicative of, or even prove, the existence of past intelligent life on the Red Planet, why is this being hidden from the public? Why are we not being told the full and complete truth? Why would anyone go to so much trouble to conceal such an obvious fact? Why the ornate coverup of an idea that many people already believe, and that the majority of others would easily and willingly accept if it was shown to be true?

The answer involves a major conspiracy theory: The scientific powers that be are well aware that many parts of Mars are covered with the climate- or war-ravaged remnants of a highly developed society that lived and died there long ago. However, just as they have done with the UFO phenomena, Sasquatch, lake monsters, ESP studies, NDE reports, and Area 51 (to name but a few examples), the elite, powerful, and secretive men and women running our world governments have decided that the public revelation of this fact would cause global unrest, threatening the world's economic, political, social, and religious stability. To this end, it has been ordered that astronomical images showing signs of extraterrestrial life on Mars, and elsewhere, be doctored in order to hide this reality from the general population. Is it possible that this theory is true?

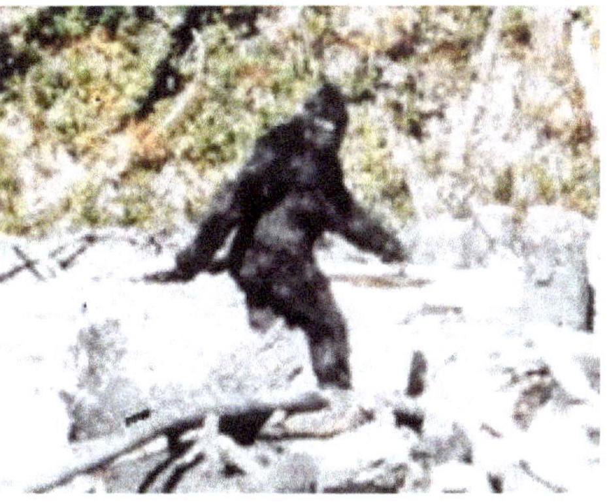

Famous frame 352 from the 1967 Patterson-Gimlin film, showing a purported female Sasquatch. Like the Martian anomalies themselves, this footage has never been proven or disproved. As far as mainstream scientists are concerned this is perfectly acceptable, for they do not seem to want the public to know the truth—whatever that might be—about either phenomenon. Image credit: Roger Patterson and Bob Gimlin.

As a lifelong historian, writer, and editor, with (currently) 80 books to my credit, I can attest to the fact that most of what we have been taught by the mainstream educational system and the mainstream media is false. Not just simply incorrect, but nearly always the opposite of the actual truth.

In other words, whether it be history, science, philosophy, sociology, politics, economics, or the arts, *nearly all* information has been manipulated by the ruling thought police to hide certain facts from the public.

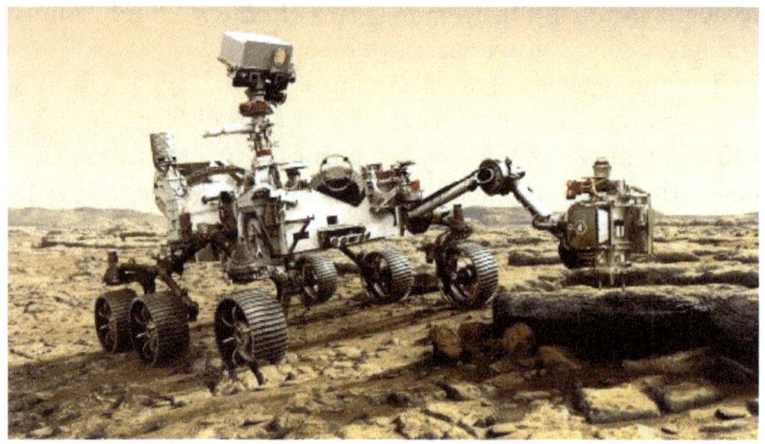

An illustration of Perseverance, which landed on the Red Planet on February 18, 2021. Will it uncover evidence of complex life on Mars? Almost certainly not. Most (if not all) of its creators do not accept the hypothesis that large intelligent life forms once inhabited Mars, and therefore they did not design their rover to search for it. Instead, one of its primary goals is to "seek signs of ancient life, particularly in special rocks known to preserve signs of life over time." The term "ancient life" as it is used here is a conventional scientific euphemism for *microbial life*. Image credit: NASA/JPL-Caltech.

Instead of the truth, we are often spoon fed mountains of both misinformation and disinformation, all, apparently, in an effort to lead us as far away from reality as possible.

In no topic is this global gaslighting campaign more obvious than the War for Southern Independence (the "American Civil War"). I should know: I have written over 50 books on this topic alone.

But the suppression of the truth affects every field of human knowledge, including astronomy. And here we are.

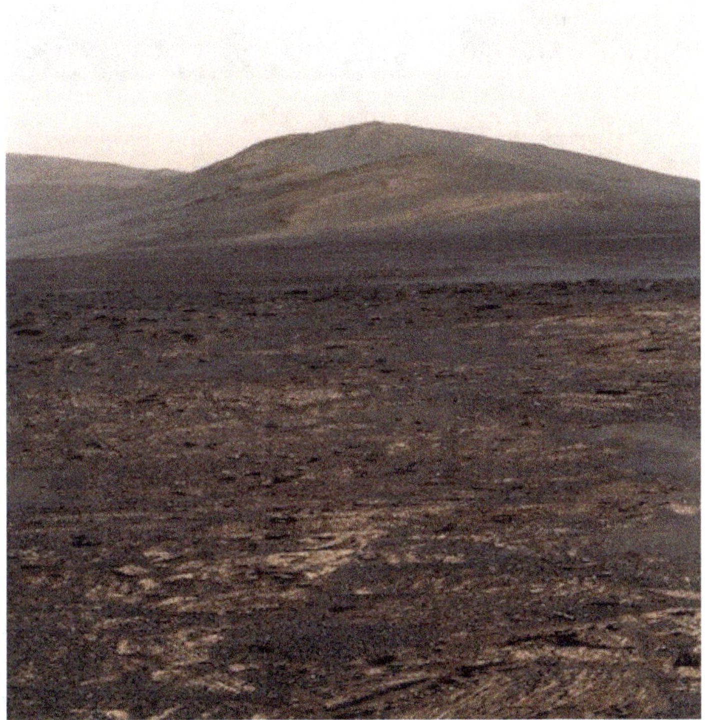

Typical Martian topography. With its thin cold atmosphere, violent "planet-encircling dust events," constant crustal movement, and ongoing meteor impacts, Mars is an ideal location for concealing evidence of probable past and present intelligent life. Image credit: NASA/JPL-Caltech/MSSS.

All that can be said with any certainty at this point is that many of NASA's Martian photos appear to contain visual fiction: areas that have been digitally manipulated in an attempt to prevent the viewer from seeing what actually lies beneath. Such image alterations include artistically applying false "sand," "rocks," "dunes," "dirt," "sky," "shadows," "clouds," "hills," "cliffs," "ripples," "haze," "light," and blurring effects to areas and objects where the genuine article or effect never originally existed. Some of these concealment efforts, both manual and automated, show the knowledge and craftsmanship of an expert hand, while others are quite primitive, clumsy, and overt (as I will show). In many cases then, in examining the NASA Mars photos we are in fact looking at simulacra: reprocessed likenesses, imitations, or representations of the real thing rather than the real thing itself.

Is this digital sleight of hand, this fake NASAian world of hyperreality, nothing more than wild speculation? Hardly. NASA has admitted that, due to what it deems image defects and camera issues, it sometimes uses software to repair or enhance its photos. But the world's largest, most powerful, and most influential space agency would rather we not focus on this. In fact, as I discovered myself, it appears that NASA may now be concealing or even deleting images that reveal potential for artificiality or which support the "ancient civilization on Mars" hypothesis.

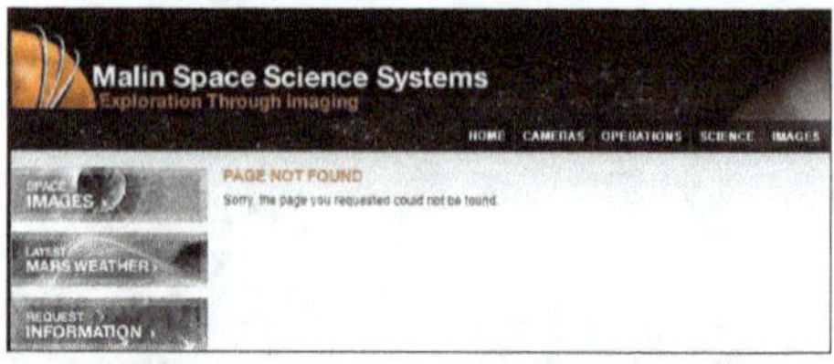

In looking for several specific images on the MSSS Website, I received this error message: "Page not found." Similar messages appeared while researching NASA's Website. Some believe the images been decommissioned, or even permanently deleted.

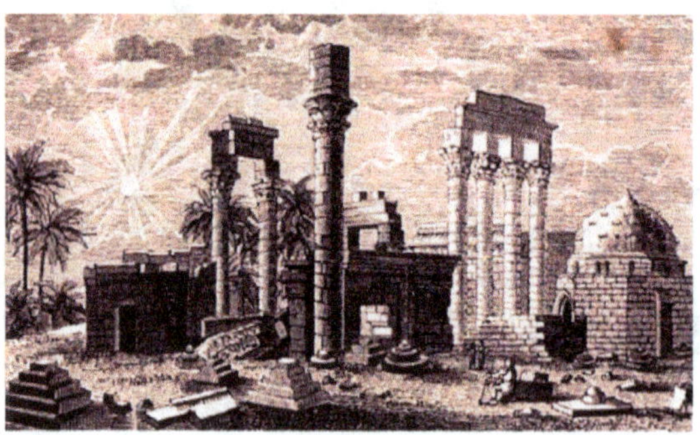

What would these ruins of ancient Egypt's Temple of Hermonthis look like if they were, not thousands of years old, but hundreds of thousands or even millions of years old? For the most part they would be buried under multiple layers of dirt and sand—faded, concretized, and fossilized. This gives us a fairly accurate picture of how the remnants of an early civilization on Mars might appear today.

Concerning NASA's Mars exploration plans specifically, the main focus is on the search for *microbial* life forms, not human-sized life forms—as the mission statement for the current Mars rover program itself states:

> "Perseverance is tasked with searching for telltale signs that microbial life may have lived on Mars billions of years ago. It will collect rock core samples in metal tubes, and future missions would return these samples to Earth for deeper study."[19]

---

19. Website: https://www.nasa.gov/feature/jpl/searching-for-life-in-nasa-s-perseverance-mars-samples

Conspiracy or no conspiracy, what is actually known about extraterrestrial life behind closed doors at NASA headquarters in Washington, D.C. (not to mention the U.S. government), is being concealed and may never be made public. Thus, this aspect is beyond the scope of this book.[20]

Suffice it to say that I have great respect for the extraordinarily important scientific work this tax-payer funded, government agency performs, as well as the worldwide space exploration community that supports it. My intent here is not to critique NASA, or anyone else, but to bring the topic of *The Martian Anomalies* to as wide an audience as possible.

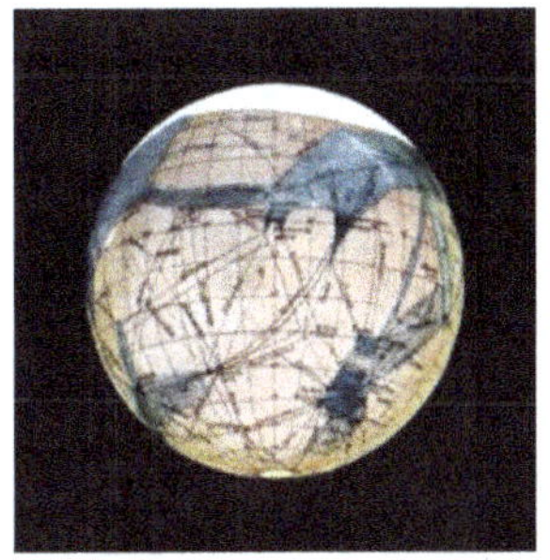

A color drawing of Lowell's view of Mars from 1905. Note the icy polar caps and "watery channels," that suggested life on Mars to early astronomers.

The arrogant, the unscientific, the unimaginative, and the know-it-alls will mock this book without ever once even glancing between its covers. This is understandable, and to be expected, for, to paraphrase Einstein: "Extraordinary ideas have always encountered violent opposition from mediocre minds."[21] But extraordinary ideas like this one do not simply disappear because a small minority dislike them, disagree with them, fear them, or do not understand them.

I have complete faith that those of my readers who possess the mind of a genuine scientist—i.e., those who approach scientific puzzles with courage, honesty, humility, curiosity, mental flexibility, and a thirst for knowledge—will be more than capable of judging for themselves what it is that is transpiring on the fourth planet from our Sun.

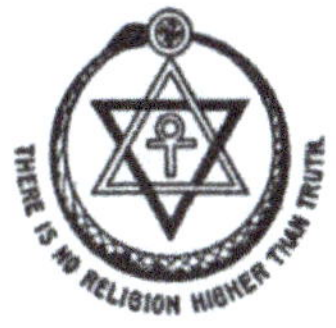

Lochlainn Seabrook
February 22, 2022
Nashville, Tennessee, USA

---

20. For those interested in digging deeper, however, I have covered this particular subject in detail in my popular book, *UFOs and Aliens: The Complete Guidebook* (foreword by Nick Pope). See bibliography.

21. Einstein's original quote was: "Great spirits have always encountered violent opposition from mediocre minds."

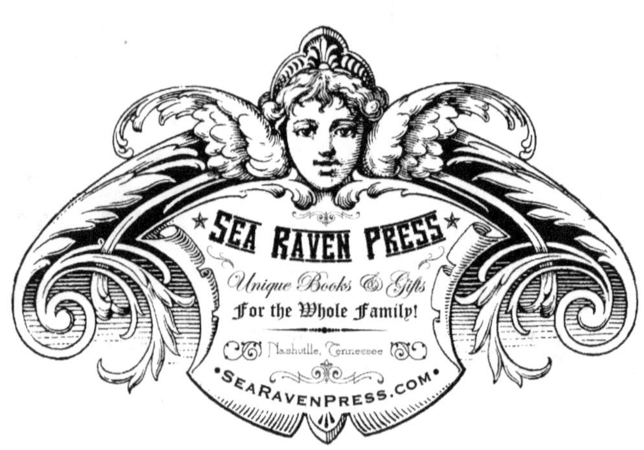

"Books invite all; they constrain none."
Hartley Burr Alexander (1873-1939)

Section One

Images From the Surface of Mars

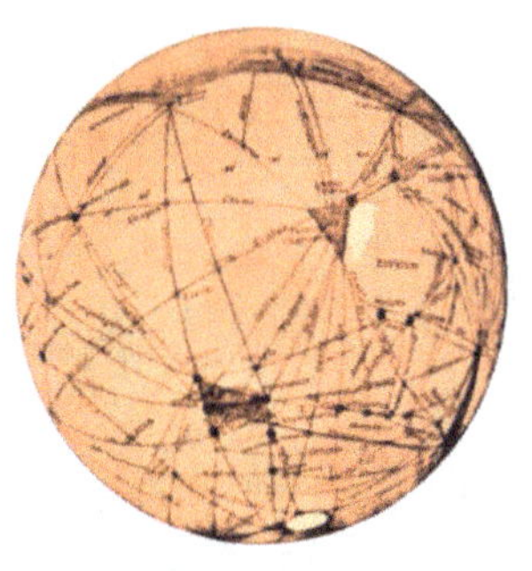

"It has been proven that microorganisms from Earth can survive in outer space. What then would prevent microbes from thriving on Mars, either today or in the distant past? And since early Mars is known to have had a more moderate atmosphere and climate, with rivers and oceans capable of sustaining life, what was there that would have prevented large complex beings from existing on Mars thousands or even millions of years ago? A lack of oxygen? It is doubtful. The fact is that ancient intelligent extraterrestrials could have manufactured their own oxygen, terraformed the surface of Mars, and protected themselves from its hazardous environment exactly the way we will when we eventually colonize the Red Planet. Nothing will stop us, and nothing would have stopped them." — LOCHLAINN SEABROOK

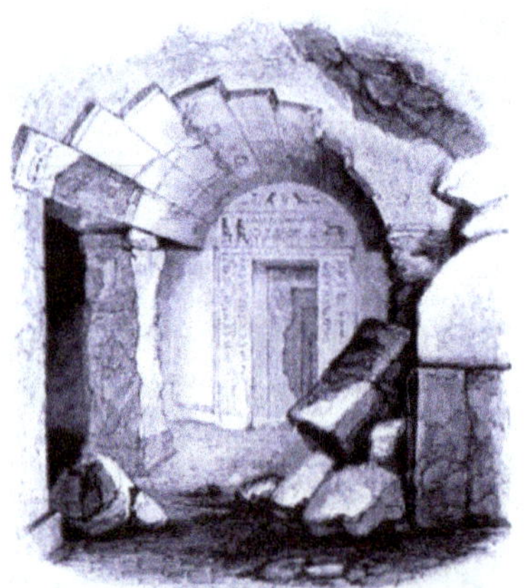

An ancient tomb at Saqqara, Egypt, from the time of the Egyptian King Psamtik II, 26th Dynasty (circa late 6th Century B.C.). Mars may contain evidence of a similar ancient society.

Chapter 1

DEBRIS FIELD

SOL 526: EARTHDATE JANUARY 28, 2014

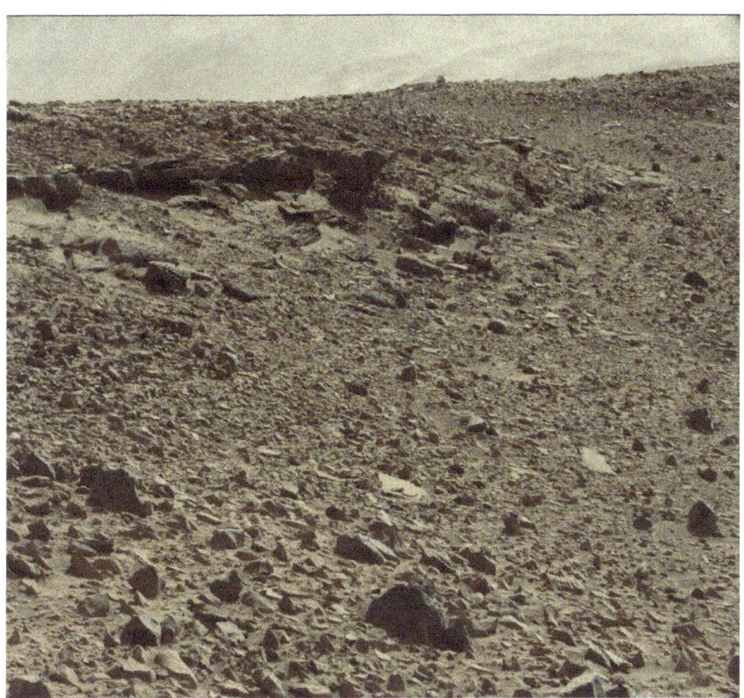

Original image. "Taken by Mast Camera (Mastcam) onboard NASA's Mars rover Curiosity on Sol 526 (2014-01-28, 15:59:52 UTC)." Image credit: NASA/JPL-Caltech/MSSS.

ANALYSIS OF DEBRIS FIELD, SOL 526

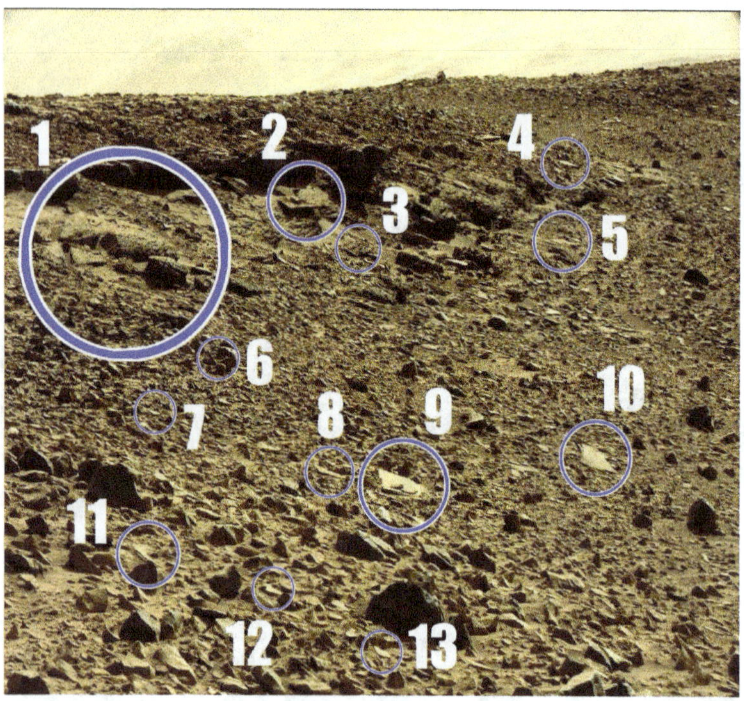

Debris Field, Sol 526, with anomalies circled. Image colored, brightened, and sharpened by the author to improve contrast. There are so many 90 degree angles in this photo it is impossible to count them all. Do 90 degree angles occur in Nature? Yes, but they are unusual, typically quite small (such as salt crystals), and rarely perfectly straight. These facts make the numerous small, medium, and large sized, square shaped objects, as well as the objects with squared off corners, in this image quite extraordinary. As the following images suggest, this may be a debris field left over from the explosion of several artificial building structures.

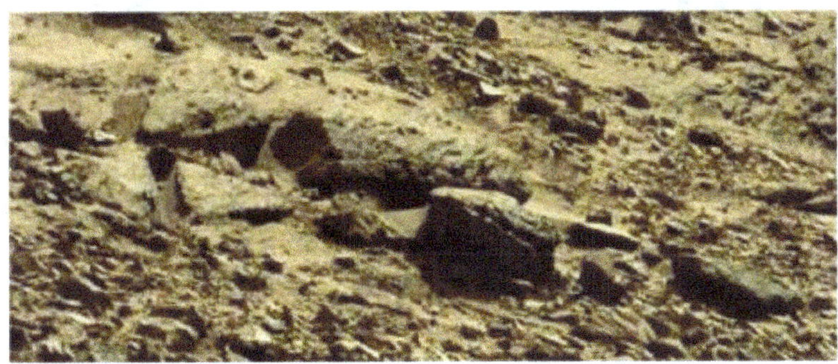

SOL 526, highlighted area 1 (above): This "rubble" may have once been a walkway or foundation. Triangular objects appear to have broken apart, or were possibly separated, by some kind of force. This "trackway" runs diagonally from upper left to lower right.

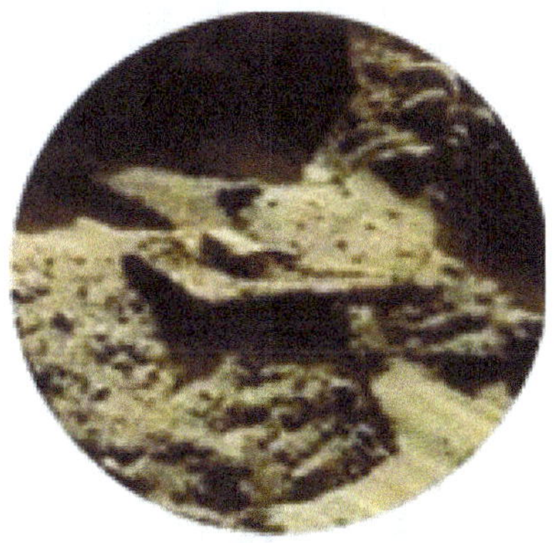

SOL 526, highlighted area 2 (above): A square or rectangular shaped object perched on a rock or dirt mound. A brick shaped object lies on top of it.

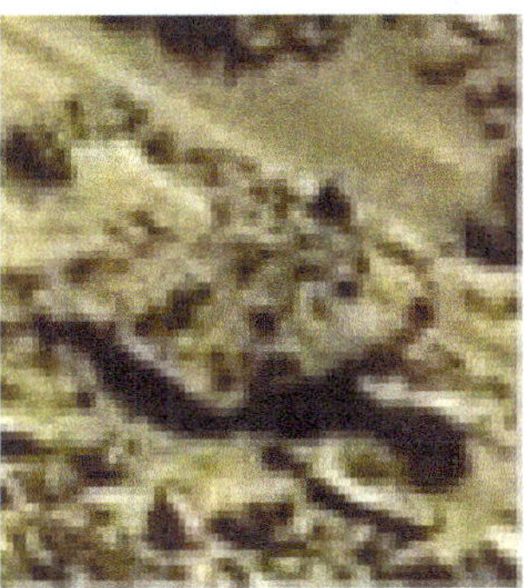

SOL 526, highlighted area 3 (above): Artificial looking square shaped objects, surrounded by assorted odd items.

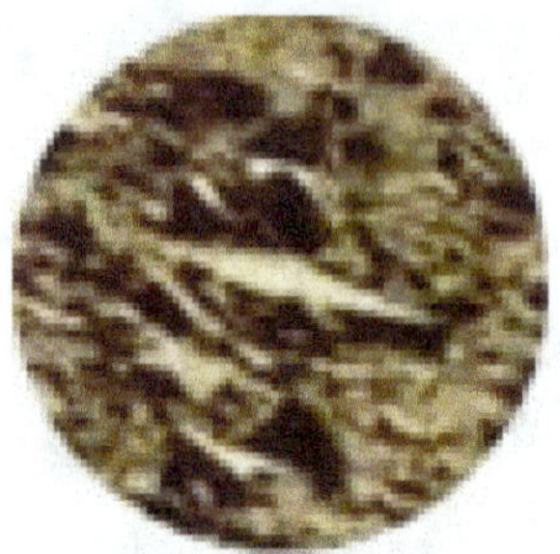

SOL 526, highlighted area 4 (above): What appears to be a perfect artificial square, possibly a tile or piece of siding. Material unknown.

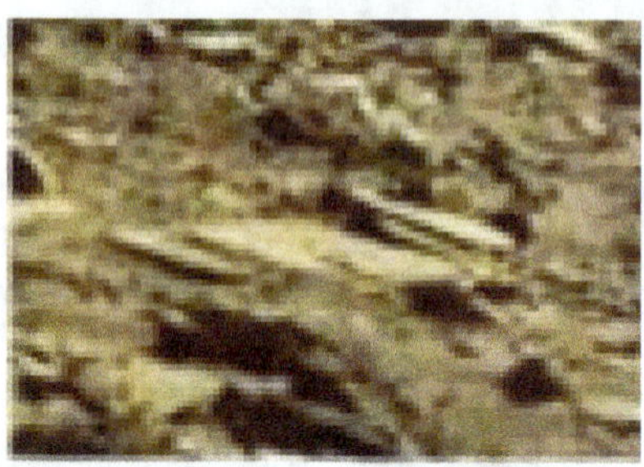

SOL 526, highlighted area 5 (above): Several flat square objects lying on top of one another.

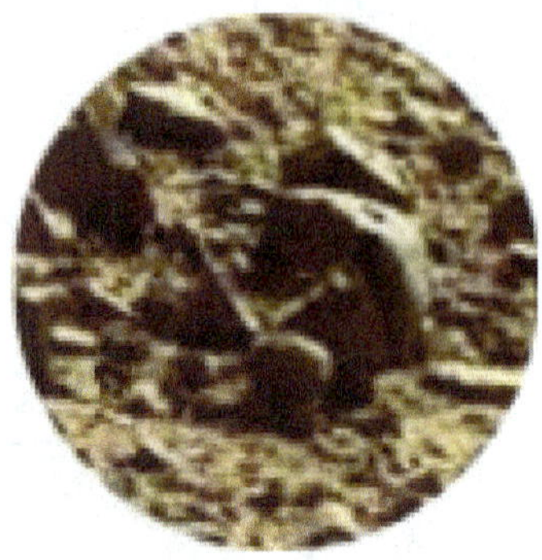

SOL 526, highlighted area 6 (above): An object that looks like a box on its side, with several sides missing.

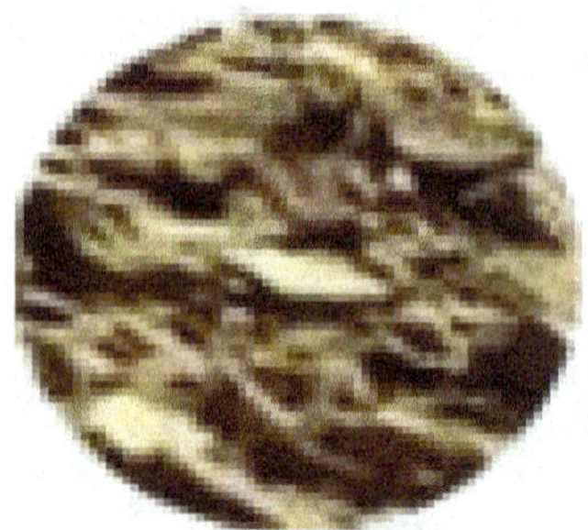

SOL 526, highlighted area 7 (above): A nearly perfect flat square. Is it a naturally formed rock or an artificial object, like a tile? Strange items and shapes surround it.

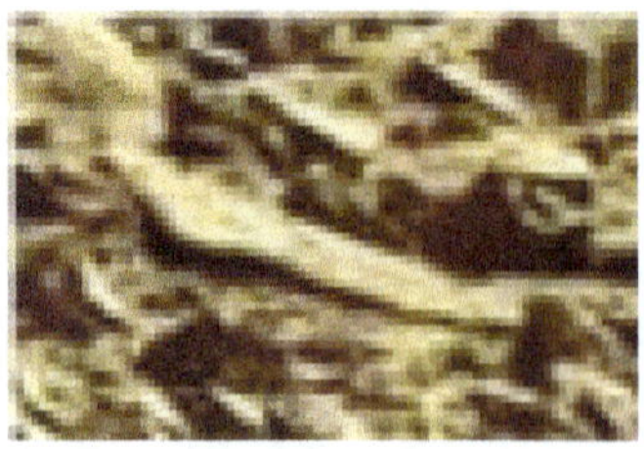

SOL 526, highlighted area 8 (above): Flat object with 90 degree corners.

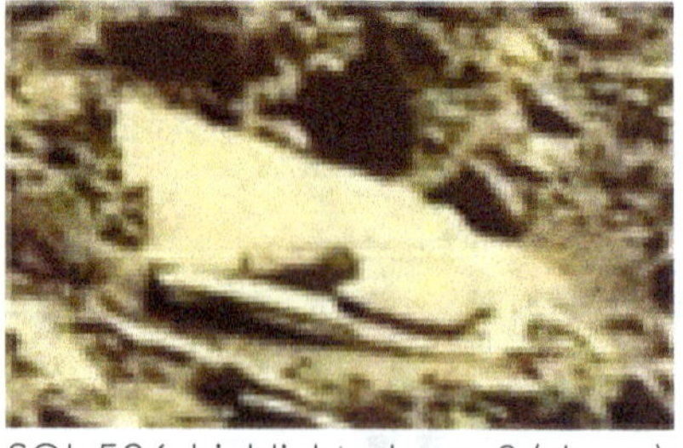

SOL 526, highlighted area 9 (above): A large squarish object embedded in dirt. An unusual notched, curved item lies atop it.

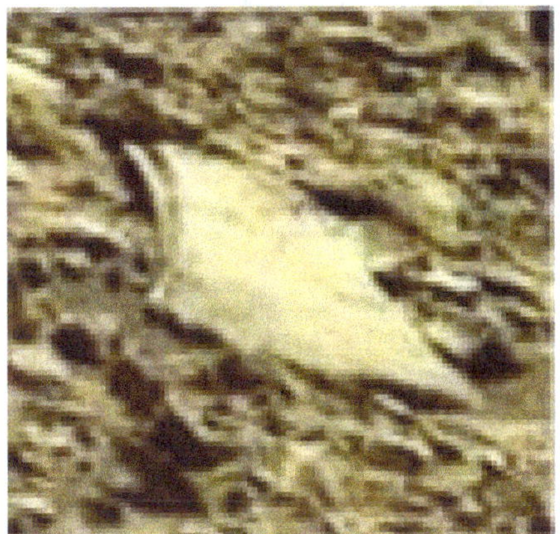

SOL 526, highlighted area 10 (above): An unusual shaped item that looks like a plaque with a beveled edge. Despite its appearance, if it is artificial, it may be a broken shard from a larger object. (Note: We will see many more objects with this exact shape and appearance.)

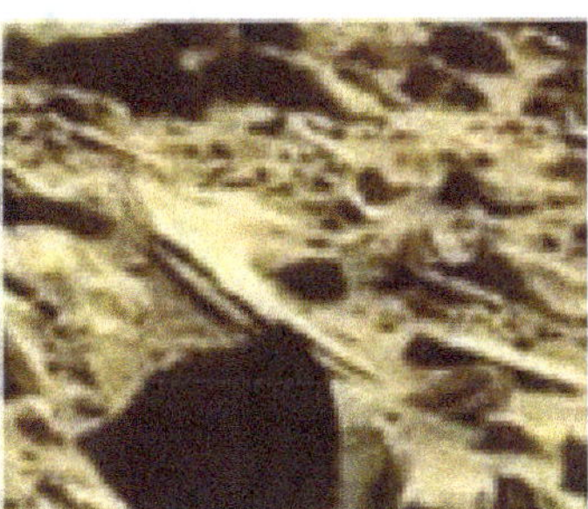

SOL 526, highlighted area 11 (above): Is this an intelligently made layered material or just a piece of Martian sedimentary rock? Look closely under magnification. Nearly all of the objects in this image appear to have been machined.

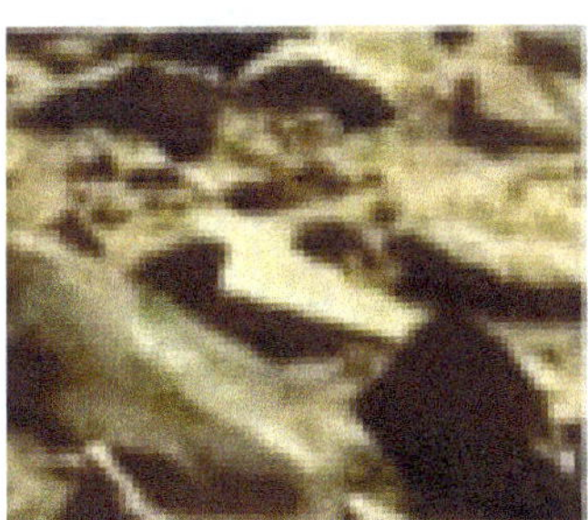

SOL 526, highlighted area 12 (above): An odd serpentine shaped object, with a myriad of other artificial looking items surrounding it.

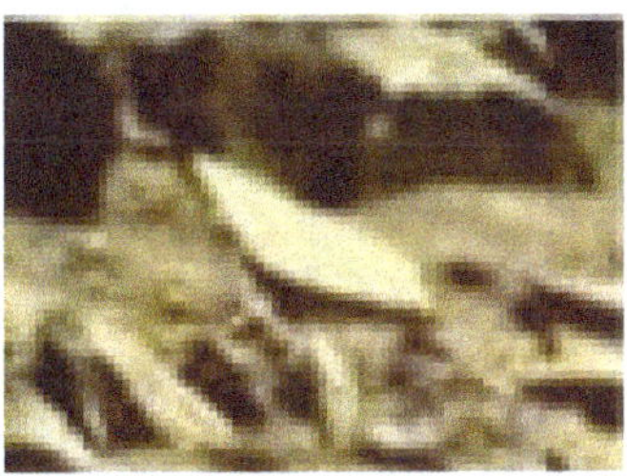

SOL 526, highlighted area 13 (above): A flat square edged object. Natural or artificial? Like so many Martian anomalies, it is located in the midst of numerous other oddly shaped items.

Chapter 2

MARTIAN PETROGLYPHS

SOL 529: EARTHDATE JANUARY 31, 2014

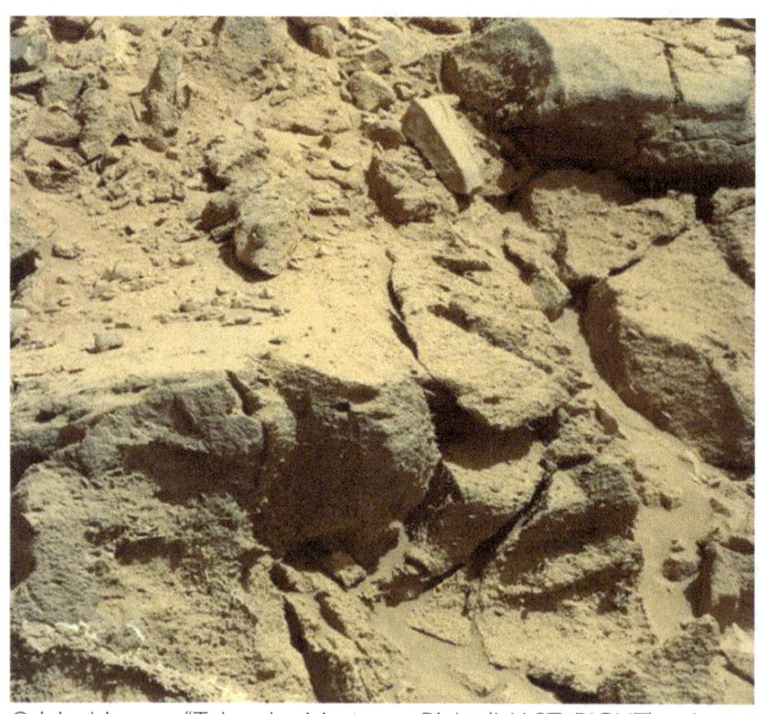

Original image. “Taken by Mastcam: Right (MAST_RIGHT) onboard NASA’s Mars rover Curiosity on Sol 529 (2014-01-31, 15:39:41 UTC).” Image credit: NASA.

ANALYSIS OF MARTIAN PETROGLYPHS, SOL 529

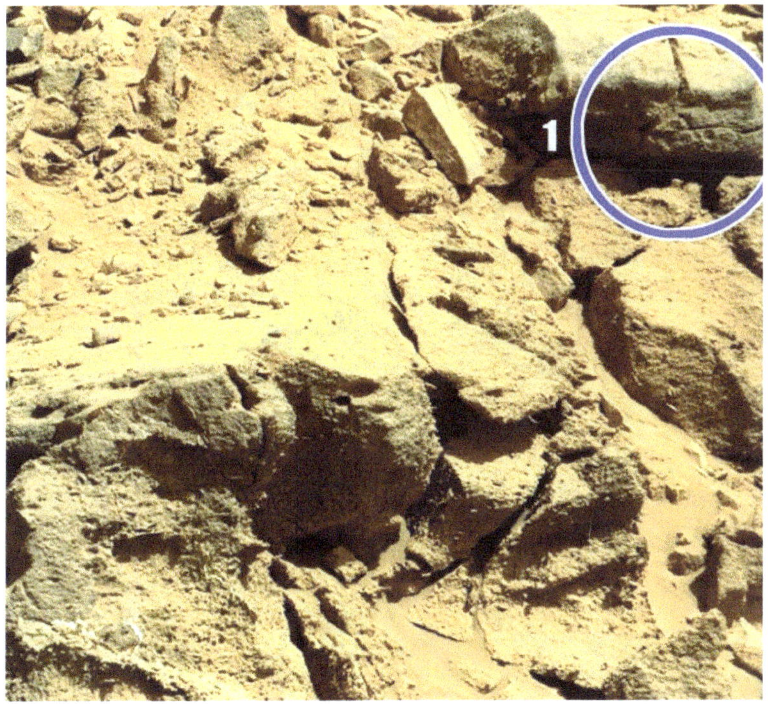

Martian Petroglyphs, Sol 529, with one anomaly circled. Image colored, brightened, and sharpened by the author to improve contrast. At first glance, this dusty rock strewn photo seems rather uninteresting for our purposes. But Curiosity's cameras caught something quite provocative in the upper right-hand corner.

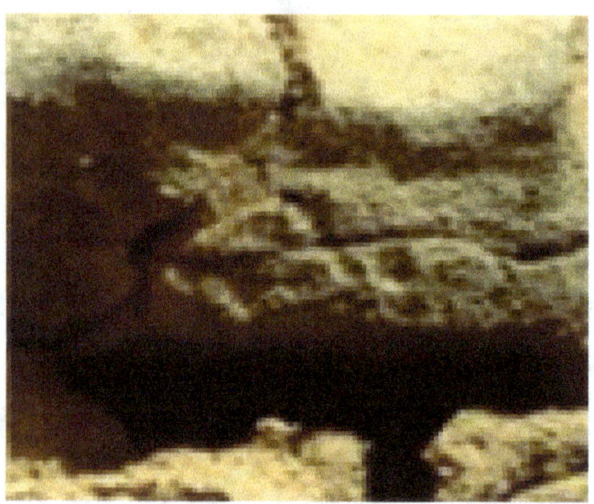

SOL 529, highlighted area 1 (above): Is this rock art, a Martian petroglyph, or perhaps even a pictograph (a pictorial form of writing), created by an advanced intelligence? Or is it nothing more than a series of markings that formed during natural geological processes, such as erosion and wind? What will ancient Egyptian hieroglyphs here on Earth look like 500,000 to 1 million years from now?

Chapter 3

DEBRIS FIELD

SOL 533: EARTHDATE FEBRUARY 4, 2014

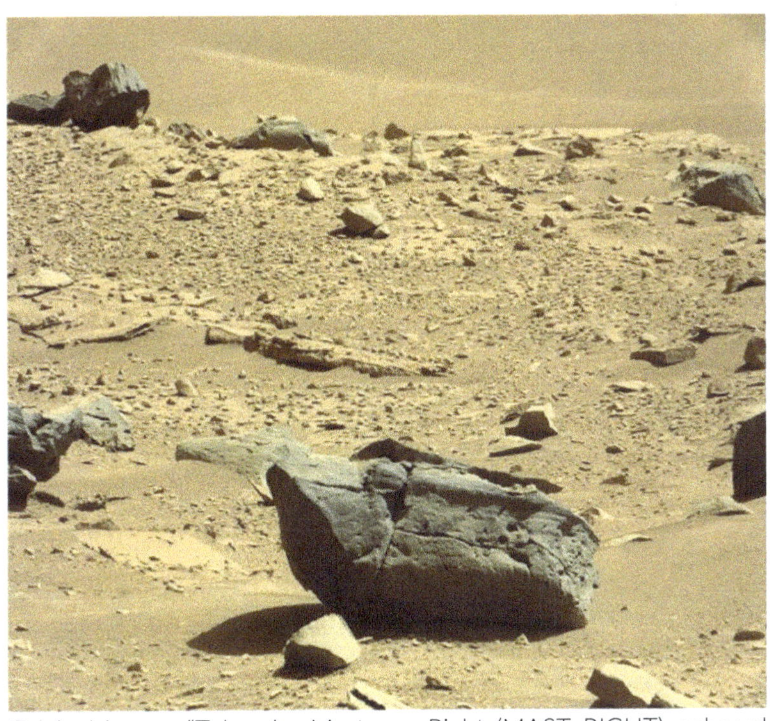

Original image. "Taken by Mastcam: Right (MAST_RIGHT) onboard NASA's Mars rover Curiosity on Sol 533 (2014-02-04, 18:16:08 UTC)." Image credit: NASA/JPL-Caltech/MSSS.

ANALYSIS OF DEBRIS FIELD, SOL 533

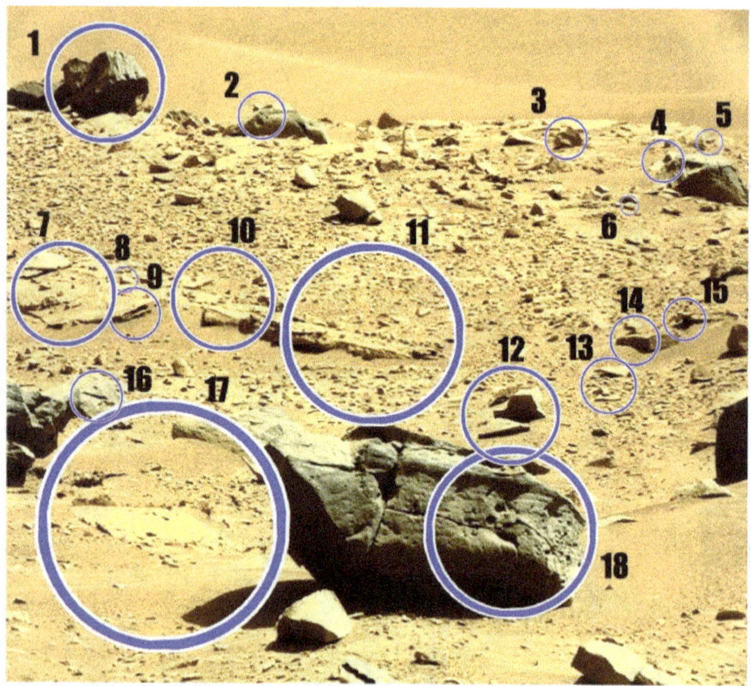

Debris Field, Sol 533, with anomalies circled. Image colored, brightened, and sharpened by the author to improve contrast. There are dozens of objects with 60 degree and 90 degree angles in this NASA image. It also contains numerous strangely shaped objects. Are any of these artificial, or are they all natural in origin? Let us take a closer look.

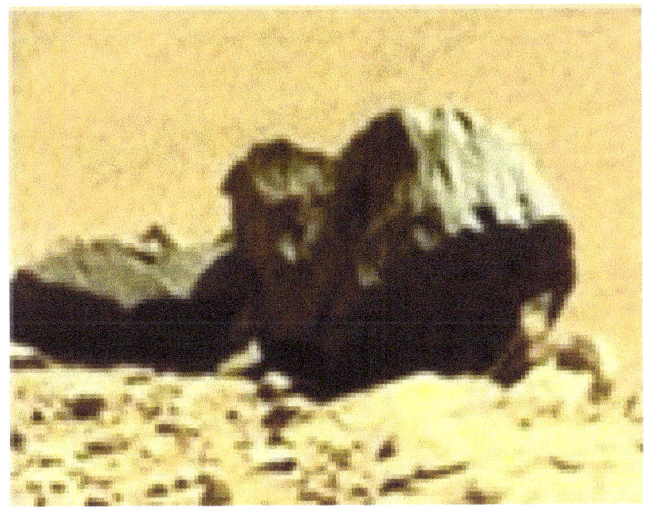

SOL 533, highlighted area 1 (above): A group of oddly shaped objects. The large rounded one tipped up on its side at center right looks something like a petrified tree trunk. A goldish brown object is protruding from underneath it that appears similar to crown molding. If the extraterrestrial hypothesis is correct, this entire pile of objects could be the fossilized remnants of a type of Martian machine, which was then digitally manipulated by modern hands to camouflage its true identity.

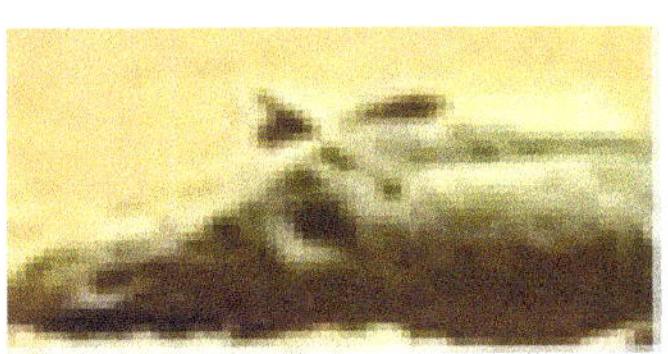

SOL 533, highlighted area 2 (above): This rock bears a square hole along with several round holes. On top there is what appears to be an open book like object. Another provocative yet unidentifiable scene.

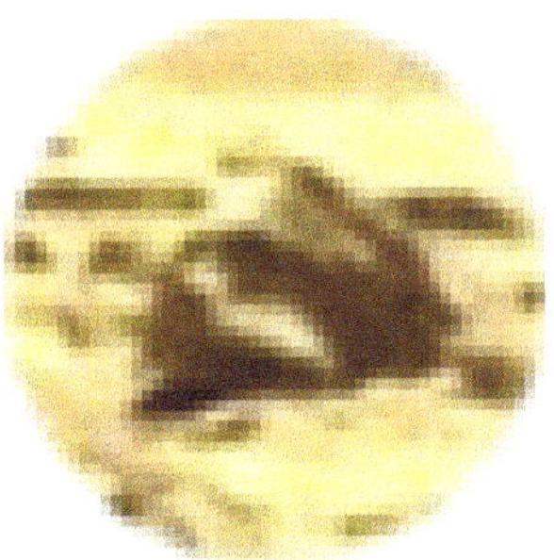

SOL 533, highlighted area 3 (above): The damaged crowned head of a demolished statue or a natural rock formation?

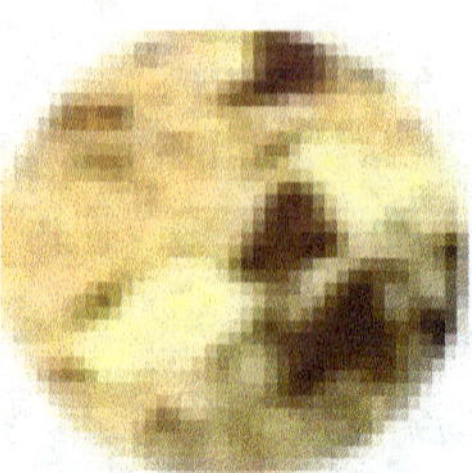

SOL 533, highlighted area 4 (above): A long triangular object and a rectangular object lie side by side. The triangular object has a honeycomb appearance on the end facing the camera, with at least three or four holes visible.

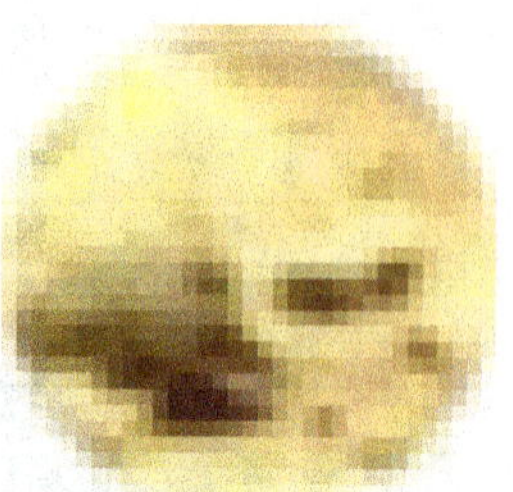

SOL 533, highlighted area 5 (above): With its elongated "skull" and huge "eye sockets" this odd "rock" has the appearance of a gray alien's head half-submerged in the sand.

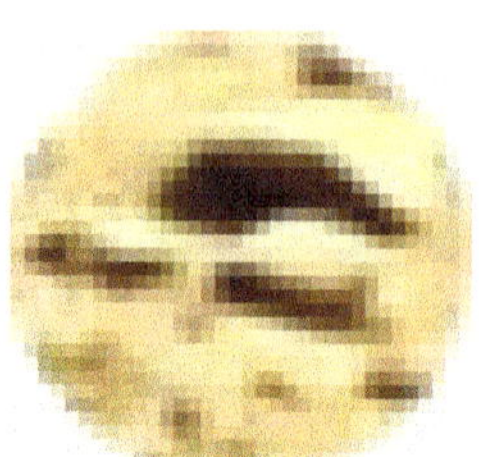

SOL 533, highlighted area 6 (above): Two nearly perfect squares lie beside an oddly shaped "rock" that "bends" downward on each end.

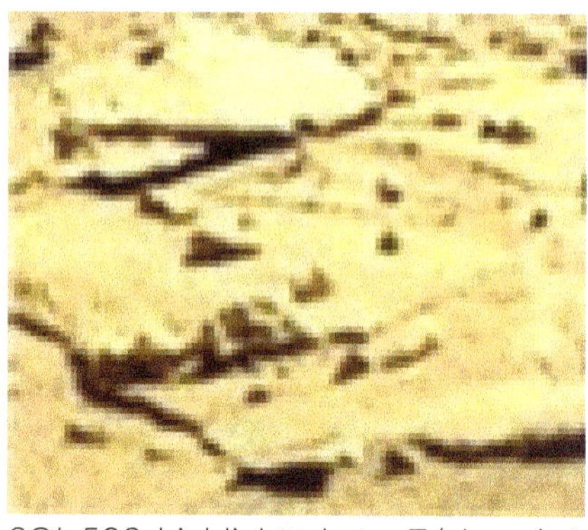

SOL 533, highlighted area 7 (above): A number of board or plate like objects seemingly linked together. Is this the remnant of an ancient cement walkway or roadway, or just a sedimentary rock formation?

SOL 533, highlighted area 8 (above): Another perfect square sitting on the surface of Mars.

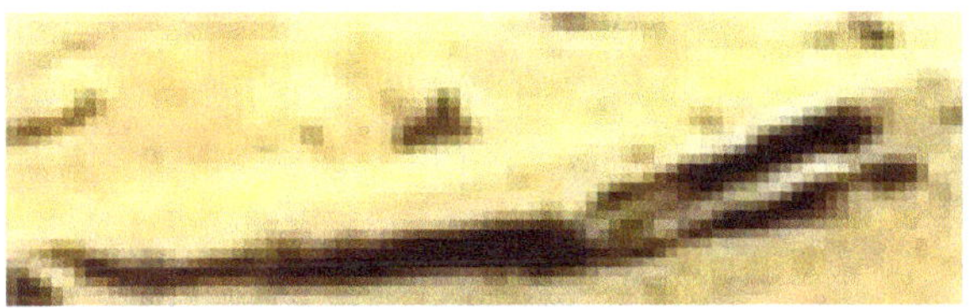

SOL 533, highlighted area 9 (above): This object has the appearance of a folded up rug or newspaper. If it is indeed artificial it could also be a type of rubber or plastic material, or more likely something unknown to our Earth sciences. Whatever its true composition, it is located in an area with other long thin seemingly artificial objects, which, together, appear to form some kind of pathway or covering on the ground.

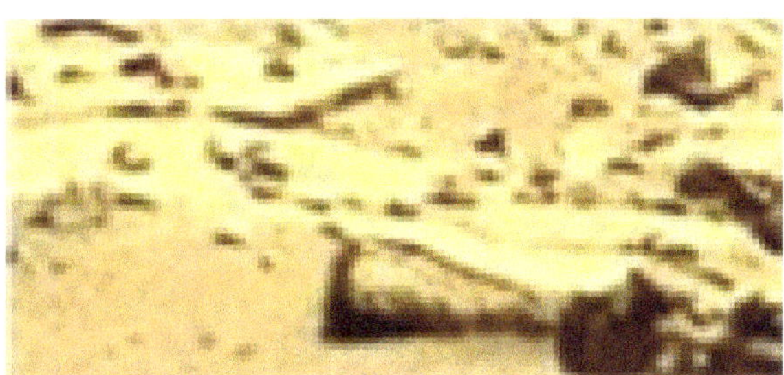

SOL 533, highlighted area 10 (above): The continuation of what I am calling the "roadway," complete with flat, squared off, board like objects that "interlock." In the upper right corner is a small pyramid with a square base. On top of the "board" in the upper left there is an "L" shaped object.

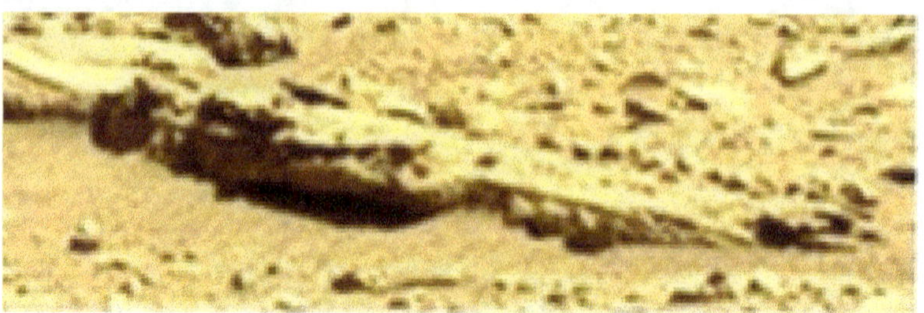

SOL 533, highlighted area 11 (above): This image has a number of strange features, including flat board like objects and small rocks "arranged" in straight lines. A rectangular object, that looks like a small piece of wood, lies at an angle on top of the debris (just left of center). In the upper left one can see the "roadway" running out of frame.

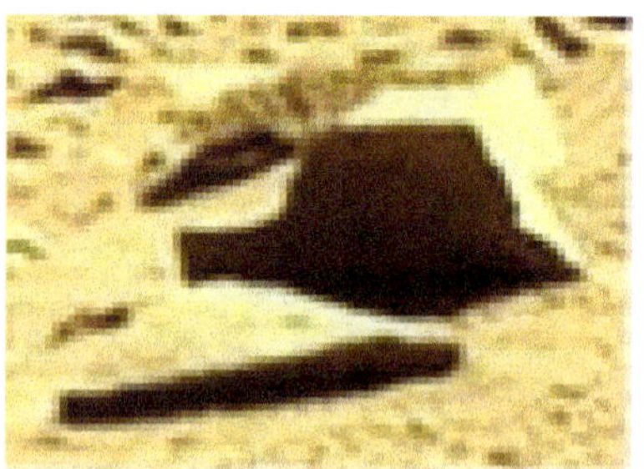

SOL 533, highlighted area 12 (above): These three "rocks" (one a near perfect triangle) could easily be natural formations. Are they? In my opinion their unusual shapes and close proximity suggest possible artificial origins.

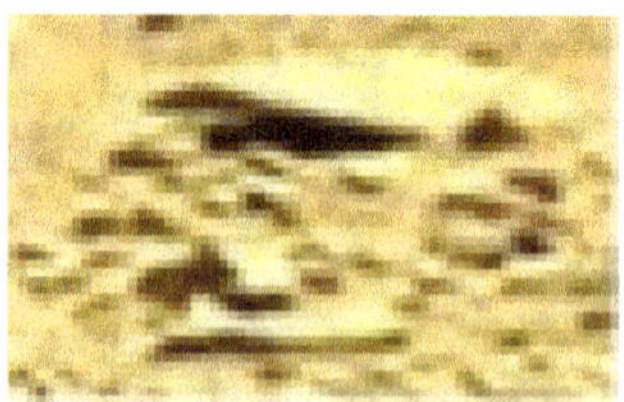

SOL 533, highlighted area 13 (above): Look closely. This pile of debris appears to contain molded or form fitted materials.

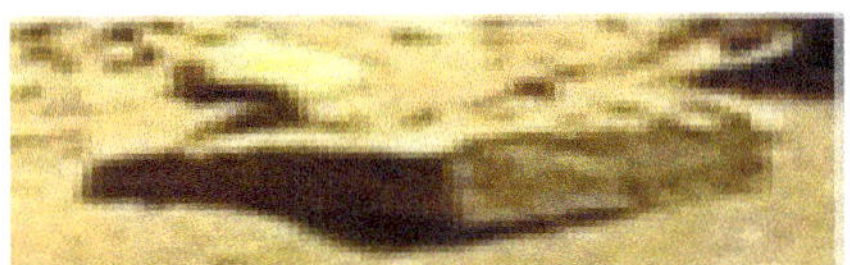

SOL 533, highlighted area 14 (above): This square shaped, multilayered (now fossilized?) object appears to be made of a flexible or spongy paper like material. And what is the multishaped, starfish like object in the far upper left corner? Skeptics say these are nothing more than rocks formed by natural Martian processes.

SOL 533, highlighted area 15 (above): Note the odd shape and squarish end shadow cast by the long object at center, as well as the serpentine objects "writhing" above and around it. A sliced avocado like object (with "pit" still intact) appears at the far left. What are we looking at here?

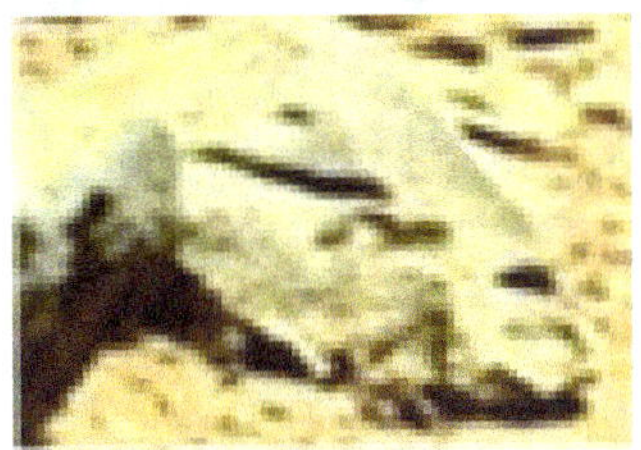

SOL 533, highlighted area 16 (above): This "rock" contains numerous artificial looking shapes, markings, and "imprints."

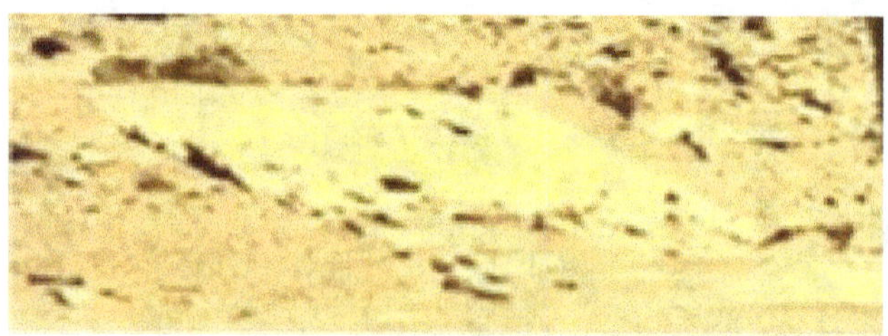

SOL 533, highlighted area 17 (above): This is one of the most piquant set of objects ever photographed by the Mars rovers. Numerous, flat, perfectly aligned, board like items overlap one another on the side of a Martian hill. Are these artificial structures, perhaps used in the construction of a walkway (seen running downward to the bottom right)? Is this debris part of a collapsed building? Note the many curious objects scattered about in the surrounding area.

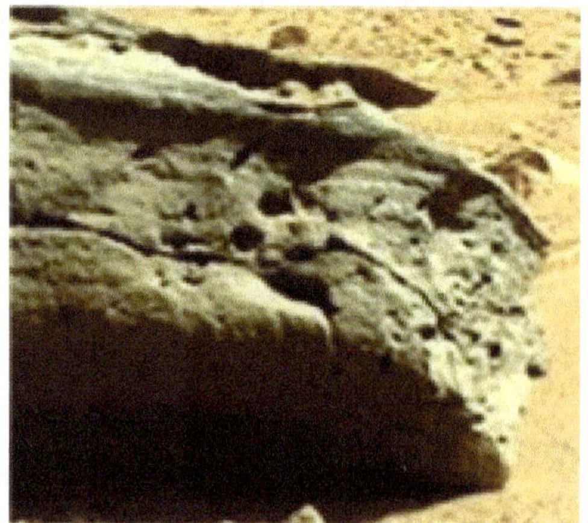

SOL 533, highlighted area 18 (above): Though this entire "boulder" is suspicious, we are focusing on the right half only, where there are enough oddities to keep both believers and skeptics occupied for years to come. Perfect circular holes, as well as "carved" or "engraved" straight lines, dots, curves, triangles, and unidentifiable shapes abound. Are these pictographs or natural phenomena?

Chapter 4

DEBRIS FIELD

SOL 532: EARTHDATE JANUARY 30, 2014

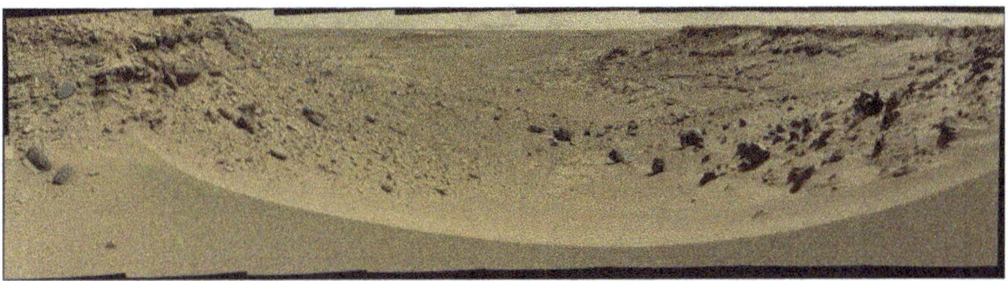

Original image. "This large composite photo combines a number of individual frames. They were taken by Mastcam's left-eye camera during early afternoon, local solar time, of the 528th Martian day, or sol, of Curiosity's work on Mars (Jan. 30, 2014). For scale, the dune in the middle, known as 'Dingo Gap,' is only about three feet tall in the center." Image credit: NASA/JPL-Caltech/MSSS.

ANALYSIS OF DEBRIS FIELD, SOL 532

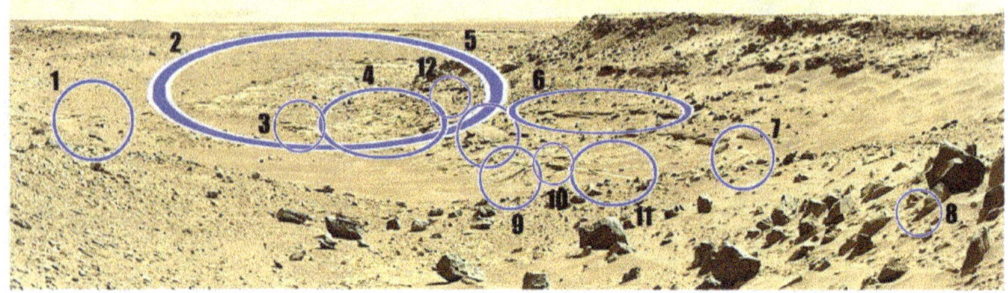

Debris Field, Sol 532, with anomalies circled. Image cropped, colored, brightened, and sharpened by the author to improve contrast. This NASA image is full of interesting anomalies, many which appear overtly artificial to both untrained *and* (many) trained eyes. If these are natural formations, as mainstream scientists claim, there seem to be few definitive explanations—and none that answer every question—about their appearance, shape, size, origins, coloring, location, and markings.

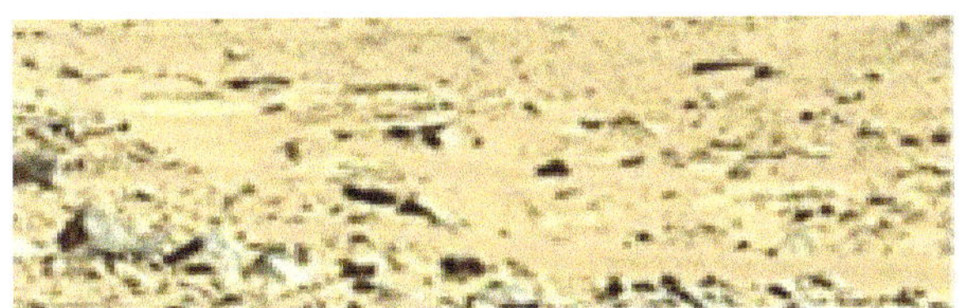

SOL 532, highlighted area 1 (above): When one looks at the full image (previous page) of this photograph, there are many indications of what appear to be paved, road like structures. Under magnification the area pictured here looks like the remnants of a road that runs across the entire (full) frame from upper left to middle right. The images below provide additional information on this mysterious anomaly.

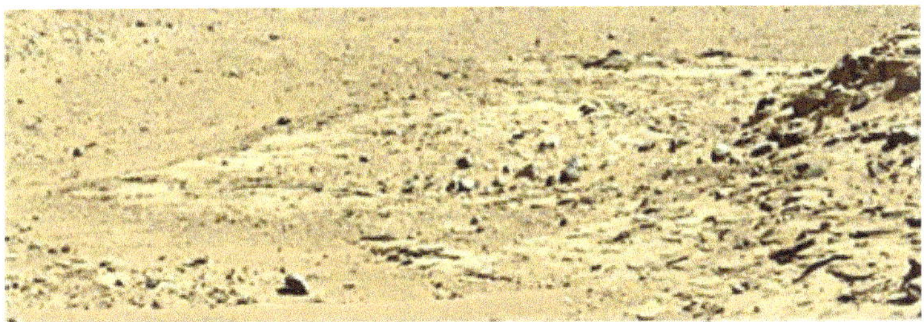

SOL 532, highlighted area 2 (above): If it were covered in grass, this large rectangular impression in the ground would look something like a small football field. If it is artificial (the overall shape, the flat cement like "plates," and the 90 degree corners suggest that it is), it may be all that is left of the foundation of a building of some kind.

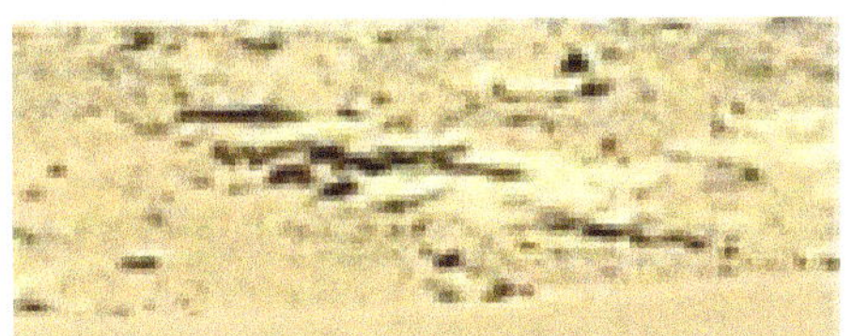

SOL 532, highlighted area 3 (above): This appears to be an upheaval of soil and rock—but caused by what? Crustal movement? Or could it have been pushed up by an underground drainage pipe, water supply pipe, or utility tunnel?

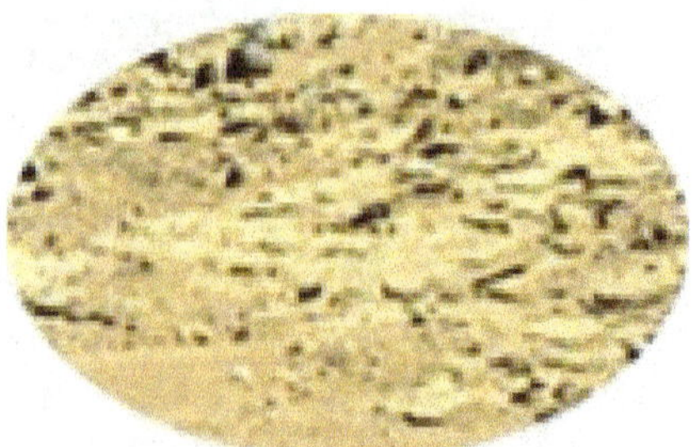

SOL 532, highlighted area 4 (above): This rectangle shape, encircled with several layers of what appears to be fitted stonework, masonry, and mortared blocks, is located next to the larger "building foundation" described in area 2. Intentional or coincidental?

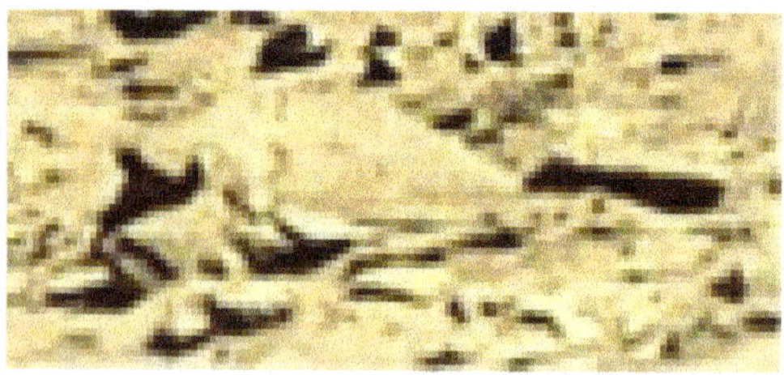

SOL 532, highlighted area 5 (above): A large, flat, square shaped object lies on its side, surrounded by irregular stacks of smaller square objects, as well as a number of very odd looking forms in the lower right foreground. Magnify them and study them carefully.

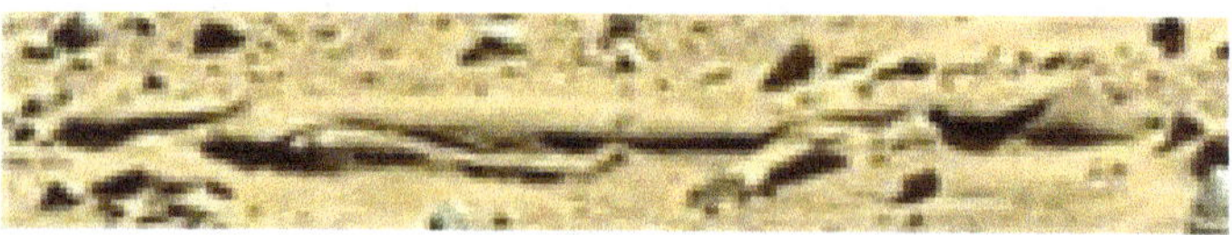

SOL 532, highlighted area 6 (above): This long, narrow, rounded object actually appears, disappears, and reappears (from left to right), through the full frame of this image. Like the image in area 5, it is surrounded by other strangely shaped objects, in this case a curious "L," as well as several small partially embedded pyramids, triangles, squares, and spheres.

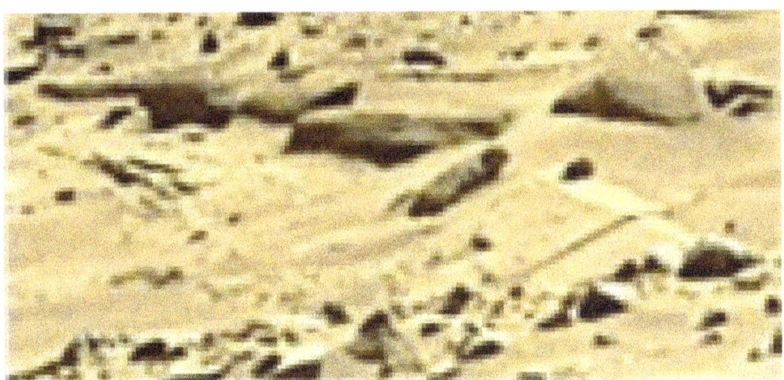

SOL 532, highlighted area 7 (above): I have included extra space on the left to provide context for this remarkable image. In the lower right is what looks like molded cement (now split down the middle). The top of the object is flat and horizontal, while its left side runs down the slope at about a 25 degree angle. To the upper left we see a continuation of the rounded object described in area 6. Here, broken into pieces, it seems to be associated with the molded cement object. Remnants of a road, an aqueduct, a parking lot? The strange "contorted" object in the upper right is only one of the many other artificial looking oddments in this photo.

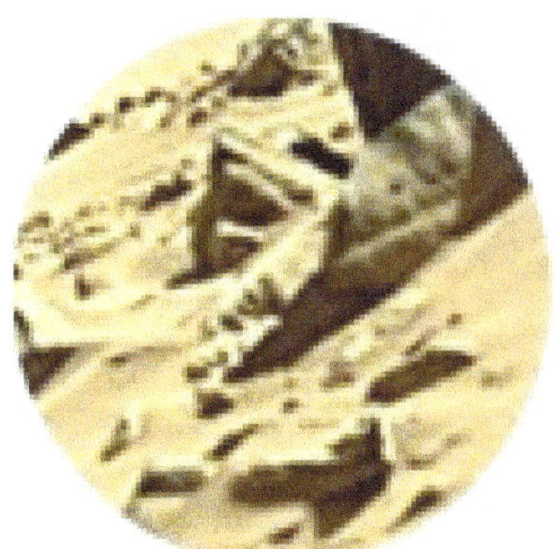

SOL 532, highlighted area 8 (above): At center is what looks like the head of an ancient statue laying on its left side in the Martian sand. This infamous "rock" has sparked much excitement and debate since the Curiosity rover captured it on its mastcam on January 30, 2014.

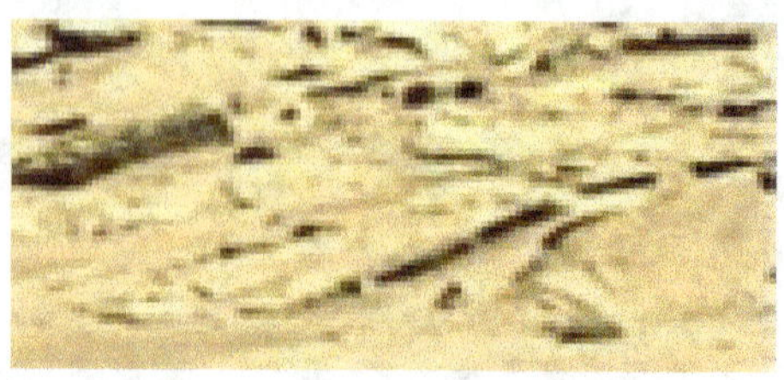

SOL 532, highlighted area 9 (above): These so-called "natural rock formations" have square corners and what look like openings for doors and windows, appearing very much like the "building foundations" described in areas 2 and 4. Additionally, the main "foundation" (lower center) seems to be surrounded by paving stones, walkways with retaining walls, and a dismantled road. Round and oval holes appear in the rocks nearby. If Mother Nature created the many odd formations and objects in this image, her talents are far more astonishing than heretofore imagined.

SOL 532, highlighted area 10 (above): I have extended this image to the left for context. The long flattish object on the middle right, embedded in the sand on its side, appears completely artificial: it has a hollow interior and is made of what looks like layers of a cement-like material. Note the "sidewalk" with "retaining walls" encircling the object to the left, as well as its proximity to the building foundation described in area 9 above. It has some similarities to the object in Sol 533, area 9.

SOL 532, highlighted area 11 (above): The whitish object in the center of this NASA image may be a "rock," but to many of us its rounded circular sides and long straight outline suggest otherwise. If this photograph had been taken in Texas, for instance, most people would immediately identify it as a drainage pipe, a sewage pipe, or an underground water supply pipe. If it is indeed artificial, it may be part of a subterranean pipe system that runs across the entire area of the full frame image above, and thus could have caused the soil and rock upheaval shown in area 3. Note the dark, flat, hollow, airplane wing like object in the upper left, which is described in area 10. It is amazing for a "rock."

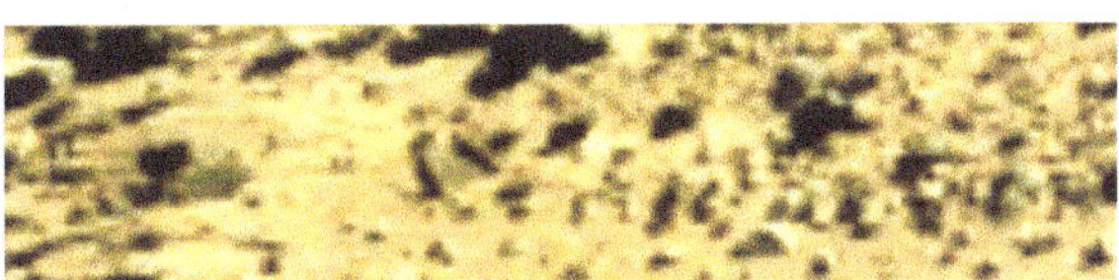

SOL 532, highlighted area 12 (above): There are several extraordinary objects in this NASA photo, all which may be natural but which are indicative of artificiality. To the left of center is an object protruding from the ground at a 45 degree angle that has the shape and look of the broken bow or stern of a small boat. To the right of it, small rocks appear quite evenly distributed over the Martian surface: their positioning next to the "boat" seems curious—but may merely be the result of weather and erosion. To the left of the "boat" is a large, whitish, circular shape in the sand with "spokes," and below that another one similar in size and appearance. Not everyone will see what I see—and some will undoubtedly see things I do not.

Chapter 5

ROCK CIRCLE

SOL 528: EARTHDATE JANUARY 30, 2014

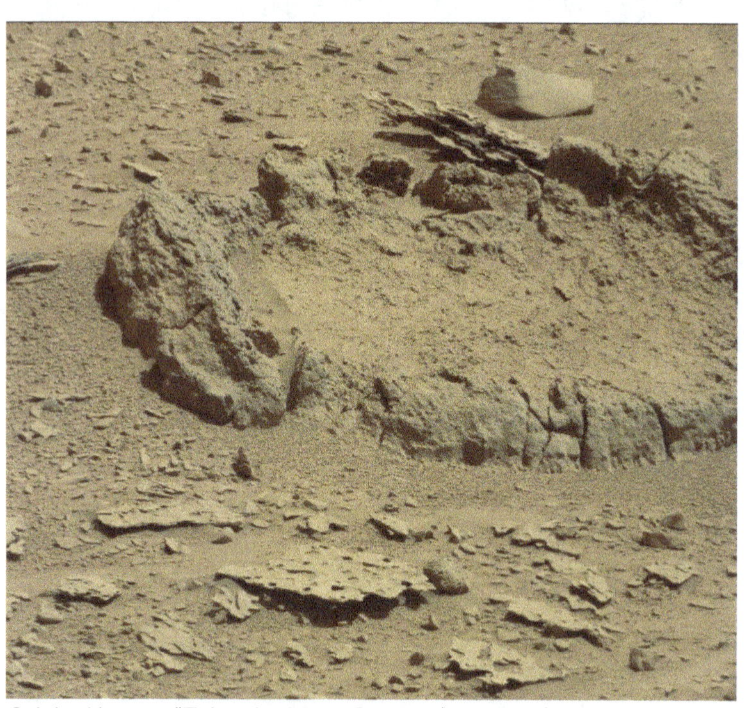

Original image. "Taken by Mast Camera (Mastcam) onboard NASA's Mars rover Curiosity on Sol 528 (2014-01-30, 13:54:31 UTC)." Image Credit: NASA/JPL-Caltech/MSSS.

ANALYSIS OF ROCK CIRCLE, SOL 528

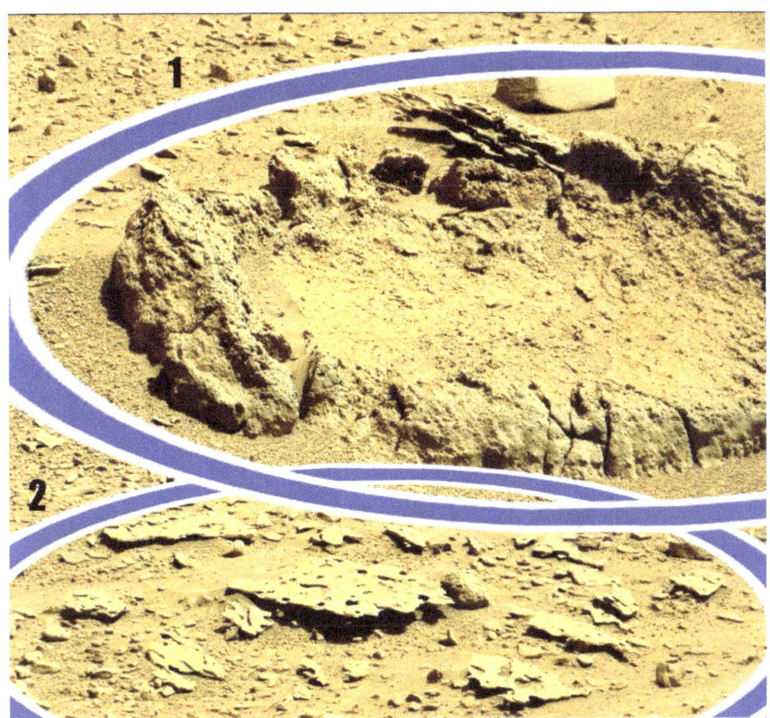

Rock Circle, Sol 528, with anomalies circled. Image colored, brightened, and sharpened by the author to improve contrast. This firepit like structure is surrounded by paper thin, cavity covered "rocks." Are we looking at astrogeology or astroarchaeology?

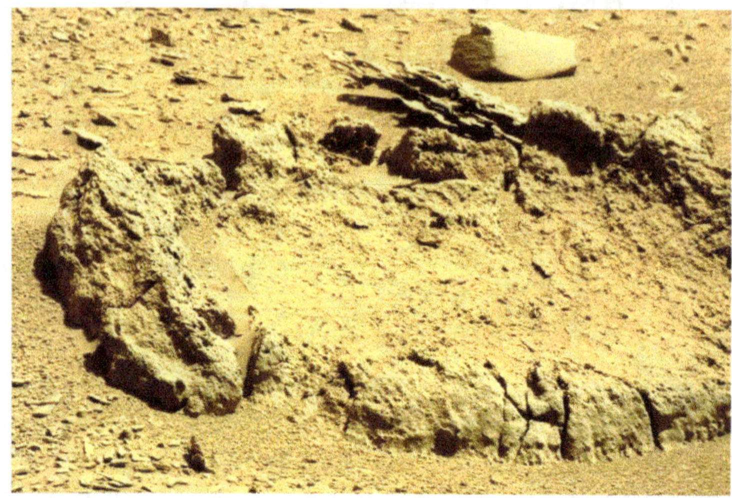

SOL 528, highlighted area 1 (above): There are indeed natural processes that could create this circular geoformation. But in this case, did they? There is no way of knowing for sure—until, if and when, human scientists examine this object in situ.

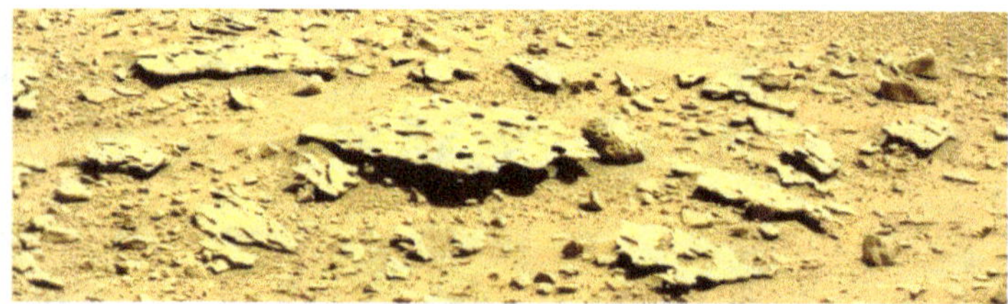

SOL 528, highlighted area 2 (above): Are the cavities and pits in these rocks the result of erosion? Are the shapes of these rocks caused by storm driven dust? Or did something artificial and more violent create their odd appearance—such as a nuclear war? Mainstream scientists have ready answers to these questions. But are they correct, or are they merely parroting one another's outdated theories? At this point everything is opinion—on both sides of the aisle.

Chapter 6
HILLSIDE DEBRIS

SOL 538-A: EARTHDATE FEBRUARY 9, 2014

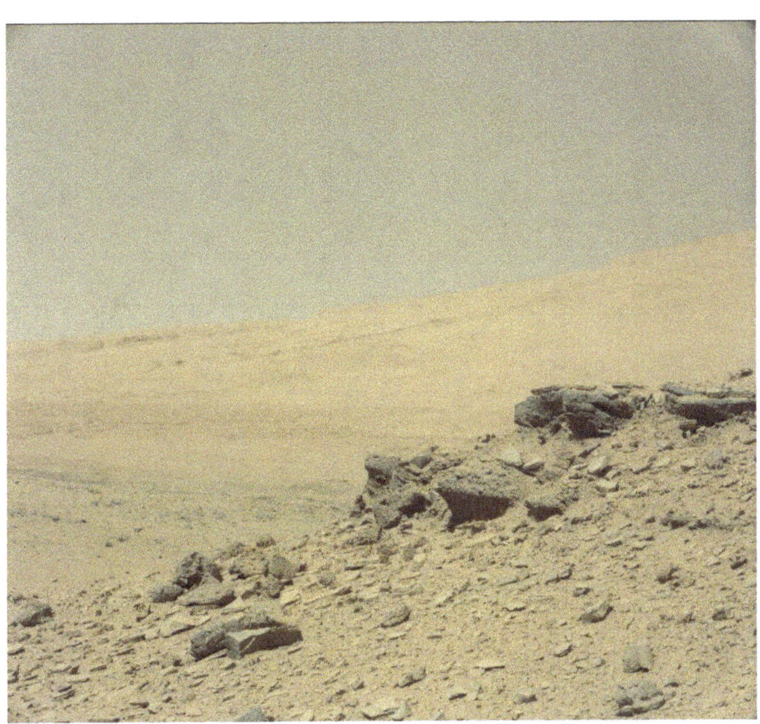

Original image. "Taken by Mast Camera (Mastcam) onboard NASA's Mars rover Curiosity on Sol 538 (2014-02-09, 22:36:47 UTC)." Image credit: NASA/JPL-Caltech/MSSS.

ANALYSIS OF HILLSIDE DEBRIS, SOL 538-A

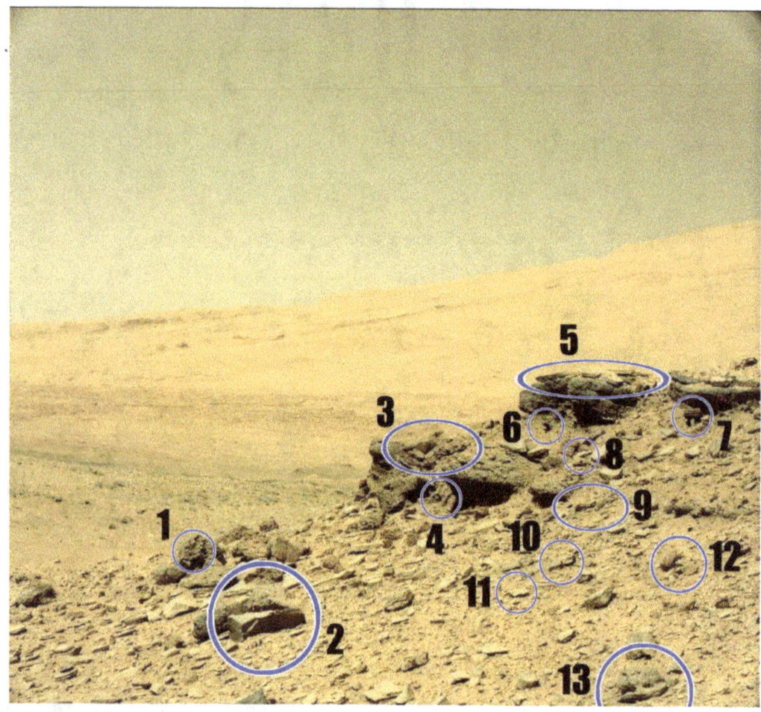

Hillside Debris, Sol 538-A, with anomalies circled. Image colored, brightened, and sharpened by the author to improve contrast. We are being told that *everything* I have highlighted in this image is a "rock," the product of natural geological Martian processes. This may be true. It may also be false, however. If a digital artist was instructed to "airbrush" away obvious artificial objects, what would we see? I believe we might see something like this NASA image, which contains not only countless stacked square edged and triangular shaped objects, but both bizarrely shaped objects and what look like mechanical pieces that have broken off larger equipment. All of these seem out of place laying on the surface of a "dead planet." If we add thousands, tens of thousands, or even hundreds of thousands of years of weathering to this scenario, it might result in an image much like this one, complete with dust covered, partially embedded, fossilized debris that is now all but unrecognizable—allowing official authorities to shrug it all off as nothing more than "geology." I believe that such images deserve a more serious look.

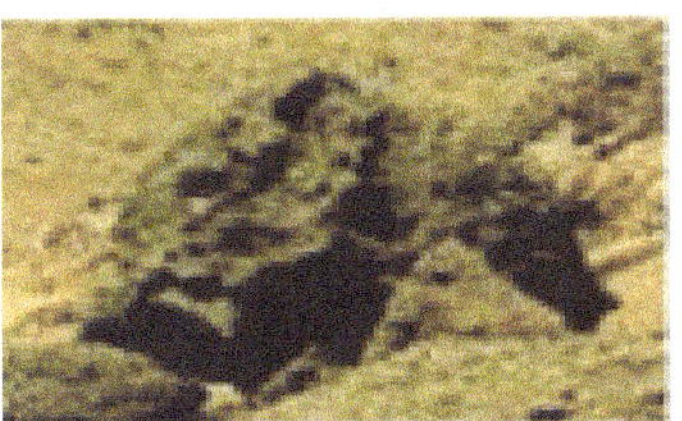

SOL 538-A (above), highlighted area 1: This "rock" sports a number of round holes (some of them near perfect circles) as well as other unusual marks, pits, and cavities.

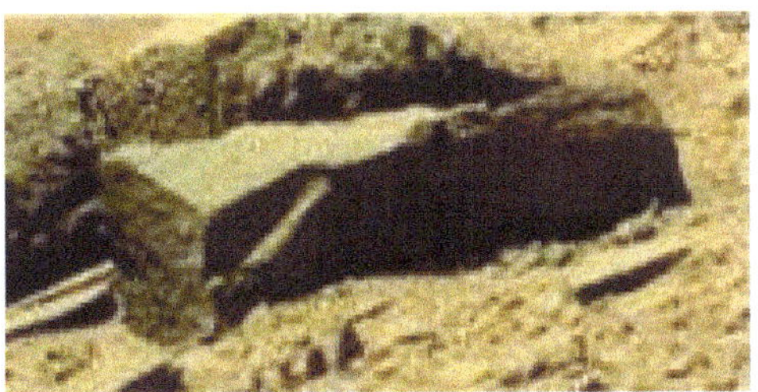

SOL 538-A, highlighted area 2 (above): This object has a box like or book like shape. Is it merely a "rock," or is this what an ancient cement block might look like after thousands of years of exposure to the arid, cold, highly radioactive Martian surface?

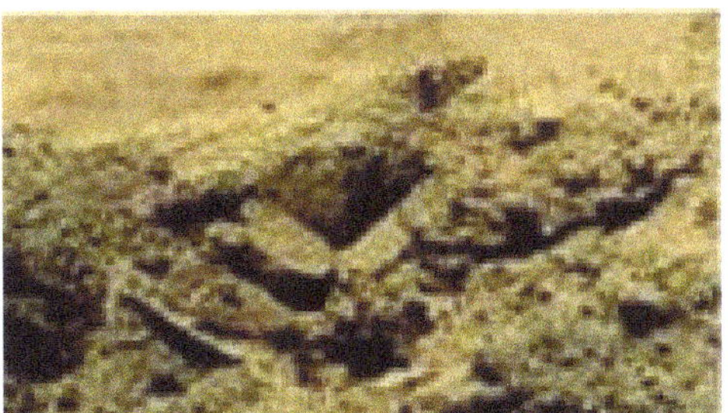

SOL 538-A, highlighted area 3 (above): There is a lot going on in this small area. Much of it has the appearance of concretized or fossilized mechanical debris.

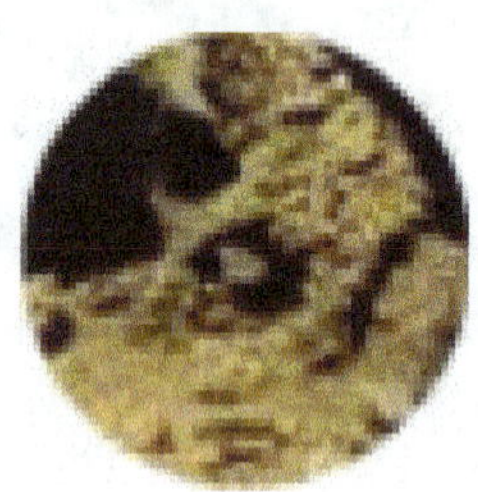

SOL 538-A, highlighted area 4 (above): At the very center of this image there is a hole "bored" all the way through an oddly shaped rock. Natural? Maybe. But the object is lying next to the jumbled pile of artificial-looking "debris" described in area 3.

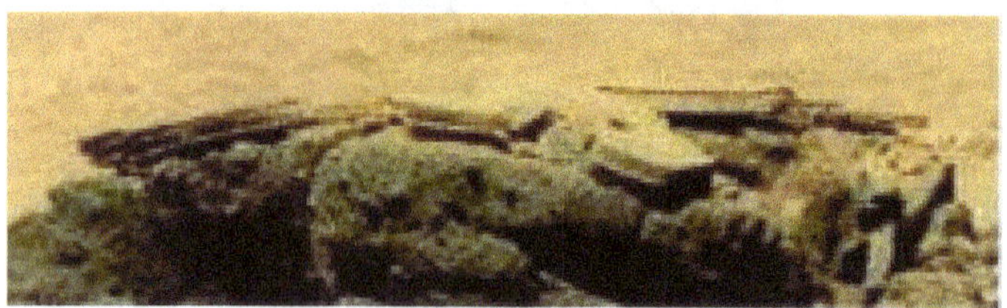

SOL 538-A, highlighted area 5 (above): The layered book like objects lying precariously on top of large grayish boulders in this NASA photo are probably naturally formed sedimentary rocks. But how can we know for sure until they are examined by geologists? There are more objects similar to these to the right (see full frame image).

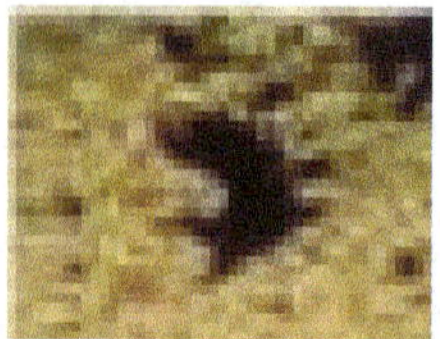

SOL 538-A, highlighted area 6 (above): What is this weird item? It is casting a pointed triangular shadow on the Martian sand.

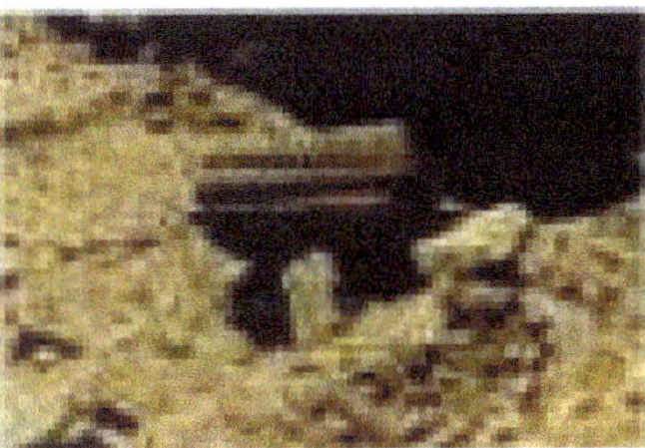

SOL 538-A, highlighted area 7 (above): This square, perfectly level, book like "sedimentary rock" is balancing delicately on a smaller object. Accidental or intentional?

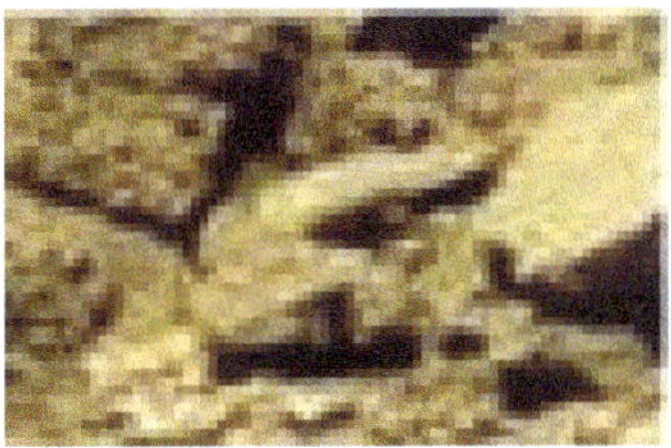

SOL 538-A, highlighted area 8 (above): There is a rounded boomerang like object lying in the center of this photo. What is it and what is it doing there?

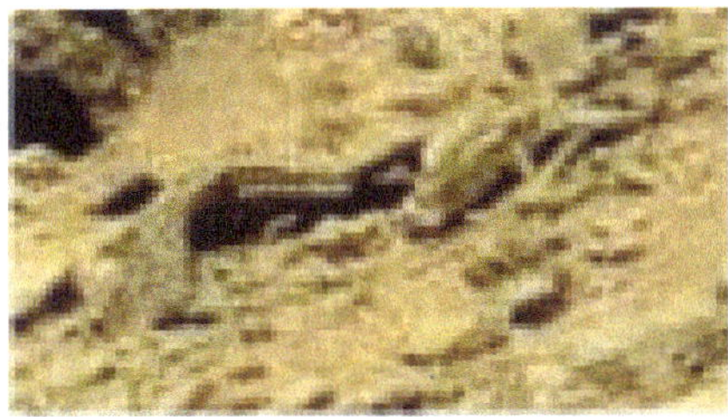

SOL 538-A, highlighted area 9 (above): This small section, which I have cut from the full frame photo, contains a number of mysterious items, including a twisted, "braided," cable or snake like object (left of center) and a pile of fragile spindly objects that look like bones or sticks. Other odd and unidentifiable items surround them.

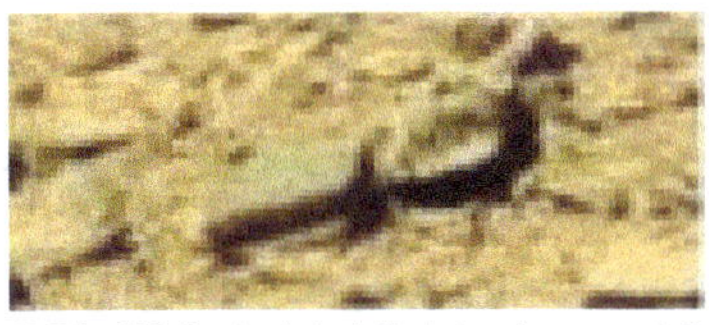

SOL 538-A, highlighted area 10 (above): Four book like "sedimentary rocks" lie perfectly stacked on top of and next to one another.

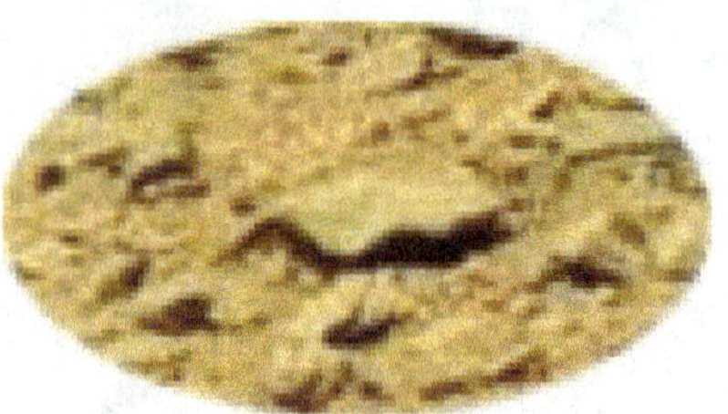

SOL 538-A, highlighted area 11 (above): There is a natural scientific explanation for why this object is shaped the way it is. However, common sense tells us that it is more likely a molded fragment that came from a building, a machine, or some type of infrastructure. Note the many artificial looking objects around it.

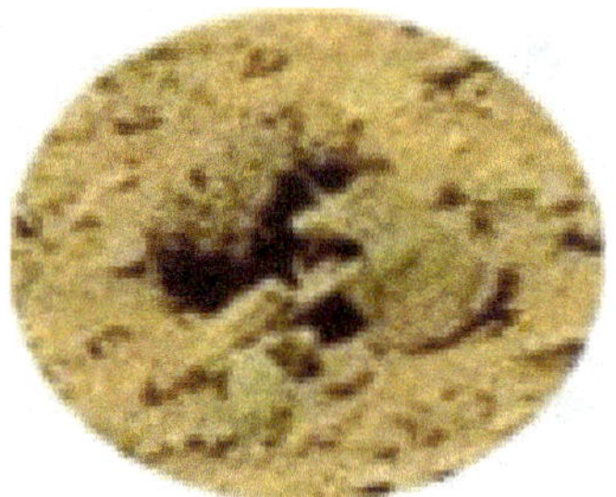

SOL 538-A, highlighted area 12 (above): A rock or the remains of an axe?

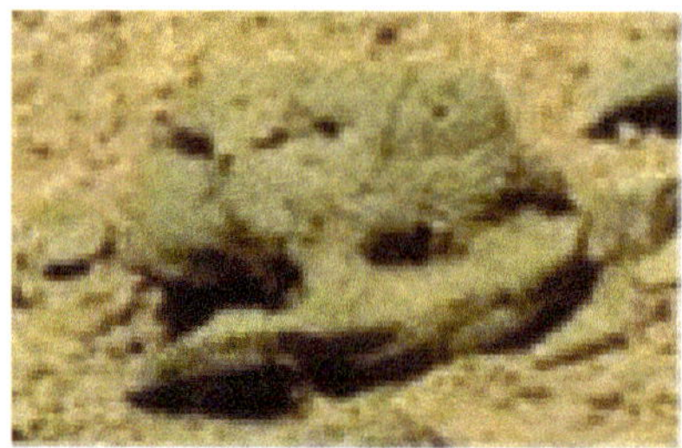

SOL 538-A, highlighted area 13 (above): This oddly shaped "rock" possesses round holes, an eye like "etching" (on its upper right surface) and a bizarre appendage (at the bottom). Remains of a statue?

Chapter 7
ODD OBJECTS

SOL 538-B: EARTHDATE FEBRUARY 9, 2014

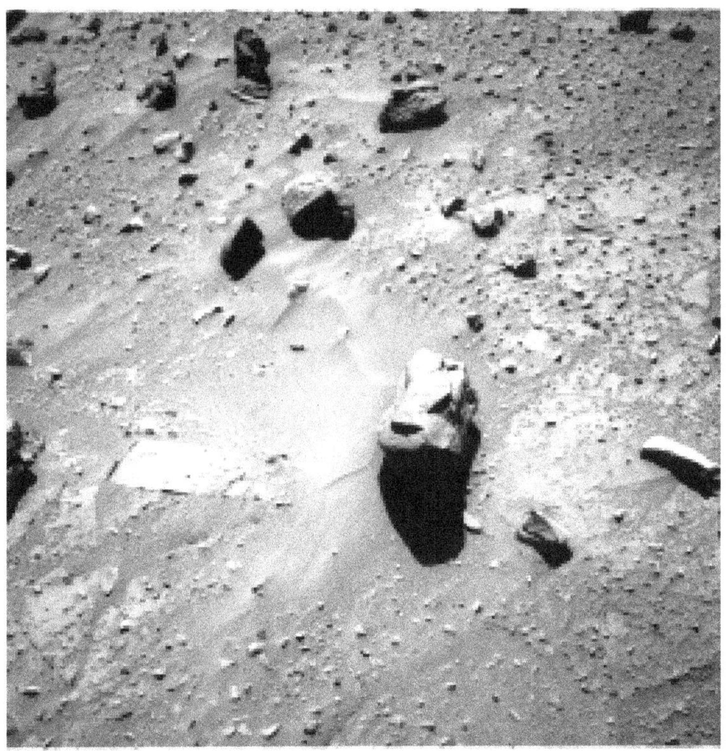

Original image. "Taken by Right Navigation Camera onboard NASA's Mars rover Curiosity on Sol 538 (2014-02-09, 22:27:10 UTC)." Image credit: NASA/JPL-Caltech.

ANALYSIS OF ODD OBJECTS, SOL 538-B

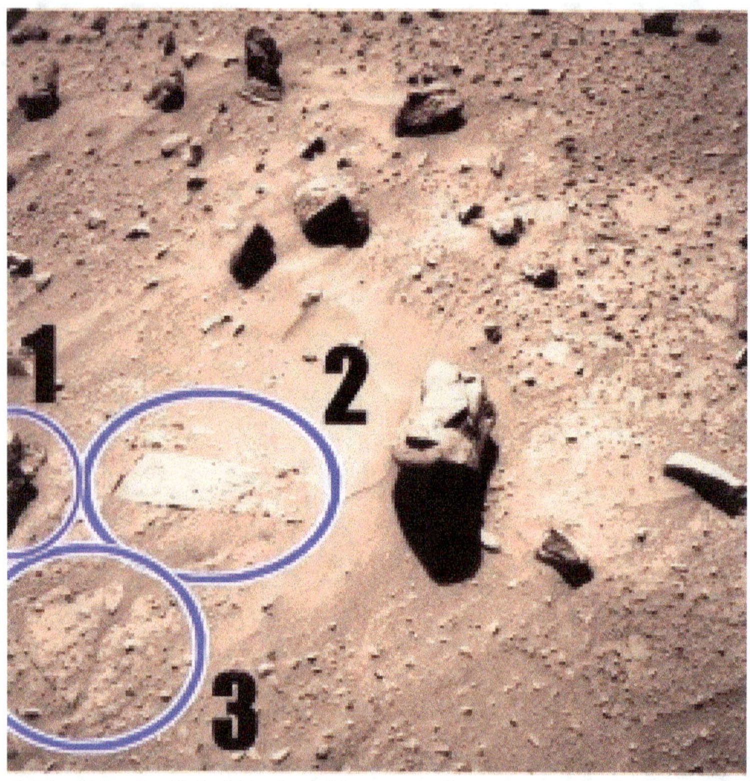

Odd Objects, Sol 538-B, with anomalies circled. Image colored, brightened, and sharpened by the author to improve contrast. Several possible anomalies cluster together in the lower left corner of this low res NASA photo.

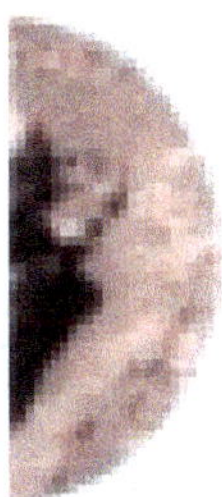

SOL 538-B, highlighted area 1 (above): Despite lying mostly outside the photo frame, this odd cluster of objects possesses what look like artificial features.

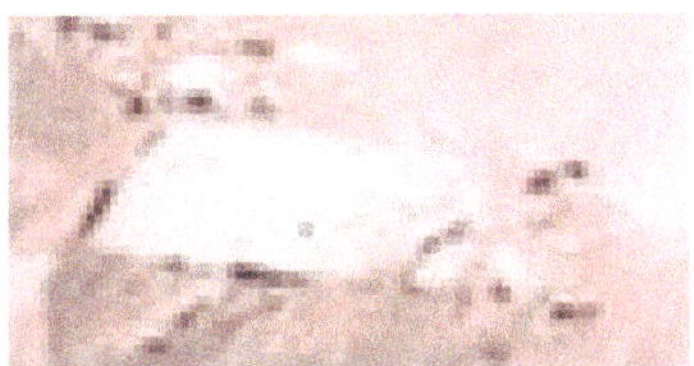

SOL 538-B, highlighted area 2 (above): This board like object with 90 degree corners is laying in plain sight on the surface of Mars.

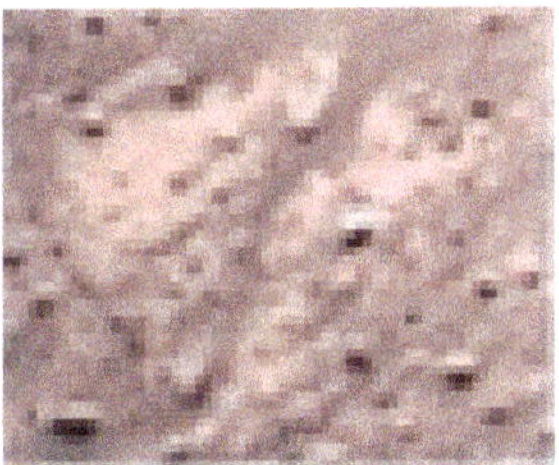

SOL 538-B, highlighted area 3 (above): There are countless thousands of triangular shaped objects on Mars. Are they natural or artificial? In my view the Red Planet hosts both types.

Chapter 8

DEBRIS FIELD

SOL 1448: EARTHDATE SEPTEMBER 1, 2016

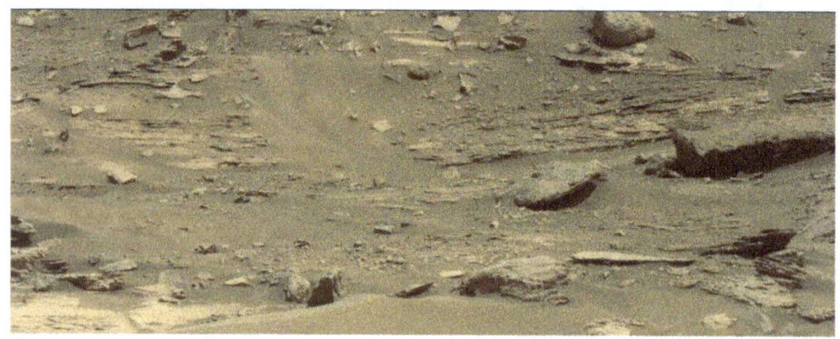

Original image. "Taken by Mast Camera (Mastcam) onboard NASA's Mars rover Curiosity on Sol 1448 (2016-09-01, 21:52:20 UTC)." Image credit: NASA/JPL-Caltech/MSSS.

ANALYSIS OF DEBRIS FIELD, SOL 1448

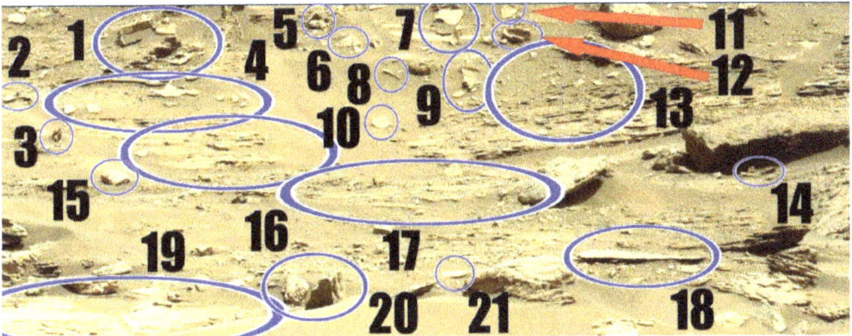

Debris Field, Sol 1448, with anomalies circled. Image colored, brightened, and sharpened by the author to improve contrast. This entire frame appears to be a debris field, with, if correct, exploded structures, sidewalks, roadways, and building furnishings, along with numerous unidentifiable but artificial looking objects.

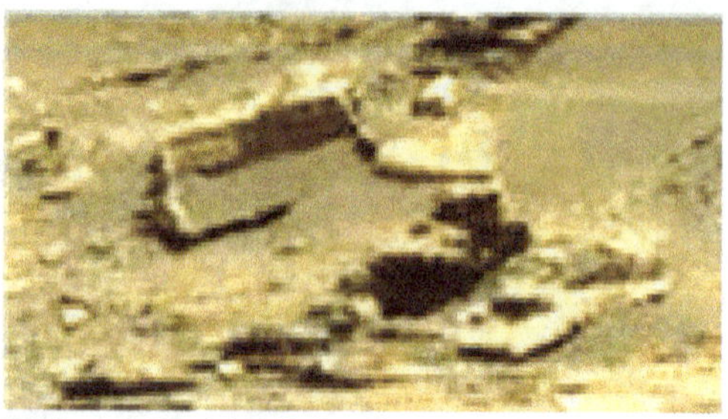

SOL 1448, highlighted area 1 (above): To many observers everything in this image looks unnatural. Among such items are the remnants of house like structures (with both their foundations and some of their brick walls still intact), paved sidewalks, roads, and embankments. What looks like an outdoor stone fireplace sits just right of lower center.

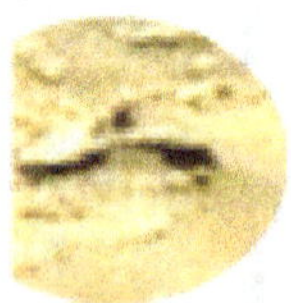

SOL 1448, highlighted area 2 (above): A "V"-shaped object laying on the surface.

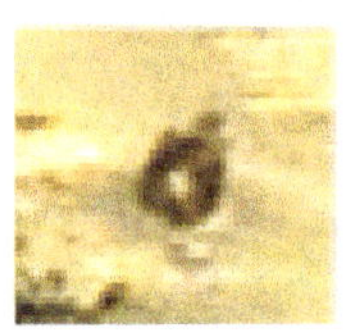

SOL 1448, highlighted area 3 (above): Besides being a different color than nearly everything else in the full frame photo, when magnified this strange object with a hole in the center is circular with what appears to be gear like appendages running along the outside "rim."

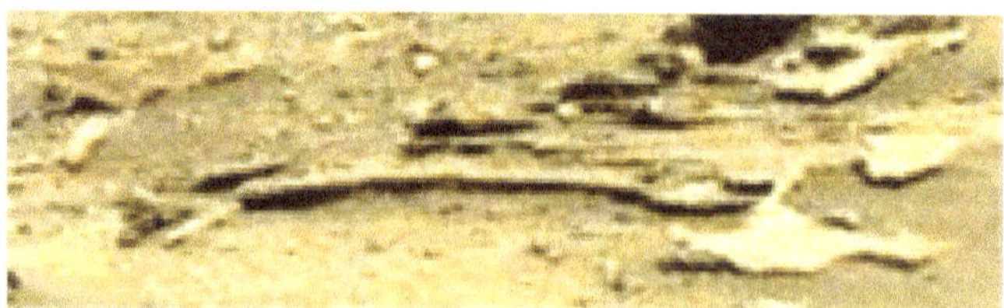

SOL 1448, highlighted area 4 (above): From a human perspective this area appears to be the backyard of a house, with steps leading down to what used to be a deck and an open space. In the upper left (10:00) a wavy rug like object lies out in the open; along the right side of this image are what look like pieces of melted metal sheeting.

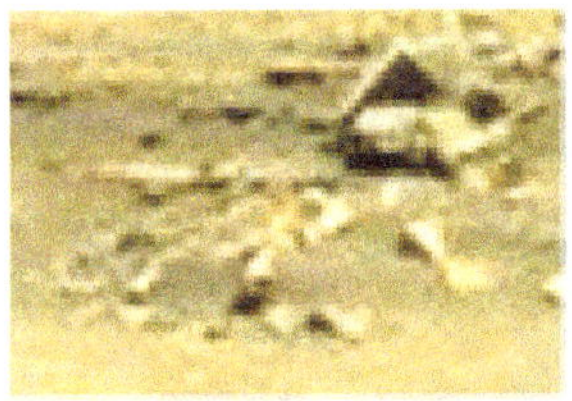

SOL 1448, highlighted area 5 (above): At first glance this section, which is probably quite small, looks like a miniature logger's camp, complete with a cabin, firepit, and log seats. On the other hand, what looks like a cabin roof could be a pyramidal structure of some kind, and the "logs," which appear hollow, could be stacked pipes used for water, electricity, or sewage. Numerous out-of-place objects lie strewn over the ground, including squares, circular items, triangles, rectangles, pyramids, blocks, and "broken" tubular objects.

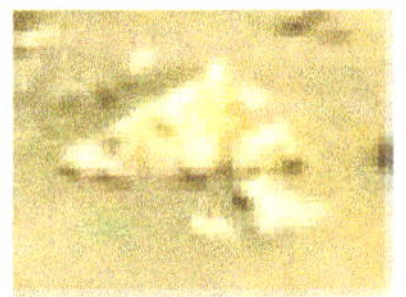

SOL 1448, highlighted area 6 (above): If this shattered triangular object was not sitting in the midst of hundreds of artificial looking objects, it might not attract our notice, and, like orthodox science, we would pass it off as a mere rock. Its location, positioning, "markings," and coloring, however, suggest it may have been intentionally made by a superior intelligence.

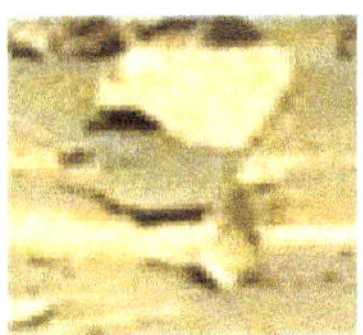

SOL 1448, highlighted area 7 (above): The low resolution of this image makes it impossible to determine what we are seeing here. What we *can* make out are three unusually shaped, flattish, bent, objects, lying next to one another on the Martian surface.

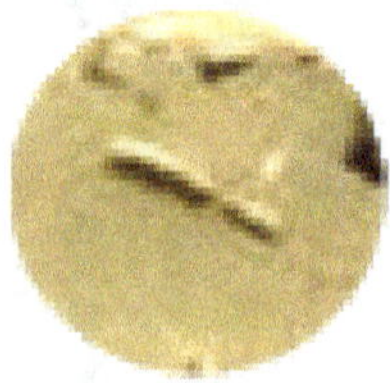

SOL 1448, highlighted area 8 (above): This hockey stick like object appears to be hooked or curved on its upper end.

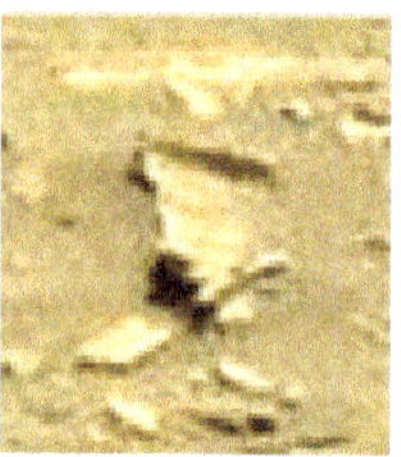

SOL 1448, highlighted area 9 (above): The large object at center appears to be made of brickwork. It has a "sculpted base," a straight side on the right, and a slanted side on the left. Did it topple over, fall forward down the hill, and break apart during a Martian war? To its lower left there is what looks like a splintered board or broken piece of stonework. At the top of this photo a number of strangely shaped objects are embedded in the Martian soil. Other "fragments" showing possible signs of intelligent design lie strewn about the area.

SOL 1448, highlighted area 10 (above): This oddly shaped object is hard to make out. Under magnification it almost appears woven together using some kind of flexible material, giving it a squarish mat like appearance.

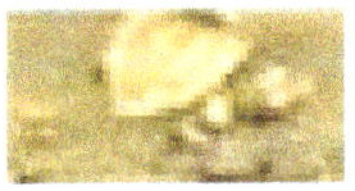

SOL 1448, highlighted area 11 (above): This object is shaped like a gravestone, a shovel blade, or a military shield. In the latter case it is probably too small to offer actual defensive protection. Depending on its actual size, perhaps it was once part of a decorative family crest. I have discovered many objects exactly like this on the surface of Mars, lending credibility to the extraterrestrial hypothesis.

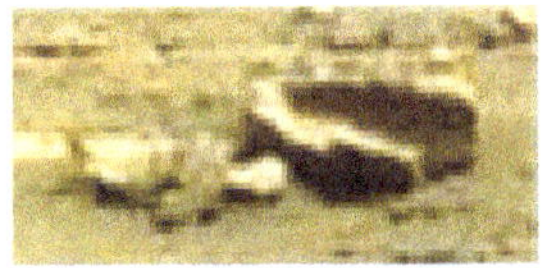

SOL 1448, highlighted area 12 (above): In my opinion this is certainly one of the more startling objects captured by any of the Mars rovers. What is it? It has the shape of the bow of a boat, but it appears to be made of stones. Thus, it is more likely the remains of some kind of memorial, gravestone, or religious statuary. Fascinatingly, it closely matches the shape of the "shield" described in area 11, and which lies directly above and next to it on the Martian surface. Are they connected? Both bear numerous traits of intelligent design. And what are we to make of the pile of debris to the left of the "boat"? One item has a hole in it and colored striations cutting across its surface. At the far left (9:00) appear two flat, perfectly rectangular objects positioned end-to-end in the sand, perhaps the remnants of a sidewalk.

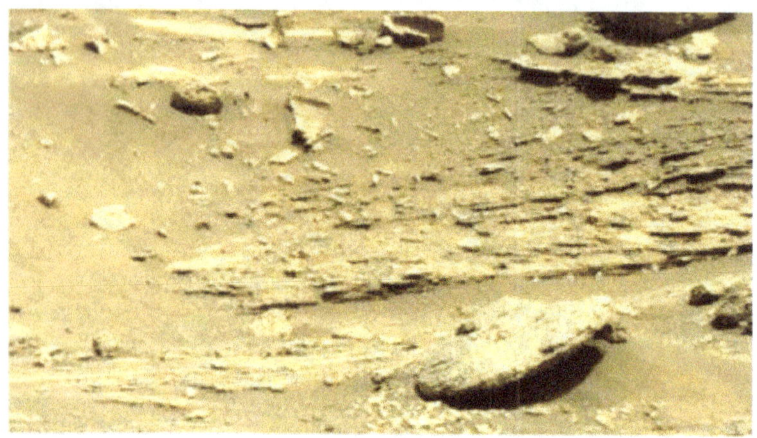

SOL 1448, highlighted area 13 (above): Here I have cut out a large section that includes some of the areas we have already analyzed. Why? To illustrate one reason why many individuals believe there was a major conflict on Mars in ancient times. This image, a bird's eye view of what looks very much like an ancient debris field created during warfare, seems to support this theory. This is not a rarity. The Mars rovers have captured many other images just like it.

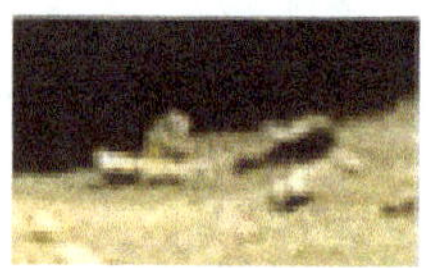

SOL 1448, highlighted area 14 (above): This curved rib like object is lying half buried in the Martian sand at the base of a large boulder. Unusually shaped and oddly placed objects lie around it.

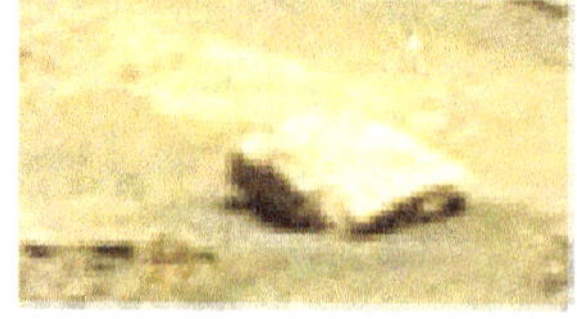

SOL 1448, highlighted area 15 (above): A square pillow like object and a board like rectangle lie next to one another, just two of the thousands of objects on Mars with 90 degree angles.

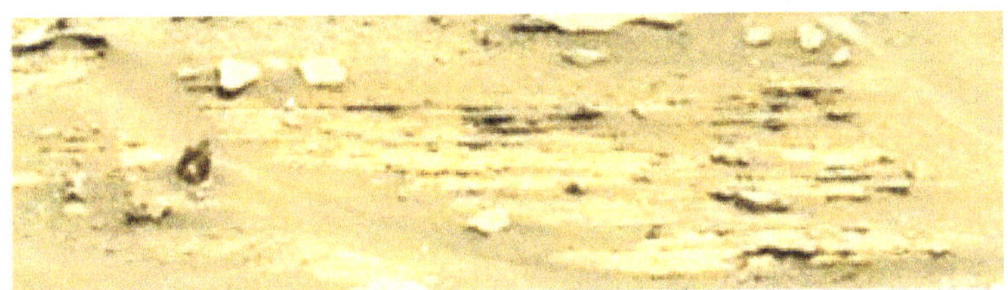

SOL 1448, highlighted area 16 (above): Another wide angle look at some previously analyzed areas. This entire image is peppered with squarish and rectangular objects. Covering most of the real estate in this photo are sweeping horizontal lines, which dominate, capturing our attention. What are we seeing here? Rocks? Or could they be the vestiges of steps, pavement, a roadway, or auditorium seating—or something else that is completely unfamiliar to us Earthlings?

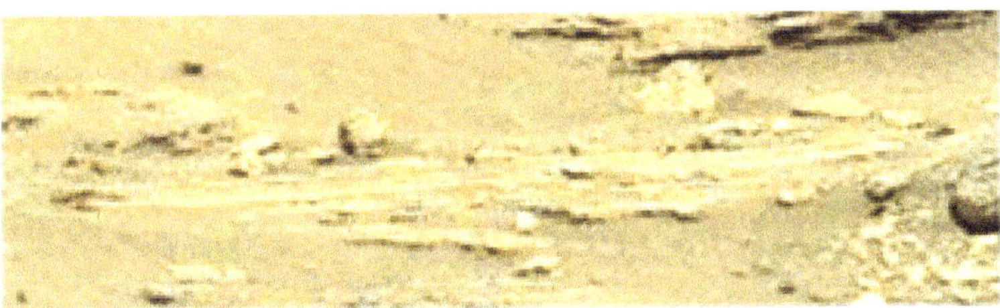

SOL 1448, highlighted area 17 (above): Conservative geologists maintain that what is pictured here is all a product of natural Martian geology. Martian anomalists hold that it is more likely the vestiges of some kind of paving, perhaps from what is left of a sidewalk or road.

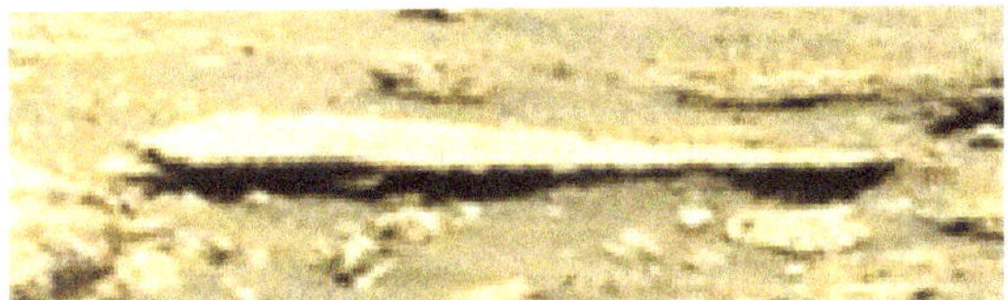

SOL 1448, highlighted area 18 (above): Is it natural for a thin piece of rock, like the one seen here, to be sitting tenuously on the surface of a planet? Yes. The real question is this: Is it common? Debate continues on this topic. Meanwhile, we must consider that this may be part of a large structure, perhaps a piece of siding, sheet metal, or insulation.

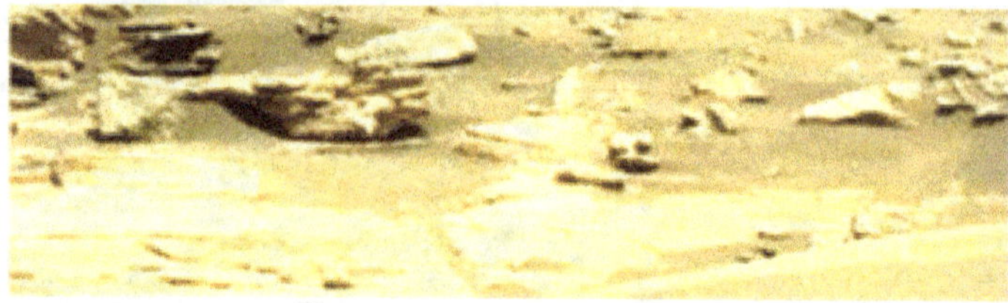

SOL 1448, highlighted area 19 (above): Are these the remains of an ancient Martian road, blown to pieces by a bomb blast? There are numerous squares, rectangles, and triangles in this image as well, inviting us to speculate on the nature of their origins. At 10:00 there is a pyramid like item with odd patterns on it. The large square flat object near the center and the oddly bent triangular piece of material in the upper right (2:00) are particularly intriguing.

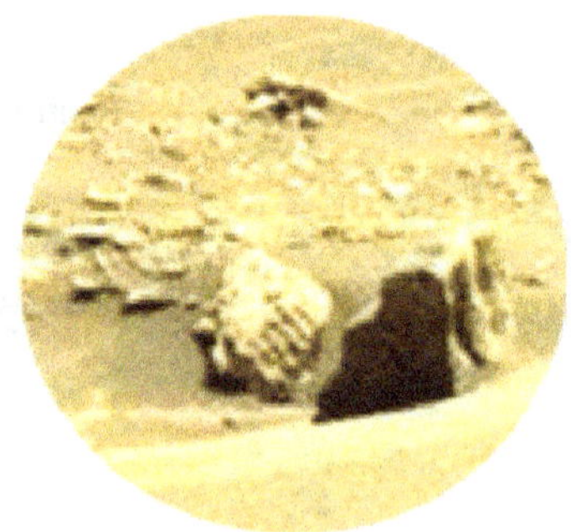

SOL 1448, highlighted area 20 (above): Almost nothing in this image looks natural. Rather, we have what appears to be the crumbled remnants of objects that were created by an advanced mind. The large shadowed object on the right (which is probably quite small in reality) looks like a piece from a statue or memorial, or perhaps it is some unknown sculpted item from a cityscape. Near the lower center is an object with what appear to be deeply engraved lines on it. To its left is a pile of debris that looks like structural elements from a building or walkway. At around the 2:00 position a small V shaped object lies in the Martian sand pointing toward the viewer. Near the top a face like object peers toward the camera. To me these are probable traits of artificiality.

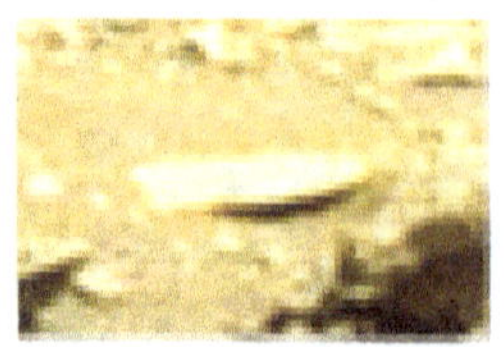

SOL 1448, highlighted area 21 (left): Another shield shaped or gravestone shaped object.

Chapter 9

THIGH BONE

SOL 719: EARTHDATE AUGUST 14, 2014

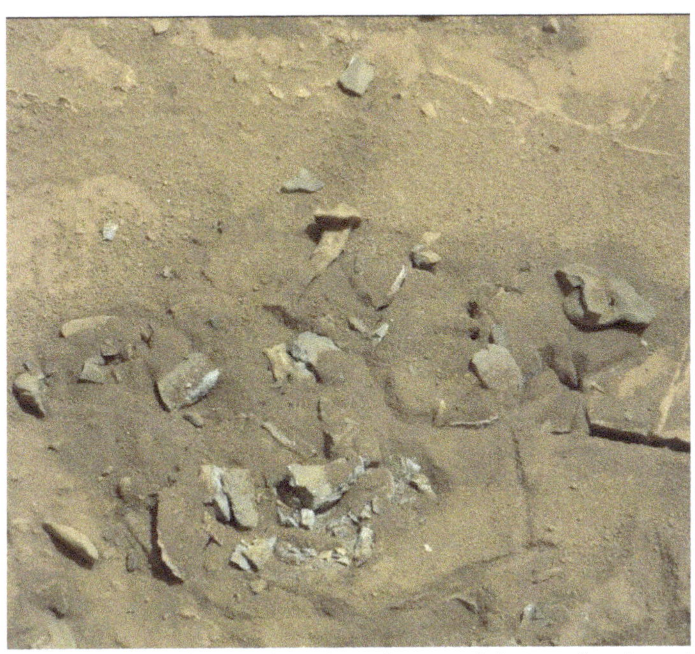

Original image. "Taken by Mast Camera (Mastcam) onboard NASA's Mars rover Curiosity on Sol 719 (2014-08-14, 21:49:28 UTC)." Image credit: NASA/JPL-Caltech/MSSS.

ANALYSIS OF THIGH BONE, SOL 719

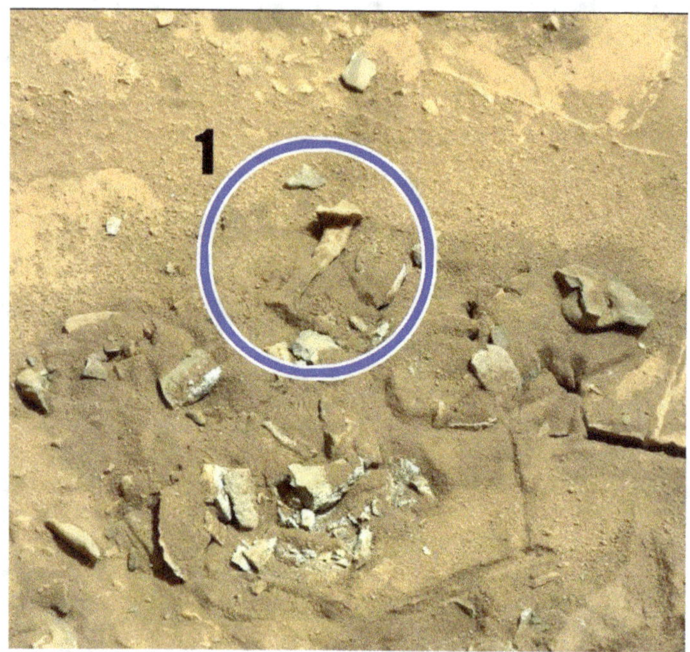

Thigh Bone, Sol 719, with one anomaly circled. Image colored, brightened, and sharpened by the author to improve contrast.

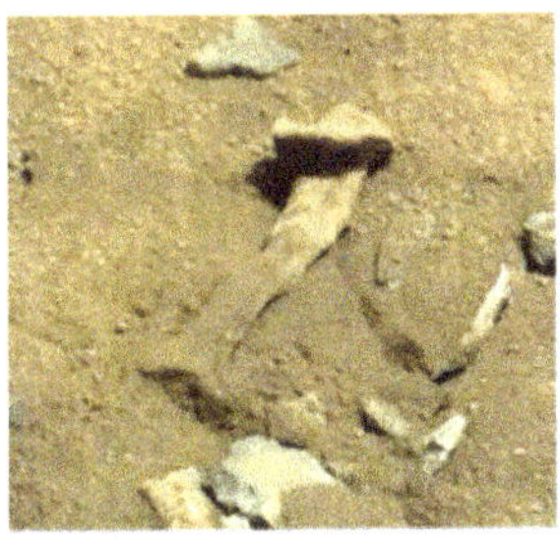

SOL 719, highlighted area 1: While Mars anomalists have long asserted that this object is a femur or thigh bone, naturally mainstream scientists claim otherwise. After the photo was released to the public, NASA itself teased believers, posting the following comments beneath it:

> "No bones about it! Seen by Mars rover Curiosity using its MastCam, this Mars rock may look like a femur thigh bone. Mission science team members think its shape is likely sculpted by erosion, either wind or water.
>
> "If life ever existed on Mars, scientists expect that it would be small simple life forms called microbes. Mars likely never had enough oxygen in its atmosphere and elsewhere to support more complex organisms. Thus, large fossils are not likely."

Let us note that there is no certainty expressed here on what this "rock" actually is or how it was formed; only opinions on how it might have been formed and what it is not. Nor is there any definitive statement on whether complex (i.e., is, large) life forms could have ever existed on Mars. Indeed, the word "expect" (i.e., an assumption without evidence) is used once while the word "likely" (i.e., pure conjecture) is used three times. I do not fault NASA for these comments. In fact, I applaud them. After all, science is not a closed but an open investigative system: science does not have all the answers and never will, and scientists who believe otherwise are not practicing authentic science.

Where the commenter comes up short, in my view, is in strongly insinuating that Mars has never had complex life forms. We cannot and do not know this for certain, making such a declaration premature. As of this writing we have only made a few shallow drill holes in the upper surface layer of Mars. Honest, open ended, fact based science demands that we continue our research of the Red Planet with a malleable yet critical mind, while preventing personal bias from interfering with what we uncover.

The fact is that many members of the mainstream scientific community are anti-extraterrestrial life: there is not only a hesitancy to search for *large* complex life forms outside Earth, but there exists an absolute refusal to even consider it a possibility. Many treat the entire subject as a joke. Why? In great part because the discovery of intelligent alien life would overturn science as we know it, every textbook would have to be rewritten, and careers built on the fragile anti-extraterrestrial hypothesis would be jeopardized. Let us recall here the words of biologist Dr. Rupert Sheldrake: "Science at its best is an open-minded method of inquiry, not a belief system."

Chapter 10

SIDEWALK & ROADWAY

SOL 108: EARTHDATE NOVEMBER 25, 2012

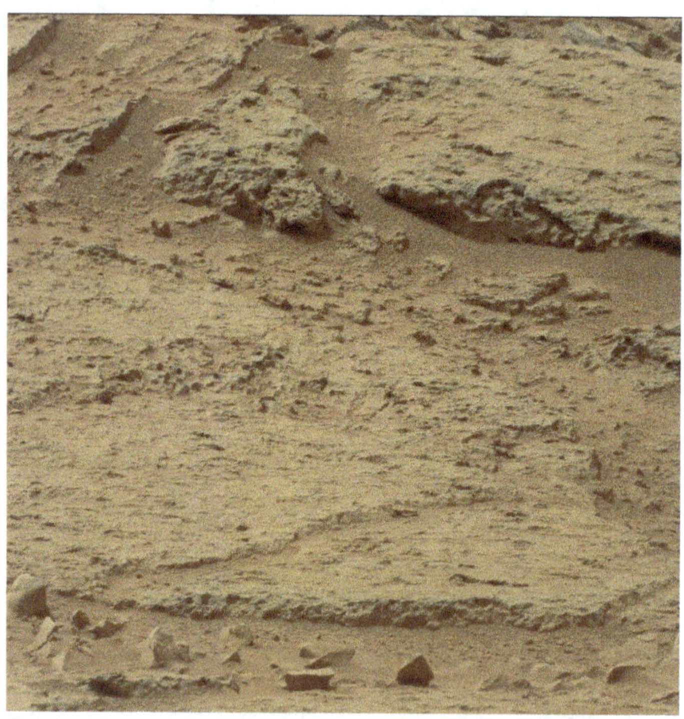

Original image. "Taken by Mast Camera (Mastcam) onboard NASA's Mars rover Curiosity on Sol 108 (2012-11-25, 03:22:07 UTC)." Image credit: NASA/JPL-Caltech/MSSS.

ANALYSIS OF SIDEWALK & ROADWAY, SOL 108

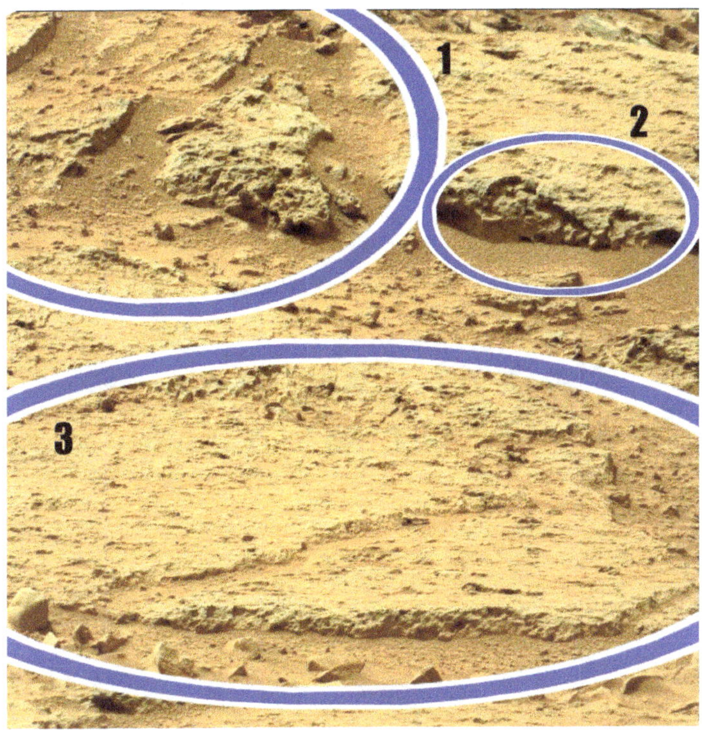

Sidewalk and Roadway, Sol 108, with anomalies circled. Image colored, brightened, and sharpened by the author to improve contrast. There are signs of numerous processes apparent here. The question, as always, is whether or not they are natural or artificial.

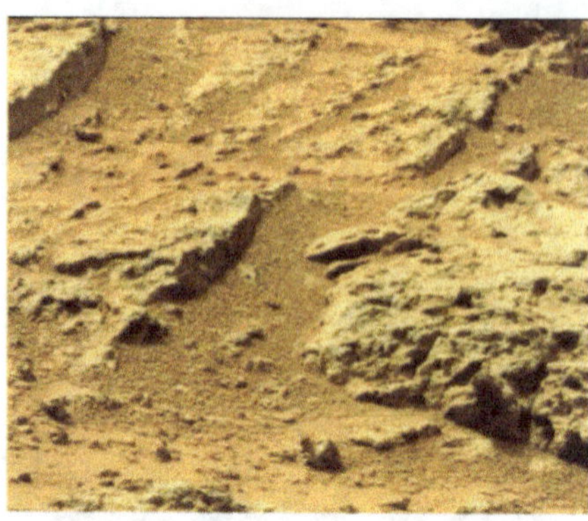

SOL 108, highlighted area 1 (above): Squared, fitted paving stones from an ancient sidewalk, or natural rock formations millions of years old?

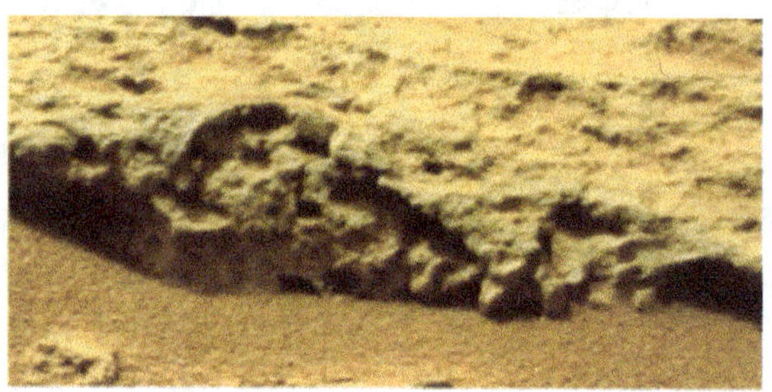

SOL 108, highlighted area 2 (above): Artificial imprints in ancient cement or natural geology?

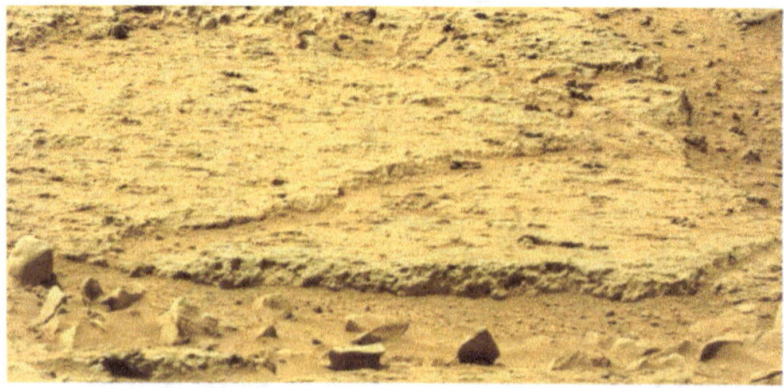

SOL 108, highlighted area 3 (above): Remnants of an ancient roadway or house foundation, or vestiges of a million year old Martian riverbed?

Chapter 11

UNUSUAL OBJECTS

SOL 1355-A: EARTHDATE MAY 29. 2016

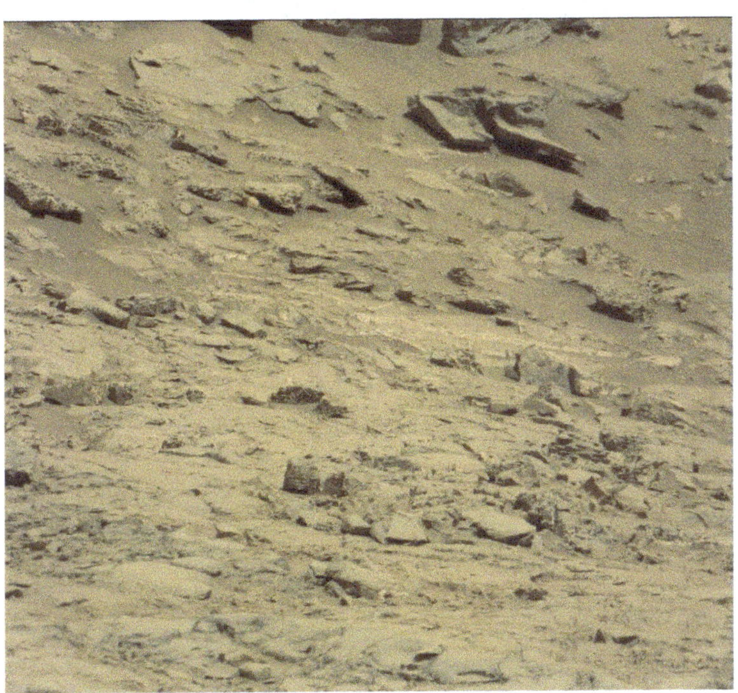

Original image. "Taken by Mast Camera (Mastcam) onboard NASA's Mars rover Curiosity on Sol 1355 (2016-05-29, 08:33:59 UTC)." Image credit: NASA/JPL-Caltech/MSSS.

ANALYSIS OF UNUSUAL OBJECTS, SOL 1355-A

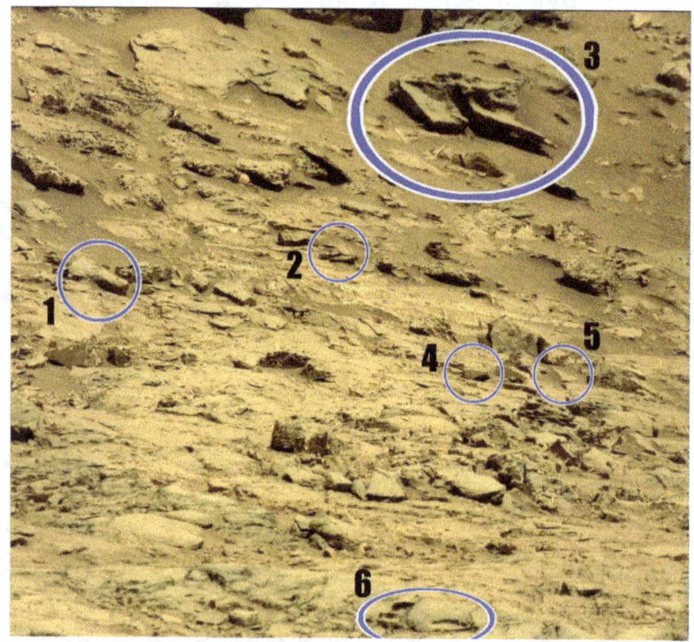

Unusual Objects, Sol 1355-A, with anomalies circled. Image colored, brightened, and sharpened by the author to improve contrast. While there are numerous obvious natural rock formations in this photo, there are a number of items that seem out of place.

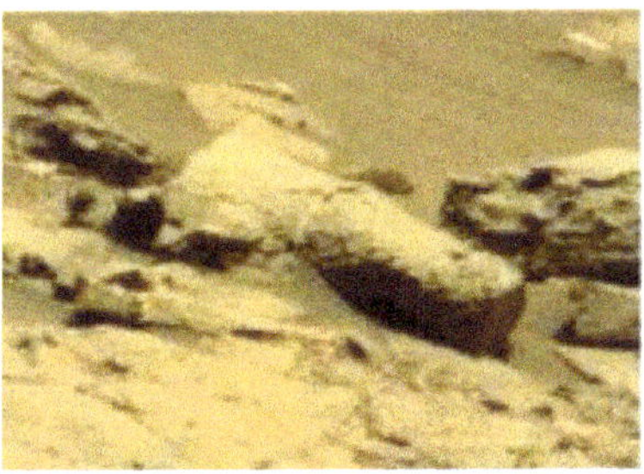

SOL 1355-A (above), highlighted area 1: Is this a pillar toppled over by an explosion, or is it a naturally formed cylindrical rock?

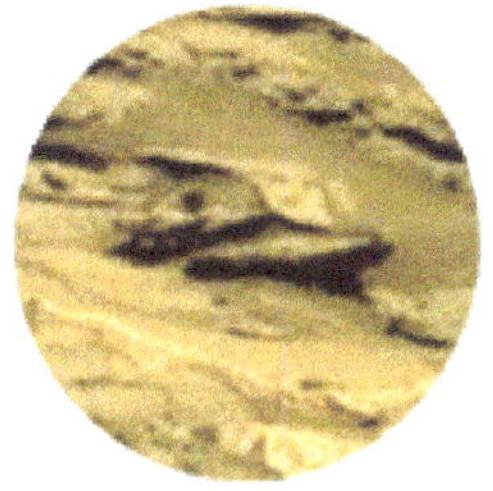

SOL 1355-A, highlighted area 2 (above): There appears to be a long, bent or flexible, rectangle shaped object draped over several rocks. On Earth this would most likely be something made of plastic, cloth, or metal.

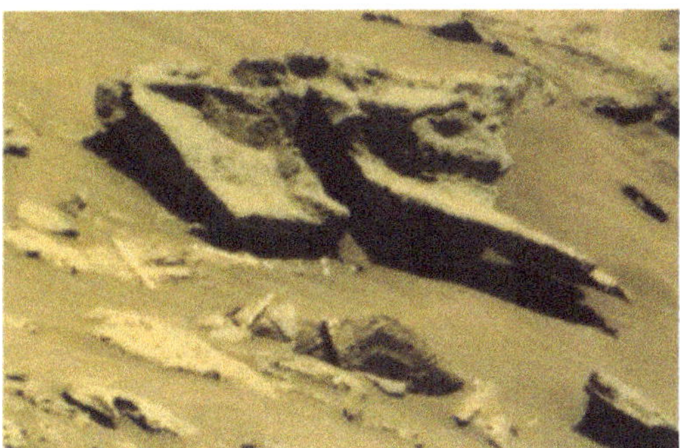

SOL 1355-A, highlighted area 3 (above): A group of oddly shaped "rocks," some looking like decorative remnants from a structure of some kind. Under magnification note the small disk shaped object with a U shaped hole in its side at the bottom left (7:00). The entire area seems digitally airbrushed, a possible attempt to cover up obvious signs of intelligent design.

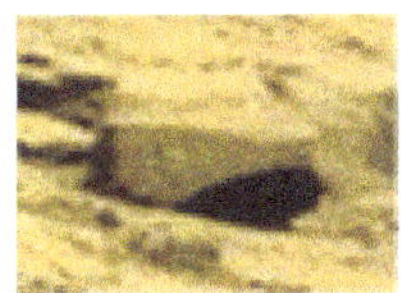

SOL 1355-A, highlighted area 4 (above): The squared and curved cuts on this "rock" look machined. Is Nature capable of this? Yes. But so is a technologically sophisticated mind.

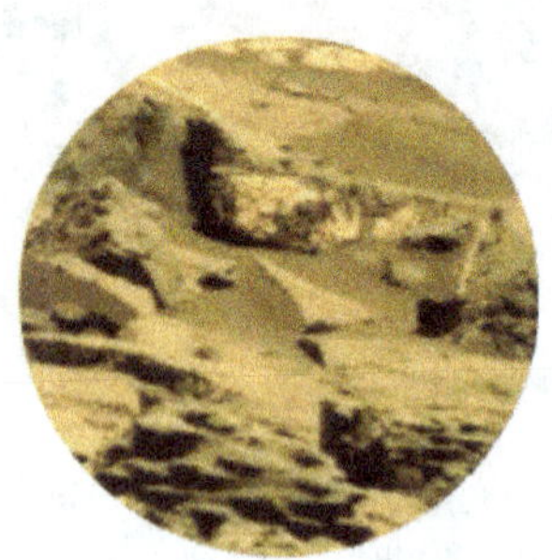

SOL 1355-A, highlighted area 5 (above): The pyramid or triangle shaped object at the center of this image is striking. Additionally, it is surrounded by a host of other peculiarly shaped "rocks," many with unusual markings.

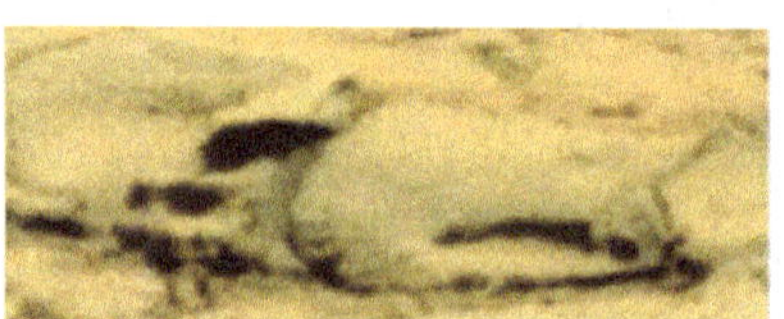

SOL 1355-A, highlighted area 6 (above): These rocks have strange indentations or cavities in them. The long one on the right looks like a fossil impression of a knife with a handle. The entire image, including everything in it, is odd.

Chapter 12

DEBRIS FIELD

SOL 1355-B: EARTHDATE MAY 29. 2016

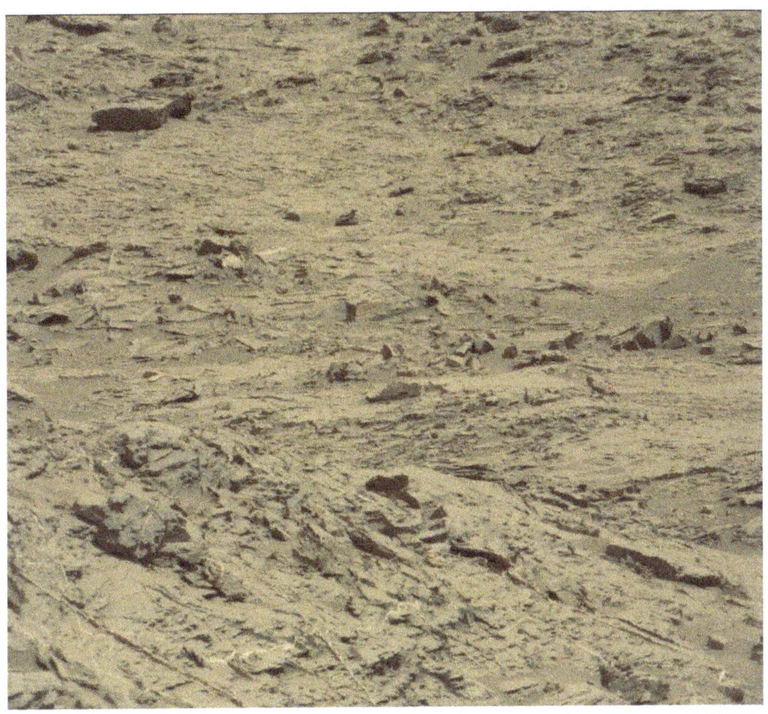

Original image. "Taken by Mast Camera (Mastcam) onboard NASA's Mars rover Curiosity on Sol 1355 (2016-05-29, 08:33:59 UTC)." Image credit: NASA/JPL-Caltech/MSSS.

ANALYSIS OF DEBRIS FIELD, SOL 1355-B

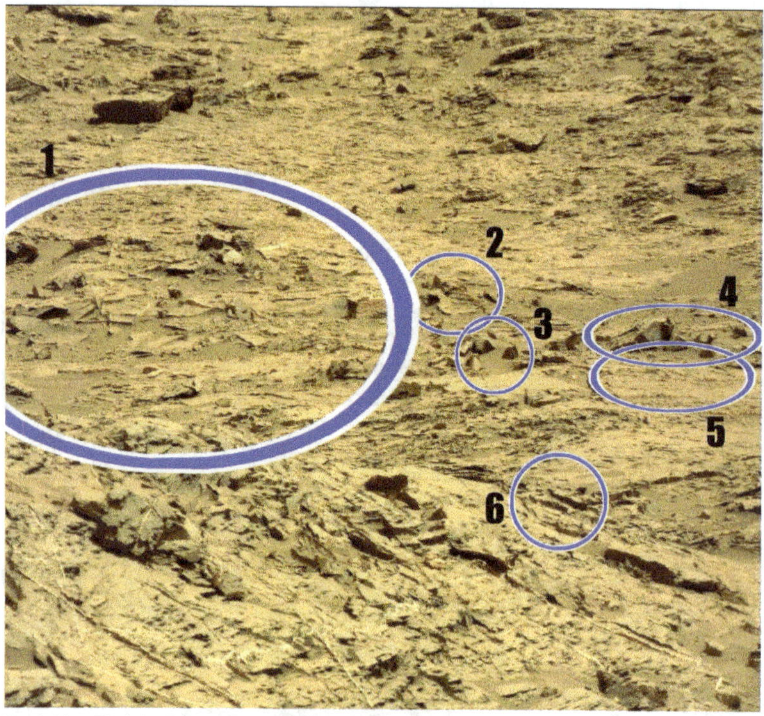

Debris Field, Sol 1355-B, with anomalies circled. Image colored, brightened, and sharpened by the author to improve contrast. Many people would consider everything in this image completely natural—and this may be true. But to others, myself included, much of it looks like a debris field, littered with the obliterated remnants of various technologically advanced structures.

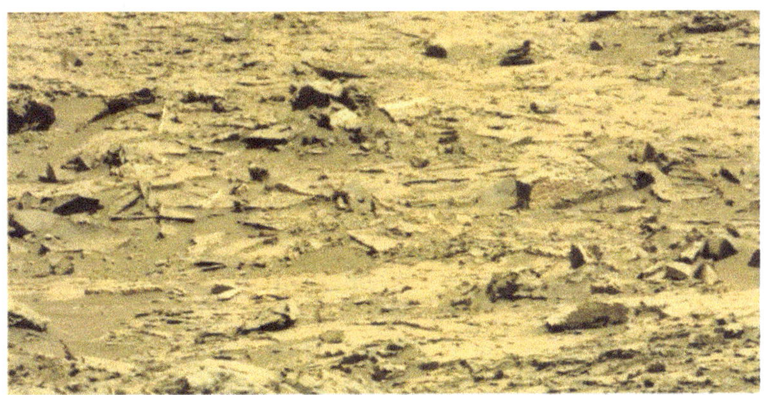

SOL 1355-B, highlighted area 1 (above): While there are no doubt natural rock formations in this photo, nearly the entire area looks like the site of a bomb blast, with exploded debris scattered in every direction. Under magnification one will see all manner of unnatural looking shaped objects, including "boards," ridged objects, "building blocks," tray like objects, "bricks," squares, rectangles, triangles, and jumbled piles of "sticks"—among numerous other unidentifiable curiosities.

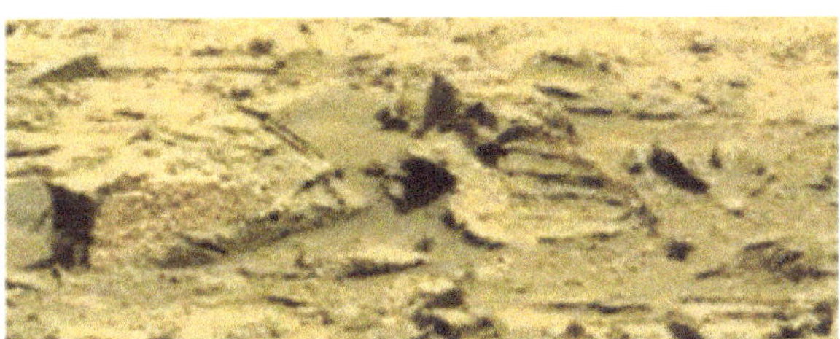

SOL 1355-B, highlighted area 2 (above): At center right, and running down to the lower right, are a series of long oval impressions in the ground. Situated side by side, they appear to be the remnants of graves. Are they? Their artificial appearance is reenforced by the fact that they are surrounded by a myriad of other strange objects, including blocks, straight lines, triangles, and 90 degree angles. Magnification will bring out details not apparent to the naked eye.

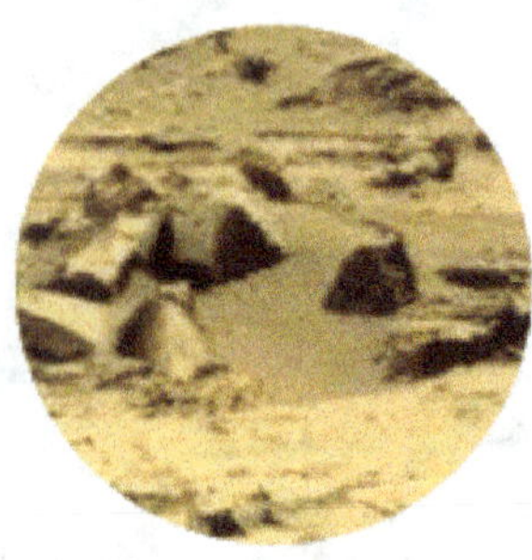

SOL 1355-B, highlighted area 3 (above): This area looks like it has been digitally manipulated (and rather poorly) by human hands. If true, this little section must have originally contained evidence of something that establishment scientists do not want us to see, and this could only be one thing: overt signs of a complex but extinct Martian civilization.

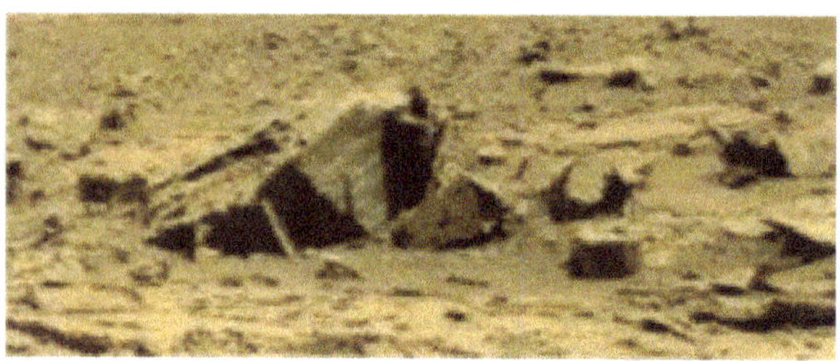

SOL 1355-B, highlighted area 4 (above): There are so many signs of seeming Martian intelligence in this image it is difficult to find a starting point. Toward the upper left, a bed like object lies collapsed in the dirt. Just to its right is a large, square, cement like "block" tilted up on one of its corners. To the right of center is an object that looks like the open end of a flare nut wrench sticking up out of the ground. Remarkably, directly in front of the "wrench" is what appears to be a six-sided "nut," or hexagon, sitting in the sand. Three of its flat, sharply defined sides are facing the camera and are clearly visible. A board like object appears to the right of the "hexagon." At the lower left is a rectangular tray like object leaning slightly to one side. Other straight lined and weirdly shaped objects abound.

SOL 1355-B, highlighted area 5 (above): Part of a circular shape can be seen on the right side of this image. At its center is a rectangular tray like shaped object. In the upper left the tray like object described in area 4 can be seen. Below it is what looks like the remnants of a wall or a rectangular building foundation. The longer you look, particularly under magnification, the more you will see.

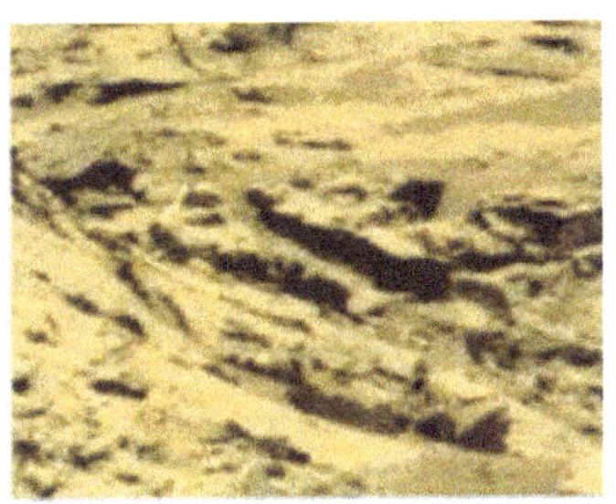

SOL 1355-B, highlighted area 6 (above): The long "carved out" object at the center of this image could be entirely natural. But if advanced life once lived on Mars it could be quite something else, perhaps part of a walkway, a trough, or an aqueduct. A host of other artificial looking items lie scattered about, including a round gear wheel like object with "teeth" just right of center, and a "brick oven" like object at 3:00.

Chapter 13

DEBRIS FIELD

SOL 1355-C: EARTHDATE MAY 29, 2016

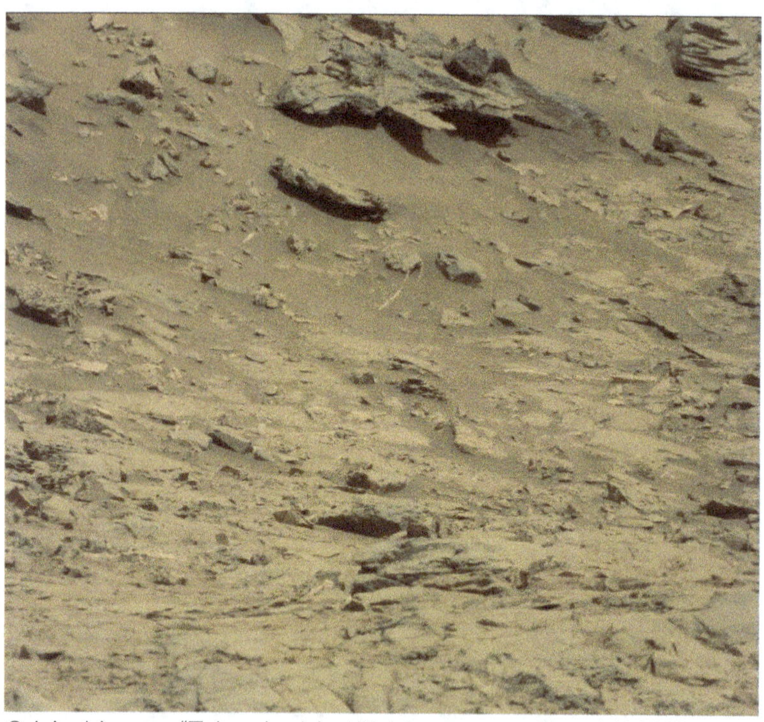

Original image. "Taken by MastRight camera onboard NASA's Mars rover Curiosity on Sol 1355 (2016-05-29, T08:41:03.000Z)." Image credit: NASA/JPL-Caltech/MSSS.

ANALYSIS OF DEBRIS FIELD, SOL 1355-C

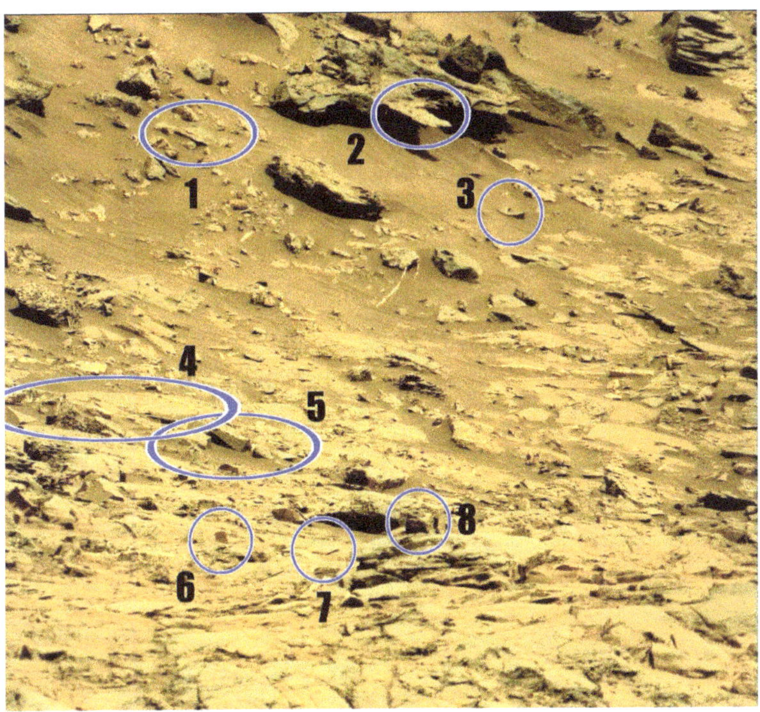

Debris Field, Sol 1355-C, with anomalies circled. Image colored, brightened, and sharpened by the author to improve contrast. This photograph appears to show artificial debris scattered across a Martian hillside.

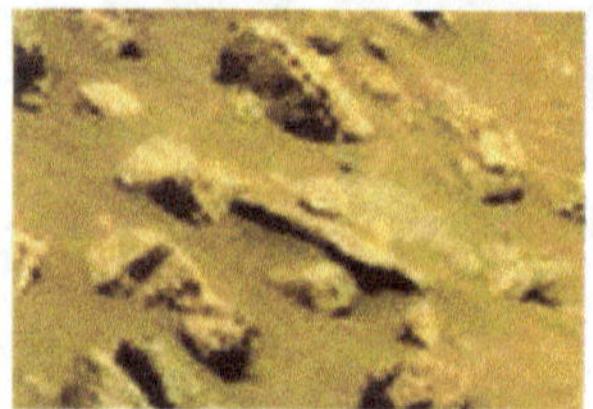

SOL 1355-C (above), highlighted area 1: At center right a large, thin, square object lies in the sand. From an Earthling's point of view it looks like a bent piece of metal sheeting. If not, it is an extraordinary rock. Study the objects surrounding it as well.

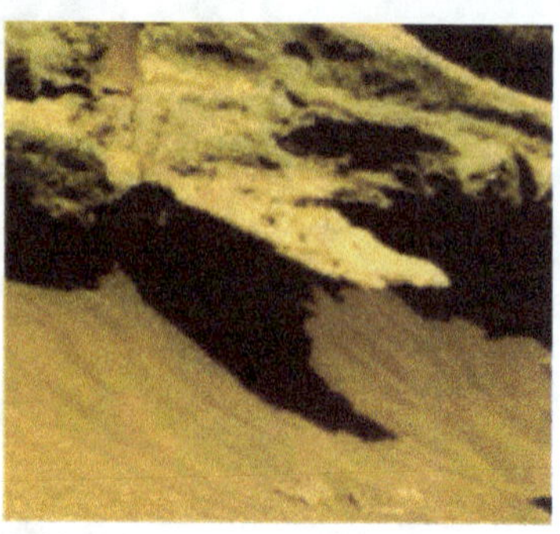

SOL 1355-C, highlighted area 2 (above): A natural formation or an artificial formation? I will side with "just a rock" on this one—but we really cannot be completely sure without examining it in person. Look around it.

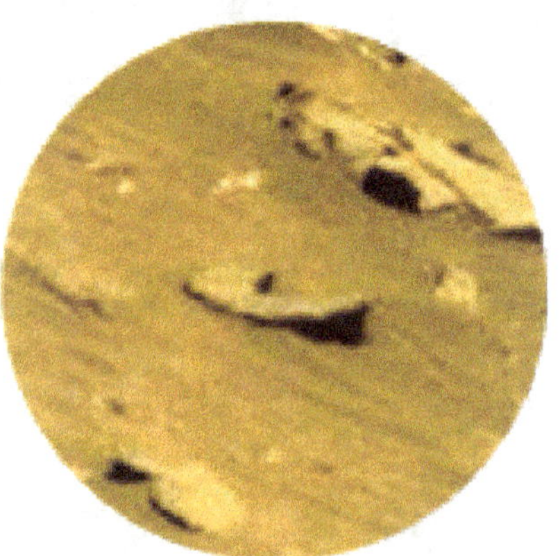

SOL 1355-C, highlighted area 3 (above): The object at center is roundish with straight sides and a deep "V" or "U" shape cut out of its far side. A host of other mysterious items lie on this Martian slope, including a large triangular object, a flattened object with a rectangular end, and a thin bent object, all at 2:00. At 3:00 there appears to be a broken "trident" laying in the sand.

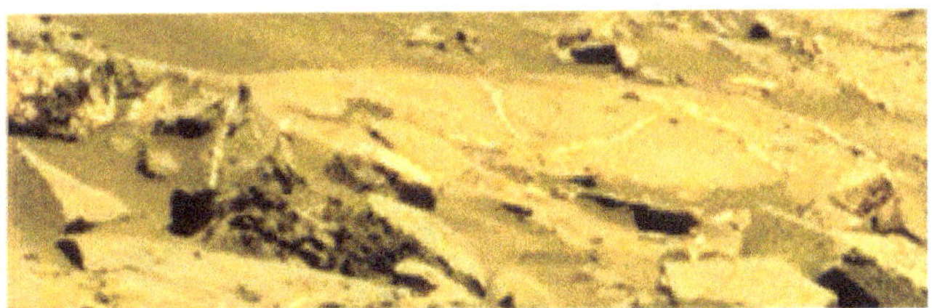

SOL 1355-C, highlighted area 4 (above): The large pale object at center right looks roundish and has a "cross" in the middle of it. On the far left (9:00) is an overtly pointed triangular or pyramidal like object embedded in the Martian sand. At around 12:00 there are two oddly shaped objects; at about 12:30 there is a large object with a square angle "cut out" of it; at 1:00 there is a small triangular object that appears to have a honeycombed interior; at about 3:30 there is a "rock" with a letter like item on it; at 4:00 and 5:00 there are a number of block like objects; just left of upper center is a serpentine like object; below it, at about 7:00, is an oddly shaped "rock" with a flat, light colored top and letter like markings on its sides, including the number "6." The entire area surrounding this image (see full frame image above) is filled with what looks like ancient bomb blast debris. Use magnification to get in close.

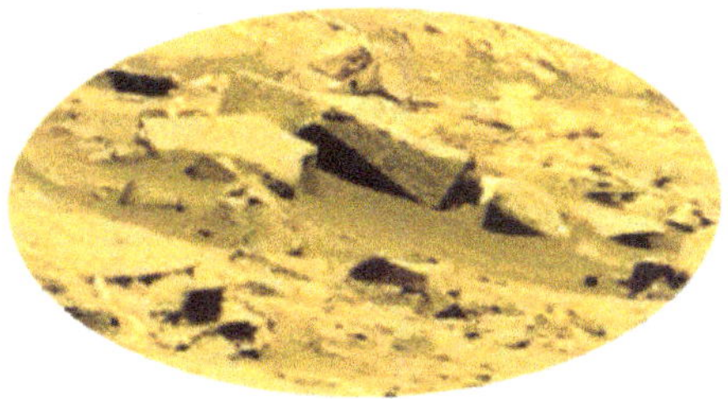

SOL 1355-C, highlighted area 5 (above): The objects in this image lie just below and to the right of the circular object described in area 4. Notice the perfectly square brick like object at about the 11:00 mark. Next to it, or perhaps attached to it, is a curved brick like object. To the right of that is a triangular or square shaped object poking out from the sand. At around 10:00 there is an oddly shaped object, appearing something like a square with a point on one end. I think even the most staunch believers can admit that all of these may be naturally formed rocks. But skeptics will never admit that they may not be.

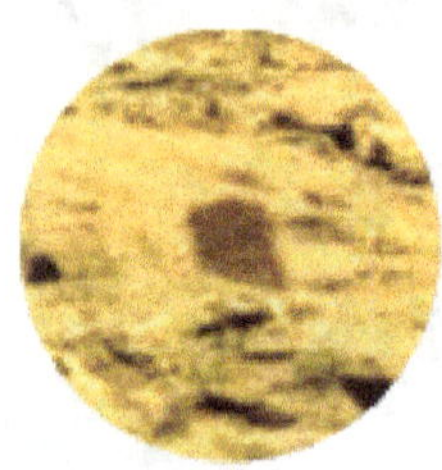

SOL 1355-C, highlighted area 6 (above): At center the shadow of a small, square, artificial looking pillar is being cast. The "pillar" casting it, however, cannot be seen. Was it digitally removed? If so, the digital censor neglected to airbrush out its shadow.

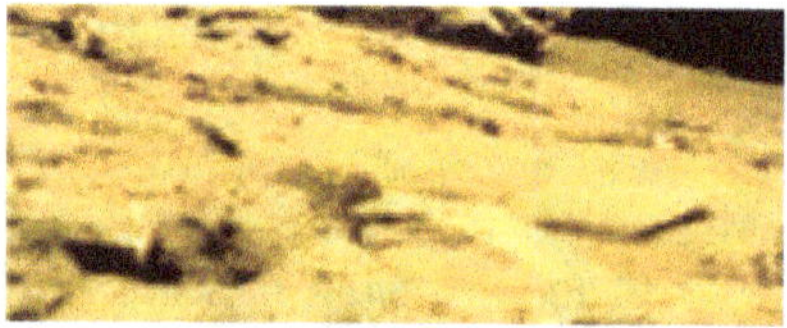

SOL 1355-C, highlighted area 7 (above): On the right a perfectly rectangular object lies in the Martian sand. To its left appear to be more of the same objects, all lined up like paving stones. Other singular objects appear around them.

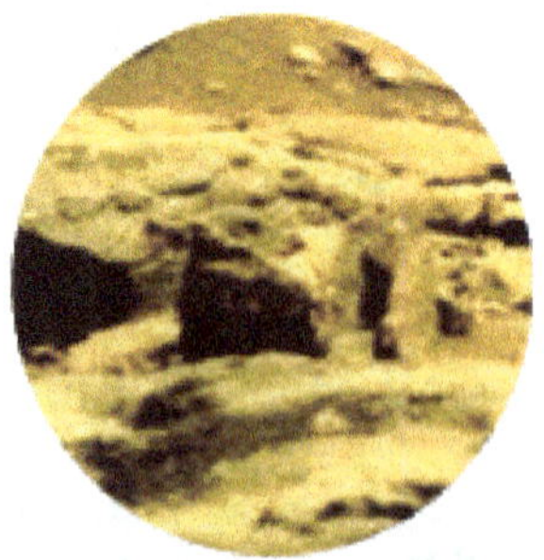

SOL 1355-C, highlighted area 8 (above): This small, square, cement like object looks like a surveyor's marker or boundary stone that has been heavily damaged by some kind of powerful force. Was the force natural or artificial?

Chapter 14

DEBRIS FIELD

SOL 1355-D: EARTHDATE MAY 29, 2016

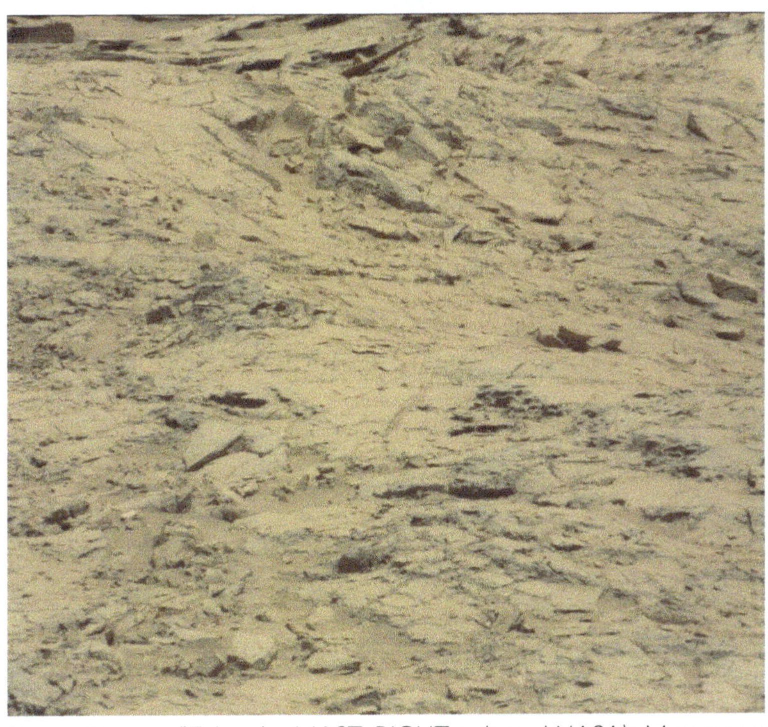

Original image. "Taken by MAST_RIGHT onboard NASA's Mars rover Curiosity on Sol 1355 (2016-05-29, T08:43:08.000Z)." Image credit: NASA/JPL-Caltech/MSSS.

ANALYSIS OF DEBRIS FIELD, SOL 1355-D

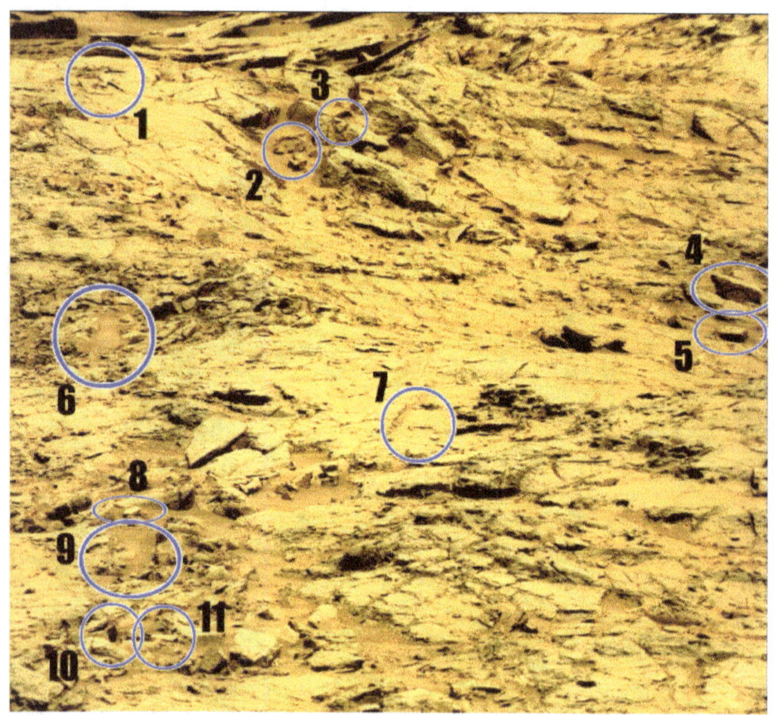

Debris Field, Sol 1355-D, with anomalies circled. Image colored, brightened, and sharpened by the author to improve contrast. There are numerous points of interest in this photo, many presenting traits of artificiality.

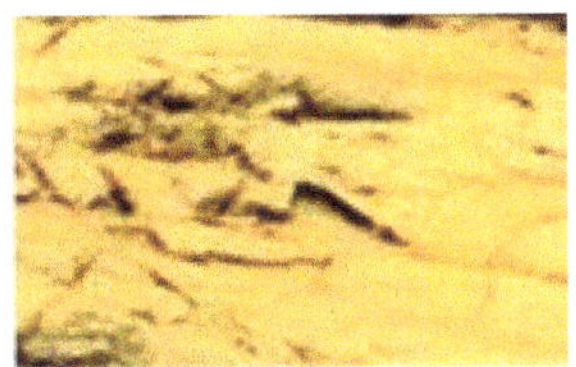

SOL 1355-D, highlighted area 1 (above): An "L" shaped object and a "Y" or "V" shaped object.

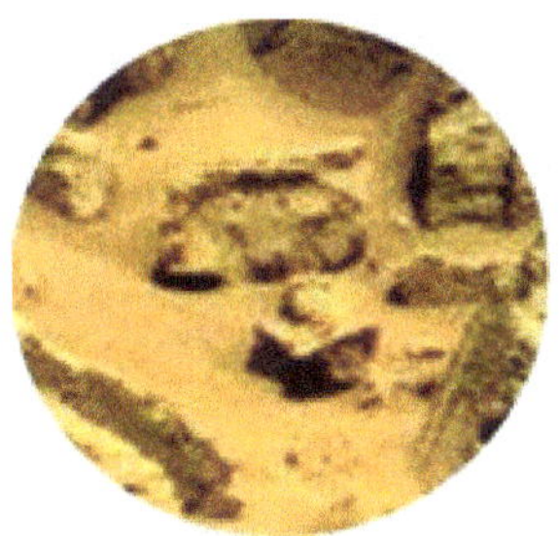

SOL 1355-D, highlighted area 2 (above): The two objects near the center bear striking signs of intelligent design. The large block like object (upper center), which I will call object A, appears to have black lettering etched across the surface of its front upper edge. The one below it, object B, has a flat base and looks something like an aquarium decoration. As this photo is low resolution, I have included extreme closeups of both objects on the next page.

Above: Closeup of front upper edge of object A in area 2 above, showing possible Roman lettering. If this is indeed from the Latin alphabet, the first two letters would be "T" and "Y"; the letters after the "Y" are illegible; the last letter seems to be an "A." Skeptics will call this pareidolia. However, the original NASA photo tells what appears to be a different story when scrutinized under magnification.

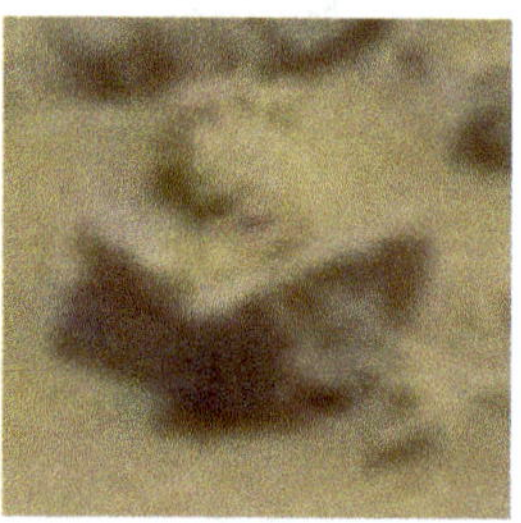

Above: Closeup of object B shown in area 2 above. Note its odd ornamental like shape and flat rounded "base."

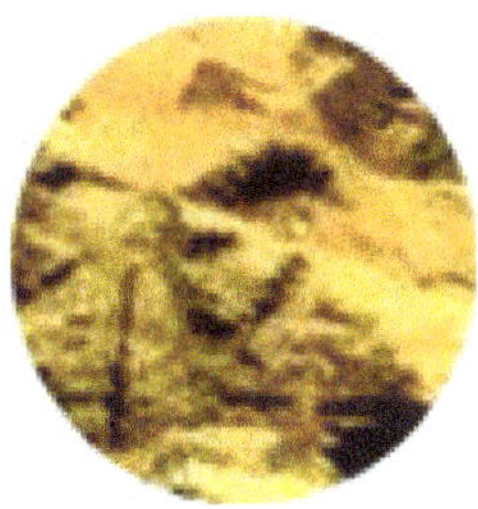

SOL 1355-D, highlighted area 3 (above): The cross hatching in this image, along with the circular gear like "wheel" buried in the sand near the top, may be nothing more than photographic artifacts—degraded pixels that result from over compression of the original image. If this is not what these marks are, then this photo is well worth closer study.

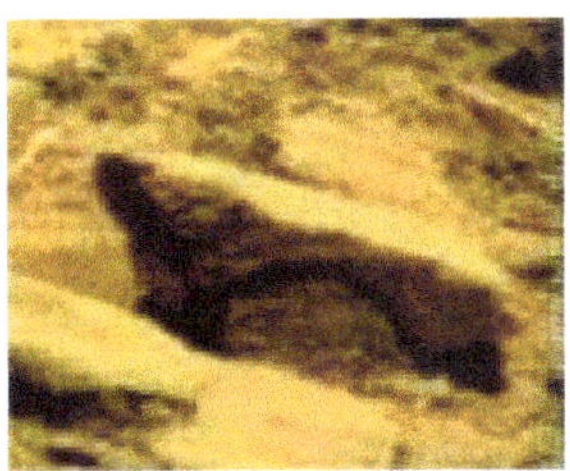

SOL 1355-D, highlighted area 4 (above): This trapezoid shaped "rock" has a clearly marked out, round "carving" in its side. What is its origin?

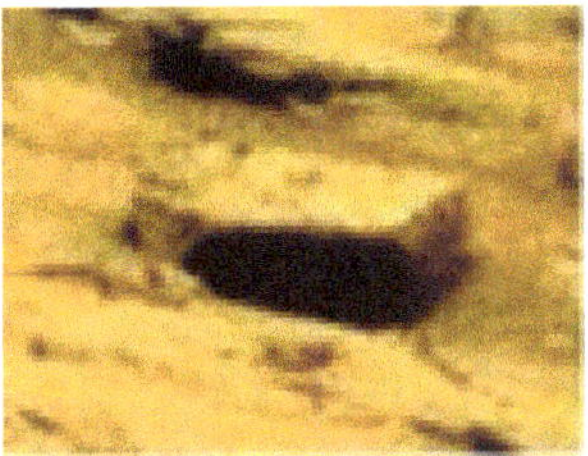

SOL 1355-D, highlighted area 5 (above): If this is not a Martian brick, what is it?

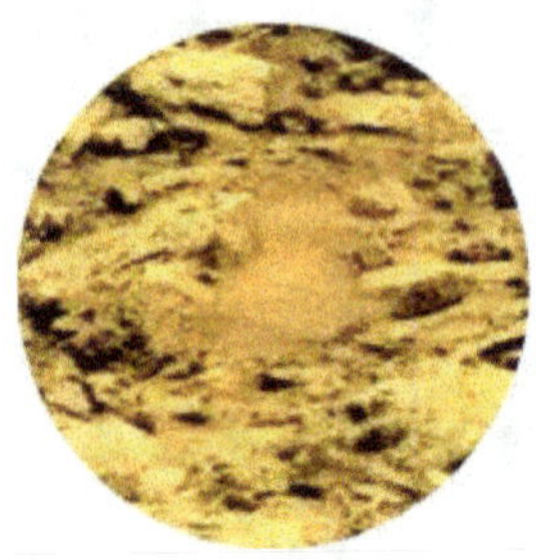

SOL 1355-D, highlighted area 6 (above): The brown wavy section in the middle of this image looks digitally airbrushed with fake "sand." If true, what is the artist trying to hide?

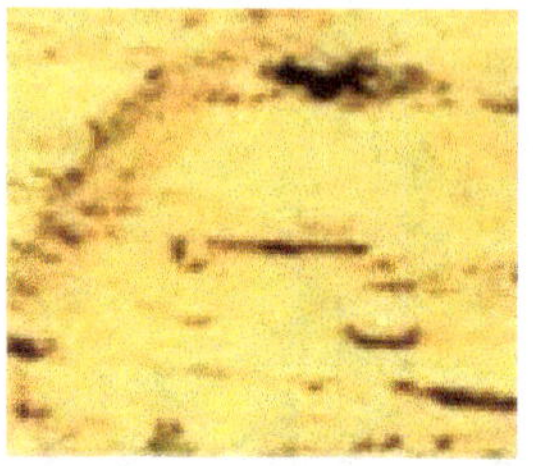

SOL 1355-D, highlighted area 7 (above): "L" shapes, like the one at center, show up repeatedly in Martian images. The full frame photo that includes this small area alone contains two that I can see.

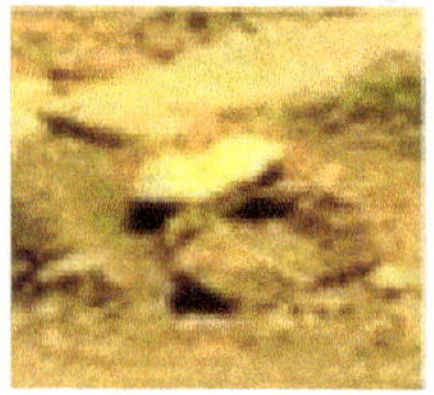

SOL 1355-D, highlighted area 8 (above): Three oddly shaped, strangely constructed objects sit together on the Martian surface.

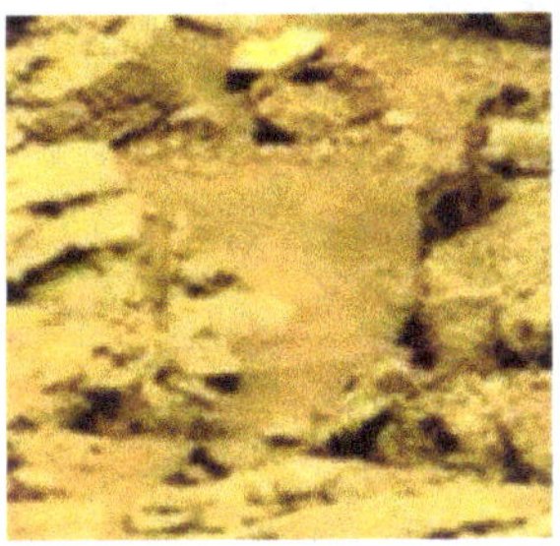

SOL 1355-D, highlighted area 9 (above): At center we see another area that looks airbrushed with artificial sand. The reason for this possible deception can be seen in the numerous seemingly intelligently crafted articles lying on the ground around it. Remember: These items could be tens of thousands of years old, making them now virtually unrecognizable.

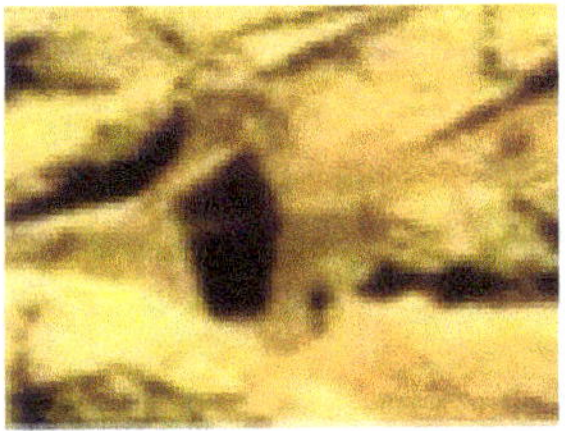

SOL 1355-D, highlighted area 10 (above): This object has a high level of strangeness. Its shape, bicoloration, and position are all quite unusual. It appears to be a semitransparent, highly polished shard of tinted glass, one that is reflecting the peculiar items to its right.

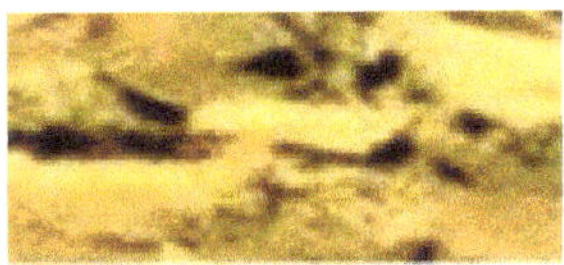

SOL 1355-D, highlighted area 11 (above): Sitting center stage is a rectangular, shingle like object surrounded by numerous other objects with 90 degree angles. All of them are situated just to the right of the "shard of tinted glass" described in area 10. We cannot identify them, but it is safe to say that they possess a number of characteristics of artificiality—that is, they seem to have been intentionally made by a sophisticated mind.

Chapter 15

DEBRIS FIELD

SOL 1355-E: EARTHDATE MAY 29, 2016

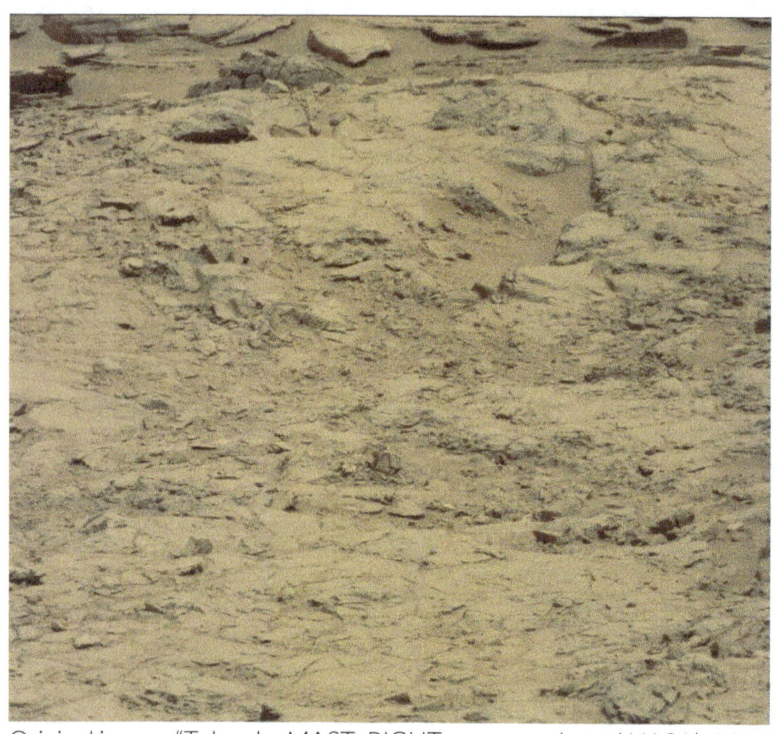

Original image. "Taken by MAST_RIGHT camera onboard NASA's Mars rover Curiosity on Sol 1355 (2016-05-29, T08:43:24.000Z)." Image credit: NASA/JPL-Caltech/MSSS.

ANALYSIS OF DEBRIS FIELD, SOL 1355-E

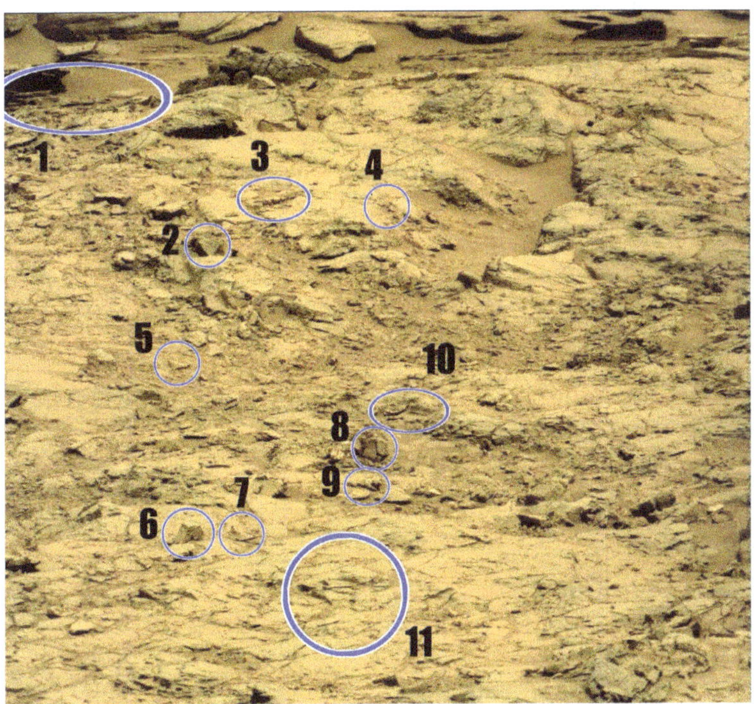

Debris Field, Sol 1355-E, with anomalies circled. Image colored, brightened, and sharpened by the author to improve contrast. If the theory that an ancient advanced society once inhabited Mars turns out to be true, its citizens seemed to have had a proclivity for living and working on hillsides, for this is where we find much of the alleged archaeology associated with them. Such is the case in this astonishing NASA photo. Let us take a closer look.

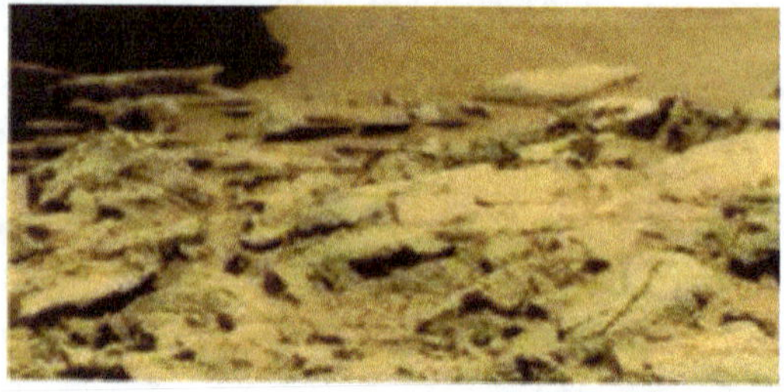

SOL 1355-E, highlighted area 1 (above): With the aid of a magnifying glass, the observant reader will quickly ascertain that there are numerous squares, squarish objects, and 90 degree angles in this image. Are they natural, the result of Martian geological forces, or artificial, the vestiges of intelligent craftsmanship?

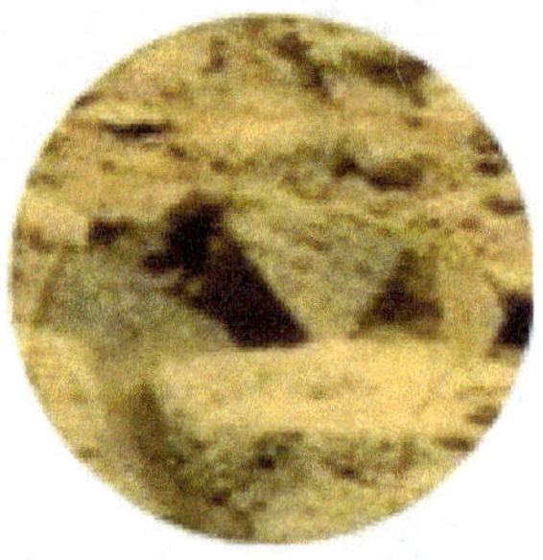

SOL 1355-E, highlighted area 2 (above): This nearly perfect "rock" triangle appears to be laying on top of another "rock" of similar proportions and shape (to its right). The object's thickness and its straight line "cuts" suggest machining.

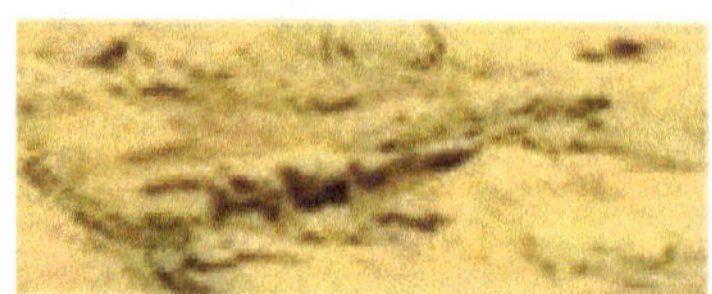

SOL 1355-E, highlighted area 3 (above): Fossilized vertebrae or natural rock formation?

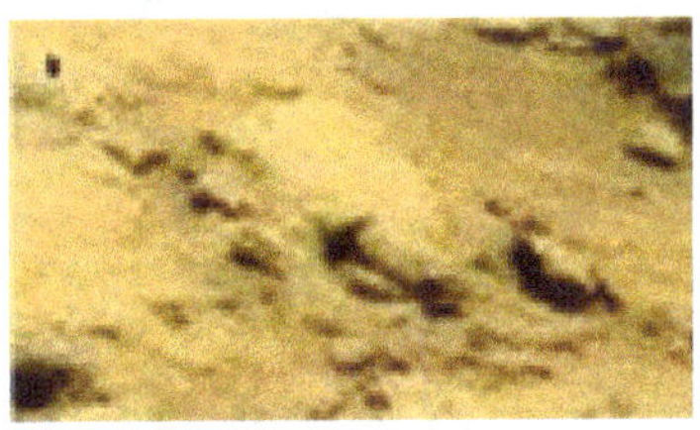

SOL 1355-E, highlighted area 4 (above): Starting in the upper left and sweeping diagonally down to the lower right, can be seen a number of star like objects. If these are not fossilized marine creatures or artificially made objects, what are they? The truth is that no one knows for sure.

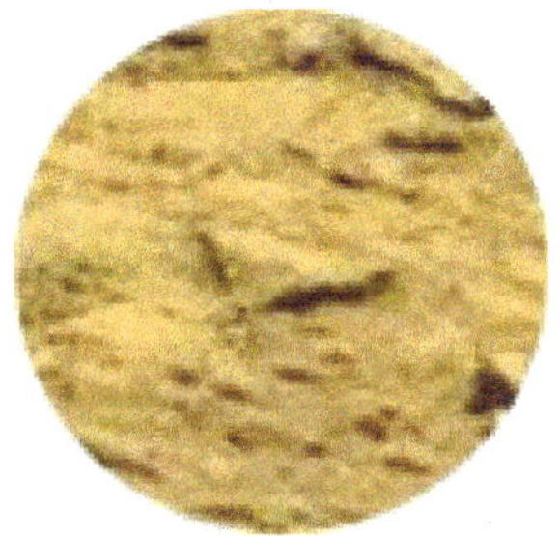

SOL 1355-E, highlighted area 5 (above): If this image was taken on Earth, this arrowhead like object could easily be mistaken for a model of a "flying wing" styled jet, perhaps similar to the fabled hypersonic "black triangle," reconnaissance aircraft known as the "Aurora," or the USAF "TR-3B," another alleged black project spy craft. Making this object even more provocative is the fact that there are several others just like it located nearby. Note the square with a triangular "slice" cut out of it at the 1:00 position.

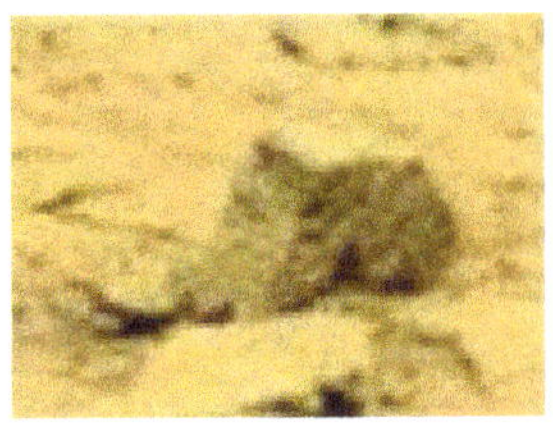

SOL 1355-E, highlighted area 6 (above): This seemingly intentionally made, delta shaped object looks like the top of a capital letter "A." Its appearance and form suggest some type of molded concrete building material.

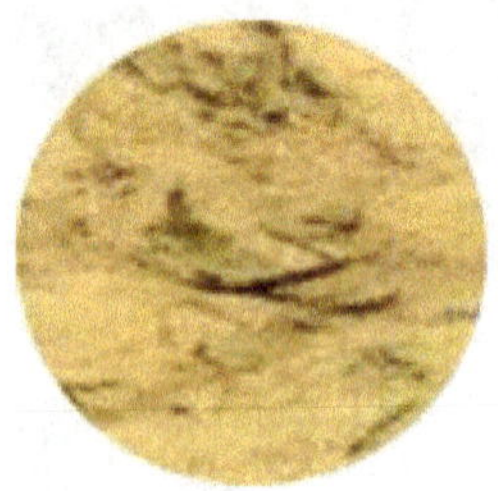

SOL 1355-E, highlighted area 7 (above): At center another one of thousands of perfectly square objects on Mars. This one appears to have a "frame." Numerous odd items encircle it.

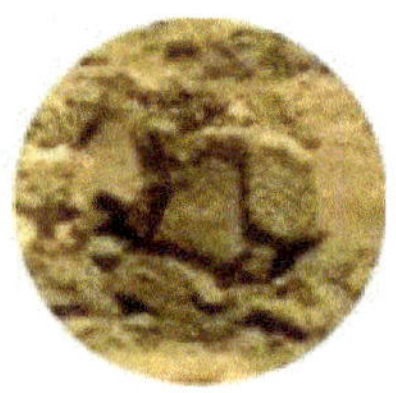

SOL 1355-E, highlighted area 8 (above): This group of unusually shaped, strangely marked, oddly clustered objects seems out of place. The resolution of the original photo is not high enough to decipher what may be going on here. It is easier for conventional science to label it a "clump of Martian rocks."

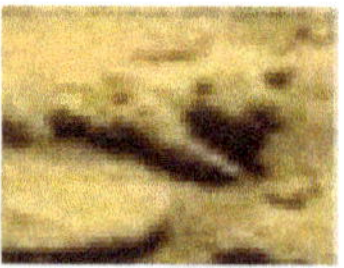

SOL 1355-E, highlighted area 9 (above): The spiral object in the center of this image has the appearance of a corkscrew; or is it just the result of digital distortion? And what about the peculiar objects on either side of it?

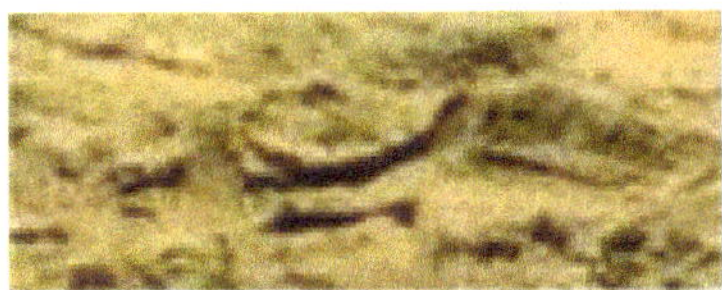

SOL 1355-E, highlighted area 10 (above): What are we looking at? Natural rocks or manufactured objects? Observing this photo under magnification reveals a scene filled from side to side with artificial looking items, shapes, forms, and structures.

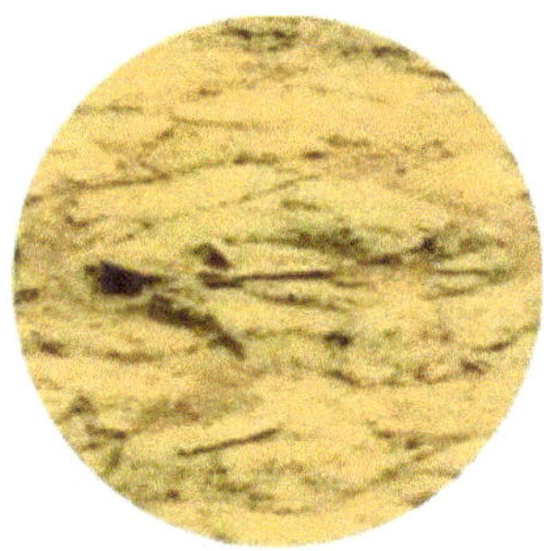

SOL 1355-E, highlighted area 11 (above): The square "cut slab" at center is slightly elevated above all of the flat "slabs" around it. Why? Several stick like or bone like objects lie on top of it and near it. Of course, everything in this image could be the result of natural geology. Conversely, none of it may be.

Chapter 16

DEBRIS FIELD

SOL 1355-F: EARTHDATE MAY 29, 2016

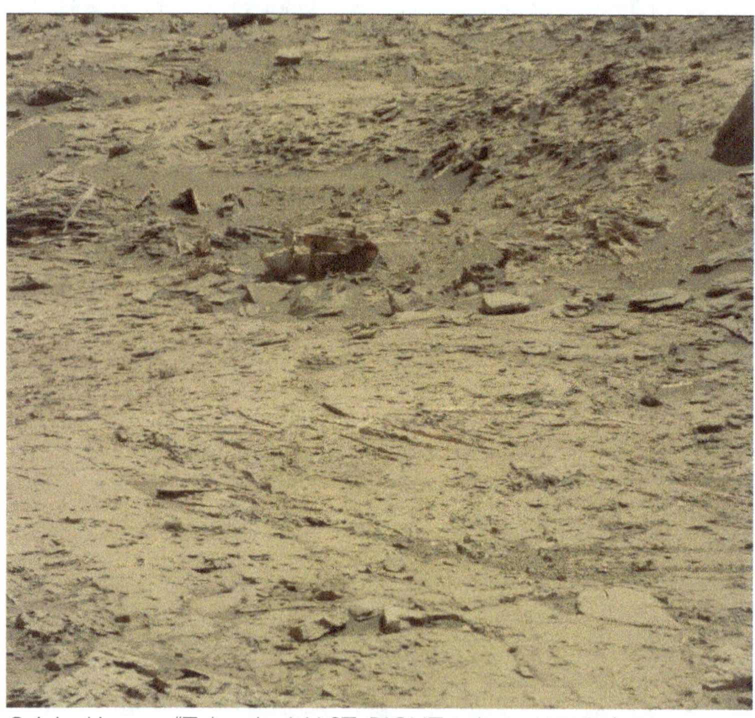

Original image. "Taken by MAST_RIGHT onboard NASA's Mars rover Curiosity on Sol 1355 (2016-05-29, T08:38:42.000Z)." Image credit: NASA/JPL-Caltech/MSSS.

ANALYSIS OF DEBRIS FIELD, SOL 1355-F

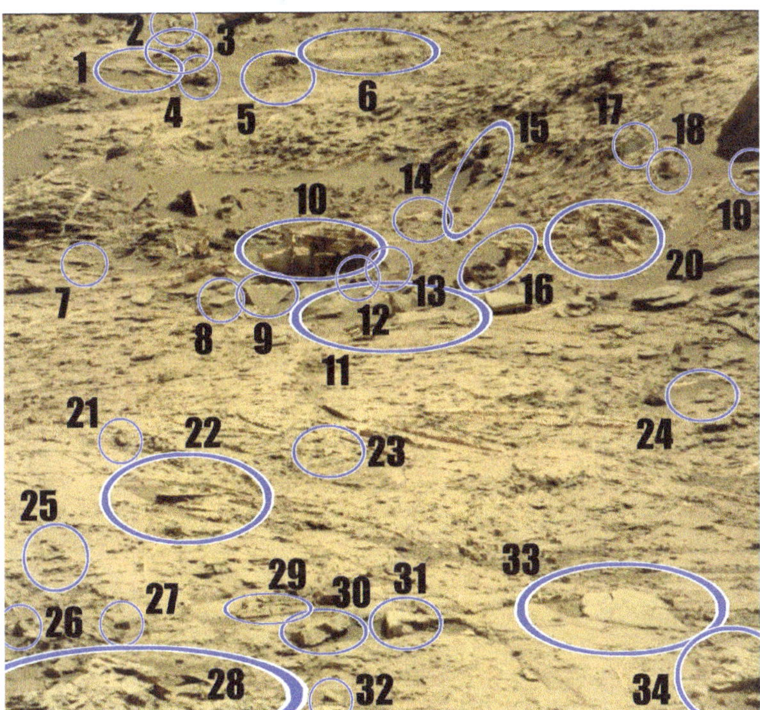

Debris Field, Sol 1355-F, with anomalies circled. Image colored, brightened, and sharpened by the author to improve contrast. This is one of the most spectacular of NASA's Mars photos, for it is littered with a veritable cornucopia of objects exhibiting traits of artificiality.

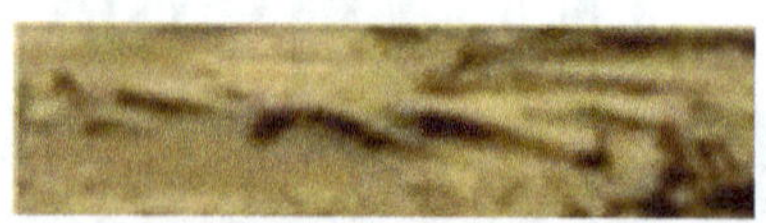

SOL 1355-F, highlighted area 1 (above): Straight cut lines and clear square cut outs (to the right) dominate this image.

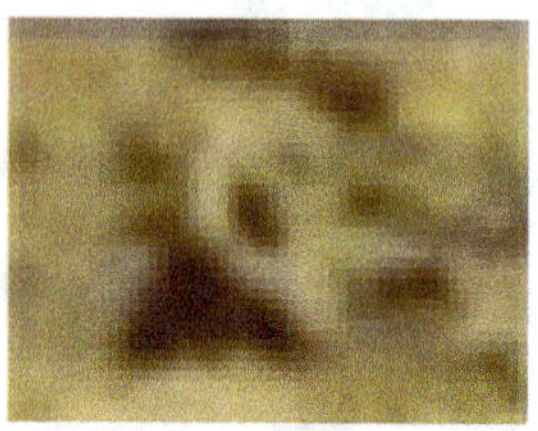

SOL 1355-F, highlighted area 2 (above): This odd rectangular object has a rectangular hole in the center of it, while a number of weird appendages are protruding along its top left edge. Also, it is leaning at an angle, as if propped up on a stand of some kind.

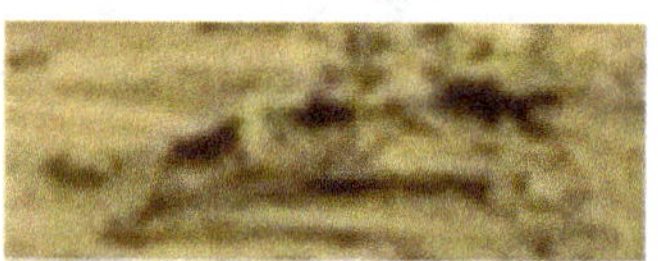

SOL 1355-F, highlighted area 3 (above): This appears to be the remains of a box like item, or possibly the remnants of a structural foundation.

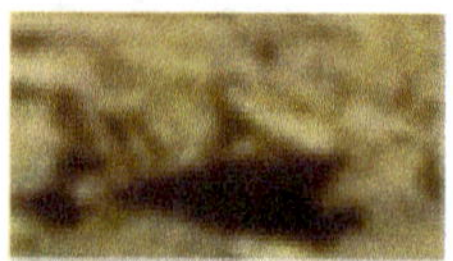

SOL 1355-F, highlighted area 4 (above): The low resolution of this image makes it impossible to see a clear outline. Despite this, this small object has all the hallmarks of intentional manufacturing. Note that areas 1, 2, 3, and 4 are all positioned next to one another.

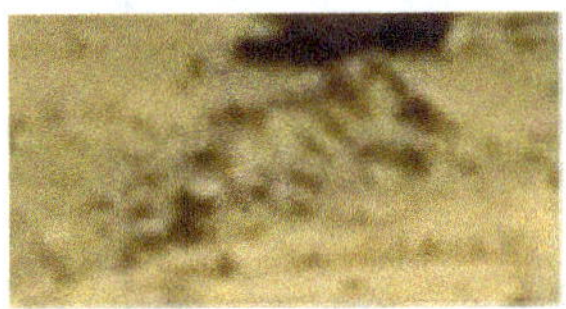

SOL 1355-F, highlighted area 5 (above): At first glance this may appear to be nothing more than a Martian dirt pile. Under magnification, however, one can see what appears to be a number of artificial objects. The whole area, in fact, looks like what is left of a collapsed structure.

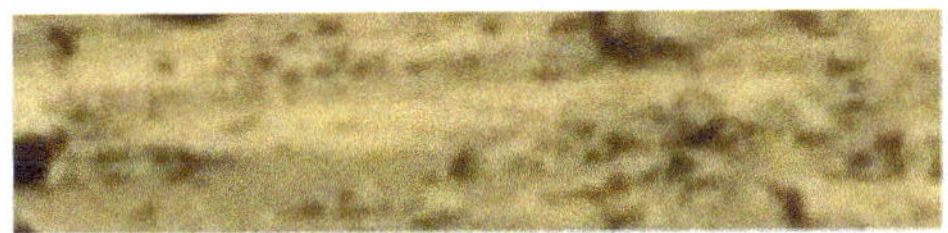

SOL 1355-F, highlighted area 6 (above): This flat, lightly colored, rectangular area stands out in the darker Martian dust. It may be naturally formed sedimentary rock. On the other hand, its close proximity to dozens of clearly artificial objects and structures raises uncertainties as to its true origins and nature.

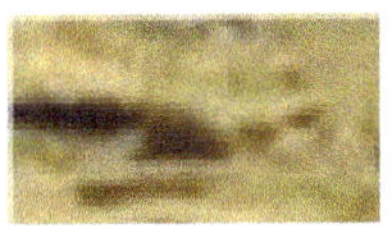

SOL 1355-F, highlighted area 7 (above): Look closely. At the center of this blurry image is what appears to be a smoothly rounded egg shaped item next to an "L" shape. It is casting an ovoid shadow.

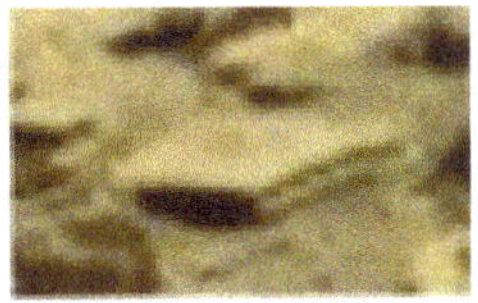

SOL 1355-F, highlighted area 8 (above): This book like object may be a Martian rock formation. But its location amid numerous objects that appear to have been created by an advanced intelligence suggests otherwise.

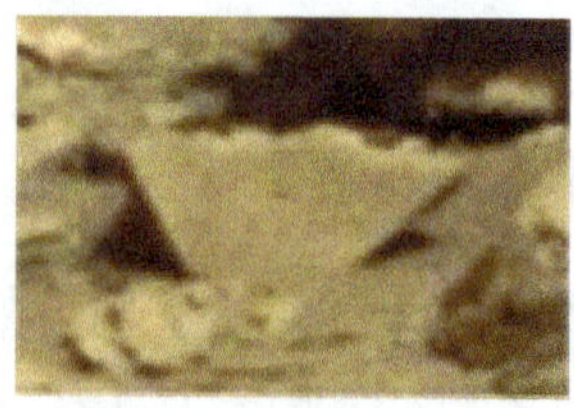

SOL 1355-F, highlighted area 9 (above): Another broken, delta shaped, cement like object half buried in the Martian soil. To believers, calling this a "rock" is laughable; to skeptics calling it an "artifact" is absurd. Study it in detail. Which side do you fall on?

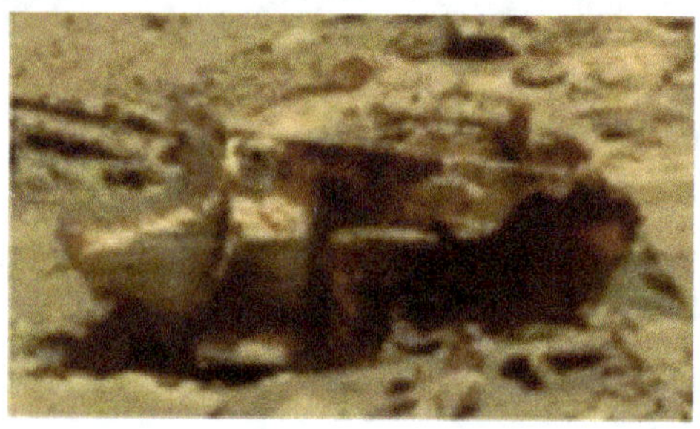

SOL 1355-F, highlighted area 10 (above): With its numerous overt square angles, there is almost nothing about this little object that looks natural. Rather, it appears to be the remnants of a small brick and mortar structure that has been digitally manipulated to conceal its true nature. Magnify it and be amazed.

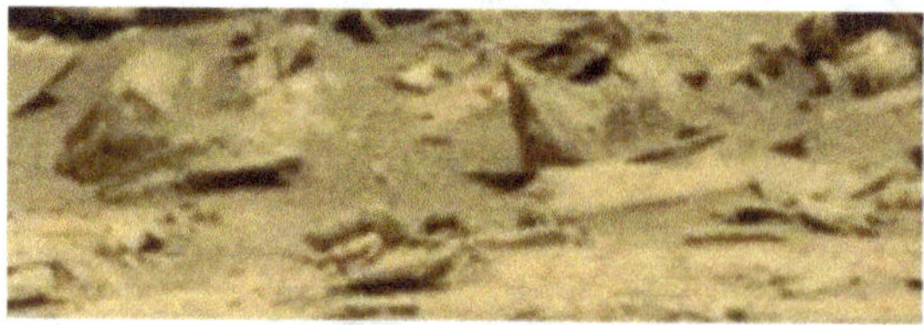

SOL 1355-F, highlighted area 11 (above): Odd shaped, unidentifiable objects—some that appear melted and bent—lie scattered throughout this image. This does not necessarily mean they are artificial. However, they are located at the epicenter of what can only be described in appearance as the site of a large ancient explosion.

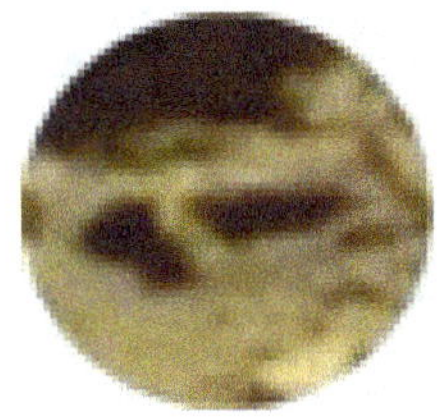

SOL 1355-F, highlighted area 12 (above): If this object were by itself in the middle of an empty Martian plain, we might ignore it. But it is surrounded by numerous other right angled structures just like it. Natural or artificial?

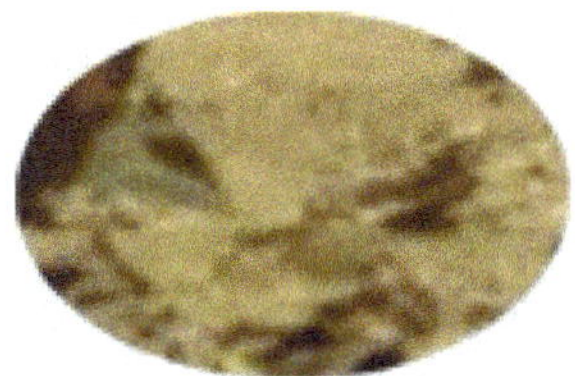

SOL 1355-F, highlighted area 13 (above): Magnify and study these oddly shaped, unidentifiable objects.

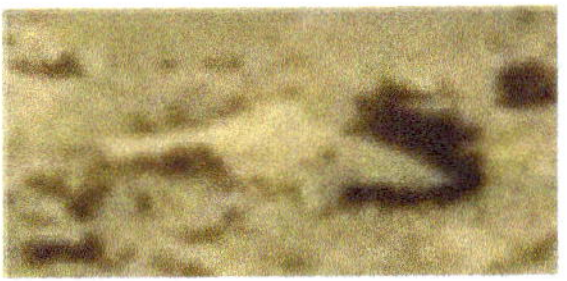

SOL 1355-F, highlighted area 14 (above): This flat pointed object appears torn and also bent in the middle. The result of geology, an "atmospheric disturbance," or possibly a bomb blast?

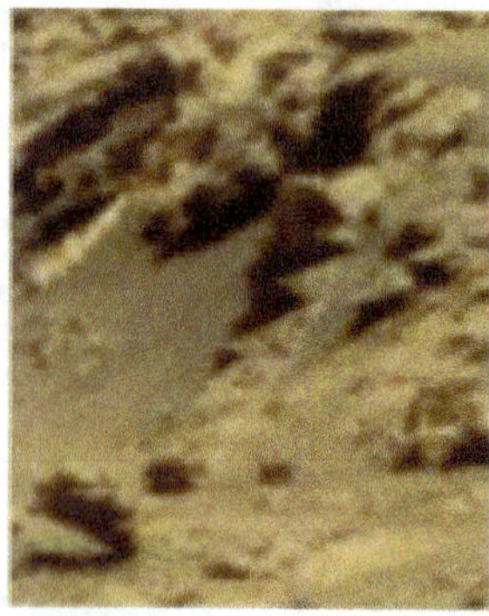

SOL 1355-F, highlighted area 15 (above): The remains of a wall running down the side of a hill, or a natural rock formation?

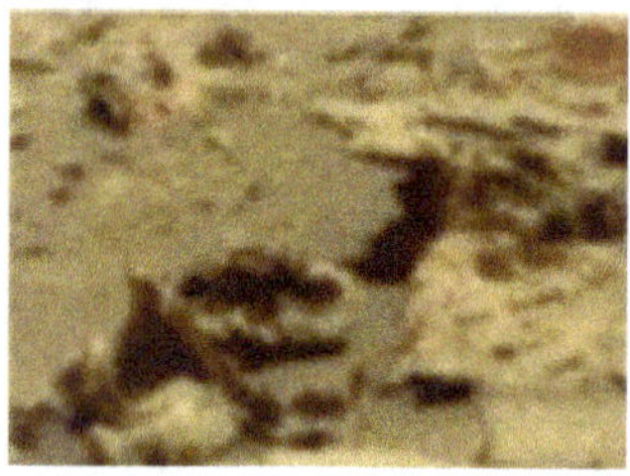

SOL 1355-F, highlighted area 16 (above): An odd parrot's bill like shape juts out from a "rock" in the upper right. Unusual shapes seem to be "carved" into its upper surface as well. Below this object and to its left is what looks like a mangled metal box that has been blown apart, partially melting from the heat created by the force of the detonation.

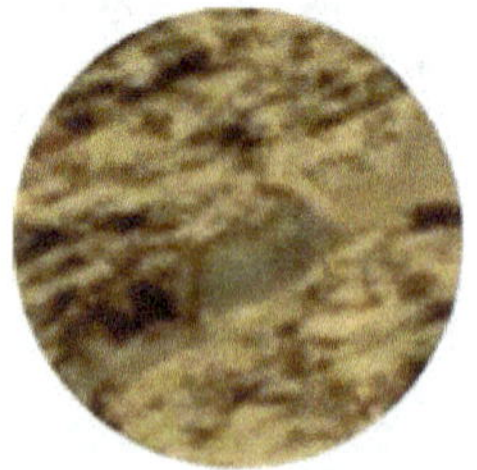

SOL 1355-F, highlighted area 17 (above): This flat, smooth, grayish object is standing vertically, and looks out of place on the reddish, rough, craggy hillside.

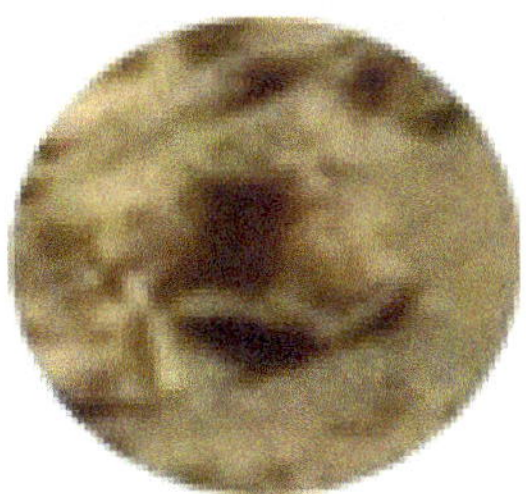

SOL 1355-F, highlighted area 18 (above): Several artificial looking objects lie in close proximity to one another.

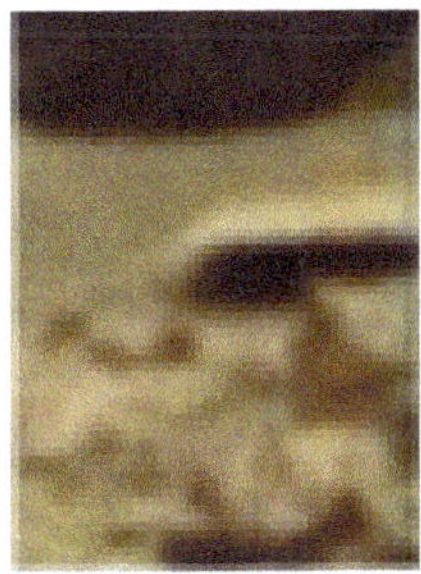

SOL 1355-F, highlighted area 19 (left): A flat perfectly square rock laying on the side of the hill.

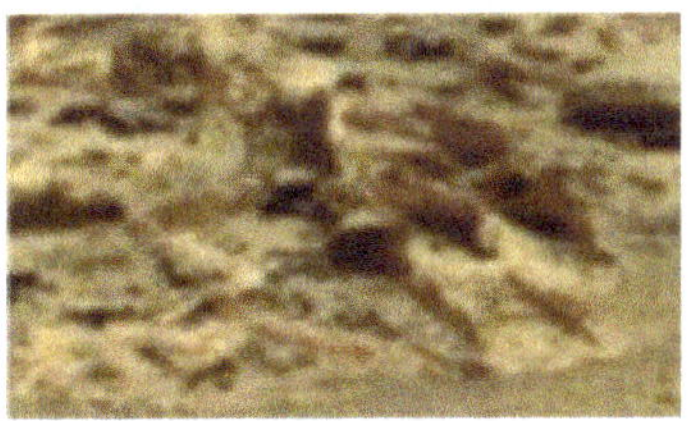

SOL 1355-F, highlighted area 20 (above): Right of upper center we find yet another boat shaped "hull" like object, strangely, like the others, seemingly made out of bricks and mortar. To its left at 9:00 and 10:00 are what appear to be small circles ringed with stones. Below the "boat hull" is what looks like a small rock lined trough leading downward to a sandy patch. Look carefully. Dozens of other oddments dot the image.

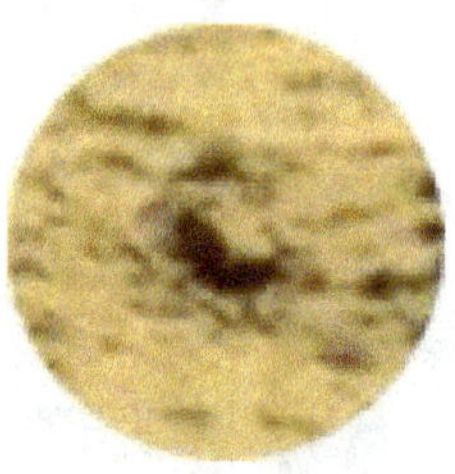

SOL 1355-F, highlighted area 21 (above): A brick or a rock? One must consider where it is located: in the midst of what looks like a large blast site, and surrounded by curious shapes and structures.

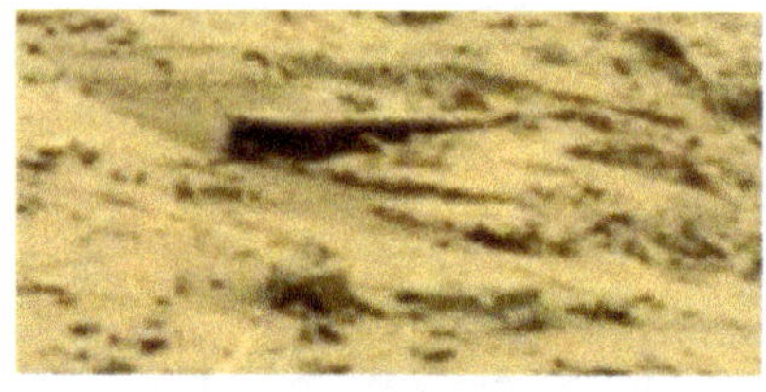

SOL 1355-F, highlighted area 22 (above): Inside a near perfect rectangle impression in the dirt lies what looks like a large triangular piece of metal. Below it another rectangle impression runs downward and to the right of frame. Additionally, "bricks," squares, "trays," ridges, straight lines, 90 degree angles, and notched objects clutter the entire landscape.

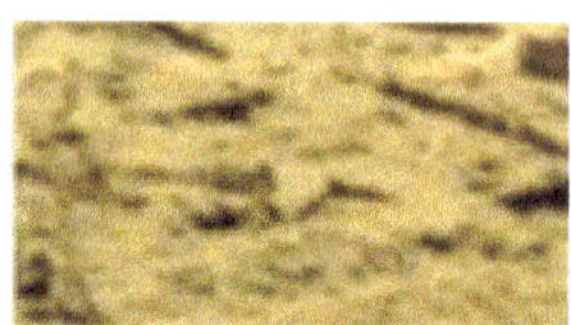

SOL 1355-F, highlighted area 23 (above): A "rock" with a triangle shape "cut" out of one side appears in the middle of this image. It is surrounded by small, straight, raised, wall like objects. Accidental mineral veins or intentionally machined objects?

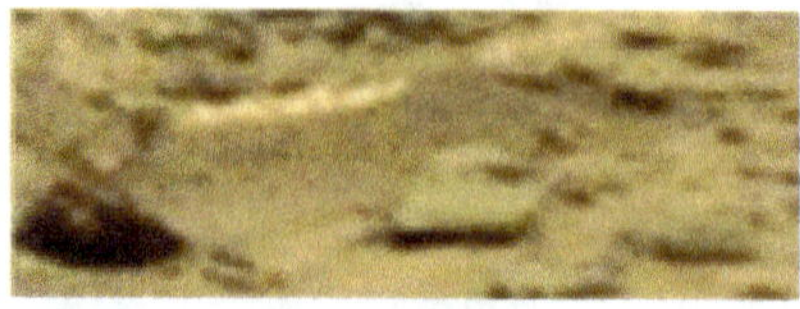

SOL 1355-F, highlighted area 24 (above): Another rectangular impression in the ground, this one with a square "rock" sitting next to it.

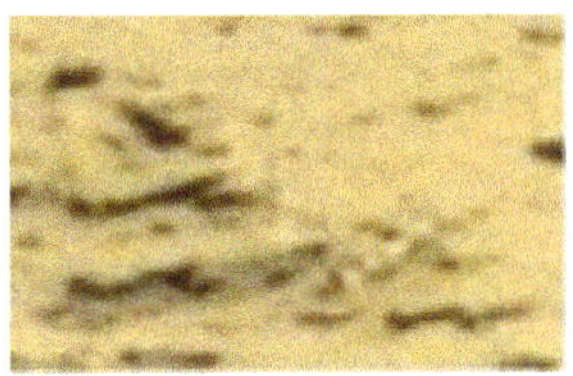

SOL 1355-F, highlighted area 25 (above): Are we looking at the remnants of a house foundation? Why are there "X"s and "crosses" in this image?

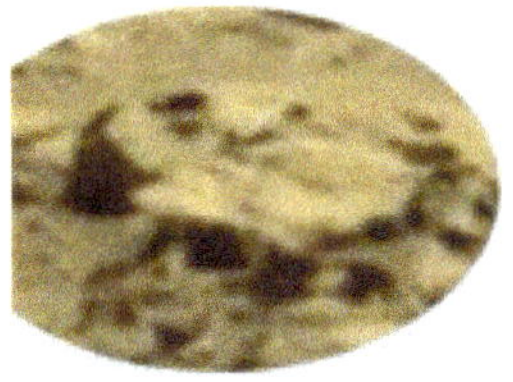

SOL 1355-F, highlighted area 26 (above): This debris seems to include a brick, pieces of broken cement, and a battered sheet of thin metal. Magnify it.

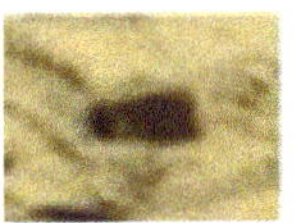

SOL 1355-F, highlighted area 27 (above): More evidence of possible Martian masonry: perhaps a brick or concrete block.

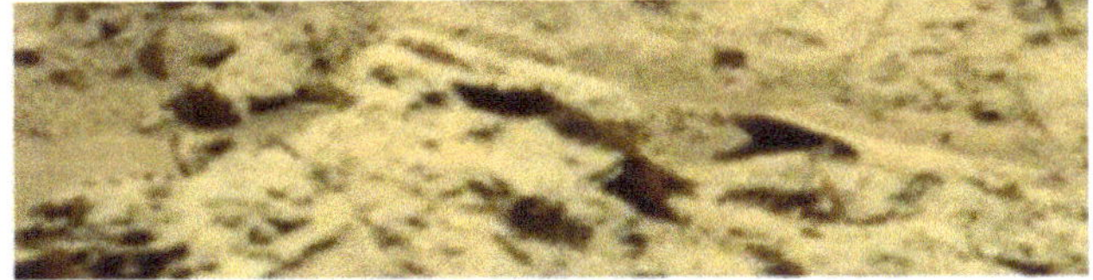

SOL 1355-F, highlighted area 28 (above): This squared off area could simply be a natural rock formation. However, it is surrounded by what appear to be bricks and tiles, as well as a host of other objects that look like they may have been manufactured by a technological culture.

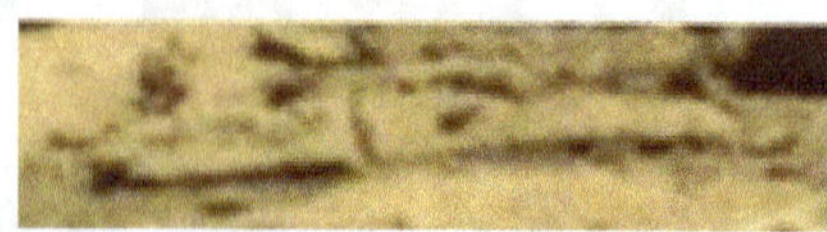

SOL 1355-F, highlighted area 29 (above): Nature could create this brick like pattern. So could an intelligent being. The question is, which is it? We must always consider the location, setting, and surroundings of such objects.

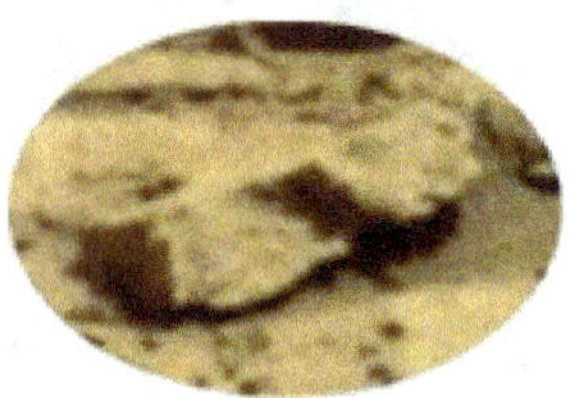

SOL 1355-F, highlighted area 30 (above): This cement like object appears to have broken off from a similar but larger object. Years of weathering, erosion, and layers of dust have given it a rock like appearance. Examine the next object (in area 31). They are lying next to one another, and seem to have both come from the same artificial structure or building.

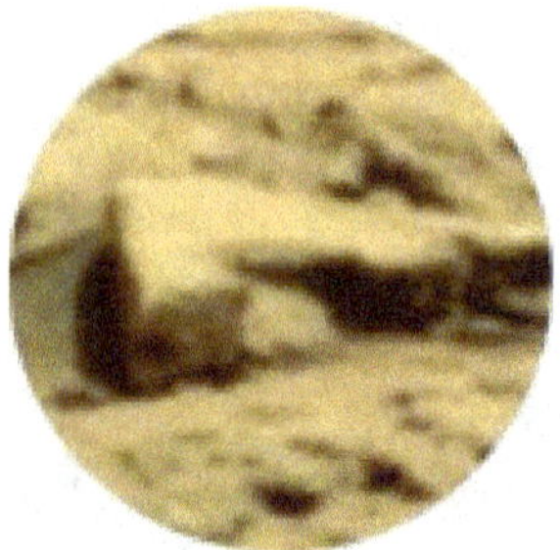

SOL 1355-F, highlighted area 31 (above): Another "molded" capital "A" like object, one that looks like it was once part of a building or ornamental structure. To its left is the object shown in area 30 above. To its right is a right angled ridge of some kind, half buried in the Martian dirt.

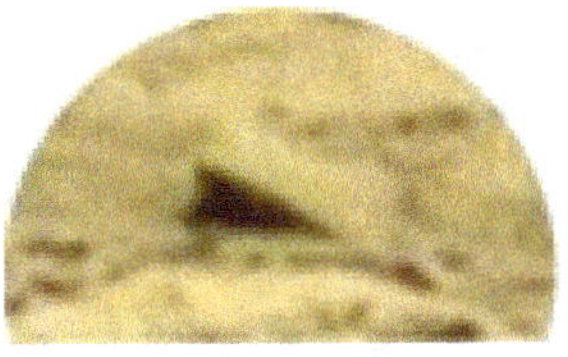

SOL 1355-F, highlighted area 32 (above): Another one of the thousands of "bricks" (or possibly ornamental "pedestals") laying on the surface of Mars.

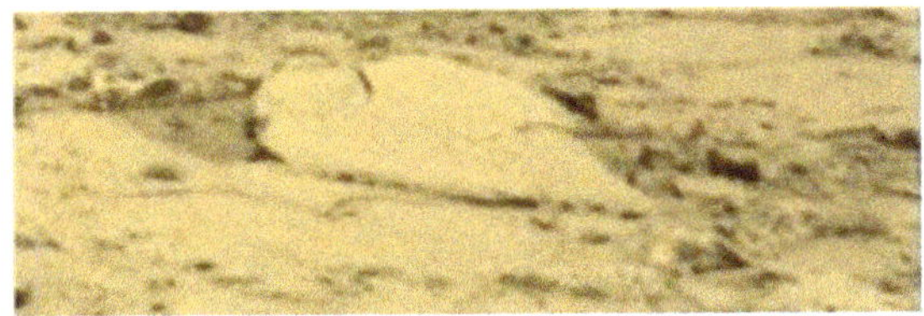

SOL 1355-F, highlighted area 33 (above): A rectangle shaped impression in the Martian soil, this one with a large, bent, arrowhead like object sitting on it. Surrounding both are numerous straight and curved objects with artificial characteristics.

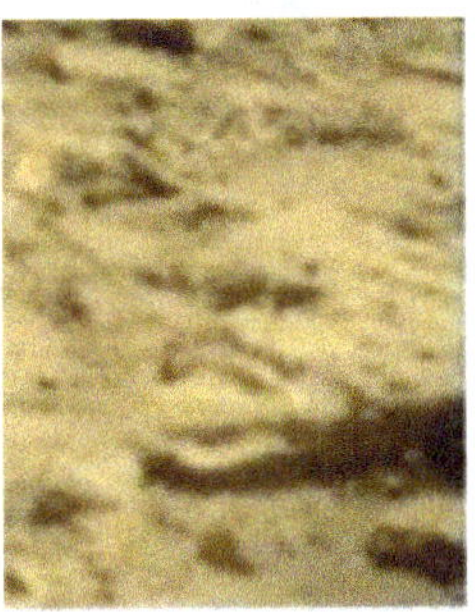

SOL 1355-F, highlighted area 34 (above): This assortment of odd but fascinating squarish objects is situated just to the lower right of area 33 described above. Although we cannot positively identify them, owing to the many other objects in the large debris field in which they lie, they could be associated with Martian building materials.

Chapter 17

UNUSUAL FORMATIONS

SOL 1355-G: EARTHDATE MAY 29, 2016

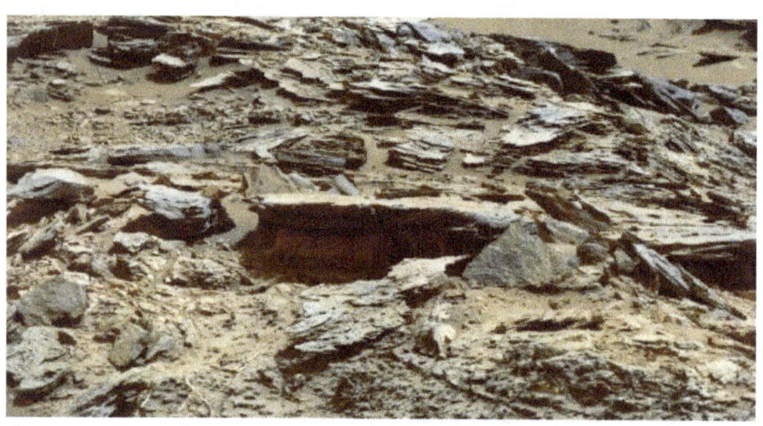

Original image. "Taken onboard NASA's Mars rover Curiosity on Sol 1355 (2016-05-29)." Image credit: NASA/JPL-Caltech/MSSS.

ANALYSIS OF UNUSUAL FORMATIONS, SOL 1355-G

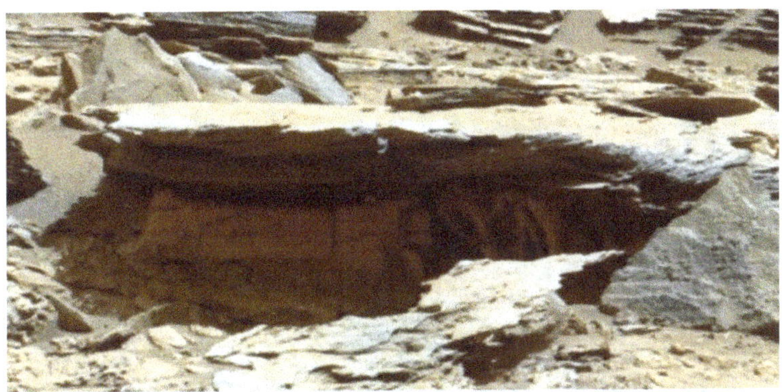

Unusual Formations, Sol 1355-G, with anomalies at center. Image colored, brightened, and sharpened by the author to improve contrast. Unusual formations beneath a Martian overhang. Use magnification to get in close. Scan the entire wall from left to right. There is a lot going on.

Chapter 18

CLEARED RECTANGLE PATCH

SOL 131: EARTHDATE OCTOBER 7, 2008

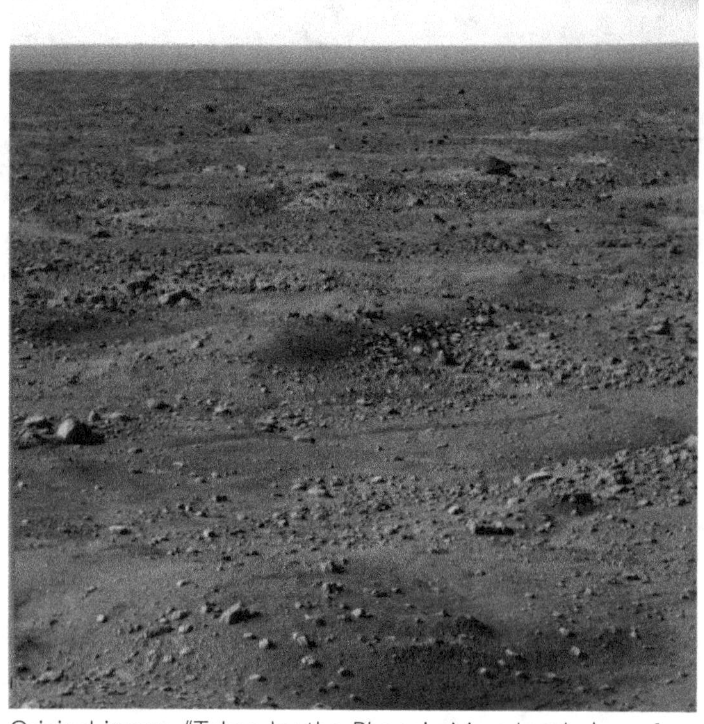

Original image. "Taken by the Phoenix Mars Lander's surface stereo imager on the 131st Martian day, or sol, of the mission (Oct. 7, 2008)." Image credit: NASA/JPL-Caltech/University of Arizona/Texas A&M University.

ANALYSIS OF CLEARED RECTANGLE PATCH, SOL 131

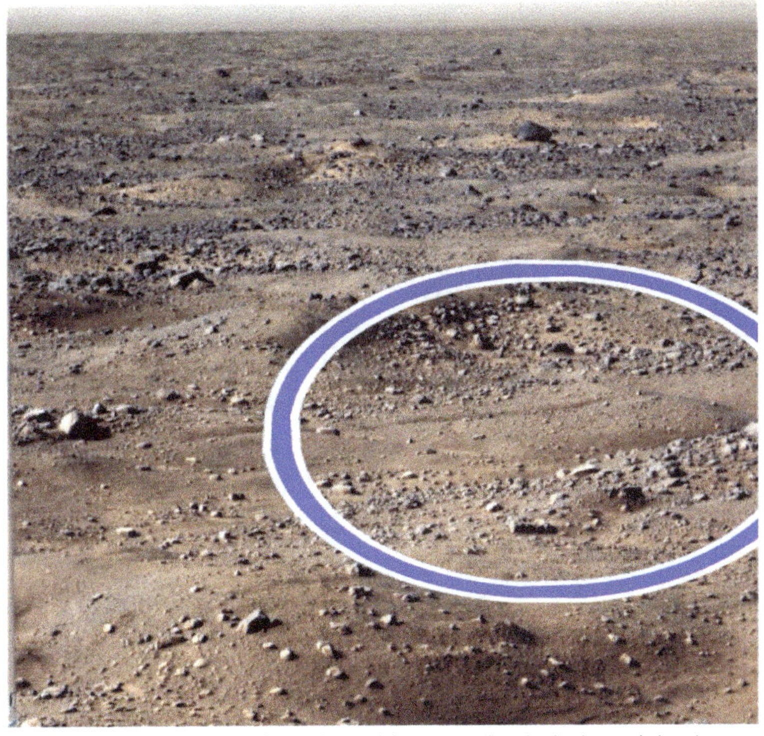

Cleared Square Patch, Sol 131, with anomaly circled on right. Image colored, brightened, and sharpened by the author to improve contrast.

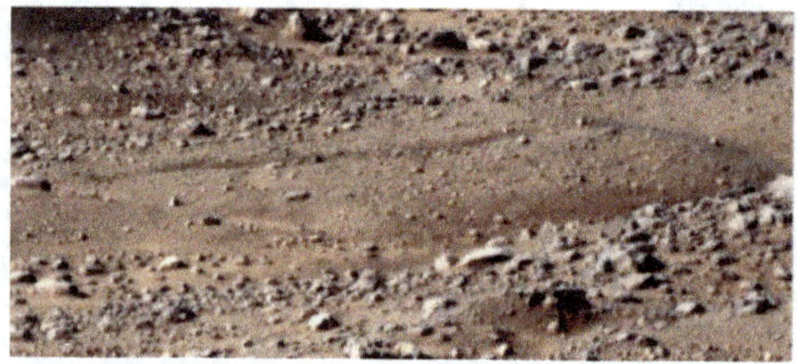

SOL 131, highlighted area: This is not the first cleared, rectangle shaped patch of ground we have seen on the Martian surface. Nor will it be the last.

Chapter 19

ODDS & ENDS

SOL 1033: EARTHDATE JULY 3, 2015

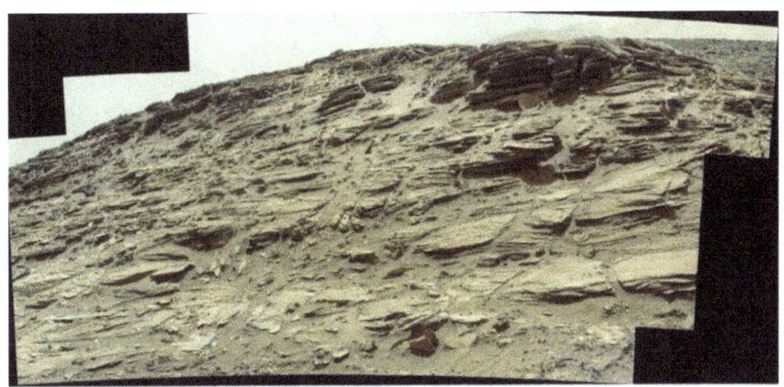

Original image. "This large composite photo combines numerous frames. They were taken by MAST_RIGHT onboard NASA's Mars rover Curiosity on Sol 1033 (2015-07-03, T13:30:45.000Z)." Image credit: NASA/JPL-Caltech/MSSS.

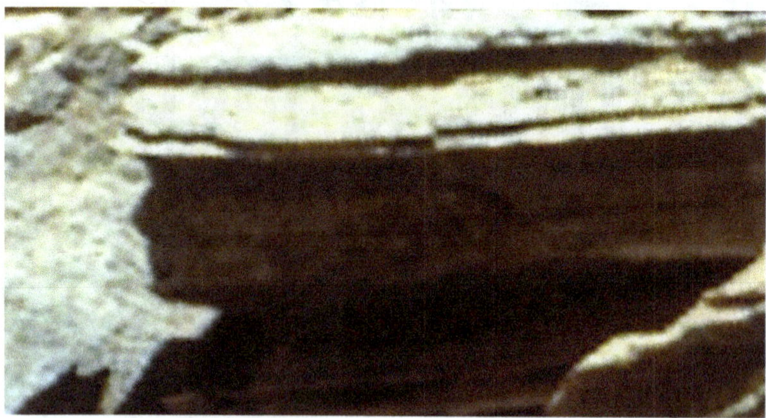

SOL 1033-1, colored, brightened, and sharpened by the author for contrast (as were all of the images in this chapter). What appear to be Roman letters have been etched or perhaps embossed into the side of this Martian overhang. Under magnification, from left to right, the letters look like "Y," "S," "O," and "A." What appears to be a parenthesis or crescent shape appears at the end (to the right of the "A"). Is this an ancient Martian word?

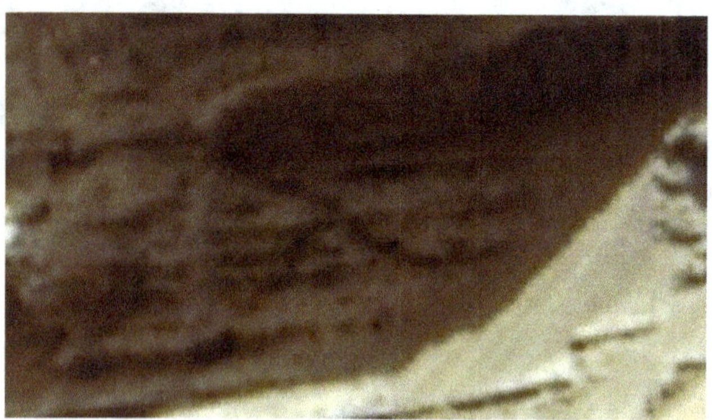

SOL 1033-2. What looks like a large "X" has been "carved" into the side of this Martian cliff face. Is this pareidolia or was this done intentionally by an intelligent being? If the latter, perhaps it is a type of "SOS" signal, used by survivors of the Great Martian Apocalypse to attract the attention of rescuers—similar to the practice of lost hikers on Earth. This heavily weathered "symbol" could be many thousands of years old.

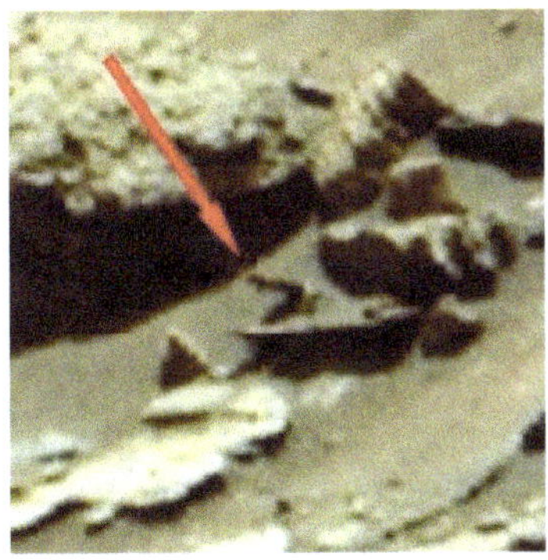

SOL 1033-3 (above). Even if we agree to label everything in this image a "rock," what is the long "rifle barrel" like object (center) sticking out horizontally from this hillside—to say nothing of the many unusual items and shapes surrounding it?

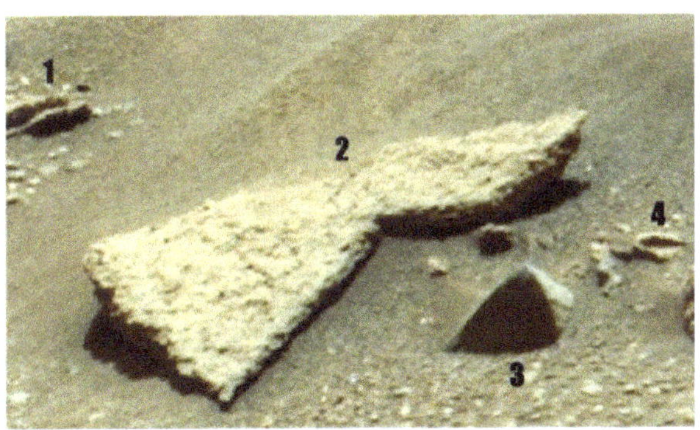

SOL 1033-4 (above). From left to right, four unusually shaped objects lie on this Martian hillside. It is possible that all four of them are natural rock formations. But how likely is that? Without evidence either way, the answer depends on your viewpoint.

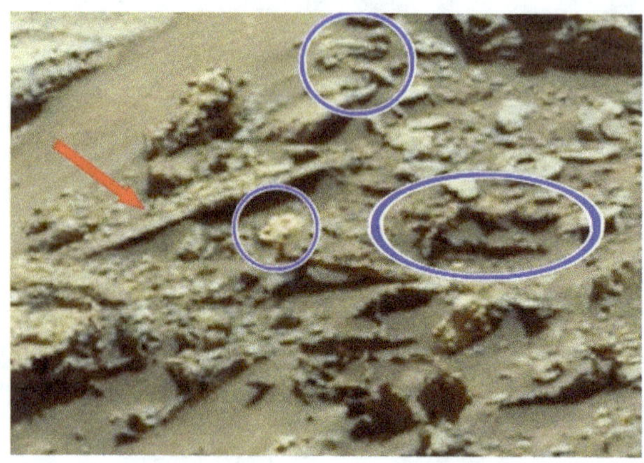

SOL 1033-5 (above). Artificiality fills this small area of the larger debris field, as my arrow and circles demonstrate. After laying on the Martian surface for perhaps thousands of years, most of these objects are not easy to identify. The circled bright gold yellow object near the center—which looks like a metal belt buckle or hinge—stands out dramatically.

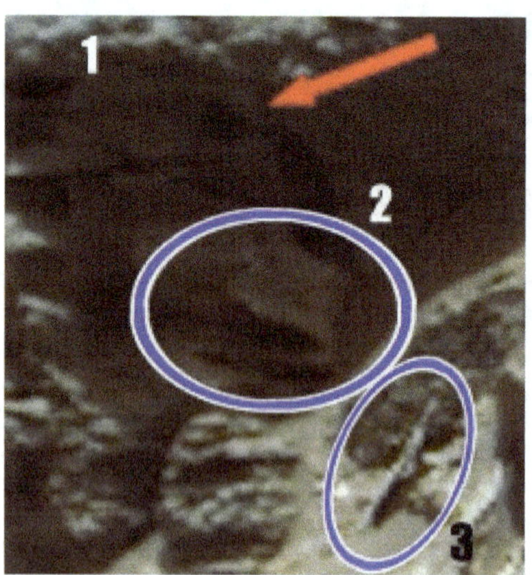

SOL 1033-6 (above). Objects 1 (left of the red arrow) appear to be flat paving stones of some kind. Object 2 looks like a piece of thin, flat metal bent nearly in half by a strong force. Object 3 may be a piece of metal like rebar (a steel rod used for reinforcing masonry). If these are artifacts we cannot ask for much better photographic evidence than this.

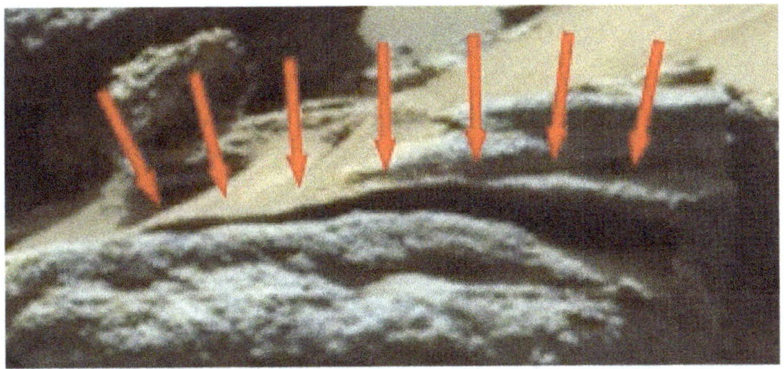

SOL 1033-7 (above). This looks like a nearly completely buried entrance to a tunnel, or perhaps a collapsed garage of some kind. If correct, the "roof" has buckled under the strain of the heavy rocks and sand above it.

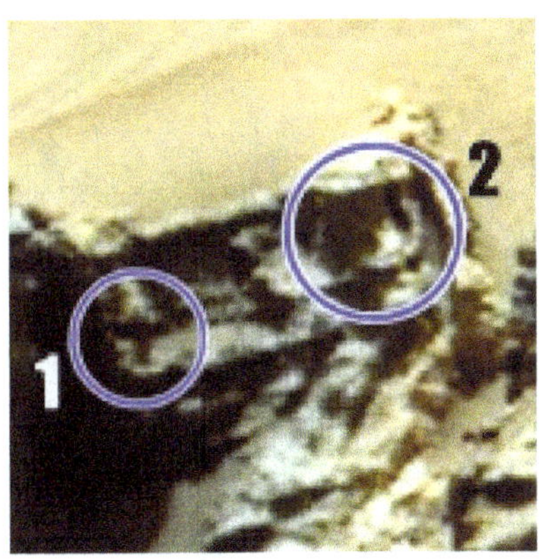

SOL 1033-8 (above). Object 1: A cross or a "T" appears to have been carved into this "rock." Object 2: This may be part of a circular object, perhaps with gears. Magnify. Other oddities abound.

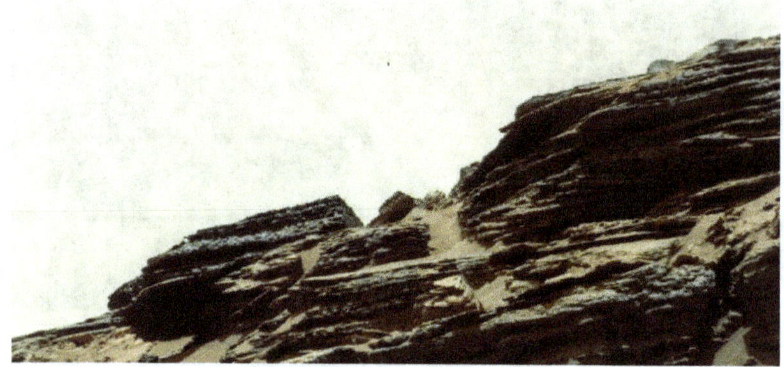

SOL 1033-9 (above). What is that small black dot in the sky above the "dead planet" Mars? Mainstream scientists like to explain objects like this away as "data processing artifacts," or as a result of other peculiarities common to digital cameras and digital photography, such as "dead pixels." See the next image (below).

SOL 1033-10 (above). Camera artifact or UFO? Here is a closeup of the "small black dot" described in Sol 1033-9 above. I have blown it up many times its original size for better viewing. This boomerang shaped object looks identical to the alien like "black triangles" seen by terrified eyewitnesses floating slowly and silently across our own skies here on Earth.

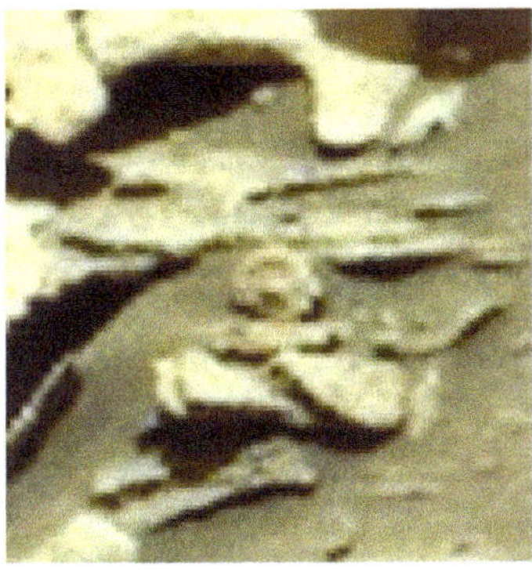

SOL 1033-11 (above). Though oddly shaped, most of the objects in this image are probably naturally formed rocks. The round "rock" at the center, however, has a number of artificial traits, including a "rim," a raised bump in the middle, and, when magnified, what appear to be broken gear teeth running along its outer edge.

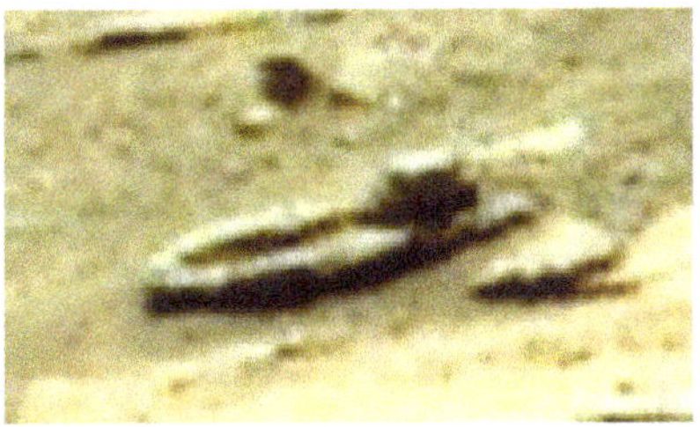

SOL 1033-12 (above). This circular or oval object has a raised ridge running around it and a small squarish object sitting on top.

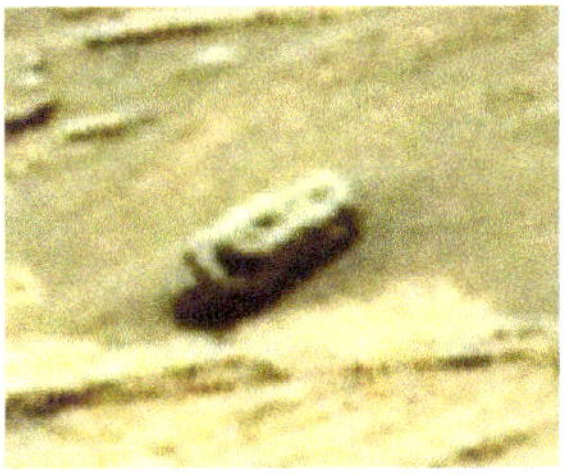

SOL 1033-13 (above). This object is laying on its side in the middle of a large debris field, filled with what some would describe as the ruins of an ancient advanced race. What is it?

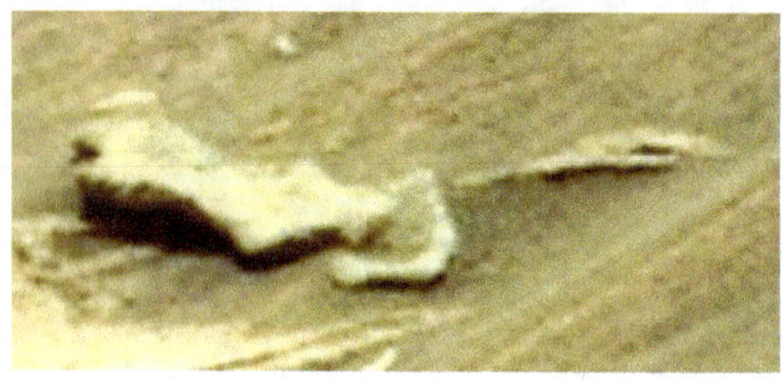

SOL 1033-14 (above). Are these fossilized bones? Sitting as they are in a debris field of some kind, amid numerous out-of-place shapes (like the perfect square edged rectangle at bottom center), they are bound to attract the attention of anomalists.

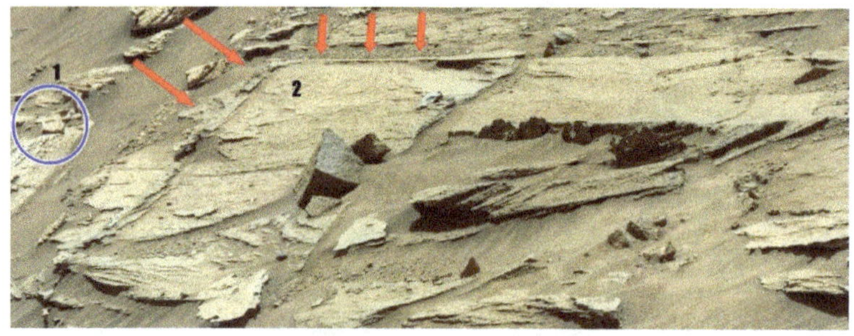

SOL 1033-15 (above). Object 1: Has a tray or trough like appearance. Object 2: The accidental conjoining of rock at a 90 degree angle, or what is left of an ancient house foundation? On close magnified inspection, dozens of other anomalies can be seen as well.

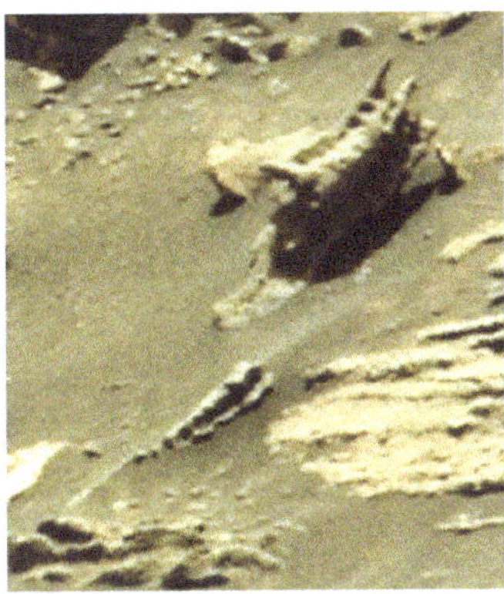

SOL 1033-16 (above). Are these the fossil remains of once living creatures? Orthodox scientists tell us that there have never been large complex organisms on Mars. If true, this is an extraordinary group of "rocks."

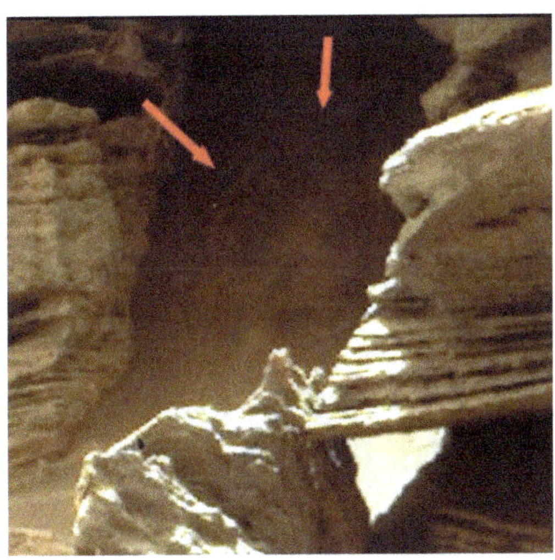

SOL 1033-17 (above). Unusual markings cover the back wall inside this cave like entrance. Pareidolia or 100,000 year old Martian petroglyphs? Magnified, the "etching" appears similar to ancient Egyptian temple art, complete with a man wearing a schenti (skirt) and carrying a heka (crook) and nekhakha (flail) staff.

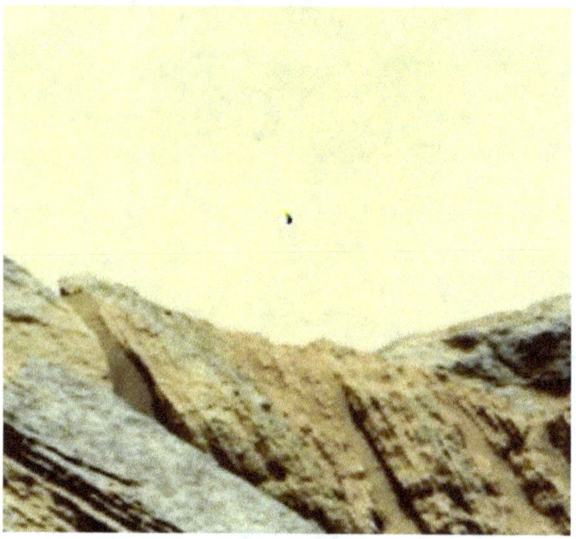

SOL 1033-18 (above). If the black speck in this image is not a "digital artifact," then apparently there were several black triangles flying in and out of the Martian mountains the day the Curiosity rover snapped these photos.

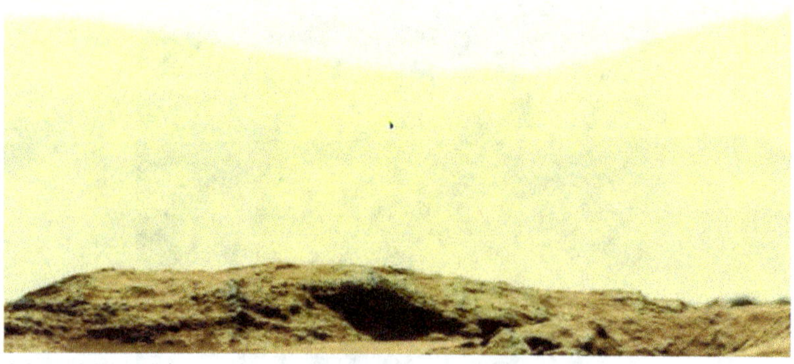

SOL 1033-19 (above). Another boomerang shaped "camera artifact" in the Martian sky. This one happens to be sailing past a large, artificial looking square opening on a mountain crest (bottom center). Is there any connection between the two objects?

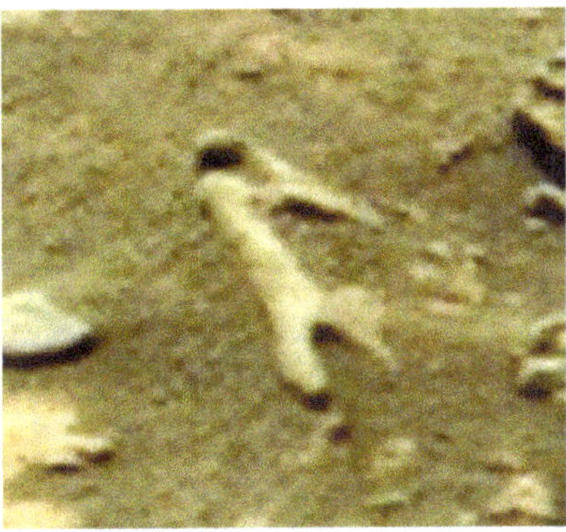

SOL 1033-20 (above). Are these bones, tools, or rocks—or something else we cannot conceive of? The truth is that no one knows for sure. Why not remain intellectually receptive to all theories then?

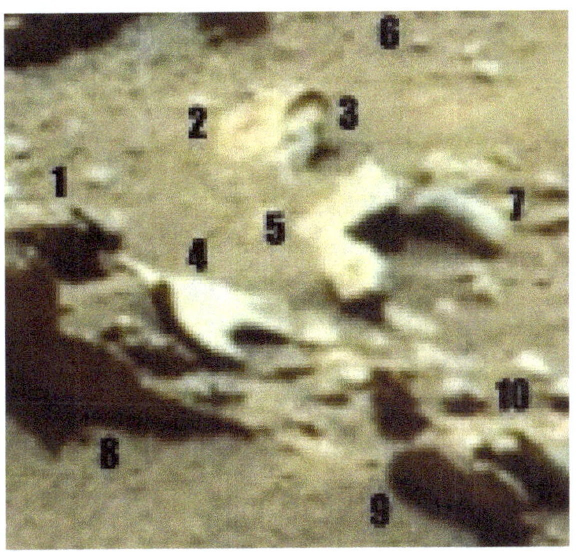

SOL 1033-21 (above). Object 1: a "rock" with a "cutout" in it. Object 2: a perfect hexagon with "designs" carved in it. Object 3: a circular shape with a mushroom like cap and gills. Object 4: an arrowhead shaped article that looks very much like the delta insignia emblem used by Starfleet on *Star Trek*. Object 5: a chevron or an "L" shape. Objects 6: several rectangular formations in the sand. Object 7: a cucumber shape. Object 8: a small "wall." Object 9: continuation of the "wall" at left. Object 10: an oddity with beetle like "pincers" (a bizarre but not uncommonly seen object on Mars).

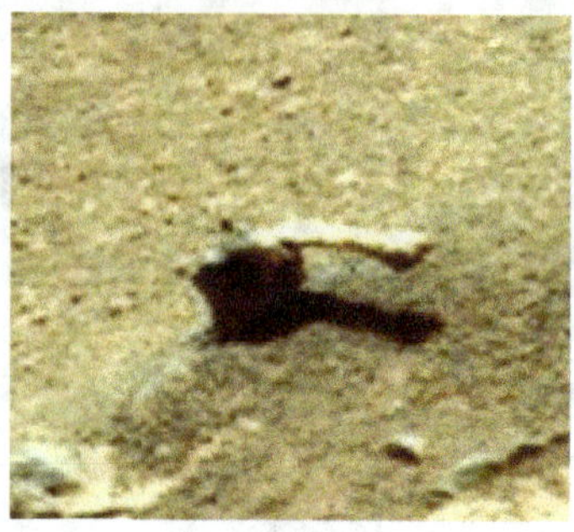

SOL 1033-22 (above). A living animal (like a mollusk), the remnant of an object made by a superior intelligence, or, as mainstream scientists posit, a naturally occurring rock? As gravity on Mars is much lower than on Earth, a delicate rock formation like this, if that is what it is, would not be impossible. However, how likely would it be, especially considering that this particular object is sitting in a massive debris field with hundreds of other objects that have all the appearance of artificiality?

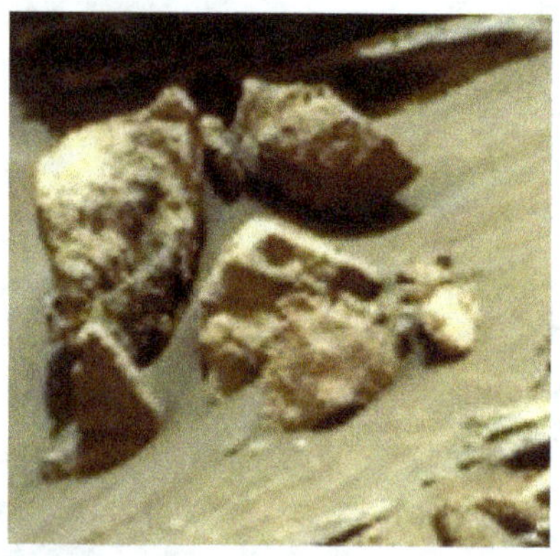

SOL 1033-23 (above). Due to the numerous "carved" surfaces, straight lines, shard like edges, triangles, and 90 degree angles on the objects in this image, everything appears artificial. If correct, this may be a group of pieces that broke off a statue during the Great Martian Apocalypse. They then slid down the hillside and sat on top of the Martian soil for millennia, where they were subjected to radiation, extreme cold, and wind and possibly water erosion. In July 2015 NASA's Mars rover Curiosity stumbled upon them and snapped this compelling image. Look carefully. Martian secrets often hide in plain sight.

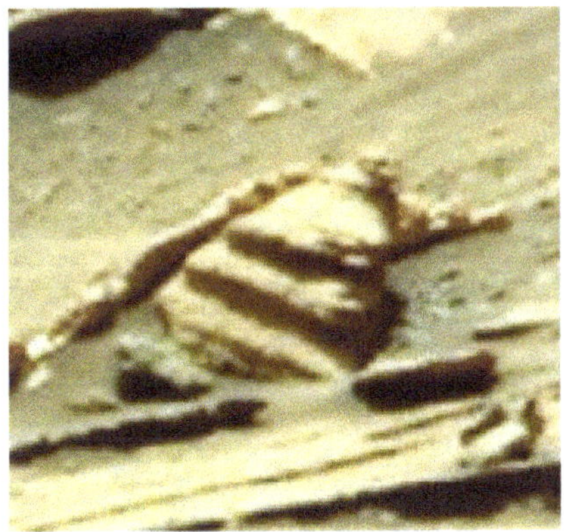

SOL 1033-24 (above). Remnants of a staircase or a small step pyramid, or just a natural geological formation? The many artificial looking ruins scattered over the large hill surrounding it offer the most likely answer. What, for instance, is the strange object at the 5:00 position?

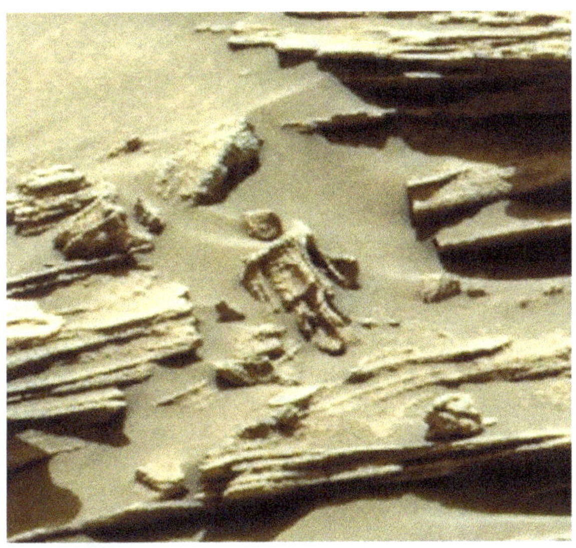

SOL 1033-25 (above). The object at the center of this image has a robotic humanoid shape. Is that what it is, or is it simply a case of pareidolia? Just below it is a gravestone like item laying in the sand. The "rock" at lower right (5:00) appears to have a "?" embossed in it. Rectangles, straight lines, and other odd shapes can be seen throughout this sandy Marsscape.

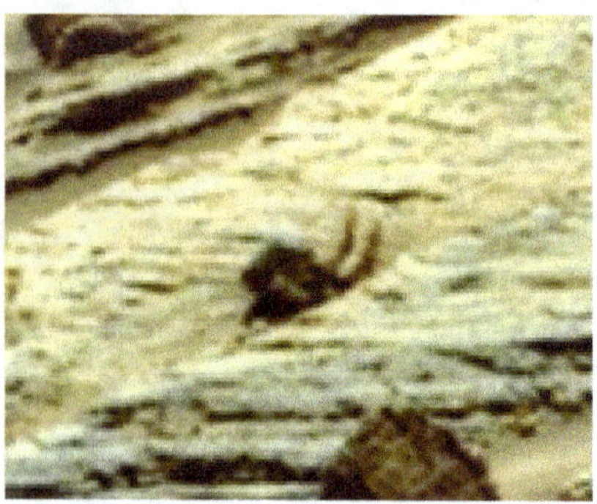

SOL 1033-26 (above). This extremely artificial looking object has a squarish gray frame or "body" and two identical appendages—similar to "horns" or "antennae"—protruding from one end. Additionally, there are at least two goldish yellow letters embossed on its side. Under magnification they appear to be, from left to right, a stylized Roman "V" and a stylized Roman "Y." The two letters, however, actually bear more similarity to the Hebrew alphabet; perhaps the letters ע ("Ayin") and צ ("Tsadeh"). See the next image (below).

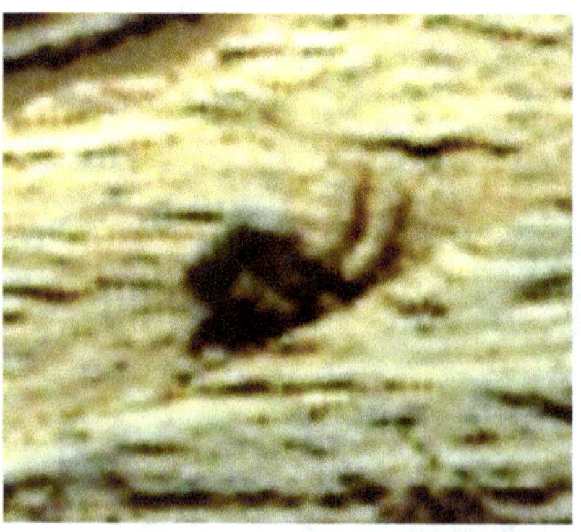

SOL 1033-27 (above). An extreme closeup of the object described in Sol 1033-26 (above). Note the Hebrew like lettering embossed on its side.

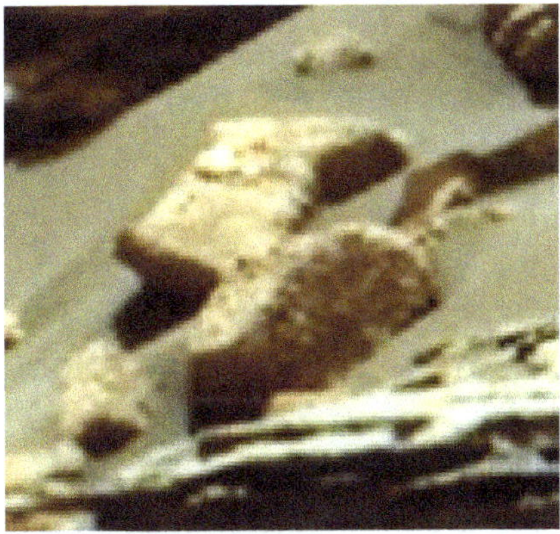

SOL 1033-28 (above). Again we have here what appear to be the remains of Martian statuary. Besides the two large "carvings" at the center of this image, there are also a number of what look like cement blocks half buried in the sand. Study the angles, edges, positions, and shapes of, as well as the markings on, these objects. There is much more going here than I can discuss in this small space.

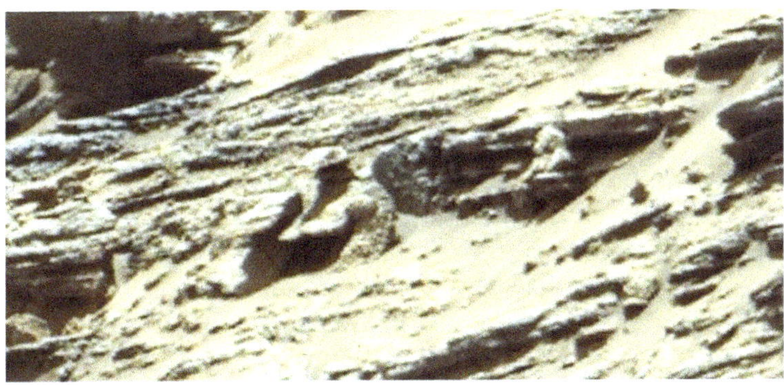

SOL 1033-29 (above). The remains of what may have been statues, ornamental architecture, and buildings dot this hillside landscape on Mars. As this particular section is quite high up on the slope, perhaps some of the statuary fragments described earlier derive from this anomaly rich area. Magnify and study.

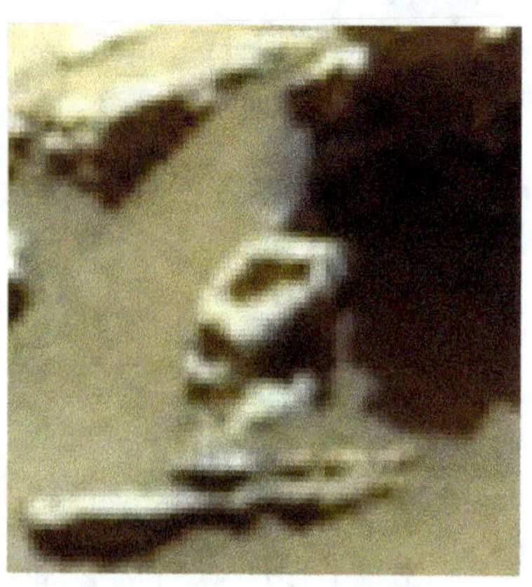

SOL 1033-30 (above). Machined looking objects are abundant in this image, including a five sided, ridged item (center), a circular item with "gear teeth" running along its edge (lower left center), the "cornerstone" of a "building" (upper left), and structural materials (bottom). An unusual rectangular shadow (lower right center) is being cast from an object out of frame. Both intuition and logic tell us we are almost certainly looking at artificiality here. Why should we ignore our cognitive faculties? Future generations will learn the Truth—whatever it turns out to be.

Chapter 20

ODDS & ENDS

SOL 153-A: EARTHDATE JANURY 10, 2013

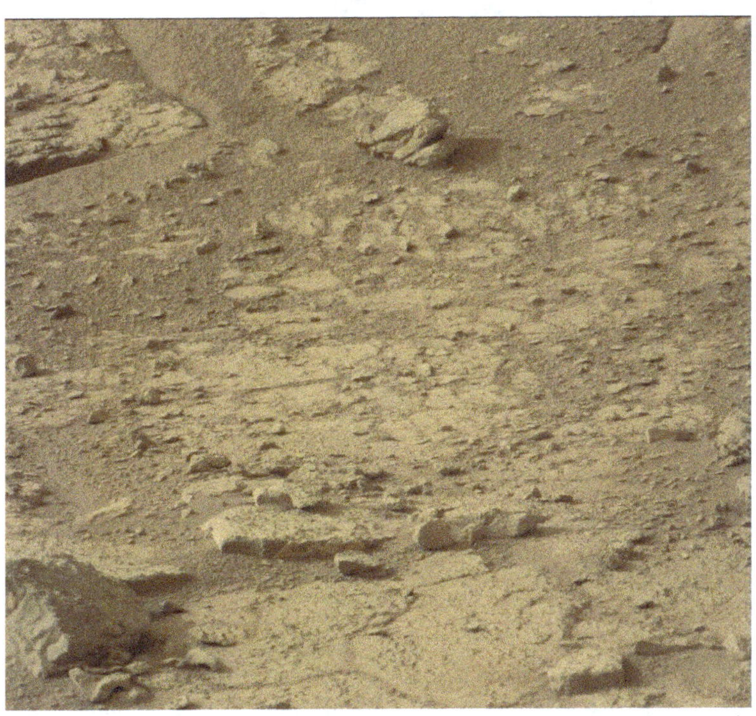

Original image. "Taken by Mast Camera (Mastcam) onboard NASA's Mars rover Curiosity on Sol 153 (2013-01-10, 10:27:56 UTC)." Image credit: NASA/JPL-Caltech/MSSS.

ANALYSIS OF ODDS & ENDS, SOL 153-A

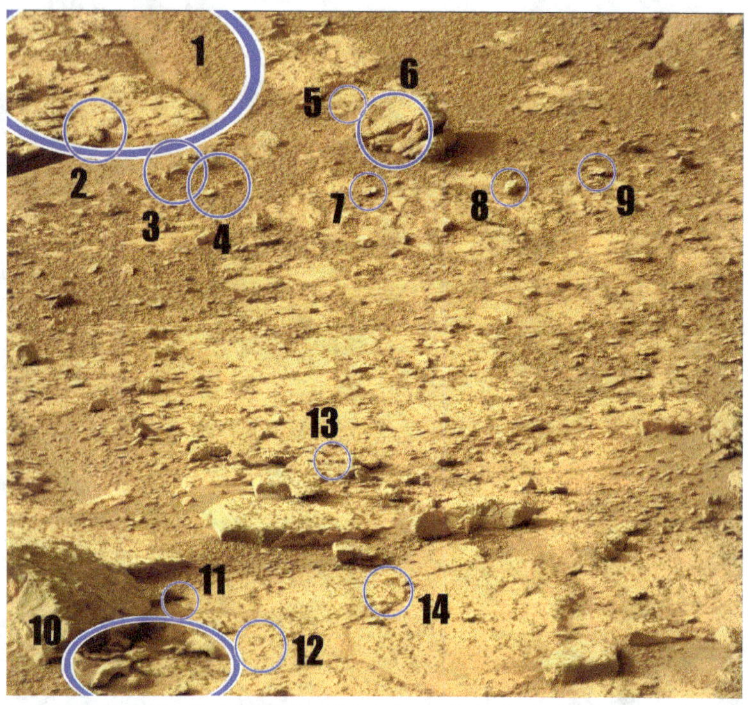

Odds and Ends, Sol 153-A, with anomalies circled. Image colored, brightened, and sharpened by the author to improve contrast. Though subtle, there are numerous potential signs of unnaturalness in this image.

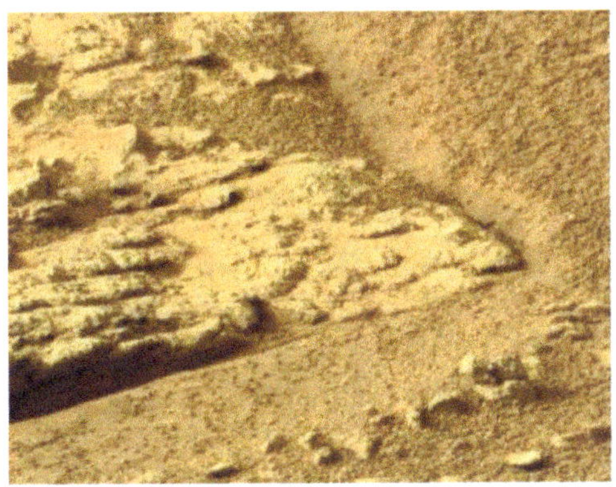

SOL 153-A, highlighted area 1 (above): A 90 degree angle in the Martian soil. This may very well be natural. However, this could also be what the remnants of a structural foundation looks like after thousands of years on the windblown surface of Mars.

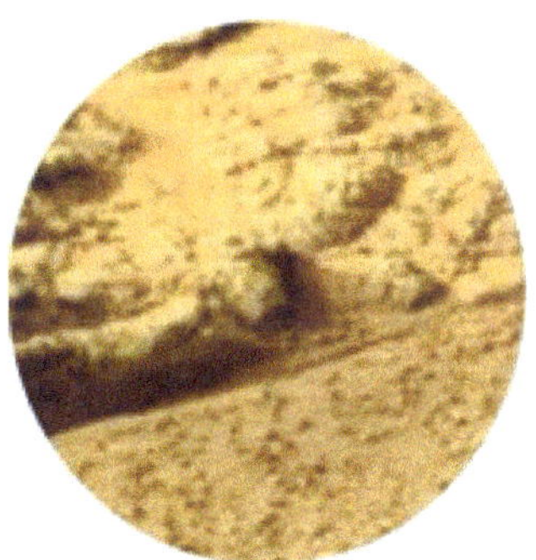

SOL 153-A, highlighted area 2 (above): There are several small donut like circular objects located just above the center of this image. If they are artificial, they seem to have been embossed on some rather odd looking "rocks."

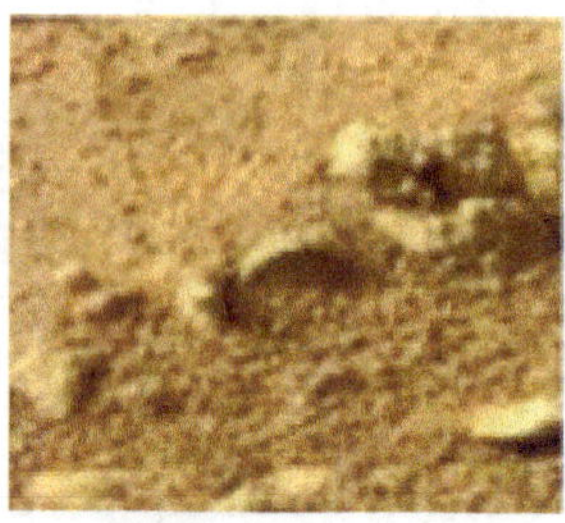

SOL 153-A, highlighted area 3 (above): A mirage at the center of this image, or a small wheel half buried in the dirt?

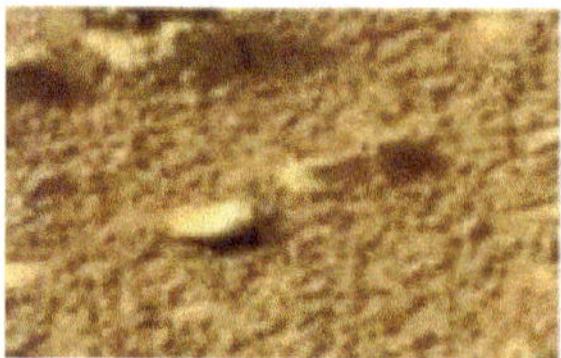

SOL 153-A, highlighted area 4 (above): These pieces of "rock" look like pottery shards.

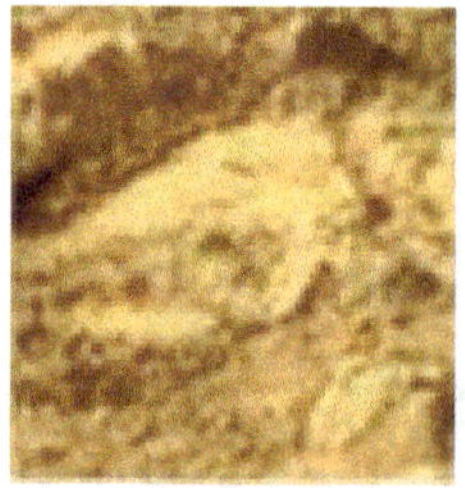

SOL 153-A, highlighted area 5 (above): At the epicenter of this image there appears to be an egg shaped badge or amulet of some kind with an arrowhead like symbol engraved into it. The "arrowhead" itself has several (decorative?) holes bored into it, as well as two "legs" at the bottom. A magnifying glass will help bring the intriguing details of this object into focus.

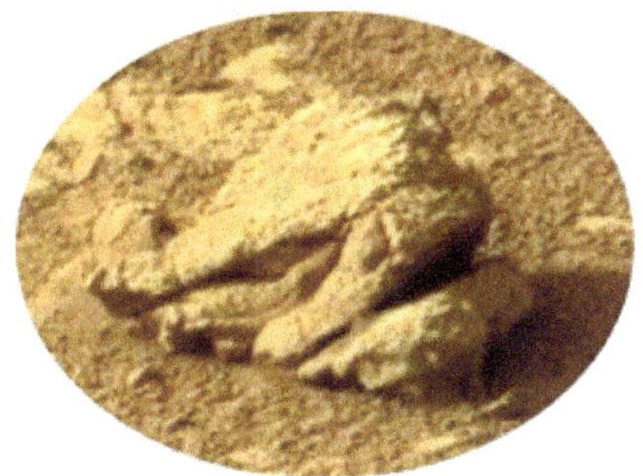

SOL 153-A, highlighted area 6 (above): These "rocks" bear numerous artificial characteristics. For the sake of illustration, let us consider this entire grouping to be vestiges of a sculpture of an alligator's head, complete with a base, upper and lower jaws, eyes, and a tongue. Under magnification there are what appear to be Roman letters carved into the upper jaw, from left to right: possibly "C," "T," "H," "Y," and a tiny vertical line with notches to the right of it. The lower jaw has a number of odd markings carved into it as well, including a backwards "L" (at the hinge) and a small vertical line with three dots to the left of it (near the outer end). The realistic looking tongue has a line "etched" into it that traces its curved shape. Keep in mind that we are quite limited by the camera's perspective, as from a different angle this object may look nothing like an alligator head. Nevertheless, the myriad of odd markings and shapes on it identify it as potentially artificial—whatever it is.

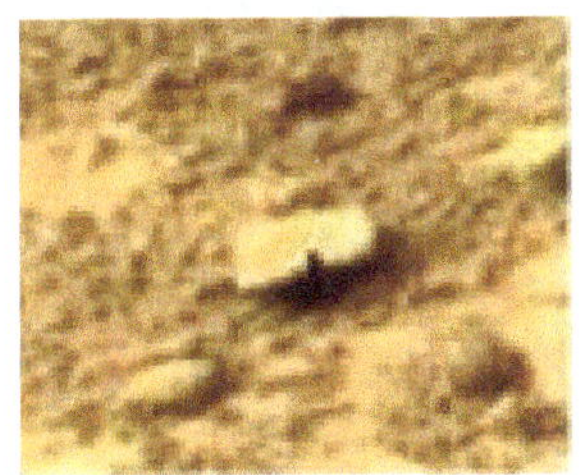

SOL 153-A, highlighted area 7 (above): This small square, of unknown material, has a perfect notch on its forward facing side. What or who created it?

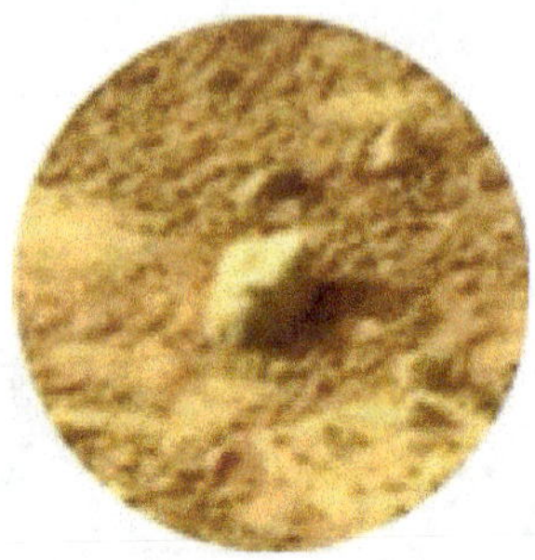

SOL 153-A, highlighted area 8 (above): This small object has an obelisk like shape and appears to be a statuette, or perhaps a piece of home or office decor—similar to what we might place on a table or desk here on Earth.

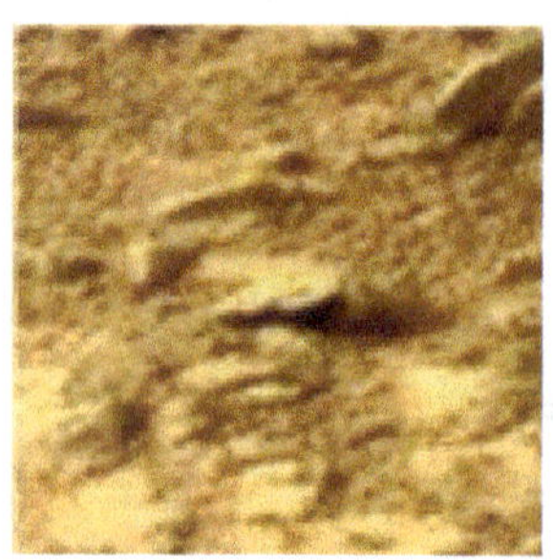

SOL 153-A, highlighted area 9 (left): Depending on one's perspective, the artificial object in the center of this image looks like a stylized capital letter "J," a capital "P," or a small "d." It is surrounded by other bizarre items.

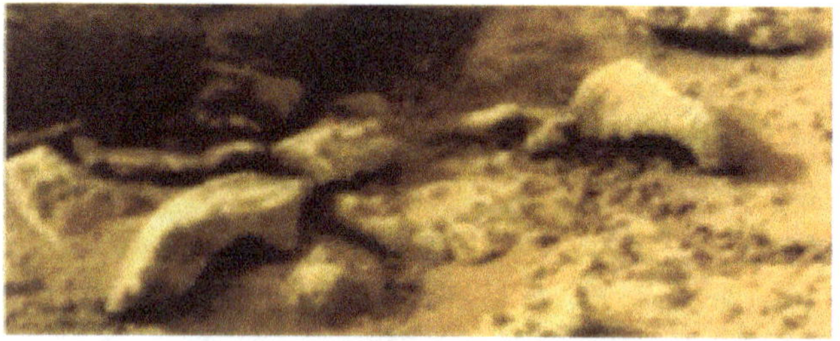

SOL 153-A, highlighted area 10 (above): This section seems to be the final resting place of numerous pieces of shattered pottery, or perhaps pieces of damaged masonry. These large chunks lie amid several brick like objects, as well as a number of oddly shaped items.

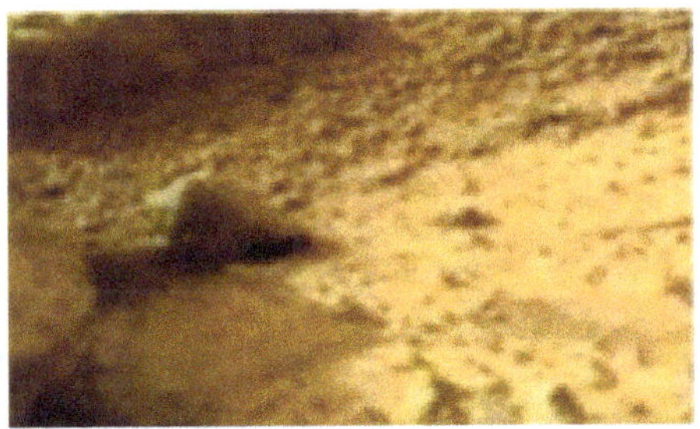

SOL 153-A, highlighted area 11 (above): Just below the pyramidal like rock at lower center left is yet another one of the countless 90 degree angles I have discovered in the Martian soil. Can *all* of them be explained away as natural geologic formations? Mainstream science says yes. What do you say?

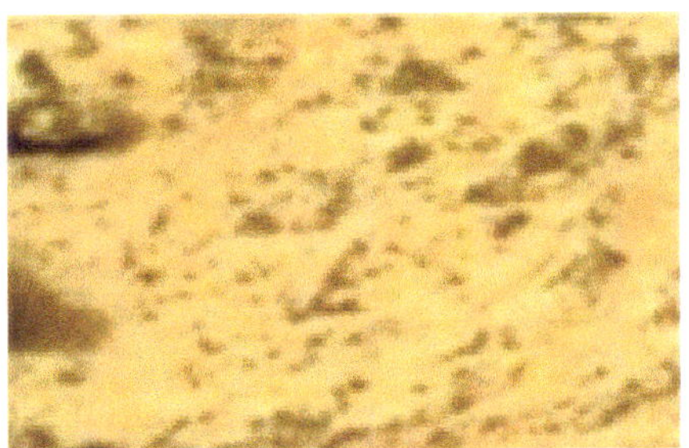

SOL 153-A, highlighted area 12 (right): To the left of center one can just make out a raised circle with an "X" or cross in it. Just to the lower right of this artificial object is a triangle shaped object. If they are artificial, it is safe to say that the two objects are somehow connected. A magnifying glass is a must for studying this anomaly packed photo.

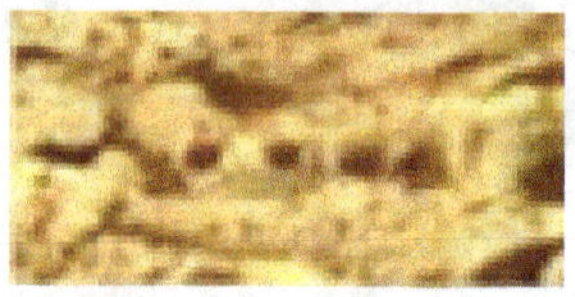

SOL 153-A, highlighted area 13 (above): These "rocks" appear to have near perfect, eye like holes "drilled" into them. If these cavities were indeed created by an advanced Martian race, what are they and what was their purpose? If we are to accept the mainstream claim that they are natural, concrete evidence must be provided in order to substantiate it.

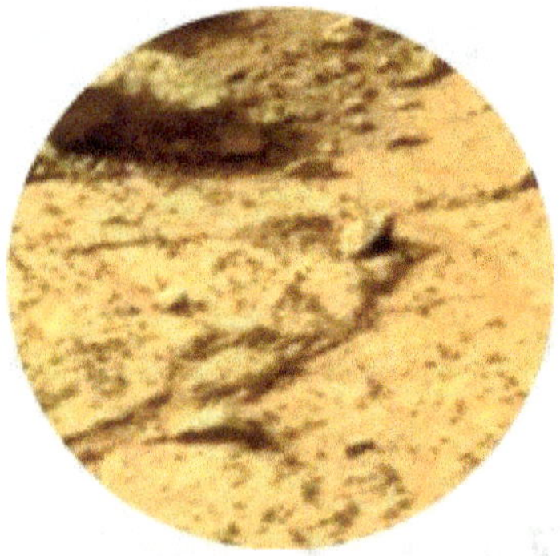

SOL 153-A, highlighted area 14 (above): Under magnification, at the center of this image another raised circle or disk with an "X" in it can be seen. To its left (at about 10:00) is a 90 degree angle; below the "X"ed circle are several unusual objects, some that look like melted fragments of metal or an unknown material.

Chapter 21

ODDS & ENDS

SOL 153-B: EARTHDATE JANURY 10, 2013

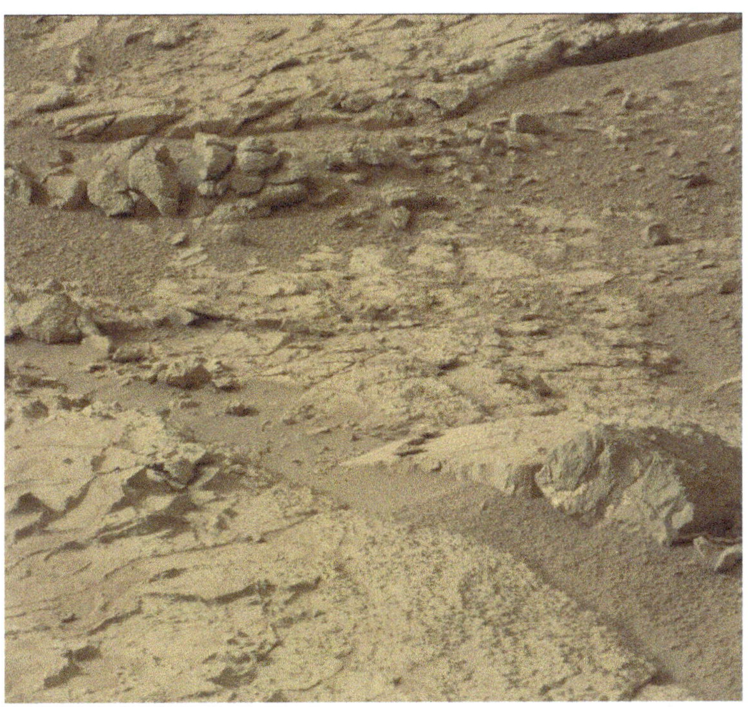

Original image. "Taken by Mast Camera (Mastcam) onboard NASA's Mars rover Curiosity on Sol 153 (2013-01-10, 10:27:11 UTC)." Image credit: NASA/JPL-Caltech/MSSS.

ANALYSIS OF ODDS & ENDS, SOL 153-B

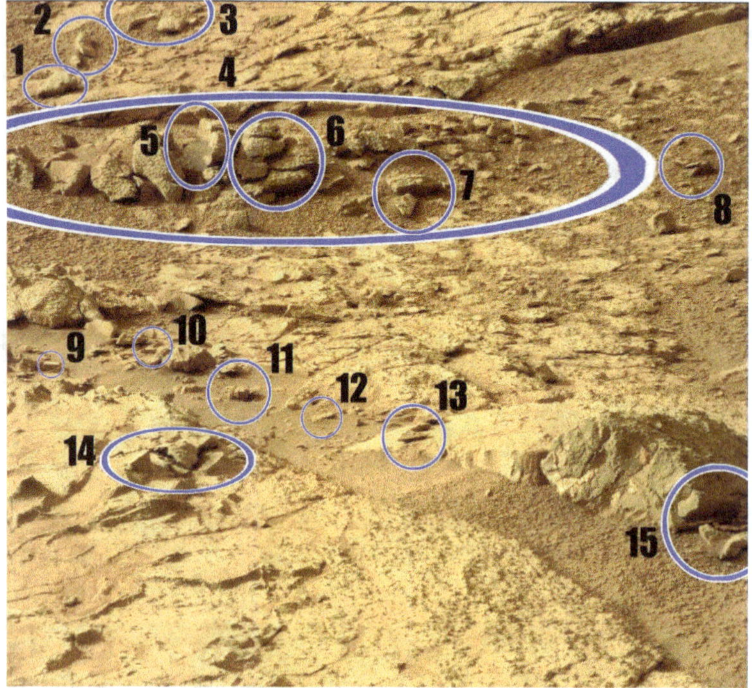

Odds and Ends, Sol 153-B, with anomalies circled. Image colored, brightened, and sharpened by the author to improve contrast. This Martianscape is dotted with artificial looking objects, providing further evidence of probable ancient intelligent design.

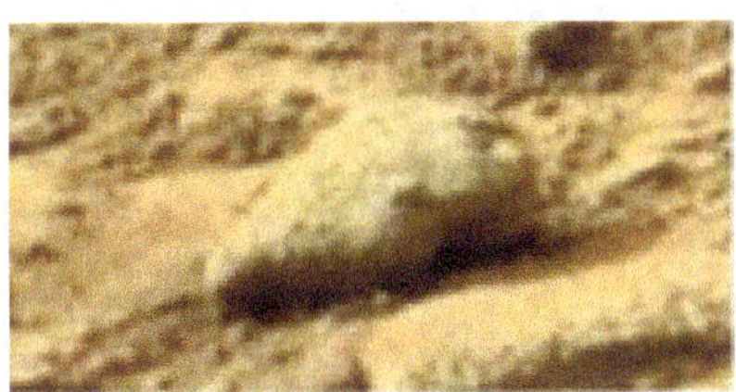

SOL 153-B, highlighted area 1 (above): Under magnification this teardrop shaped "rock" has a small hollowed out square in its upper right side. The square itself has a hole "drilled down" into its center. Can Nature do this?

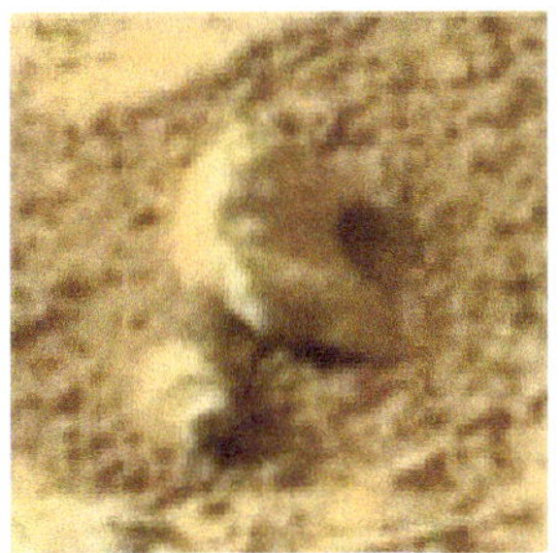

SOL 153-B, highlighted area 2 (above): These two rounded objects look like household items. Is this what is left of an ancient Martian mortar and pestle? Or perhaps they come from some type of machinery—a wheel and axle?

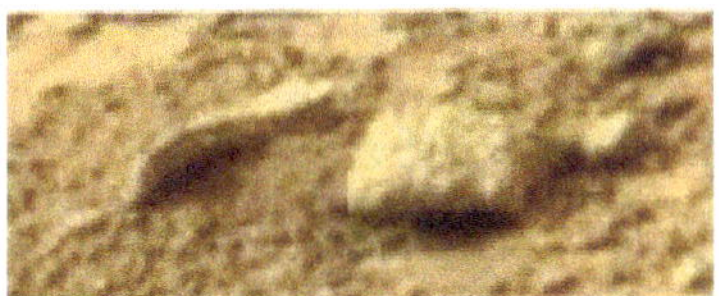

SOL 153-B, highlighted area 3 (above): The object on the left appears to be a pottery shard, or perhaps a broken piece of cement. The object on the right, which could be anything from a shingle to a paving stone, appears to be encrusted with Martian concretion.

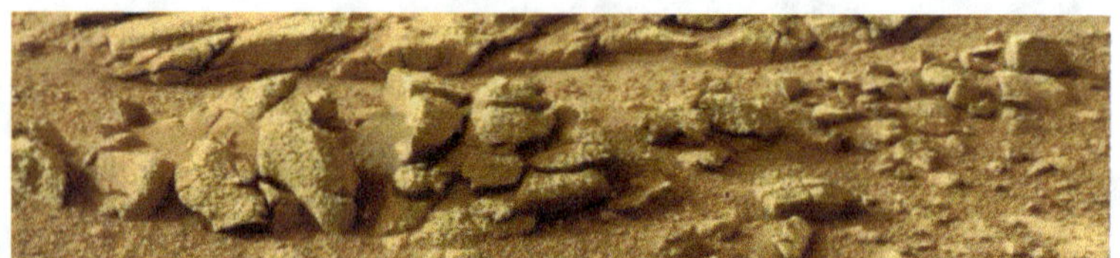

SOL 153-B, highlighted area 4 (above): This long row of "rocks" contains a number of curiously shaped objects, including rounded items and perfect squares, at least one of them with writing on it (see next image, area 5). Is this the remnants of a stone wall? Or perhaps the rubble of a crumbled statue? Look at the straight trajectory of the rocks, which appear to have fallen from right to left, and which then smashed upon impact.

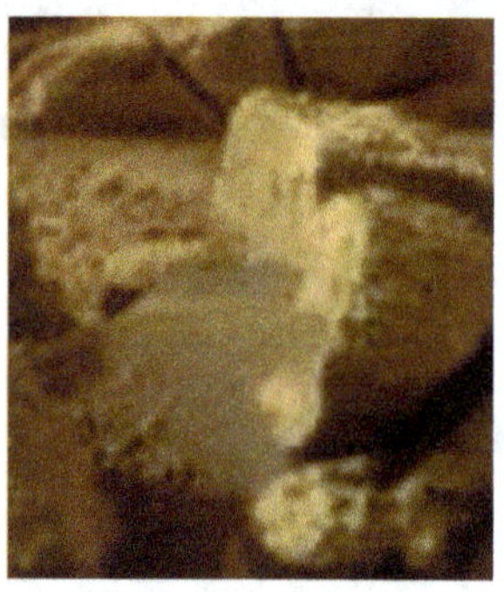

SOL 153-B, highlighted area 5 (above): Under magnification the flat side of the perfectly square "rock" on the upper right has Roman like writing on it. What does it say? The letters appear to be, from left to right: "B," "J," and "t." There are also letters or symbols on the second "rock" to the right of it, but they cannot be distinguished due to the camera angle. At the base of both "rocks" there is what looks like heavy digital manipulation, with fake "sand" airbrushed over the Martian soil. If correct, whatever lay beneath may have been deemed too "dangerous" for the public domain.

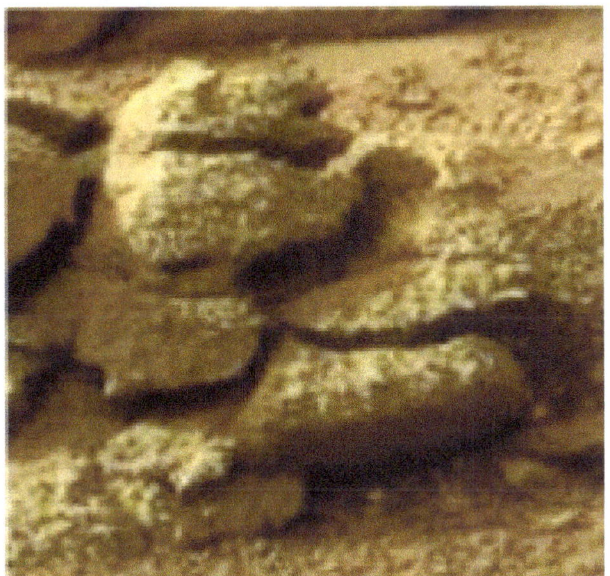

SOL 153-B, highlighted area 6 (above): Under magnification these heavily encrusted "rocks" appear to have writing on them, this time in a Hebrew like alphabet, or more probably a Martian alphabet we have yet to acknowledge let alone decipher. Among the alleged lettering here are also a number of seeming Martian symbols. We cannot discount the possibility of natural geologic processes. However, *observe closely.* Not only do these not look like rocks, whatever they are, they are located in the middle of a debris field occupied by numerous other objects that also appear to be artificially created.

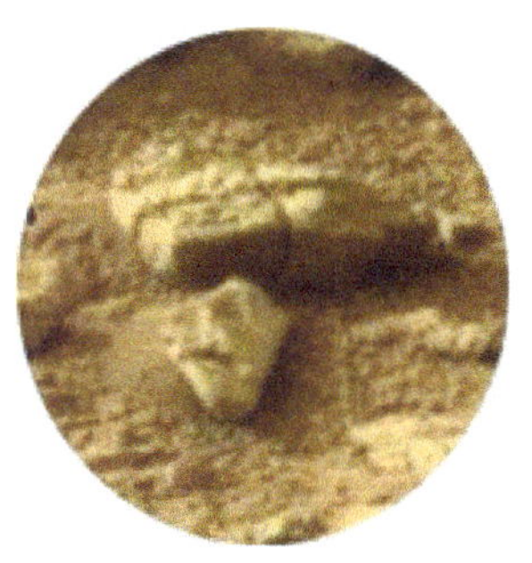

SOL 153-B, highlighted area 7 (left): The object at the upper center looks like a fossilized paper package or a loaf of bread tied up with brown string. It is obviously neither of these things. What is it then? Masonry work? Below it a possible broken clay shard appears to have a (partial) face with an open mouth etched into it (note: the face is upside down). Pareidolia or ancient Martian artwork?

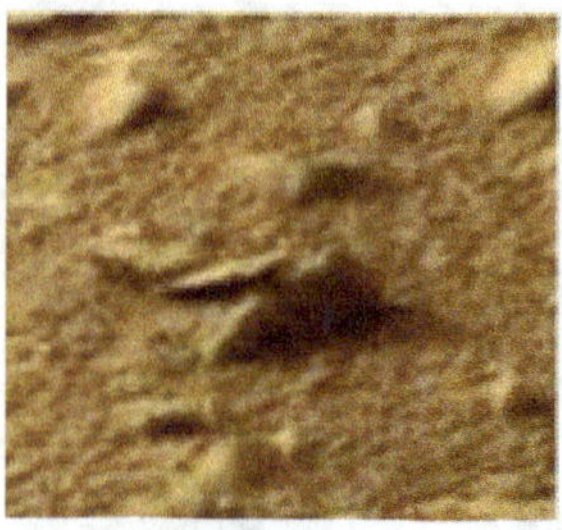

SOL 153-B, highlighted area 8 (above): At center a small, rectangular, tile like object is propped up on a rock. It may be natural. But it may also be artificial. Considering its location in the midst of a large area filled with items that look like artifacts, which is it more likely to be?

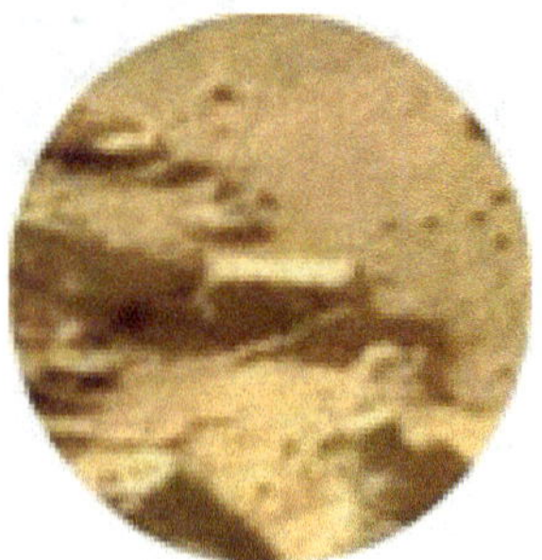

SOL 153-B, highlighted area 9 (above): Most people would call the object at center a "Martian brick."

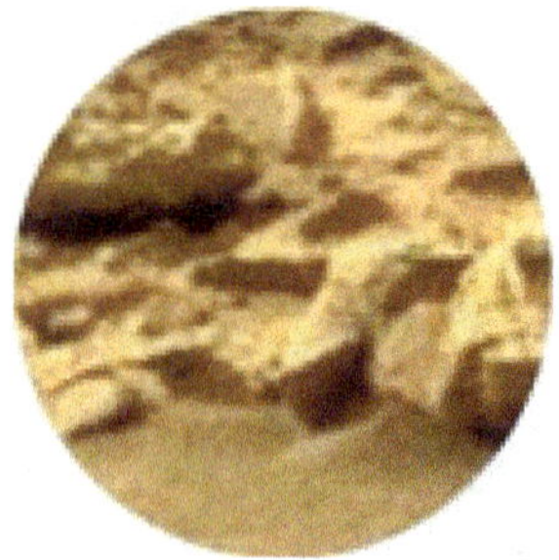

SOL 153-B, highlighted area 10 (above): In my opinion, multiple oddly shaped objects, as well as several pointed triangular objects at the center of this image, demonstrate probable evidence of a superior intelligence. The item at the upper center looks similar to *Star Trek's* delta insignia, not the first time we have come across this familiar emblem on the Martian surface—nor will it be the last.

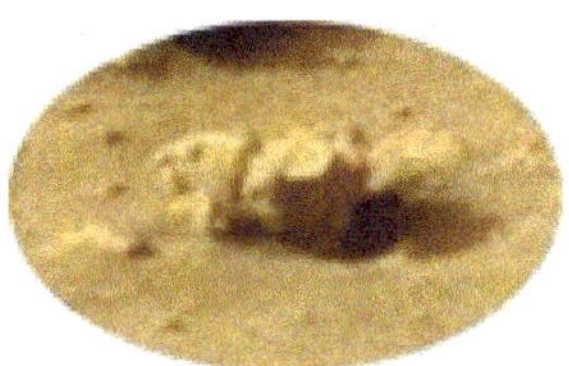

SOL 153-B, highlighted area 11 (above): I have no idea what this is, but I think a majority of my readers will agree that almost nothing about it looks natural.

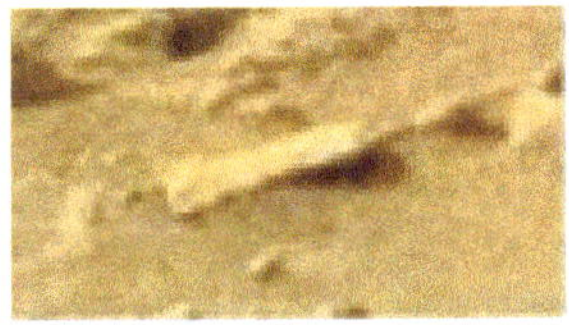

SOL 153-B, highlighted area 12 (above): A naturally formed rock, or the remnant of a small statue? A piece of house decor, or shattered brickwork?

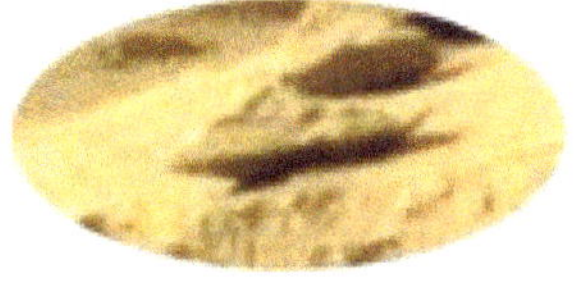

SOL 153-B, highlighted area 13 (above): This arrowhead or chevron like object is sitting on top of a large rock. It possesses odd letter like markings on its upper surface. What is it, and how did it get there? Conventional explanations for Martian anomalies are often not convincing, satisfying, or realistic—or irrefutable.

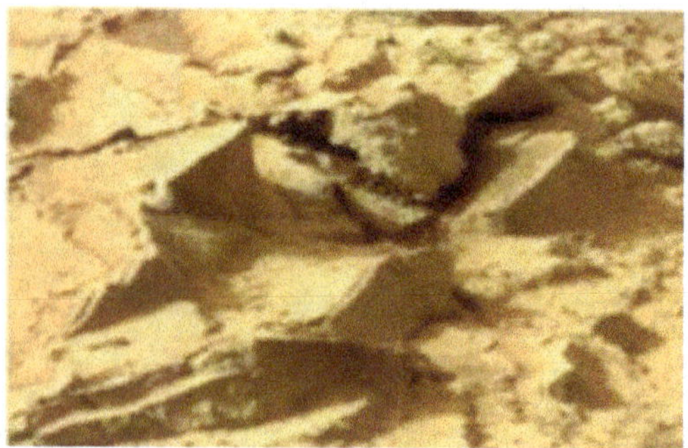

SOL 153-B, highlighted area 14 (above): This pile of what appears to be bombed cement blocks is identical to the war torn, weathered city ruins we see everyday on our own planet. Ninety degree angles are ubiquitous.

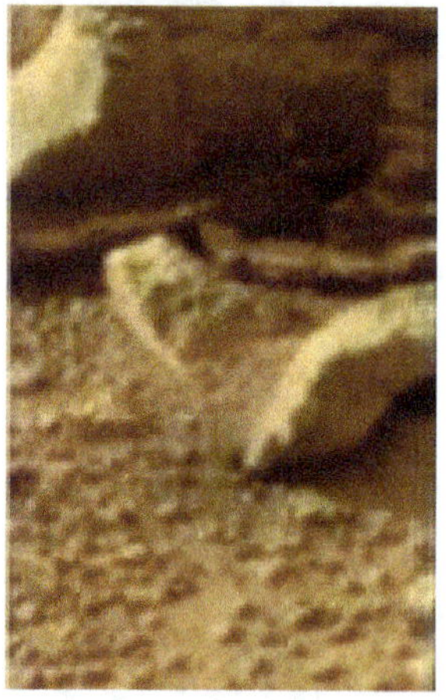

SOL 153-B, highlighted area 15 (above): From top to bottom, unnatural looking "rocks" fill this image, including an odd half circle on the right. Could we be looking at the remains of an ancient, sophisticated, and now extinct Martian civilization?

Chapter 22
BLAST SITE

SOL 153-C: EARTHDATE JANURY 10, 2013

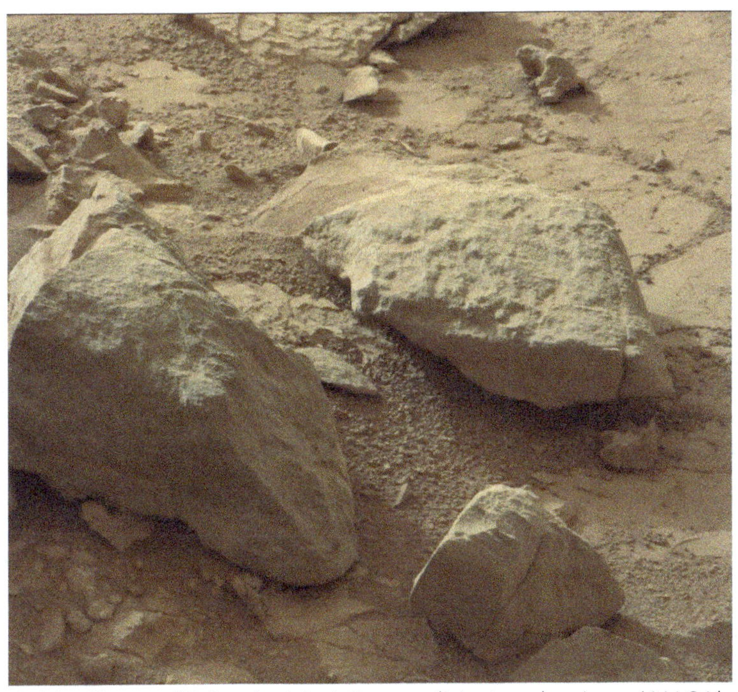

Original image. "Taken by Mast Camera (Mastcam) onboard NASA's Mars rover Curiosity on Sol 153 (2013-01-10, 10:27:11 UTC)." Image credit: NASA/JPL-Caltech/MSSS.

ANALYSIS OF BLAST SITE, SOL 153-C

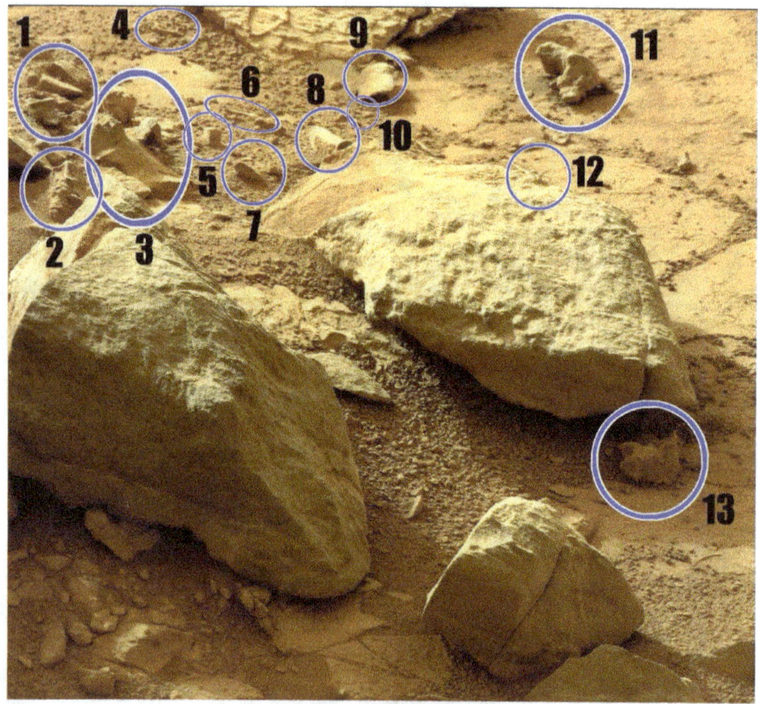

Blast Site, Sol 153-C, with anomalies circled. Image colored, brightened, and sharpened by the author to improve contrast. As this photo shows, by paying close forensic attention to the location of the many anomalies on Mars, we can formulate an idea of the "crime scene"; that is, the direction and force of the alleged bomb blasts or "environmental disturbances" that might have taken place during what I have coined the hypothetical "Great Martian Apocalypse."

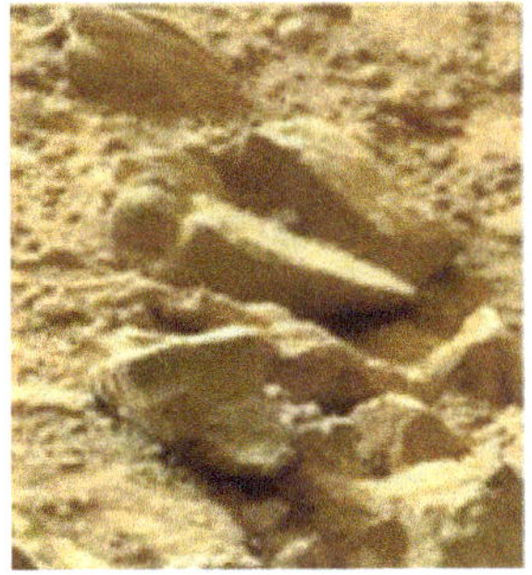

SOL 153-C, highlighted area 1 (above): The anomalist's view: This is not natural geology. These are complex, aesthetic, artificially created objects whose functions and identities we can only guess at. The skeptic's view: All rocks.

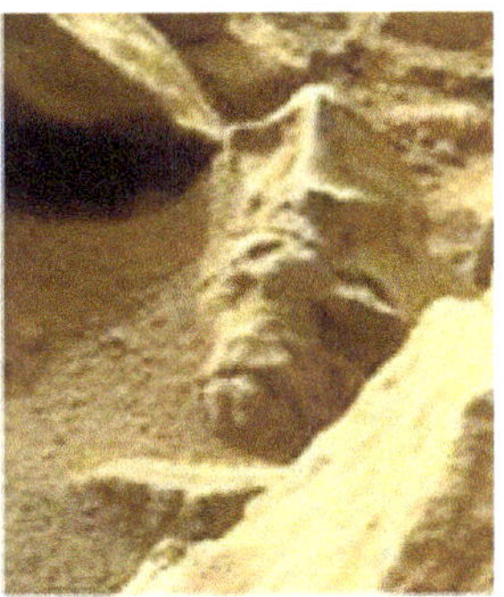

SOL 153-C, highlighted area 2 (above): What appears to be an ornamental feature of some kind is lying on the Martian sand between dozens of other objects that look like they have been intelligently made. Study it slowly under magnification.

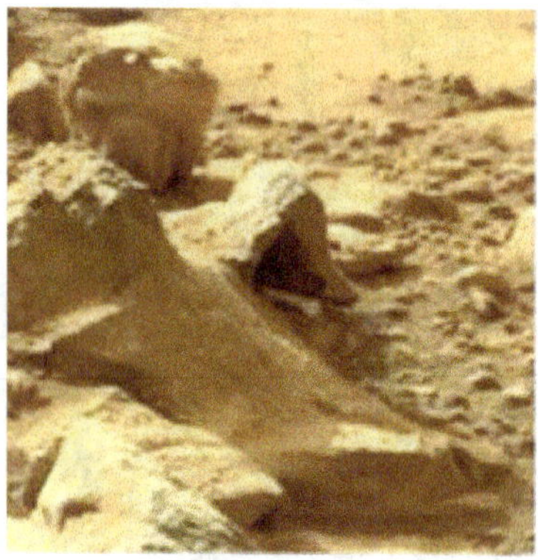

SOL 153-C, highlighted area 3 (above): Is this rubble from collapsed buildings or bombed statues, or possibly something we cannot conceive of as Earthlings?

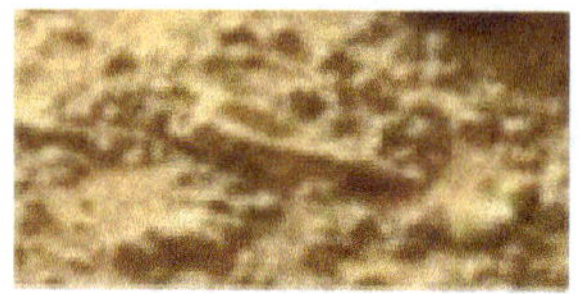

SOL 153-C, highlighted area 4 (above): At center a ring or horseshoe shaped object sits in the middle of several other roundish items with similar shapes.

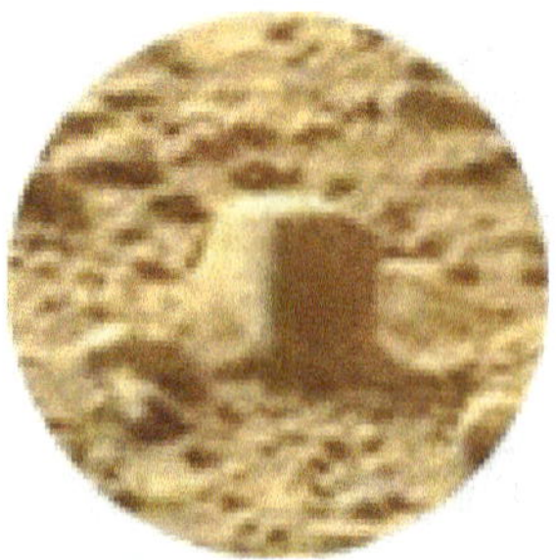

SOL 153-C, highlighted area 5 (above): Another surveyor's boundary stone? A grave marker? A brick?

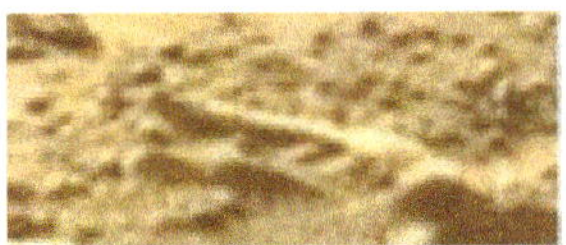

SOL 153-C, highlighted area 6 (above): Skeletal remains? Fossilized coral? Artificial objects? Natural rock formations?

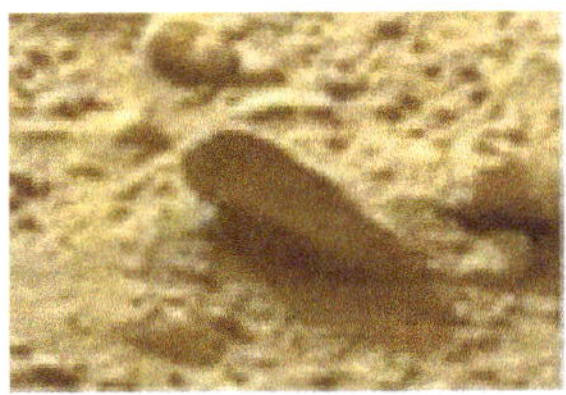

SOL 153-C, highlighted area 7 (above): From the perspective of the camera this object looks like the sole of a leather shoe curving upward out of the sand.

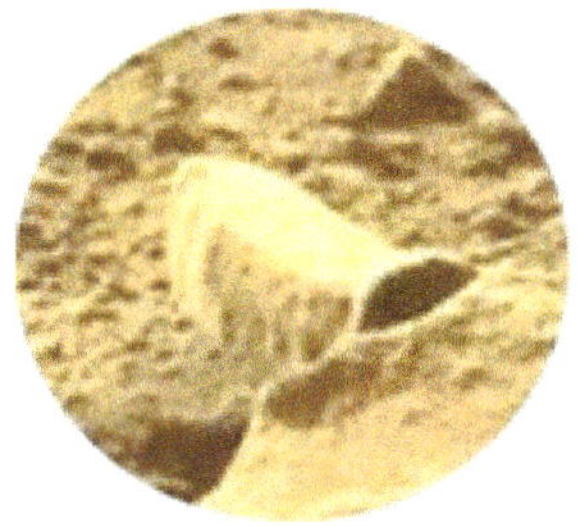

SOL 153-C, highlighted area 8 (above): This pointed, rectangular object with 90 degree corners appears to have been intentionally chiseled (note what look like weathered "chisel marks" along its sides). Moreover, the object has a flat "base" and a pointed "top." Is this piece of possibly carved stone from a statue or perhaps a stone structure, or was it once a house decoration? Calling such a structure a "rock" not only insults the intelligence of a majority of people (including many scientists), but it does nothing to help advance human knowledge. Is that not why we are exploring Mars?

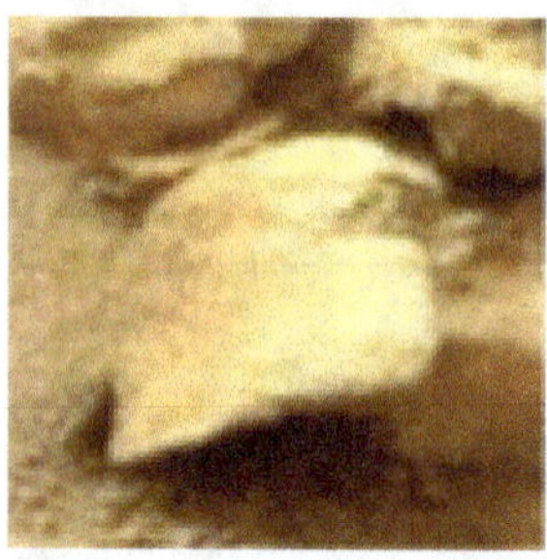

SOL 153-C, highlighted area 9 (above): Another strange, square tile like object encrusted with Martian soil.

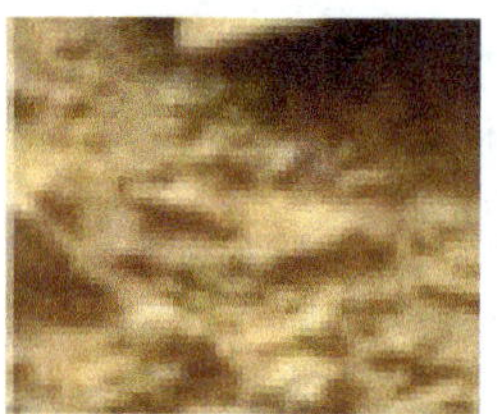

SOL 153-C, highlighted area 10 (above): At center another "Star Trek" emblem.

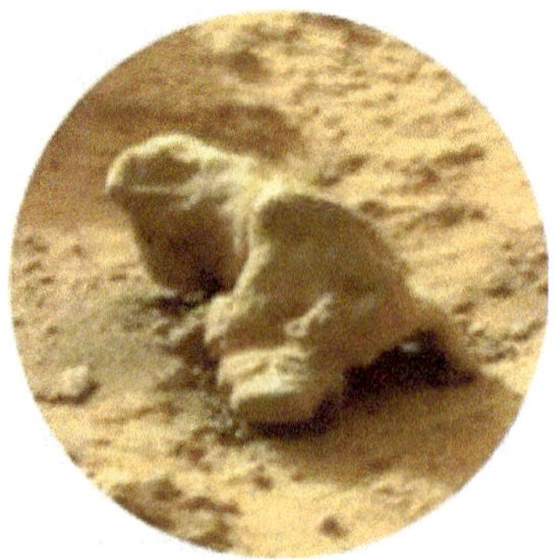

SOL 153-C, highlighted area 11 (above): This famous object is known as the "Mars iguana." Is it a sculpture of an ancient, now extinct Martian reptile, an actual living life form, or just a strange naturally formed rock formation?

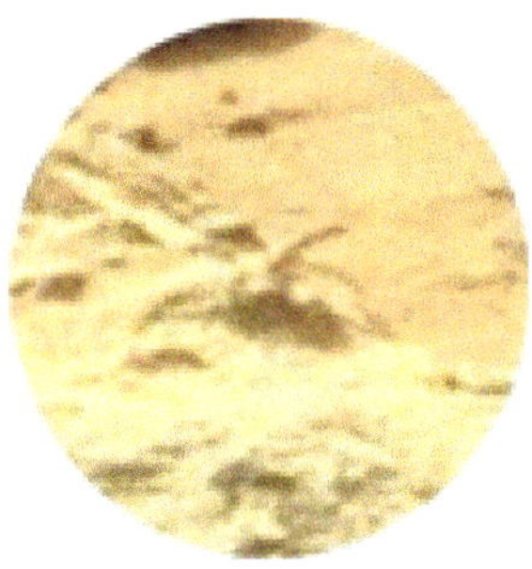

SOL 153-C, highlighted area 12 (above): What is the thin, fragile looking object at center? It is very small and is protruding from the top of a large rock. Can we wholly dismiss the idea that it may be a living creature? Due to the lack of oxygen, as well as the great amount of unfiltered cosmic radiation that bombards the surface of Mars, perhaps. However, it is not as easy to disregard the artificial looking objects around it, and to which it appears to be attached or possibly emerging from. Another image to study closely.

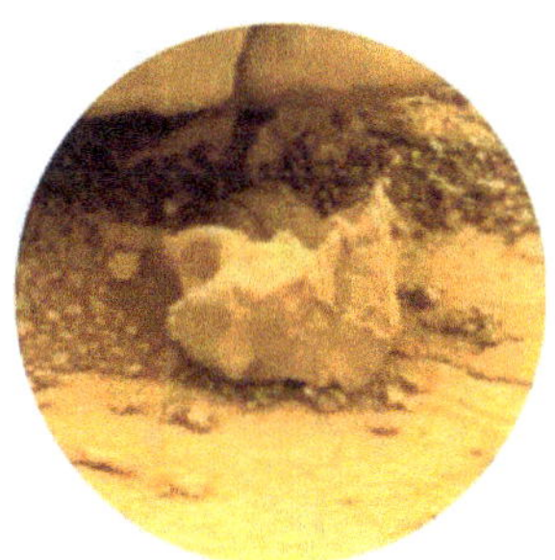

SOL 153-C, highlighted area 13 (above): Is this the fossilized remnant of a once living creature, or something made by an advanced race of beings? Or is it simply a curiously shaped rock?

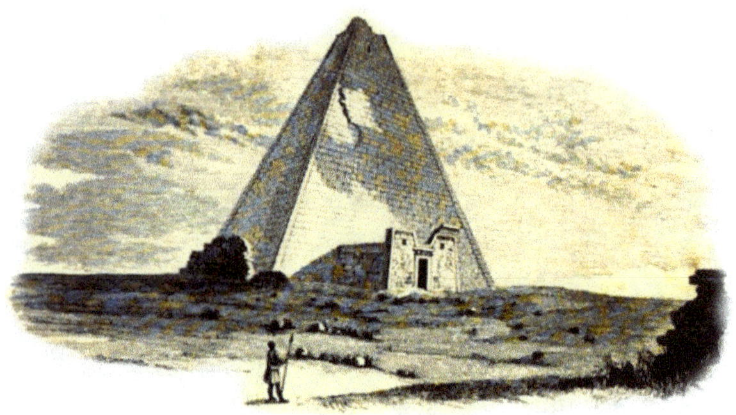

Above: Ancient Earth pyramids, like this relatively small one in Sudan, may have mathematical and ritualistic associations with what appear to be pyramids on Mars—particularly, as we will see, in the Cydonia region. Only future Mars missions focusing on astroarchaeology will be able to sort out fact from fiction.

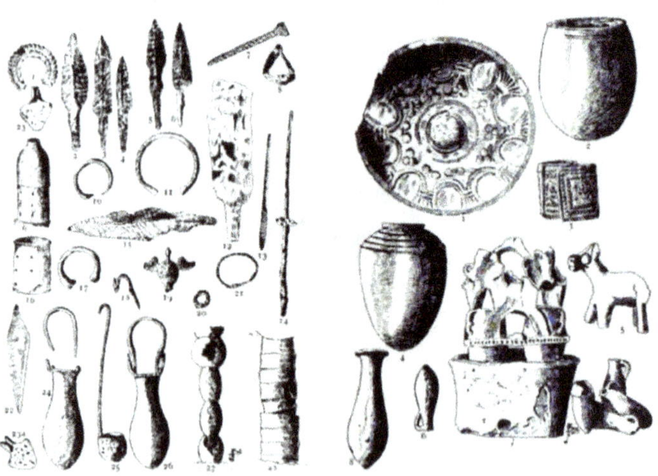

Above: Thousands of personal and household items, similar to these ancient artifacts uncovered in the city remains of Babylonia, appear to rest on or just below the surface of Mars. If this is what they turn out to be, the question every rational individual will want answered is, who made them?

Chapter 23

LETTERS & SHAPES

SOL 153-D: EARTHDATE JANURY 10, 2013

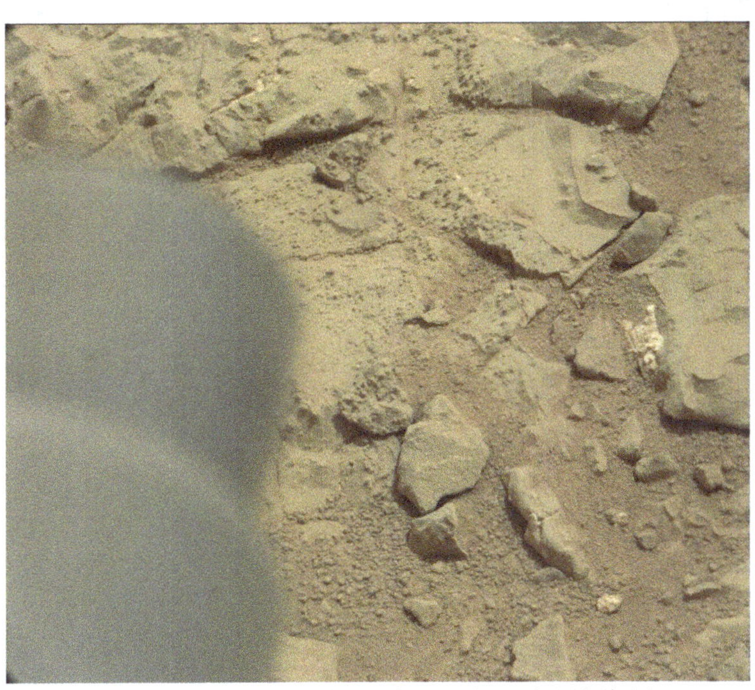

Original image. "Taken by MAST_RIGHT onboard NASA's Mars rover Curiosity on Sol 153 (2013-01-10, T07:46:49.000Z)." Image credit: NASA/JPL-Caltech/MSSS.

ANALYSIS OF LETTERS & SHAPES, SOL 153-D

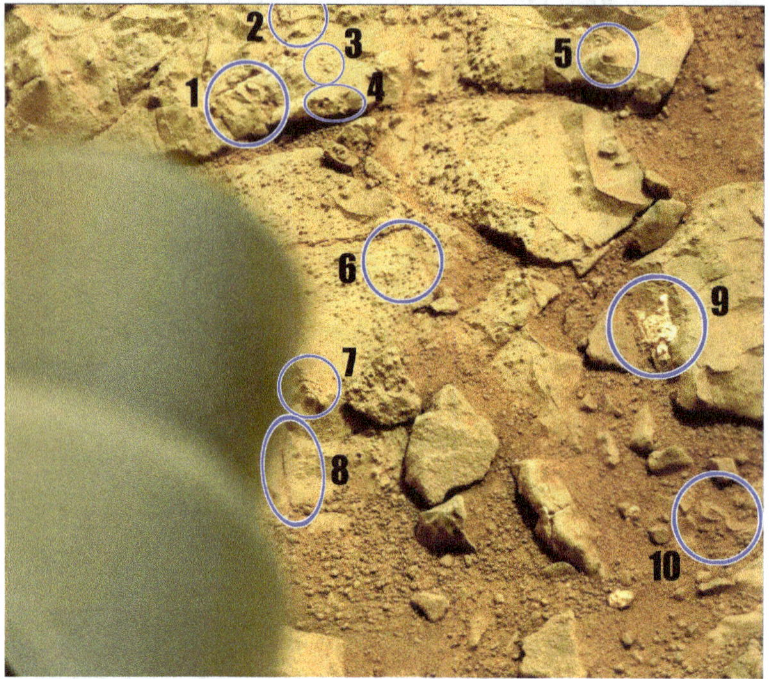

Letters and Shapes, Sol 153-D, with anomalies circled. Image colored, brightened, and sharpened by the author to improve contrast. Unfortunately Curiosity blocked a portion of its own field of vision when it took this photo on January 10, 2013. Otherwise we can be sure that many other anomalies would be visible.

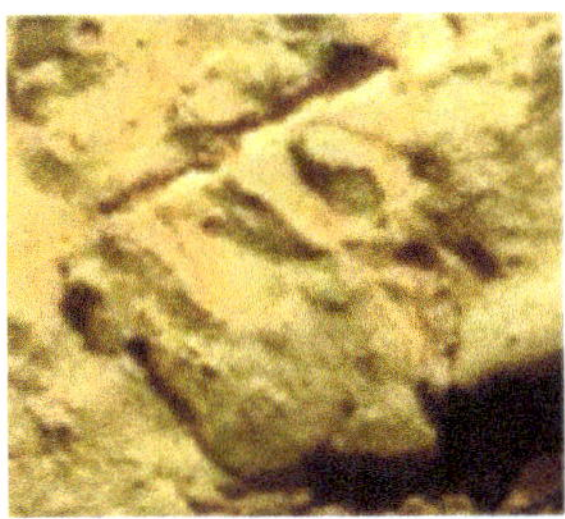

SOL 153-D, highlighted area 1 (above): A squared off shape with what looks like Roman letters embossed into its surface. If true, the letter on the left is either a "W" or an "M," depending on which side is right side up.

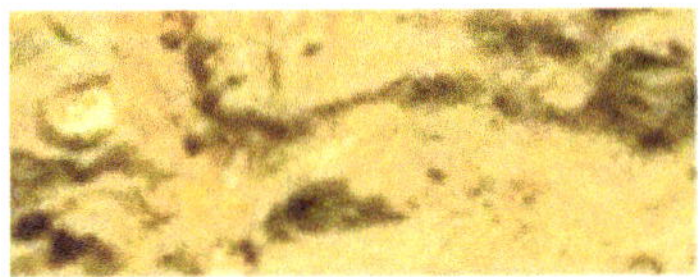

SOL 153-D, highlighted area 2 (above): A myriad of unidentifiable objects and shapes. Products of mindless geology or intelligent biology?

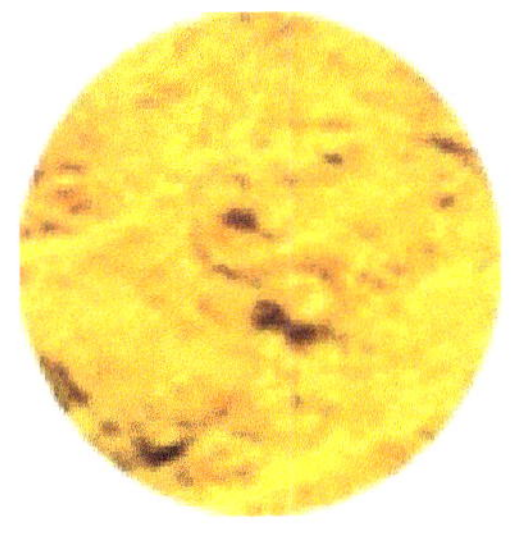

SOL 153-D, highlighted area 3 (left): Study this astounding object with a magnifying glass. Establishment scientists tell us it is a "natural geological Martian feature." If this is true, we know next to nothing about the aesthetic talents of Mother Nature. The rest of us see either the Roman letter "S" or a beautifully carved pictograph representing a serpent: note the curled "tail" and the enlarged "head" with an "eye" in the center of it at the top. If it is an artifact, we may be looking at a fragment from a work of Martian art; or perhaps it is a Martian physician's sign, similar to our caduceus—the snake entwined rod of Asclepius, the ancient Greek god of medicine and healing.

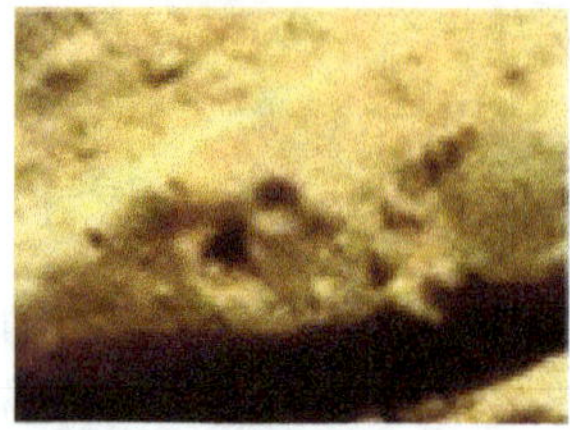

SOL 153-D, highlighted area 4 (above): This "rock" is covered in odd markings. They may be nothing more than the result of wind or water erosion. Since they are located within inches of a host of what look like artistically designed letters, objects, and shapes, however, they could also be the products of an advanced race of beings. Just because we cannot identify them at the moment does not mean they are "natural formations." That is not how real science operates.

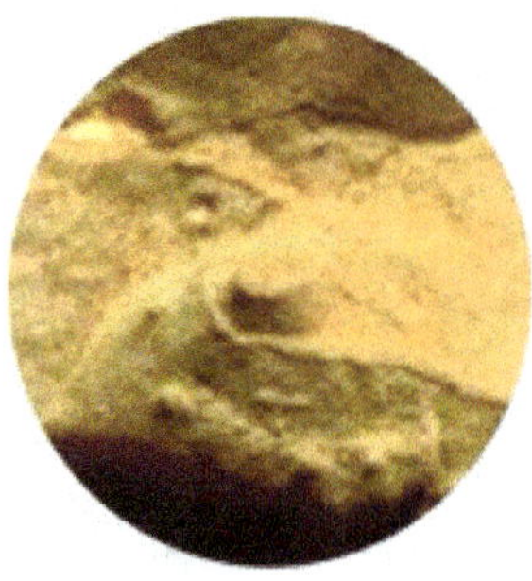

SOL 153-D, highlighted area 5 (above): Why is there a perfectly round nodule on the top of this rock? Below it, on the front rock face, we see what appears to be a sideways "y," or perhaps a Hebrew like letter. Below that, under a layer of what looks like concretion, we see some raised unknown forms. In other words, all of the objects in this image look manufactured.

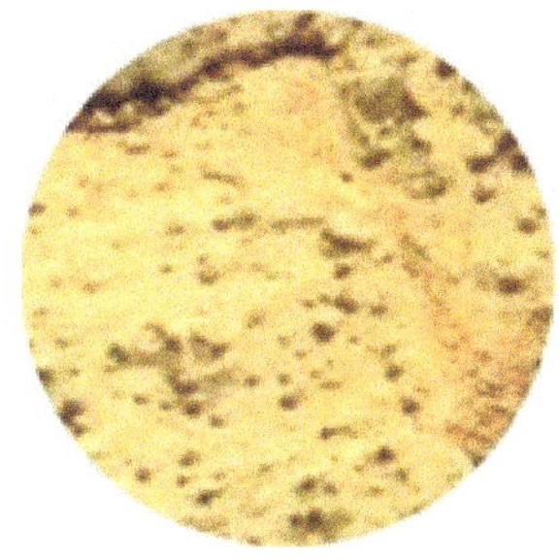

SOL 153-D, highlighted area 6 (left): Under magnification a plethora of curious angular shapes and "designs" can be seen in this image, including 90 degree angles, triangles, and circles. Can all of them be attributed to natural processes?

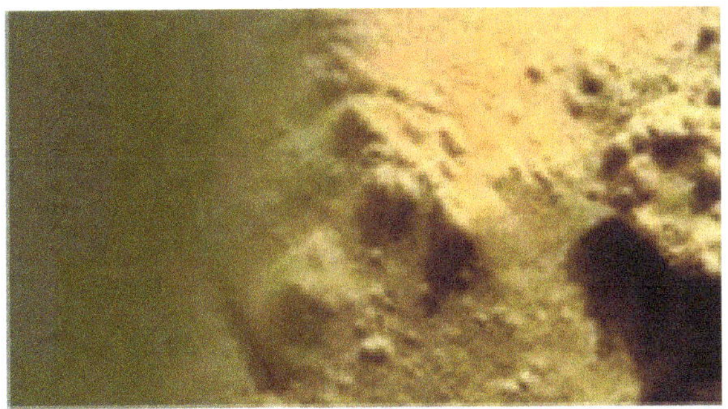

SOL 153-D, highlighted area 7 (above): Under magnification we see more odd right angles and letter like shapes, including an "n," a "u," and an "L."

SOL 153-D, highlighted area 8 (right): A spear tip or boat shape embossment dominates this image. What is it? What was it used for? Who or what created it, and when? These are questions that can only be answered by future astrogeologists, astrochemists, and astrobiologists visiting the Red Planet.

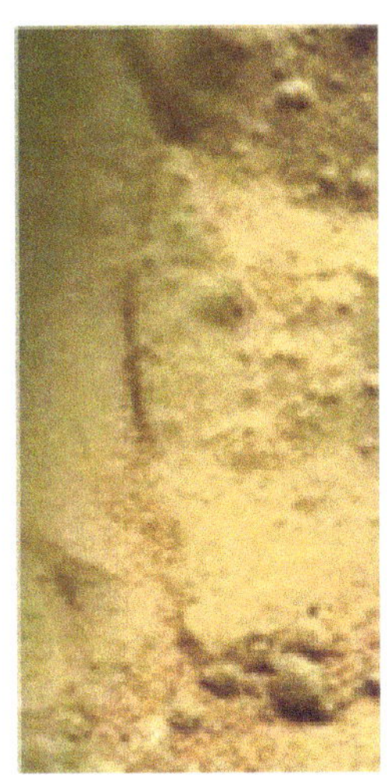

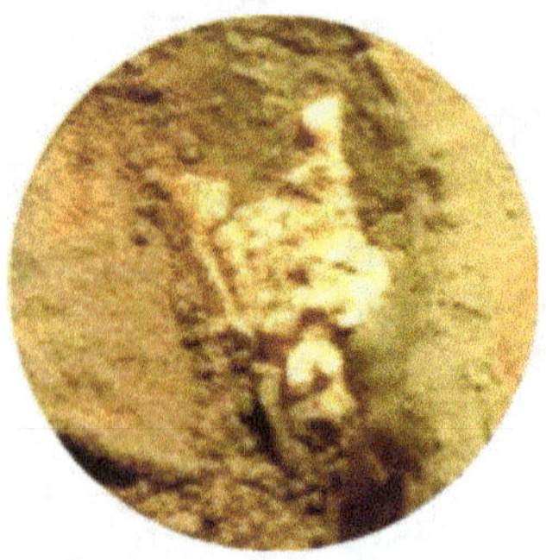

SOL 153-D, highlighted area 9 (above): The statue like shape and bright out-of-place white yellow color of this object half buried in the red Martian sand suggests that it is wholly artificial. A gold religious icon? A metallic child's toy? Look at the objects around it.

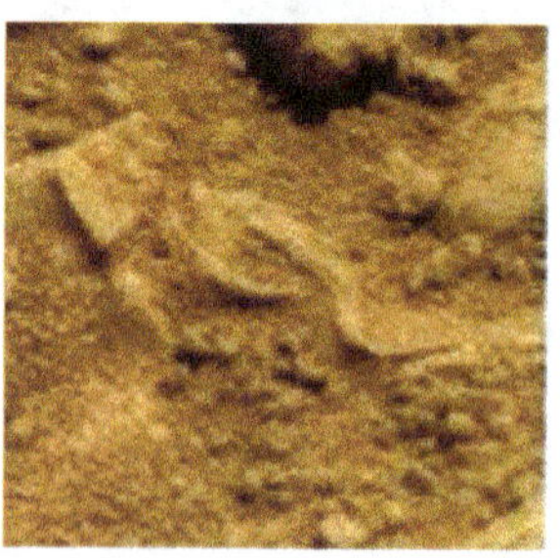

SOL 153-D, highlighted area 10 (above): Though unrecognizable, the totality of these unusual items grouped together in this one small spot—next to numerous other unnatural looking items—argue strongly for intelligent design. If they are the shredded fragments of objects that survived a Martian war, or perhaps a catastrophic "atmospheric disturbance" (as described by the U.S. government remote viewer in my introduction), one could hardly be expected to identify them. Again, this uncertainty does not necessarily make these objects products of natural forces. It makes them Martian anomalies.

Chapter 24

ANCIENT DEVASTATION

SOL 1452-A: EARTHDATE SEPTEMBER 6, 2016

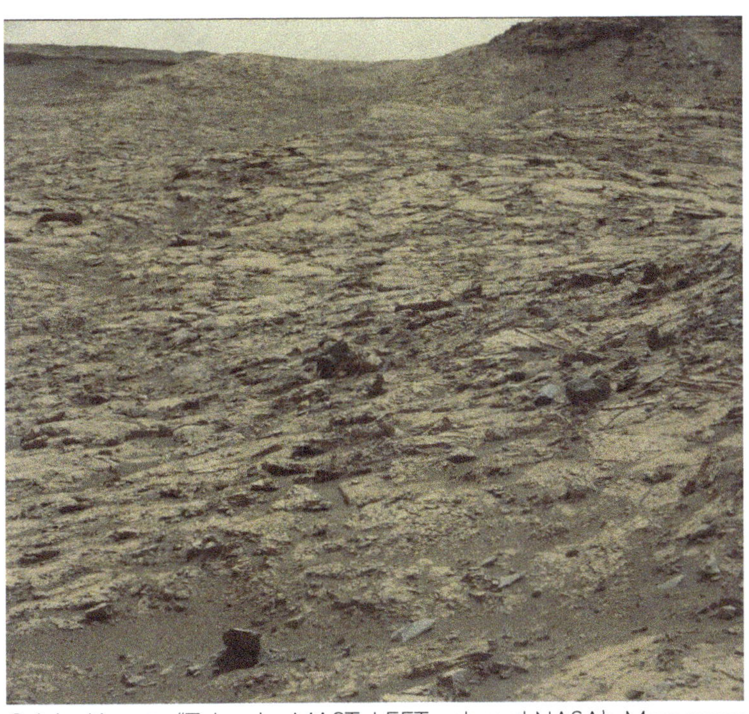

Original image. "Taken by MAST_LEFT onboard NASA's Mars rover Curiosity on Sol 1452 (2016-09-06, T02:13:50.000Z)." Image credit: NASA/JPL-Caltech/MSSS.

ANALYSIS OF ANCIENT DEVASTATION, SOL 1452-A

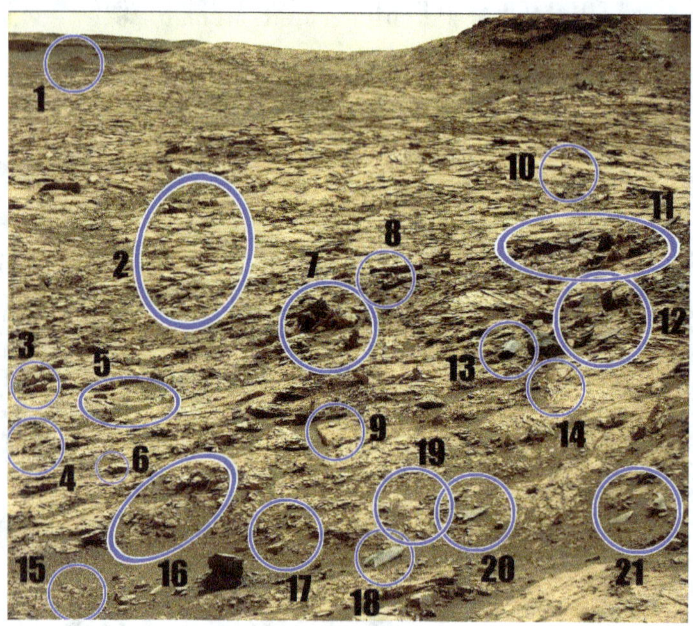

Ancient Devastation, Sol 1452-A, with anomalies circled. Image colored, brightened, and sharpened by the author to improve contrast. In this area we have what appears to be the scene of some kind of ancient devastation—frozen in time by Curiosity's cameras. Objects and structures of nearly every description appear to have been blown to smithereens, and now lie widely scattered over the Martian surface, gathering layers of dust and concretion with the passage of millennia. Due to both the camera distance and the poor resolution of this NASA photo, I cannot detail all of the seemingly countless anomalies that appear in it, especially those that are located toward the back half of the landscape. In what we *can* see, however, there is still much to study.

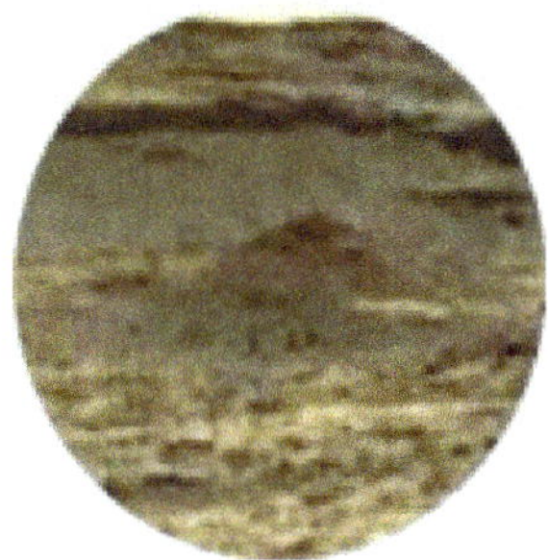

SOL 1452-A, highlighted area 1 (above): In the distance there appears to be a pyramid or pyramidal like structure or hill. It is impossible to make out precisely what it is at such a great distance and in such low resolution. Strangely, there appears to green "grass" and "plants" growing just below the "pyramid." Or perhaps we are seeing digital manipulation (artificially colored pseudo sand) both in front of and behind the "pyramid," intended to hide signs of possible former or present intelligent life.

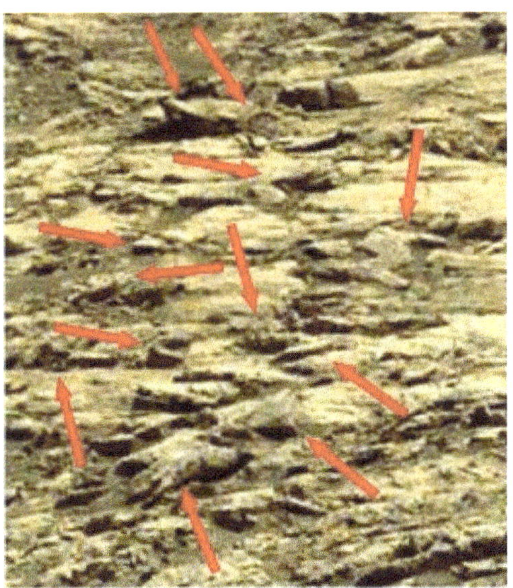

SOL 1452-A, highlighted area 2 (above): As a magnifying glass will show, this image alone is comprised of dozens of anomalies, a few which I have marked with red arrows. These include circles, crescents, squares, stars, triangles, "walls," and "wheels," along with an assortment of odd letter like objects, such as "U," "M," "Y, and "V."

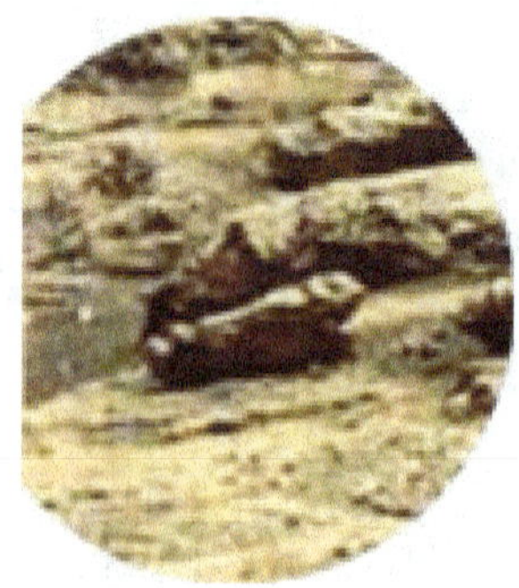

SOL 1452-A, highlighted area 3 (left): Under magnification numerous peculiar objects can be seen from top to bottom across this image. Most look quite unnatural.

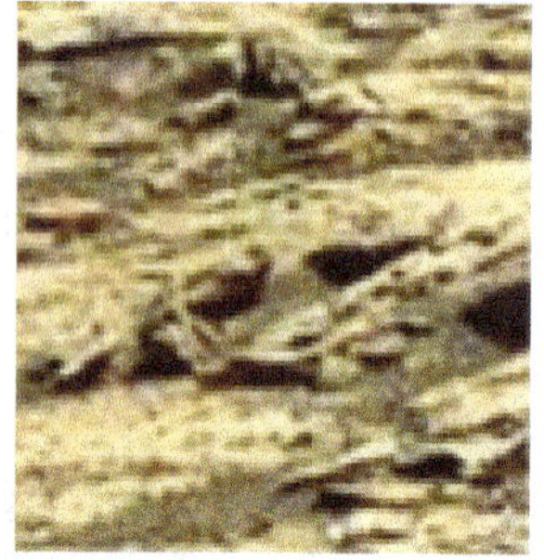

SOL 1452-A, highlighted area 4 (right): At center a perfectly square hole or pit in the Martian ground is surrounded by what looks like rubble; perhaps shattered masonry and twisted cables. A wide assortment of other bizarre looking items dot the landscape.

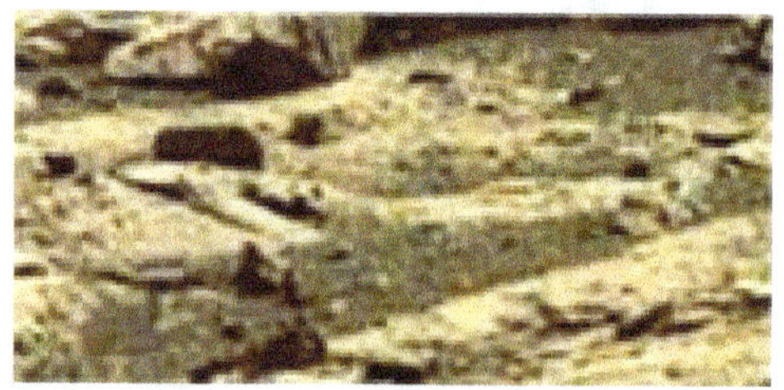

SOL 1452-A, highlighted area 5 (above): What is going here? At the left center is an odd, flat, cement like or metallic like structure laying on the ground. It is a completely different color than its surroundings. Its top end (left) is round, while one of its "arms" (on the right) seems to have been heavily damaged (note the folds, bends, and holes in it). In the upper right we can see a number of brick like or paving stone like objects. This entire area may have once been the site of a building. While artificiality appears to be evident, the resolution is so poor that anything beyond pure speculation is impossible.

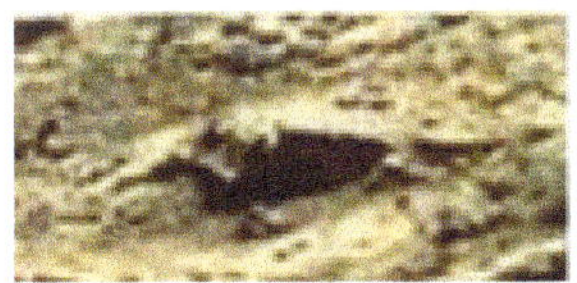

SOL 1452-A, highlighted area 6 (above): At center, a building block or a natural rock formation? Look closely at what lies around it.

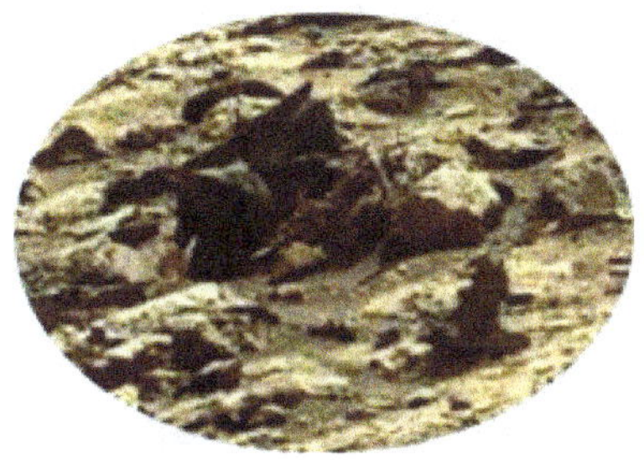

SOL 1452-A, highlighted area 7 (above): Under magnification this area contains a number of weirdly shaped objects. Additionally, at left of center there is a large pile of rubble surrounded by what appears to be sheet metal, melted objects, broken masonry, and twisted cables. The entire scene looks, in fact, like an ancient junkyard, similar to what one might see here on Earth. Note the "shredded" upright object at the 4:00 mark.

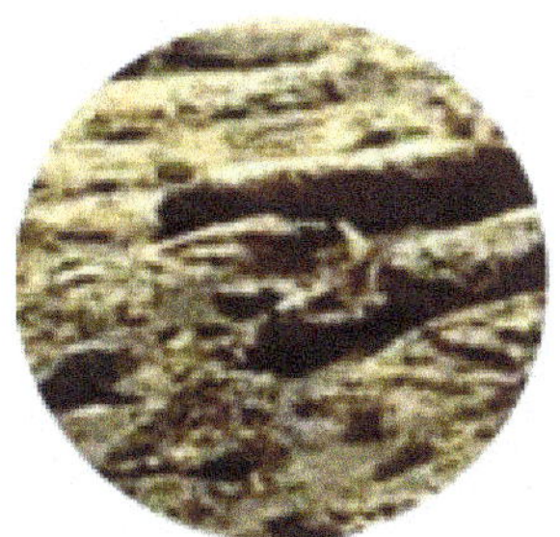

SOL 1452-A, highlighted area 8 (above): More unusual objects possessing what appear to be artificial configurations. Unfortunately, there are too many to describe and discuss in this small space. The reader is advised to explore this image in detail, methodically and critically.

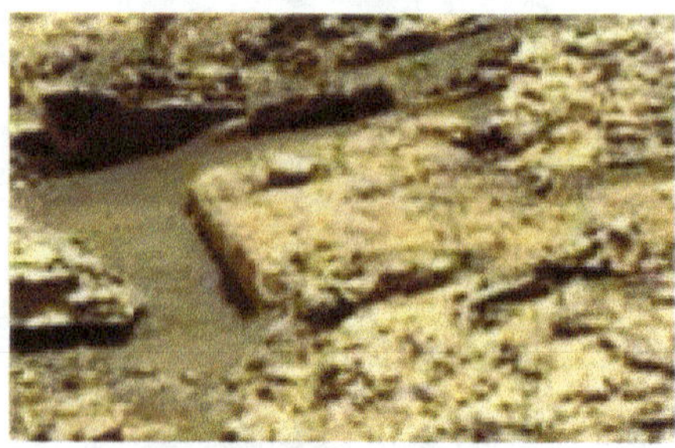

SOL 1452-A, highlighted area 9 (above): What have we here, ancient natural rock or an ancient building foundation? Closely examine the unusual markings on the "rocks" around it and decide for yourself.

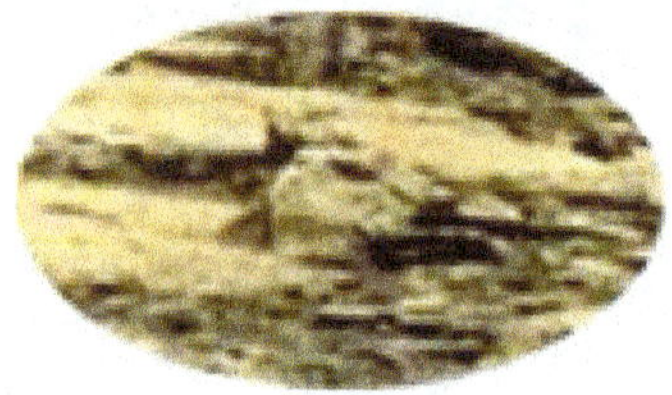

SOL 1452-A, highlighted area 10 (above): Under magnification what look like bizarre artificial shapes can be seen both at center and in the distance. Conventional scientists have tried but failed to definitively, or even adequately, describe their origins and nature.

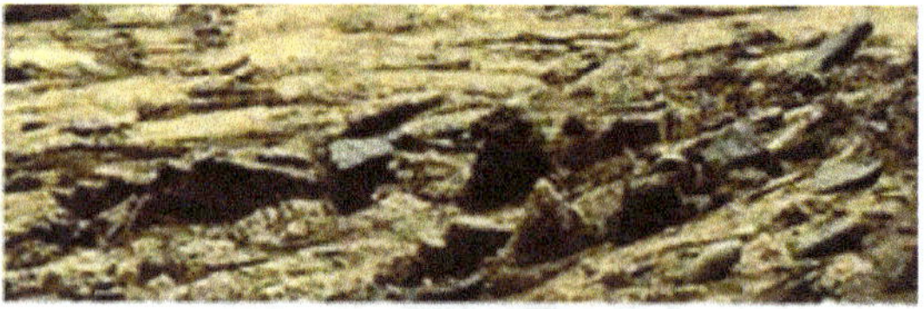

SOL 1452-A, highlighted area 11 (above): Upturned angular objects amid triangles, squares, rectangles, blocks, and various odd thin flat items. Are we looking at the house and temple ruins of an early but now extinct Martian civilization?

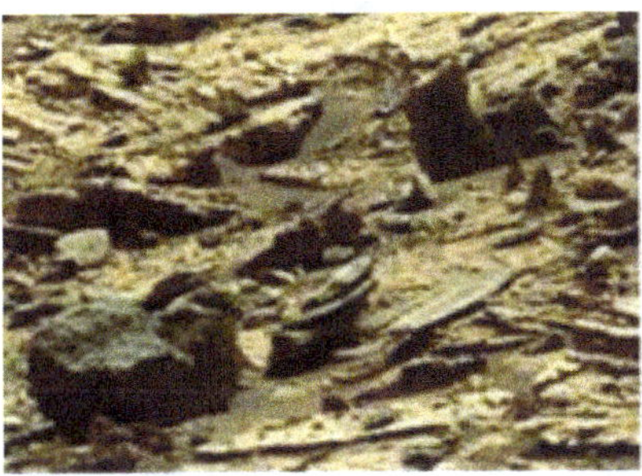

SOL 1452-A, highlighted area 12 (above): Odd "boulders" and "rocks" dot what can best be described as some kind of artificial debris field, many with coffin like, shield like, box like, and surfboard like shapes. Provocatively, these are located amid numerous half circles, rectangles, and squares, as well as strangely marked objects. Magnification will aid the discerning.

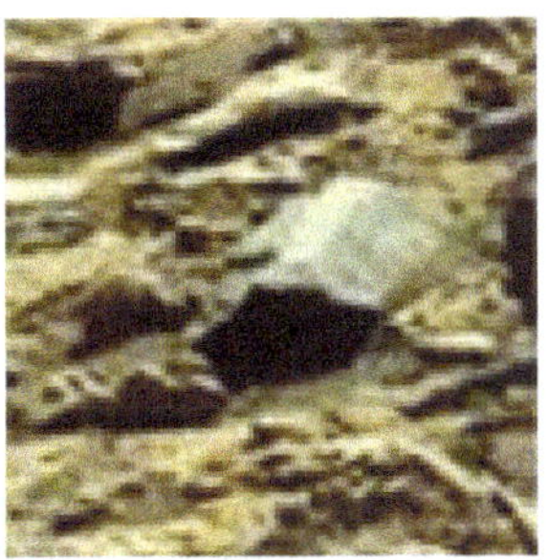

SOL 1452-A, highlighted area 13 (above): This blue gray greenish "box" or "container" seems to be made out of metal and is very different in color, appearance, and texture than nearly everything else in the full frame image. It sits in the midst of piles of twisted "wreckage": countless objects (seemingly artifacts) that appear to have been torn, smashed, and hurled by some terrific force.

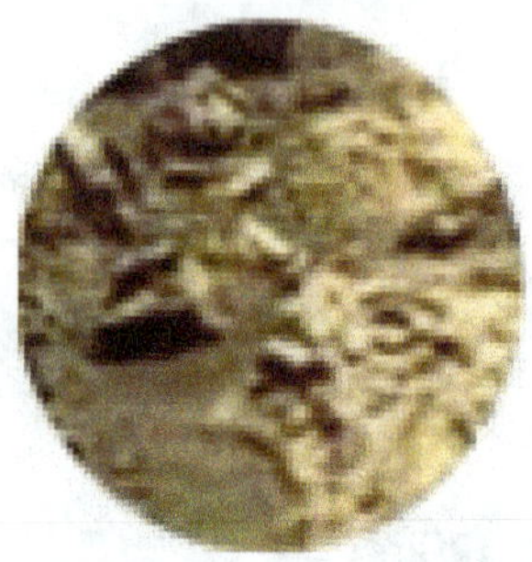

SOL 1452-A, highlighted area 14 (above): Magnify every object in this image. If there have never been intelligent beings on Mars, as mainstream scientists unscientifically claim, who or what made these intricate "designs," "sculptures," and "structures"—and why?

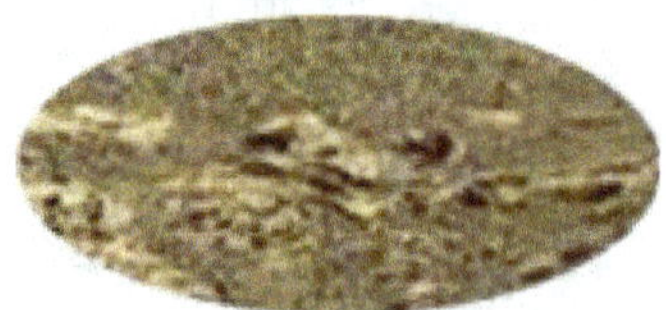

SOL 1452-A, highlighted area 15 (above): Rocks, bones, sticks, or artifacts, or possibly something we do not yet understand?

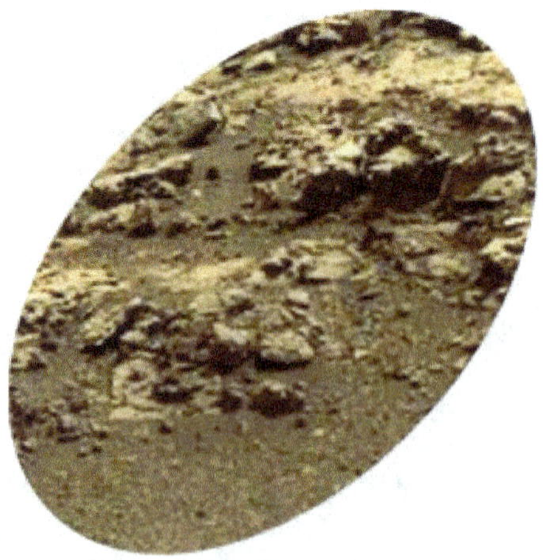

SOL 1452-A, highlighted area 16 (above): At upper right we see what looks very much like the remains of an exploded square structure, with demolished walls and odd objects strewn about.

SOL 1452-A, highlighted area 17 (right): At the center of this debris field is what is left of a small, white, square structure. It appears to have been made out of bricks and looks like it has a hollow rectangular interior. Surrounding it are hundreds of "demolished," possibly artificial objects of nearly every description.

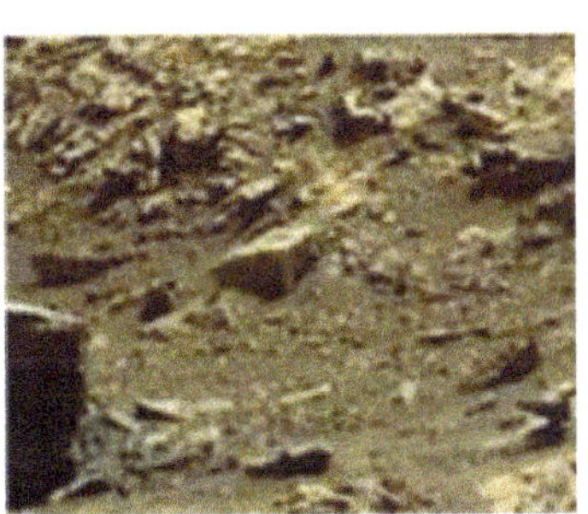

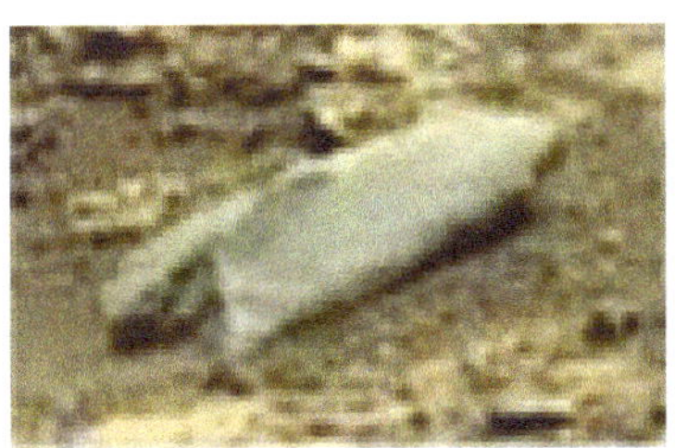

SOL 1452-A, highlighted area 18 (left): This large piece appears to be part of a Martian aircraft, or perhaps a watercraft. (See areas 20 and 21 below.)

SOL 1452-A, highlighted area 19 (right): Under magnification this image will astonish most individuals. If the objects in it are not artificial, what are they? Can every single one of the thousands of strange anomalies I (and others) have discovered on Mars all be explained away as "rocks" or the product of "natural Martian forces"?

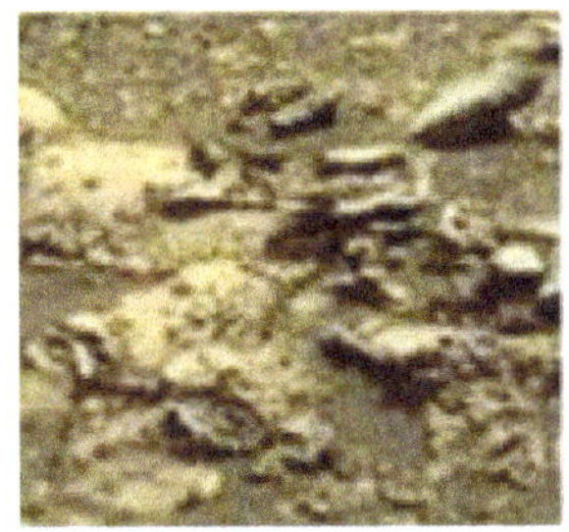

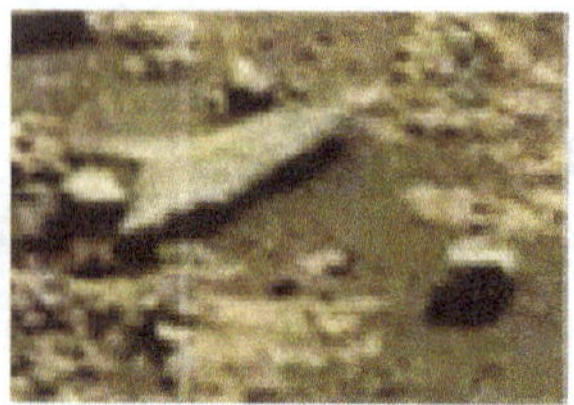

SOL 1452-A, highlighted area 20 (above): More blue gray metallic like objects on the Martian surface. Is the large arrowhead like piece (left of center) part of the wrecked craft described in area 18 above? They are located near one another.

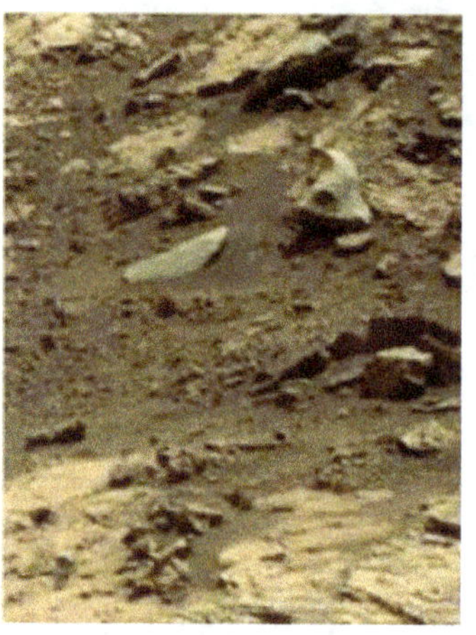

SOL 1452-A, highlighted area 21 (above): What appear to be blue gray metallic objects stand out on this Martian hillside. In addition, they are surrounded by a fantastic array of what look like damaged structures, walls, and artifacts, as well as a spectacular variety of unknown items. Magnification will increase perception as well as the mystery.

Chapter 25
CATASTROPHIC HILLSIDE

SOL 1452-B: EARTHDATE SEPTEMBER 6, 2016

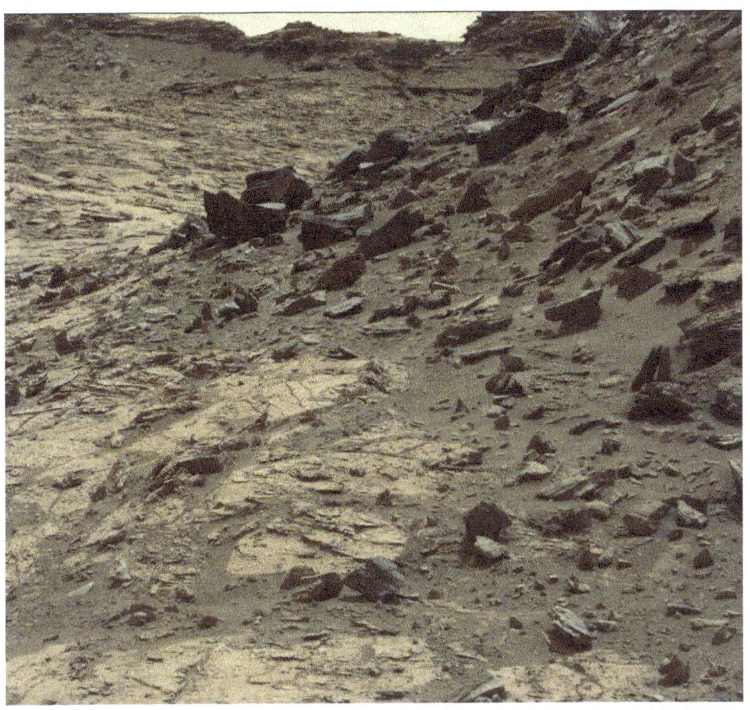

Original image. "Taken by Mast Camera (Mastcam) onboard NASA's Mars rover Curiosity on Sol 1452 (2016-09-06, 02:14:08 UTC)." Image credit: NASA/JPL-Caltech/MSSS.

ANALYSIS OF CATASTROPHIC HILLSIDE, SOL 1452-B

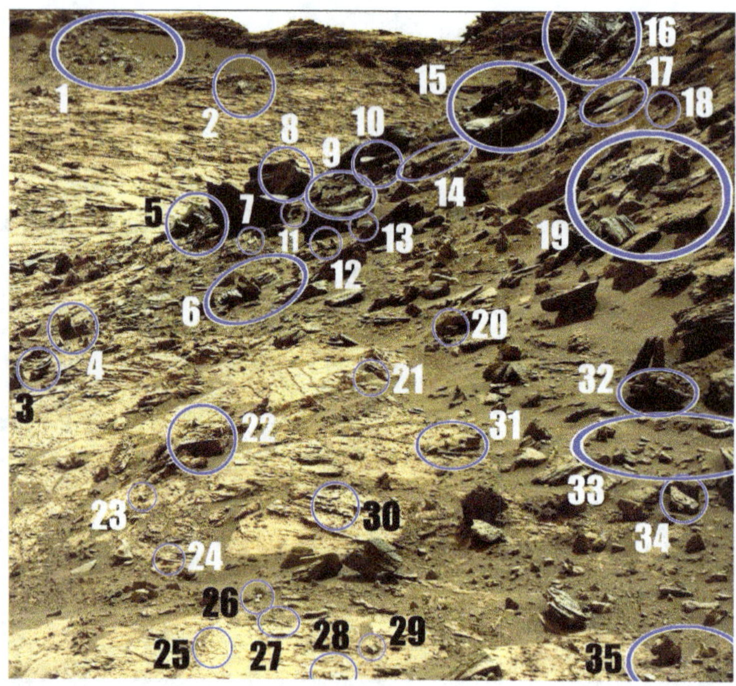

Catastrophic Hillside, Sol 1452-B, with anomalies circled. Image colored, brightened, and sharpened by the author to improve contrast. Mainstream scientists tell us that we are looking at nothing more than a group of rock slides here, with broken stone and shattered natural rock formations strewn over the hillsides. Under magnification, however, a completely different scenario seems to come into focus: a vista depicting some kind of devastation, one that contains countless obliterated artificial objects. As with other extra large NASA images, this one simply contains too many anomalies to analyze. And in any event, when zoomed in the photo resolution is so poor we cannot be sure of what we are actually seeing, making even general identification impossible.

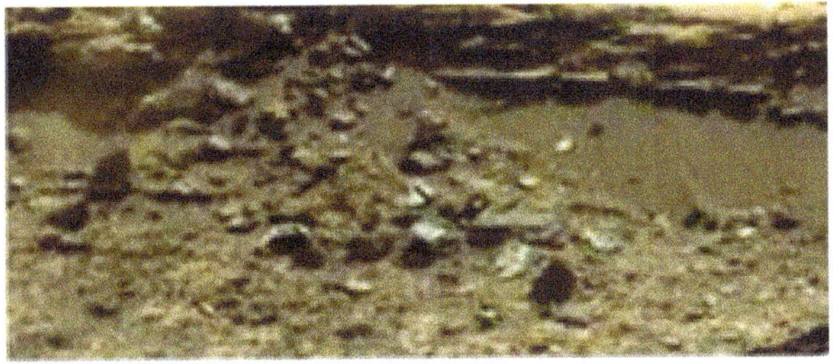

SOL 1452-B, highlighted area 1 (above): A natural small landslide or the remnants of a bombed hillside? Though the great distance and low resolution make it impossible to understand precisely what we are seeing here, numerous damaged, artificial like objects in the debris field tell us it is more likely the latter.

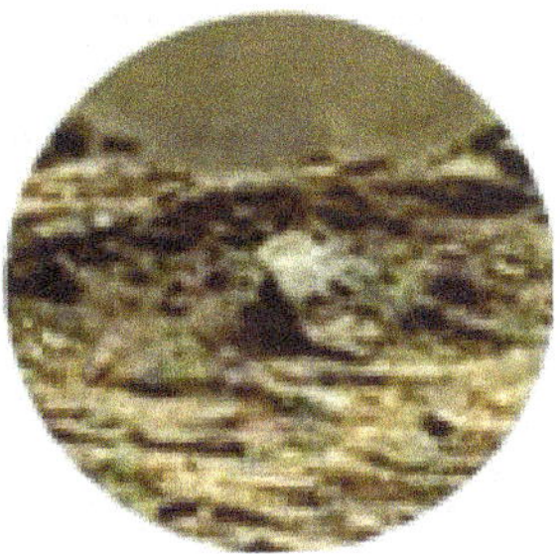

SOL 1452-B, highlighted area 2 (above): What is the oddity at center? It resembles a damaged bucket from an excavator arm. As is common on Mars, a host of other peculiar items encircle it.

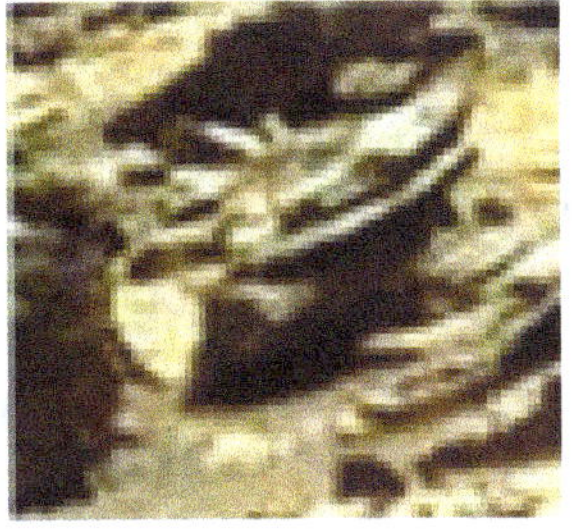

SOL 1452-B, highlighted area 3 (above): This rounded object looks very much like the front half of a broken wooden rowboat. If correct, what appears to be timber planking and the bow frame are both clearly visible. Numerous other equally outlandish objects are plentiful throughout.

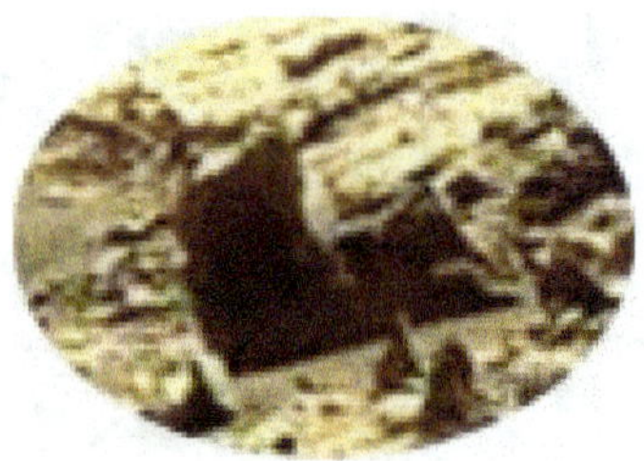

SOL 1452-B, highlighted area 4 (above): This wrecked machine like object seems to have a squarish chute at one end (3:00)—perhaps with some kind of material coming out of it.

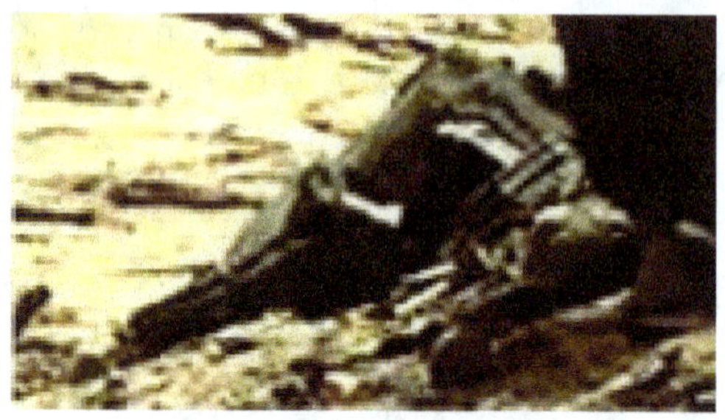

SOL 1452-B, highlighted area 5 (above): This appears to be a large machine that has toppled over due to a powerful but unknown force. We can see what look like metallic pieces, cables or pipes, and a bizarre, squarish "false teeth" like object, along with numerous broken objects scattered over the ground from top to bottom.

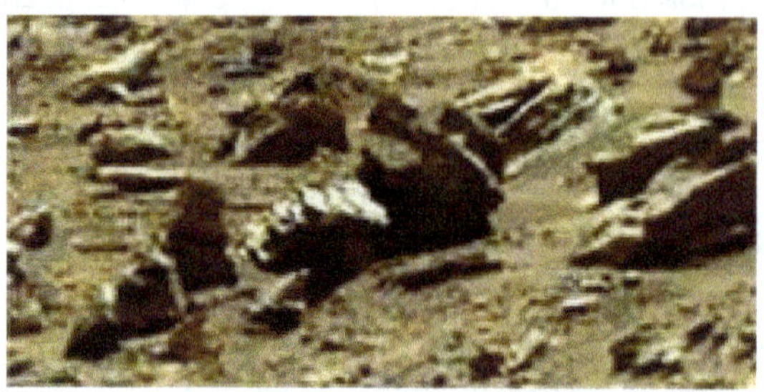

SOL 1452-B, highlighted area 6 (above): Multiple artificial objects seem to decorate this image, including a triangular well, pit, or chamber of some kind (about 1:00); a ruined rectangular stone enclosure (about 2:00); what is left of a small wall or aqueduct (about 3:00); pieces of machinery (center); and some strange colored debris (about 11:00). The closer you look the more you will see. Unluckily for us, parts of this image appear to have been intentionally digitally altered.

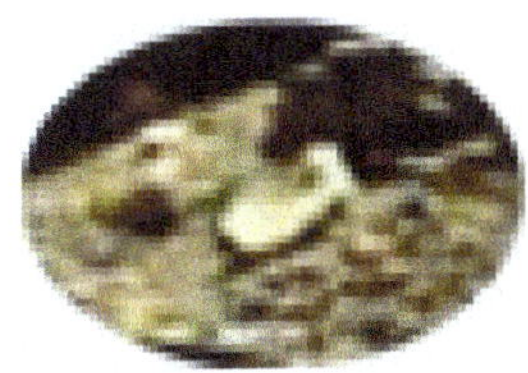

SOL 1452-B, highlighted area 7 (above): This white, queerly shaped object stands out unnaturally. Whatever it is, it looks heavily damaged. And what are the items around it?

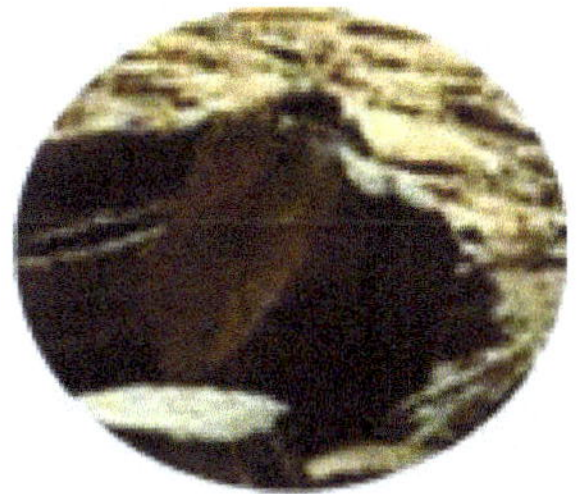

SOL 1452-B, highlighted area 8 (above): This cylindrical object looks like a piece that has broken off from a statue; if so, perhaps it is part of a thigh or torso.

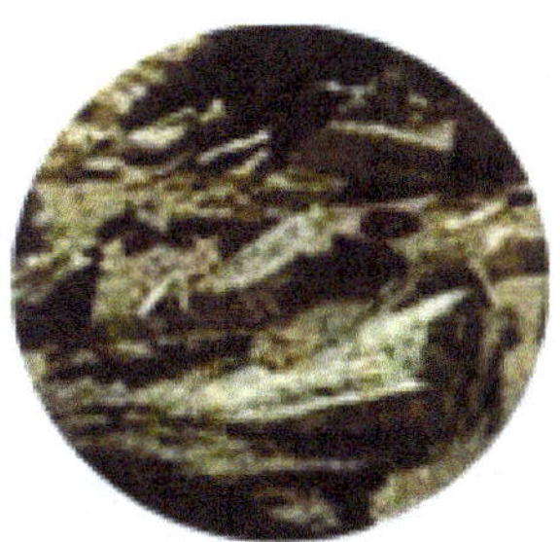

SOL 1452-B, highlighted area 9 (above): A broken "wall" (about 2:00); remnants of an orangish "brick box" (right of center); an oddly shaped whitish greenish object with an "X" carved into the exposed end (about 4:00); a circular "rock" and a strange impression in the ground (center); a small triangular shaped enclosure (left upper center); what look like bones lie scattered about; an unusual shaped squarish object sits at the 9:00 position. Artificial looking right angles, squares, rectangles, circles, bows, ovals, and triangles command the scene.

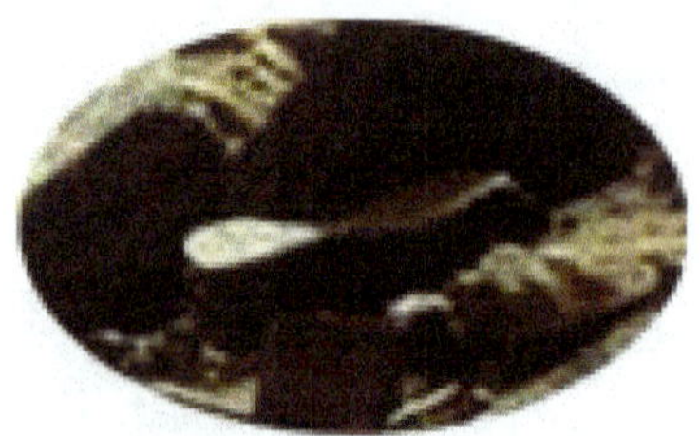

SOL 1452-B, highlighted area 10 (above): This is not the only boat shaped object in the full frame image of Sol 1452. In addition, a number of artificial looking oddities surround it.

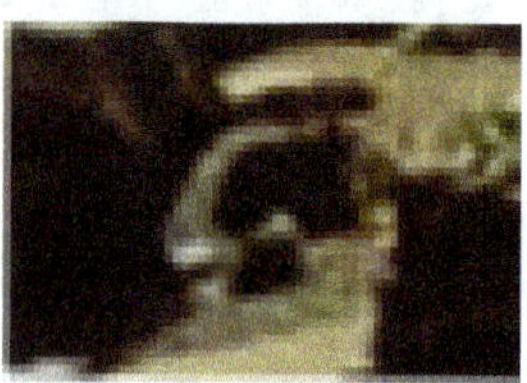

SOL 1452-B, highlighted area 11 (above): This appears to be what is left of an underground well made of stone or cement; or perhaps it is the entrance to some kind of subterranean chamber or shelter. To call such objects the result of "natural Martian geology" beggars credulity.

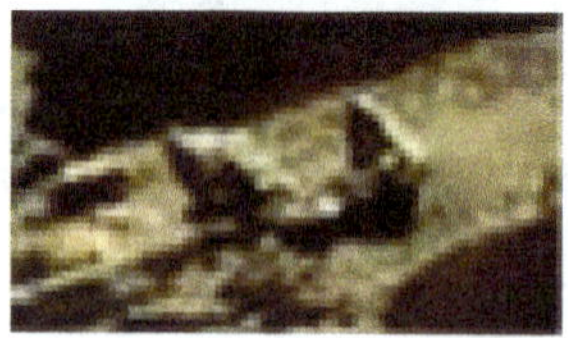

SOL 1452-B, highlighted area 12 (above): A broken sink or bathtub? A damaged piece of factory machinery or agricultural equipment?

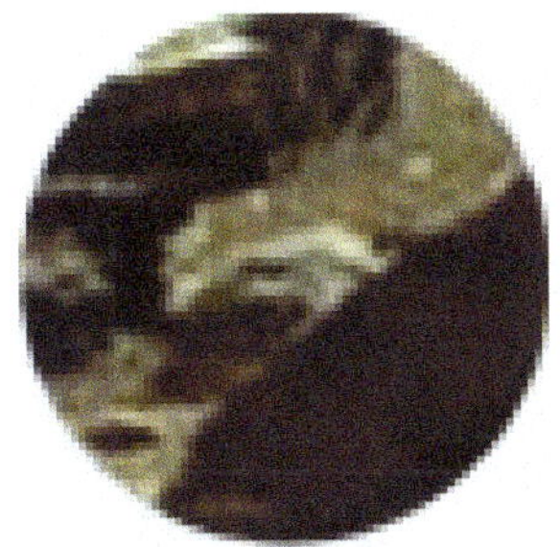

SOL 1452-B, highlighted area 13 (above): Under magnification, at the center of this peculiar image, is what looks like a large, beautifully sculpted eye, complete with an upper and lower eyelid, a pupil, an iris, a sclera, and a tear duct. If this is what it is, it seems likely that it is from the same statue as the one described in area 8.

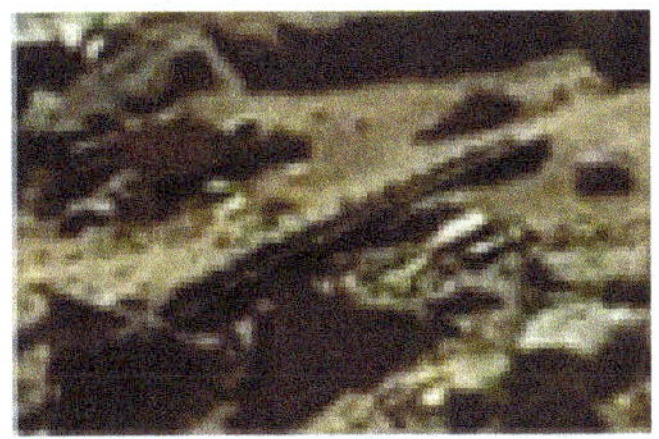

SOL 1452-B, highlighted area 14 (above): The strange, hollowed out object at center looks like a long, narrow wooden boat, perhaps something similar to our dugout canoes here on Earth. If so, it has sawtooth edges along the gunwales running from bow to stern. Around it are a number of startling, artificial looking objects of various shapes and sizes.

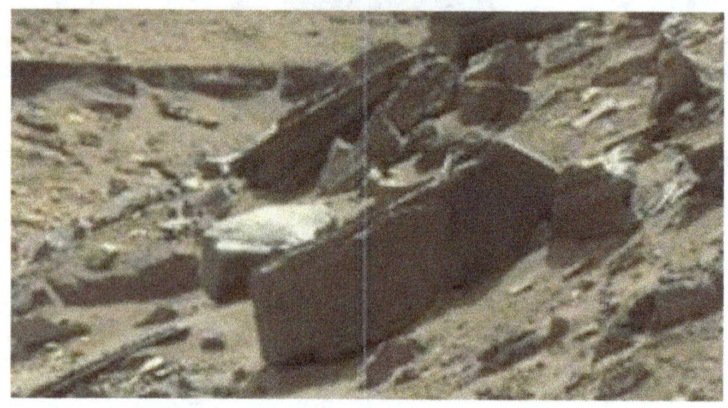

SOL 1452-B, highlighted area 15 (above): Unidentifiable objects displaying probable intelligent design permeate this photo. At around the 1:00 position a rounded "V" shaped object is visible. To the left of it (12:00) is a coffin shaped object. To its left are several large roundish objects. At the 11:00 mark there are what look like the remnants of retaining walls cut into the hillside (like many of the features on Mars, these could, of course, be naturally formed sedimentary rocks). Strange "Y" shaped objects appear at the 9:00 and 2:00 positions. Filling most of this image, at the lower center a large dark "rock" can be seen. Its round shape, identical curved sides, flat top, and "hollowed out" interior betray what seems to be artificiality. Above it and to its left (at about the 8:00 mark) there is what appears to be a large shiny chunk of metal with carefully cut angles. Odd shapes continue at the 3:00 position, one that looks similar to a large wisdom tooth that has been extracted (with roots still attached); another, a large star like object; and also a crab like object. The odd dugout canoe like object with the serrated "gunwales" (described in area 14) can be seen at the lower left.

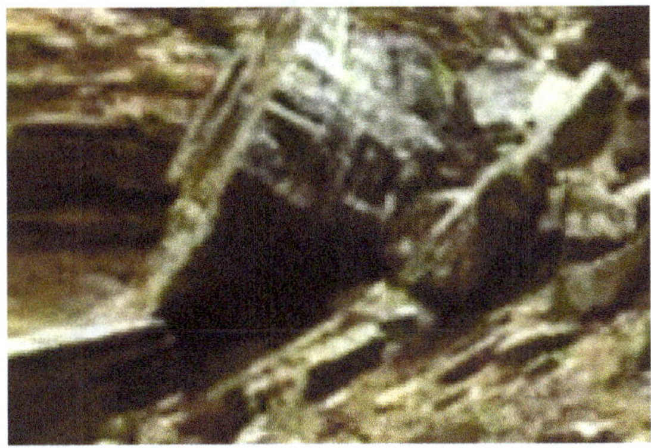

SOL 1452-B, highlighted area 16 (above): Not knowing anything about ancient Martian culture, one can only hazard random guesses as to what this fascinating artificial looking object might be. I would describe it as large and round with a circular hollowed out center. Horizontal lines encircle it and it has unknown markings on its sides, including oddly shaped squares with peculiar lettering or symbols inside them. Was it a military turret of some kind, or perhaps a piece of factory machinery? It seems to be surrounded by long squarish posts or pillars constructed of cement or some similar material.

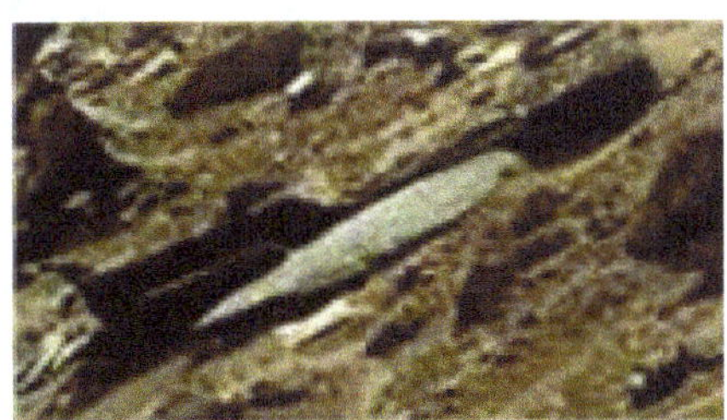

SOL 1452-B, highlighted area 17 (above): The surfboard shaped object at the center of this image is quite possibly just a naturally formed rock. However, its peculiar shape, along with the odd objects its laying next to, as well as its location just yards from the seemingly artificial objects described in areas 15 and 16, raise doubts.

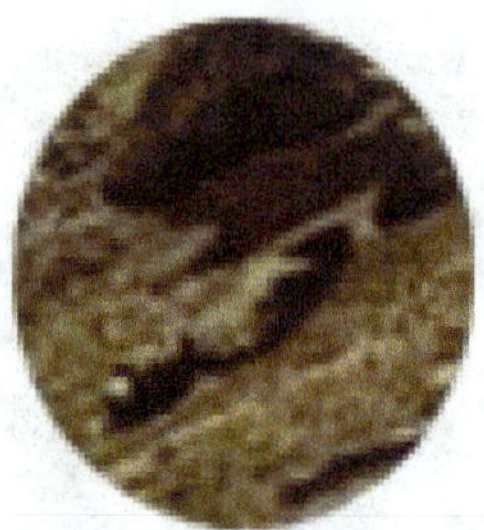

SOL 1452-B, highlighted area 18 (above): Low resolution makes the probable artificial objects in this image impossible to identify or even understand. An educated guess might be that these are pieces from some type of machine. At the 2:00 position we see a damaged object that looks similar to a vertebra; at the 8:00 position there is a roundish cylindrical object with what appears to be a shiny metal ball attached to one end. It looks wholly unnatural. What could it be?

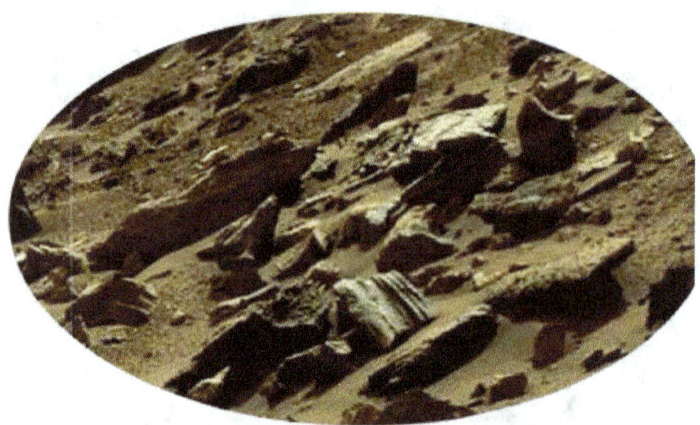

SOL 1452-B, highlighted area 19 (above): A small section of the massive debris field shown in the full frame photo above. At 12:00 another surfboard shaped object. At 1:00 odd artificial shapes that appear broken. At 3:00 the probable remnants of a square box or container. At 4:00 a board like object made of some type of conglomerate material pokes out from the Martian dirt. At 5:00 and 6:00 are several pieces that look sculpted and decorated with lines and symbols. At 8:00 a "rock" with what look like tines or teeth. At 9:00 another curved object that looks like the bow or stern of a small boat. At 10:00 a strange circular clearing or pit that is lined with rocks. At 11:00 an object with several large cavities. Around the center of this image a group of unusual and extremely artificial looking objects can be seen. Magnify them.

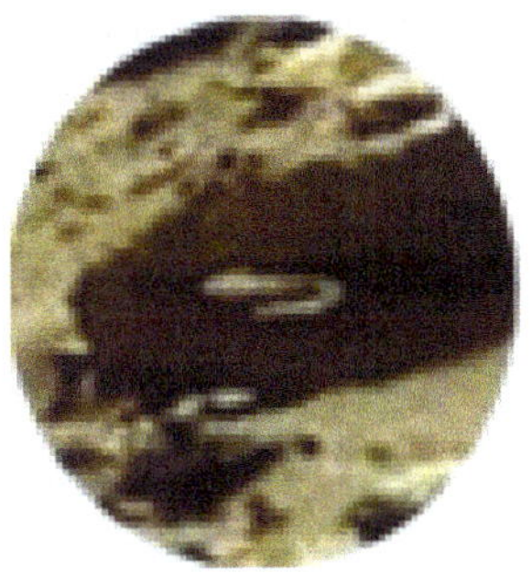

SOL 1452-B, highlighted area 20 (above): Why is there a bright white, sideways "J" emblazoned on the rock face beneath this overhang? Or is this nothing more than minerals leaching out of the stone? Is it part of the fossilized remains of a prehistoric Martian life form? Is it a serpentine petroglyph from an ancient Martian mural? Is it an optical illusion caused by sun and shadows? Is it an example of digital manipulation, an effort to coverup what lies beneath, possibly lettering?

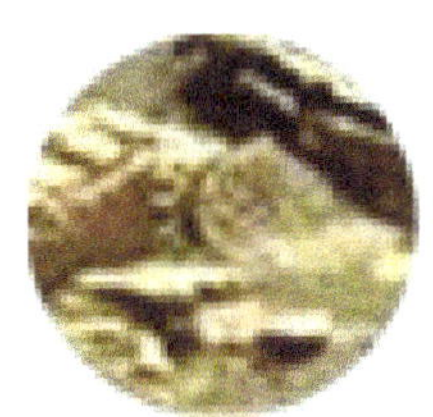

SOL 1452-B, highlighted area 21 (left): The rounded object at center looks like a stone tablet with symbolic engravings on it. Is it? Look closely at what is lying around it.

SOL 1452-B, highlighted area 22 (right): There is a lot going on in this small piece of Martian real estate. The large rectangular object at center appears to have debris on and around it, markings on its sides, and a box like object on the top right side. It sits in a cleared space lined with rocks. A kiln like object made of "bricks" can be see at the 2:00 position. The roundish object at the 12:00 mark may be what is left of an ancient Martian wrecking ball, complete with a lifting hook.

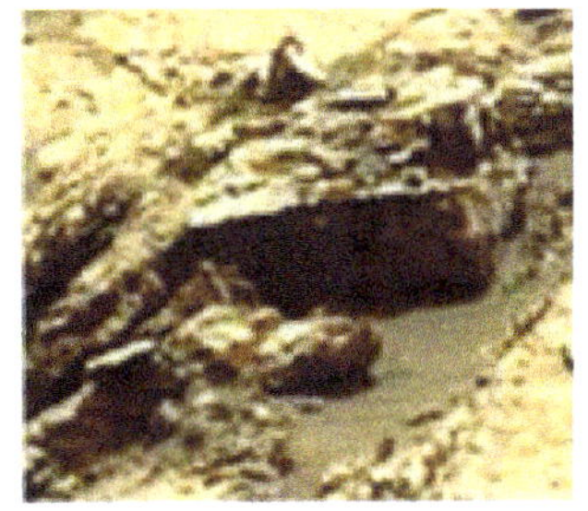

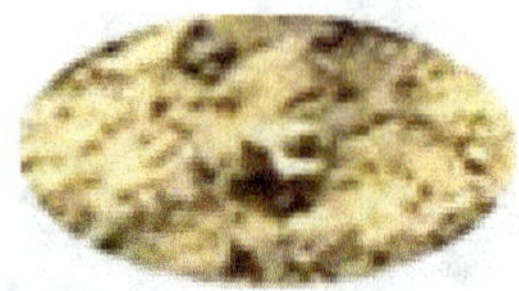

SOL 1452-B, highlighted area 23 (above): A naturally formed rock, or the broken remnants of an object made by a superior intelligence?

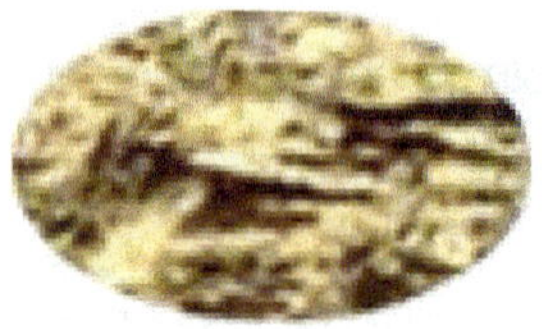

SOL 1452-B, highlighted area 24 (above): At center another *Star Trek* insignia like object.

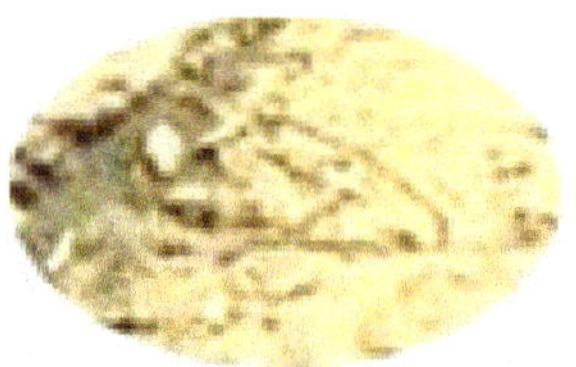

SOL 1452-B, highlighted area 25 (above): Artificial engravings and objects or natural geological processes?

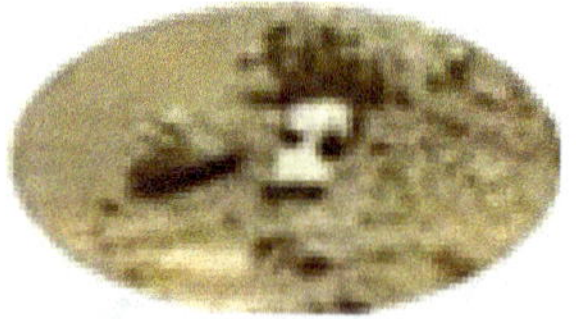

SOL 1452-B, highlighted area 26 (above): This bright, oddly shaped object appears metallic. Look at the objects nearby.

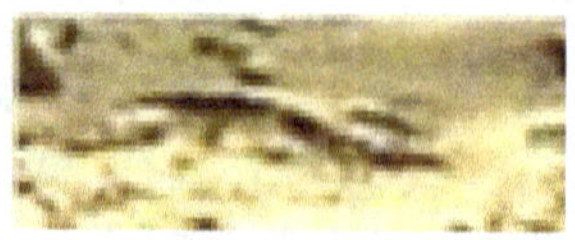

SOL 1452-B, highlighted area 27 (above): At center, nothing but a "rock."

SOL 1452-B, highlighted area 28 (above): The "V" shaped "rock" at lower right center has a notch in it, and is surrounded by a group of other unusually shaped "stones."

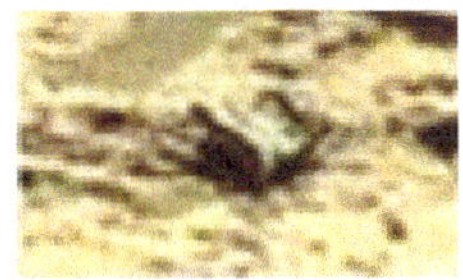

SOL 1452-B, highlighted area 29 (above): While we cannot positively identify this object at the moment, it has the appearance of a crushed metal can or a rolled up rug that has been sitting on the dusty surface of Mars for hundreds if not thousands of years.

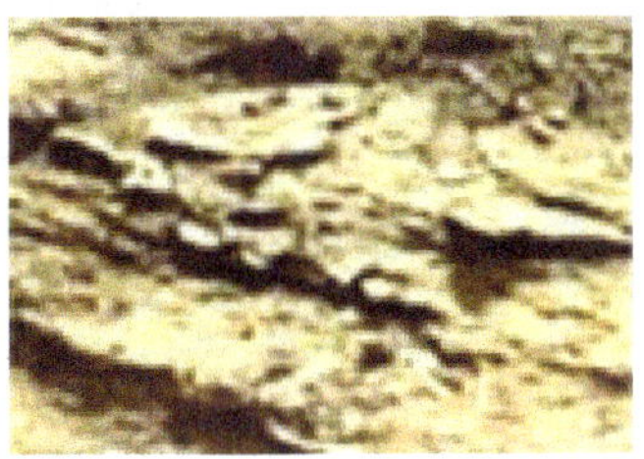

SOL 1452-B, highlighted area 30 (above): Odd shapes, rocks, and formations. Possibly all result from natural geology. Considering their appearance and location, possibly not.

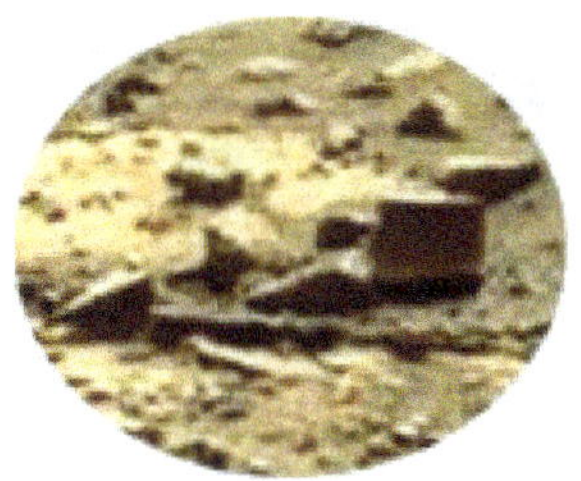

SOL 1452-B, highlighted area 31 (above): Peculiar shapes, along with squares, rectangles, blocks, straight lines, curvatures, and what appears to be a small pointed pyramid laying on its side (at the 8:00 position), fill this image.

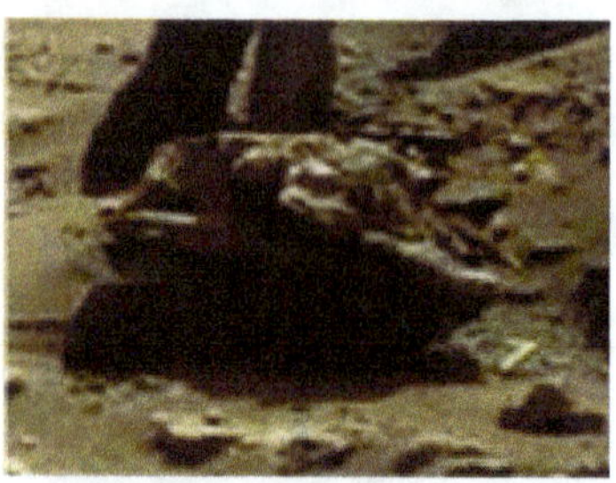

SOL 1452-B, highlighted area 32 (above): This strange object appears to be comprised of conglomerate materials, mostly artificial. Has it undergone a period of intense heat?

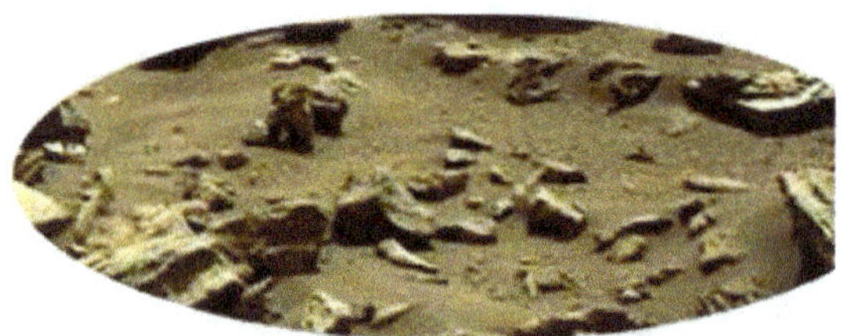

SOL 1452-B, highlighted area 33 (above): Bones, artificial debris, or rocks? Look closely.

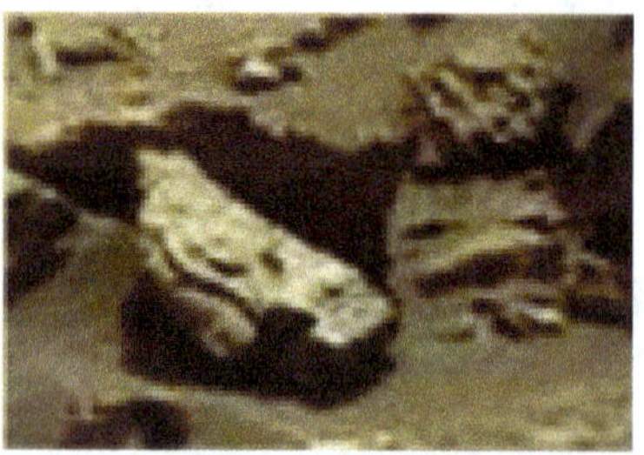

SOL 1452-B, highlighted area 34 (above): Remnants of Martian hieroglyphs and artwork or natural rock formations? Magnify.

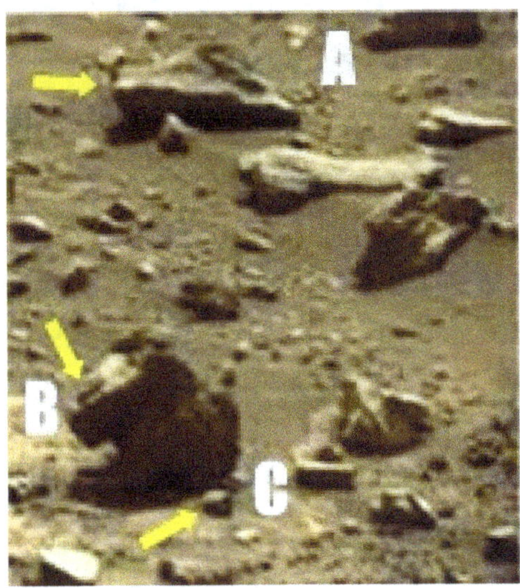

SOL 1452-B, highlighted area 35 (above): Very few objects in this image look geologic. Blocks, triangles, 90 degree angles, rectangles, "boards," "bricks," "shingles," and other unnatural looking articles fill the frame. For example, look closely at the hollowed out triangular object I have marked "A." It looks intelligently constructed. What about squarish object "B"? It has a square opening or cavity on top. Then there is object "C." This small, cylindrical item has a round hole in it (or all the way through it), and resembles a pipe or plumbing fitting. Numerous other objects in this photo also look out of place on Mars, a so called "lifeless planet." Does Nature create these sorts of things? More to the point, how do conventional scientists explain them? They do not have unequivocal answers. They can only guess based on their conventional school training, a highly restricted worldview that does not allow for an extraterrestrial explanation. Thus, their perspective not only defies both intuition and logic, but it violates the very foundation of science itself, the *scientific method*, which is defined as: "The principles and procedures for the systematic pursuit of knowledge involving the recognition and formulation of a problem, the collection of data through observation and experiment, and the formulation and testing of hypotheses." Has the scientific method been applied to these particular objects? Absolutely not. We must then seriously consider the possibility that they are what they appear to be, and most likely are: the ruins of an ancient Martian society.

Chapter 26
BROKEN ARTIFACTS

SOL 1454-A: EARTHDATE SEPTEMBER 8, 2016

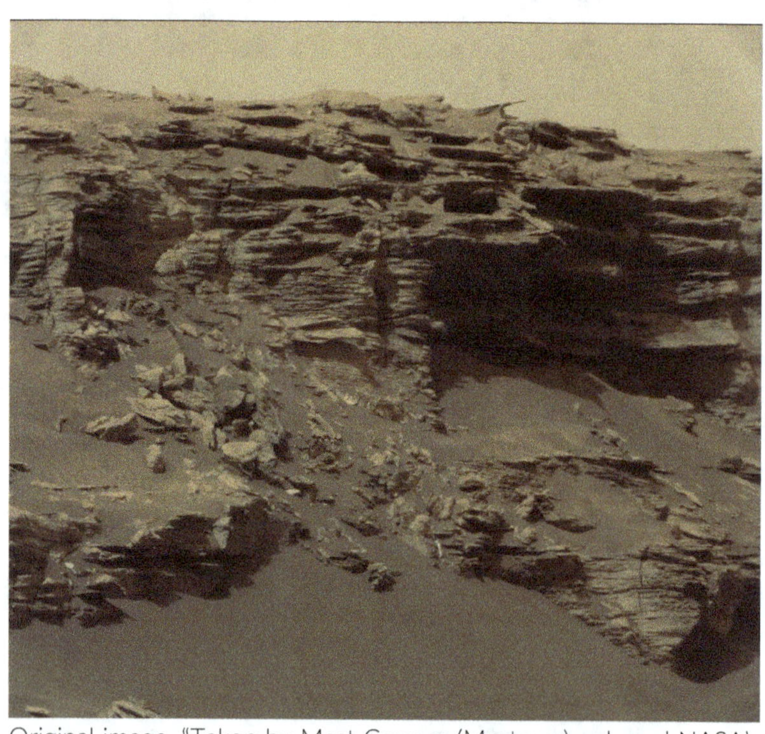

Original image. "Taken by Mast Camera (Mastcam) onboard NASA's Mars rover Curiosity on Sol 1454 (2016-09-08, 03:03:41 UTC)." Image credit: NASA/JPL-Caltech/MSSS.

ANALYSIS OF BROKEN ARTIFACTS, SOL 1454-A

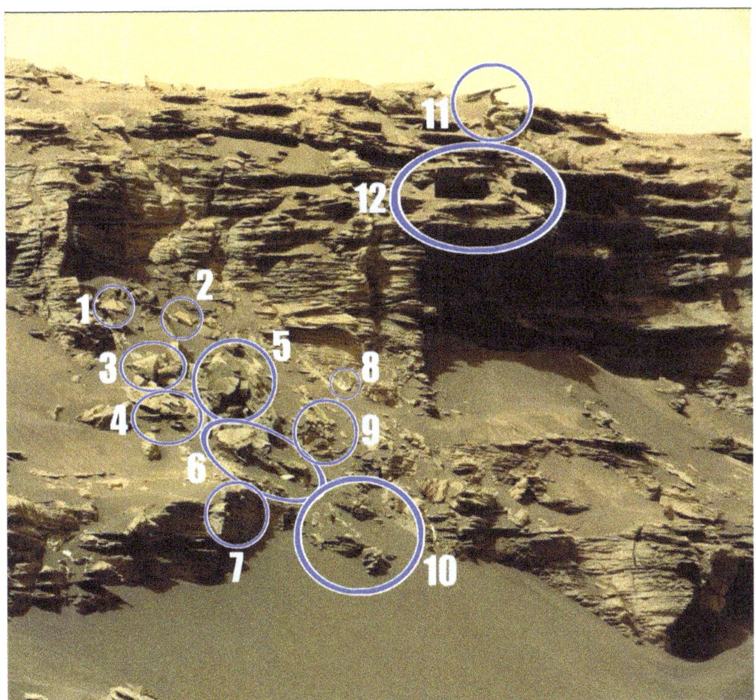

Broken Artifacts, Sol 1454-A, with anomalies circled. Image colored, brightened, and sharpened by the author to improve contrast. From a distance, on the lower left, we see what looks like a natural rockslide of mostly sedimentary material—and this is how orthodox scientists view it. However, under magnification we see what looks more like a debris field filled with ancient broken artifacts. Is this what it is? We cannot know positively until humans examine these objects in situ. Until then we must use a combination of common sense, intuition, speculation, and rational thinking to make a determination.

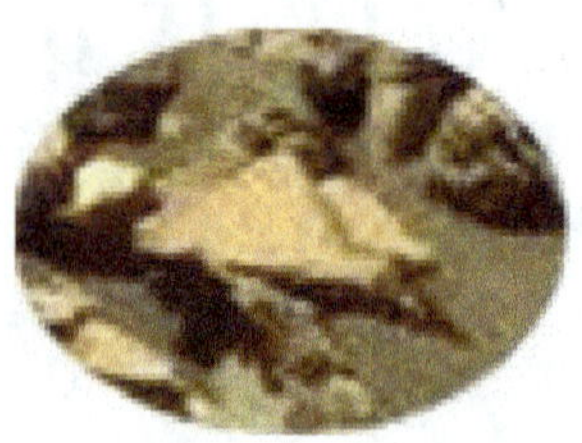

SOL 1454-A, highlighted area 1 (above): Natural or unnatural?

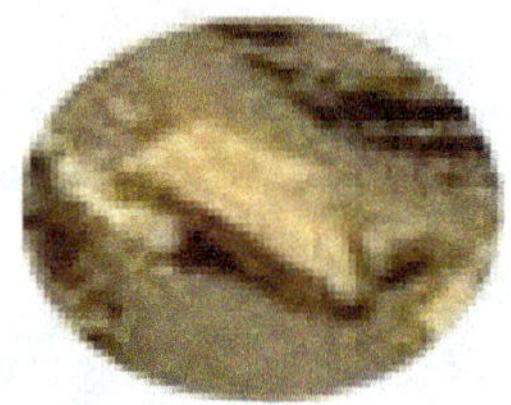

SOL 1454-A, highlighted area 2 (above): This object looks like it originated from the same source as the object shown in area 1. They are both the same color and bear the same unusual straight edges. What are they?

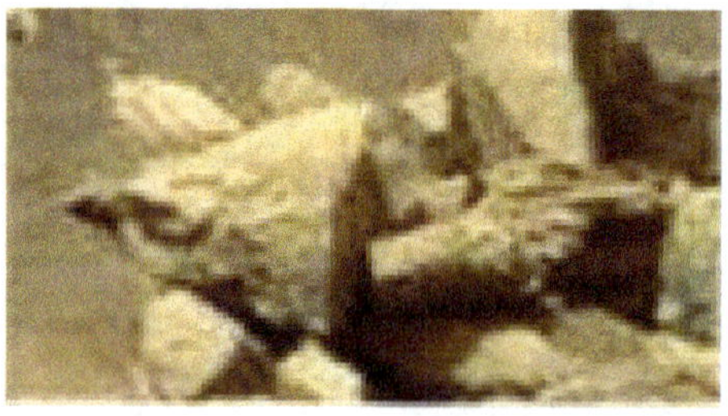

SOL 1454-A, highlighted area 3 (above): Does anything in this stupendous image look natural? The large triangular shaped object on the left has what appears to be raised writing on it, which, from left to right, looks like a "G" and the numeral "2" (see next image below for a closeup). To its right is what appears to be the top of a round Corinthian styled column that has toppled over. Above the lettered triangle (at 11:00) is an odd item with squares carved into it, giving one the impression that it is the bottom half of a small step pyramid or ziggurat. Square "cement" blocks lie tumbled about in disorderly fashion. The crumbled ruins of a temple courtyard?

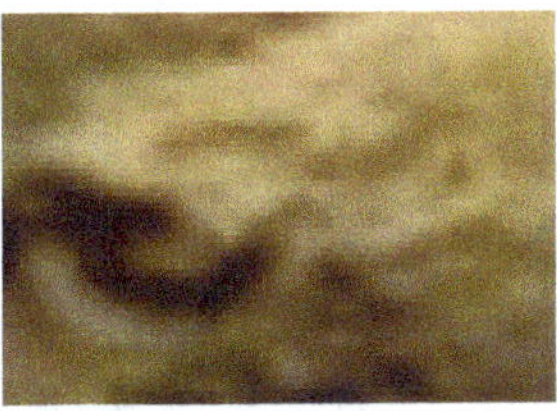

SOL 1454-A (above): An extreme closeup of the raised letter "G" (at left) and the raised number "2" (at right) described in area 3 above. Interestingly, these are similar in style to many of the letters and numbers we have examined in earlier images. The arm at the top of the capital "G" (if that is what it is) appears to have a stylized cross on the end of it. At close range the "2" stands out in bold relief.

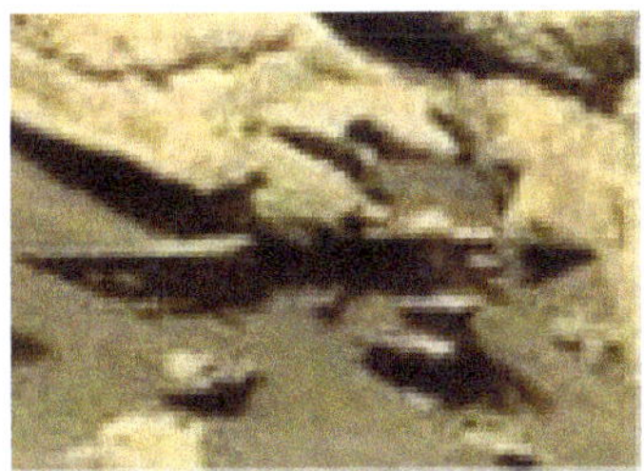

SOL 1454-A, highlighted area 4 (above): I would be willing to call many of the items in this photo "natural rock formations"—if they were not sitting in the middle of a debris slide filled with hundreds of artificial looking objects. Study them under magnification.

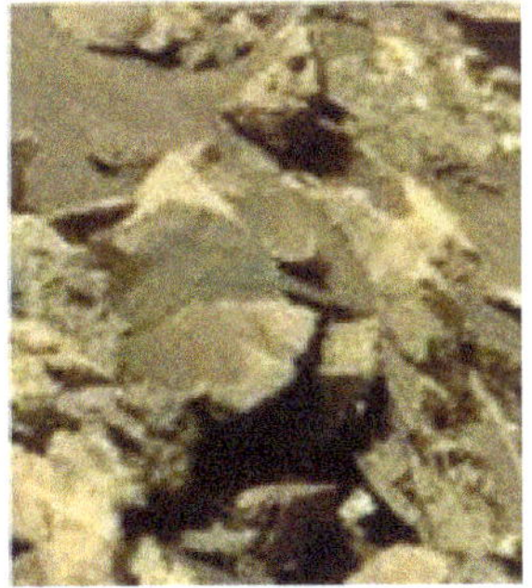

SOL 1454-A, highlighted area 5 (above): More "cement" blocks, squares, rectangles, triangles, and unusually shaped objects—many with circles and marks seemingly cut into them—piled on top of one another. I will call the partial "face" on the rock at 12:00 pareidolia.

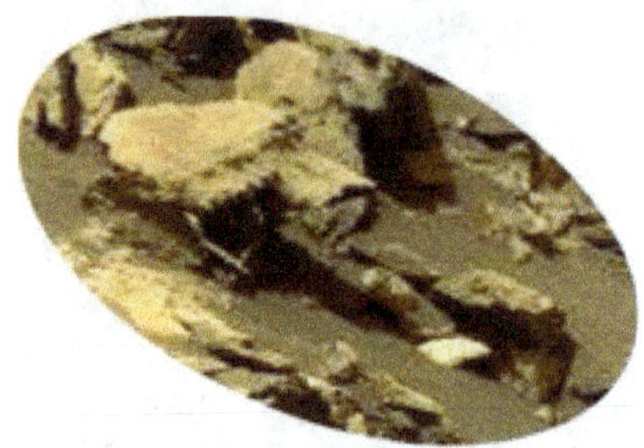

SOL 1454-A, highlighted area 6 (above): If this object is actually what it appears to be from an Earthling's point of view, all of our history books will have to be rewritten. Combining the anomalous objects in this book with the fact that there were rivers, lakes, and even oceans on early Mars, only the unimaginative would be surprised to find the encrusted, barely recognizable remains of a fossilized ship on the surface of the Red Planet. This one, if that is what it is, has retained most of its "hull" and "foredeck," with the "planking" and remnants of the "stern" still clearly visible.

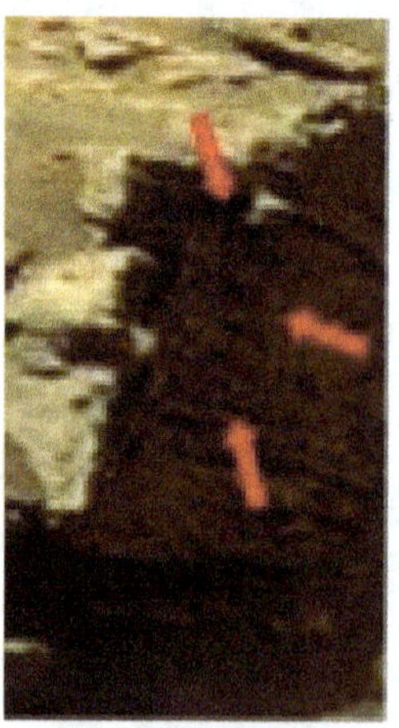

SOL 1454-A, highlighted area 7 (above): This "rock face" appears to be the remnant of a wall that has tumbled over on its side, for it has what looks like clear Roman letters engraved on it. Turn this book to the right until the page is horizontal. Though the resolution makes it impossible to be definitive, from left to right there are three letters that appear to be "I," "N," and perhaps "P," "F," or "T."

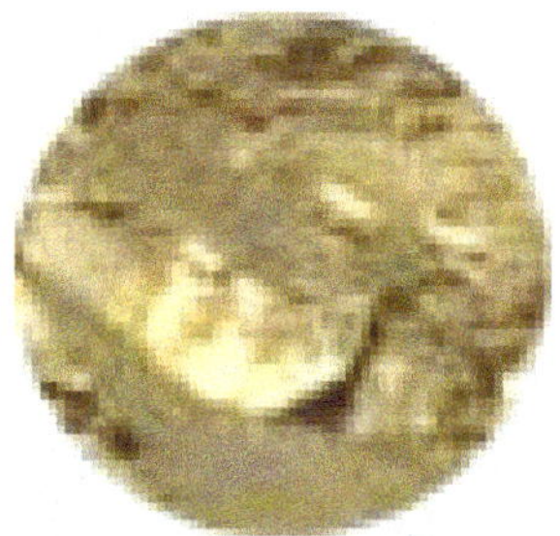

SOL 1454-A, highlighted area 8 (above): This white tablet like or book like object could be a rock. However, not only does it possess traits of artificiality, it is located in an anomaly rich area where an ancient catastrophic event of some kind seems to have taken place—one that destroyed nearly everything around it. This increases the possibility that it was created by intelligent thought.

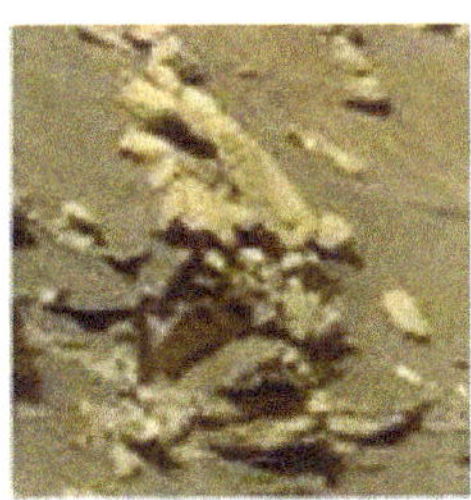

SOL 1454-A, highlighted area 9 (left): A number of unnatural looking objects appear in this pile of hillside debris. Possible artifacts with curves, straight lines, and 45 and 90 degree angles can be seen. A large arrowhead like object sits upper left of center (pointing uphill). Magnify it.

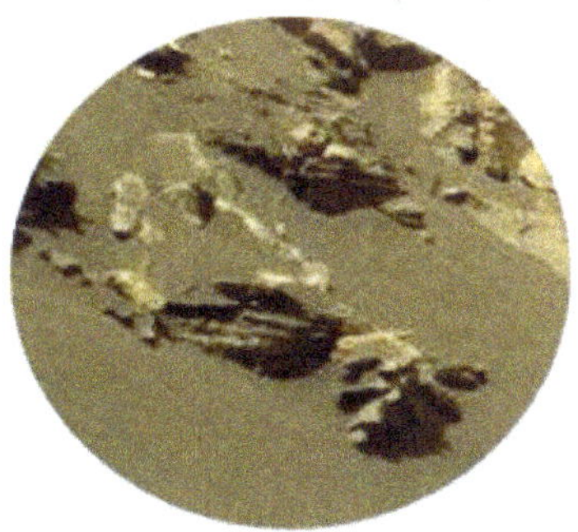

SOL 1454-A, highlighted area 10 (above): At first glance this appears to be a group of natural rock formations. Is it? Due to the number of square and curved objects visible it is more likely a collection of both geological features and artificial objects.

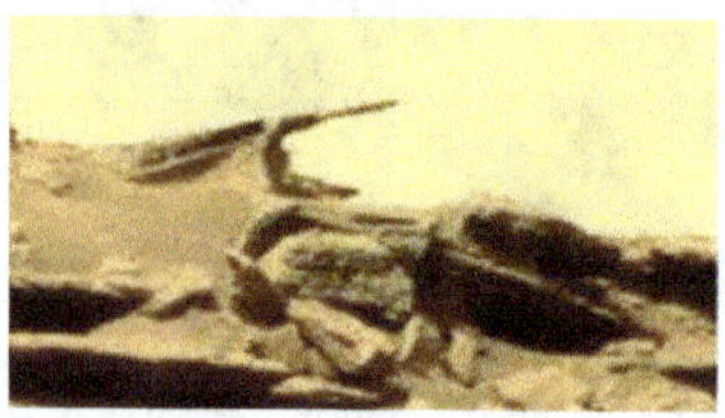

SOL 1454-A, highlighted area 11 (above): What is the long narrow object at the top of the hill? A curious rock formation, or a gun covered in centuries of concretion? Look at what is below it. Are these the devastated remains of a Martian hill fort?

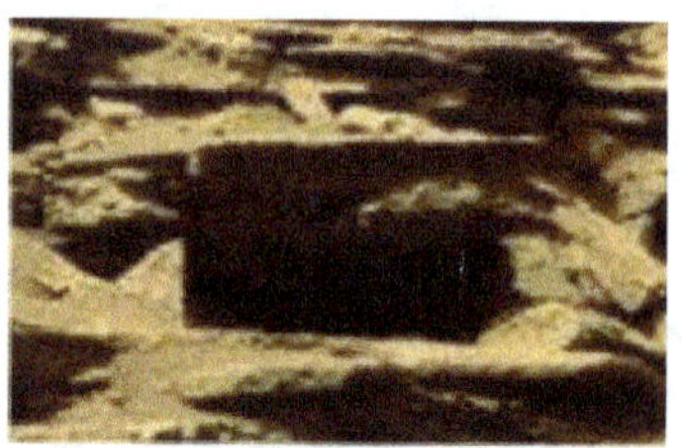

SOL 1454-A, highlighted area 12 (above): The squarish, straight lined shadow at center is peculiar. What is casting it? A rock, or the remnants of a floor or tumbled wall?

Chapter 27

DEBRIS FIELD

SOL 1454-B: EARTHDATE SEPTEMBER 8, 2016

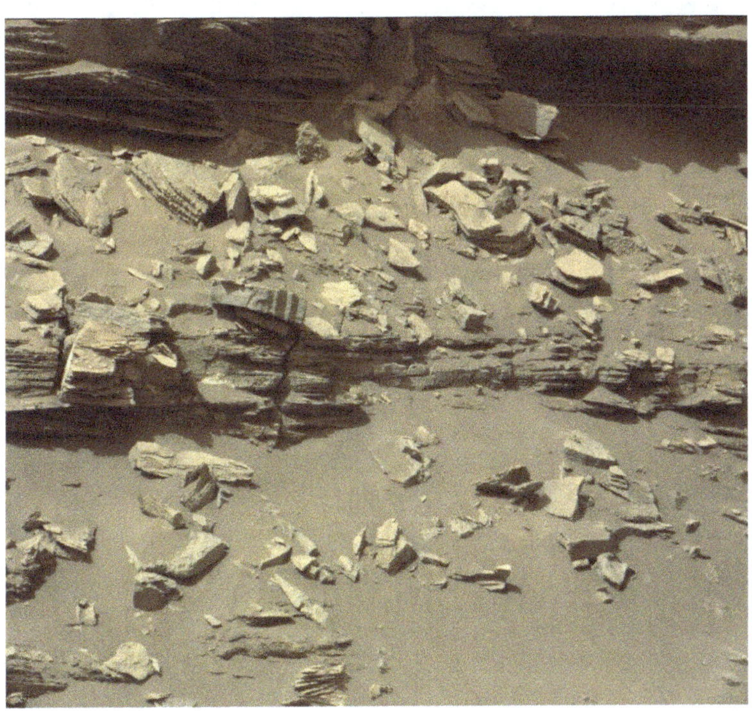

Original image. "Taken by Mast Camera (Mastcam) onboard NASA's Mars rover Curiosity on Sol 1454 (2016-09-08, 03:08:52 UTC)." Image credit: NASA/JPL-Caltech/MSSS.

ANALYSIS OF DEBRIS FIELD, SOL 1454-B

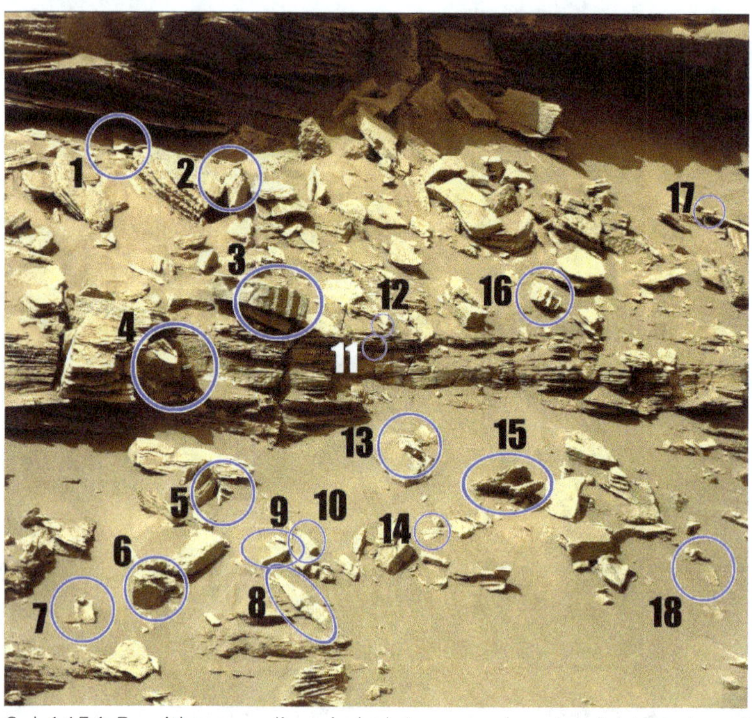

Sol 1454-B, with anomalies circled. Image colored, brightened, and sharpened by the author to improve contrast. Another large debris field with possibly hundreds of artificial looking objects strewn about. If true, most of them would not be visible because they are buried under layers of sand, dirt, and rock. As always I will be examining only those that I believe are most likely to have been made by an advanced intelligence.

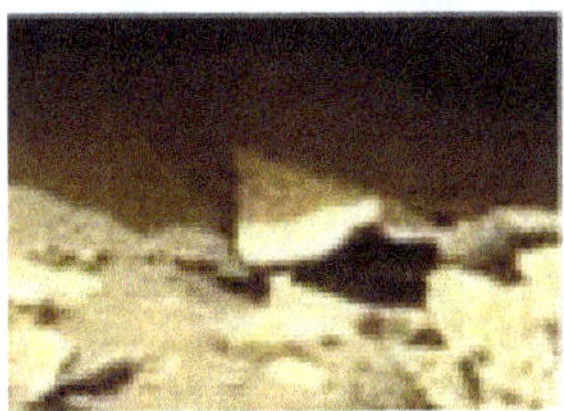

SOL 1454-B, highlighted area 1 (above): An arrowhead like object. It appears to be rumpled on the trailing end, suggesting that it may be made of metal or plaster, or more likely a material unknown on Earth.

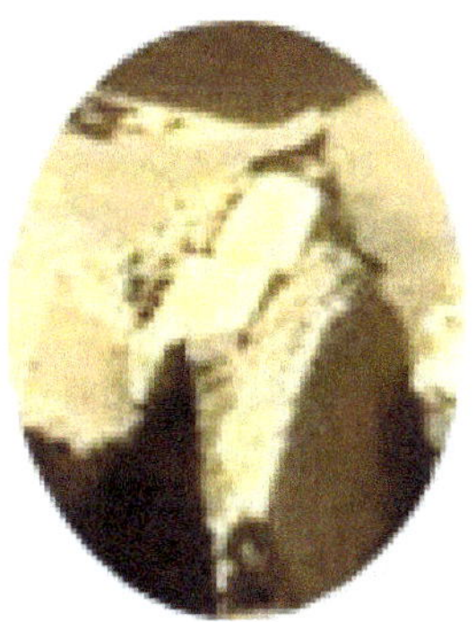

SOL 1454-B, highlighted area 2 (above): The brightly colored "rock" at center appears to have a sharply defined, perfectly cut "lightning bolt" like design along its upper side. Is it accidental or intentional?

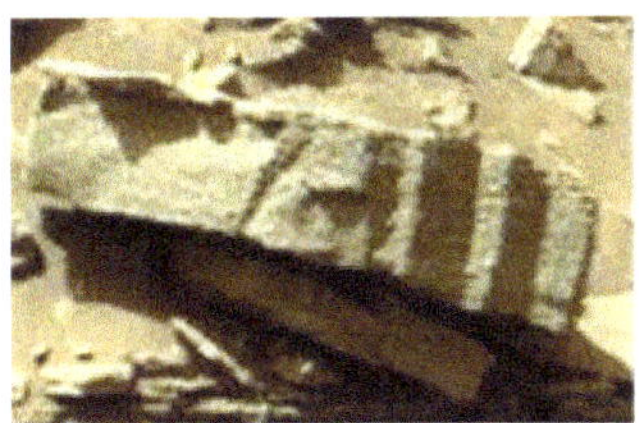

SOL 1454-B, highlighted area 3 (above): More jagged "lightning bolt" cuts. This object has the appearance of cement and may have once been part of a decorative wall. It is only a few feet below the similar object described in area 2. Are the two related?

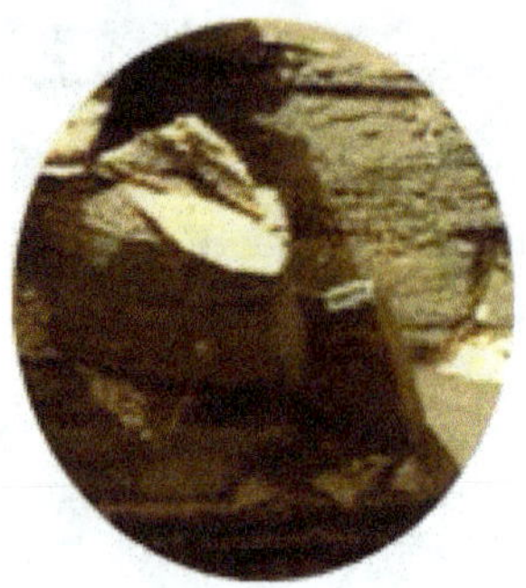

SOL 1454-B, highlighted area 4 (above): Two odd objects: At 11:00 there is a bright gravestone like or spade like object that does not match the color or shape of anything around it. Below it and to its right (4:00) a small square chimney like object can be seen. It is tall, narrow, and hollowed out. Strange detritus lies at its base. Other oddments abound. Magnify them for clarity.

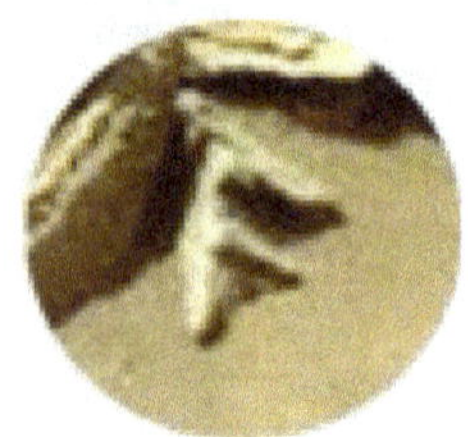

SOL 1454-B, highlighted area 5 (left): This artificial looking structure, possibly a piece that has broken off a Martian building, lies half buried in the sandy hillside.

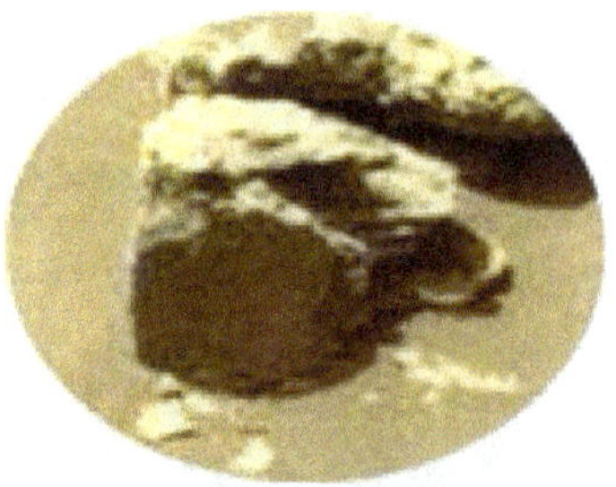

SOL 1454-B, highlighted area 6 (above): This looks like a rock, but with its straight square base and flat sides, it seems more likely that it is an artificial structure made from a conglomerate material. Notice the two basin like structures on its right side. The bottom one is rounded, the one above it is angled. Can geology be as creative as biology?

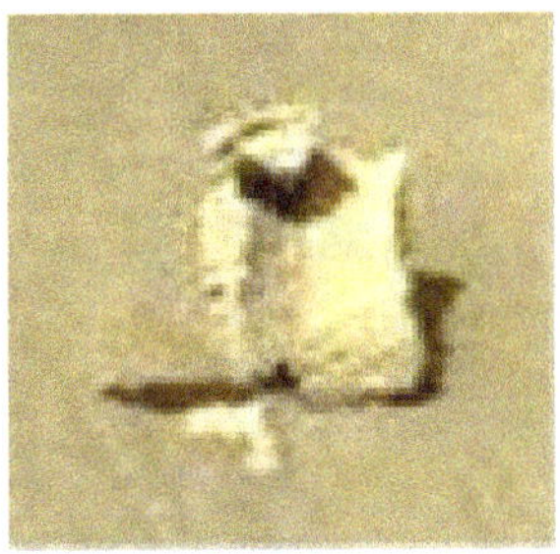

SOL 1454-B, highlighted area 7 (above): There is nothing natural looking about this strange "woven" mat like object. It even appears to have lettering or symbols along the left side.

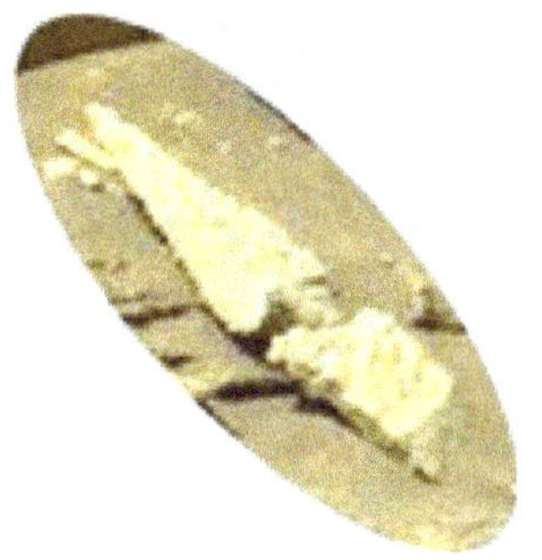

SOL 1454-B, highlighted area 8 (above): A shattered spear tip like object. If artificial, it may have originally been a decorative piece attached to a statue.

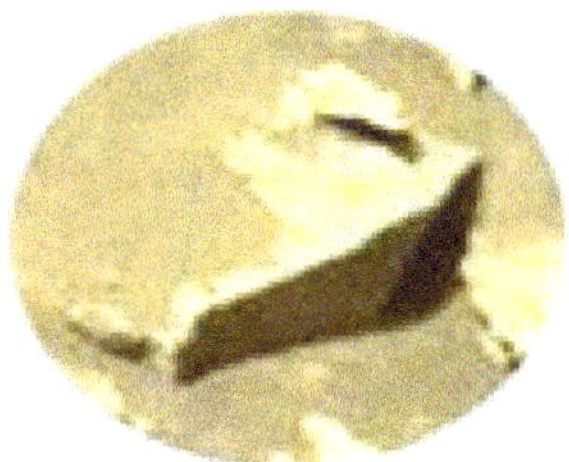

SOL 1454 B, highlighted area 9 (above): This small object looks a like a brick retaining wall of some kind, or perhaps it once served as the base of a statue.

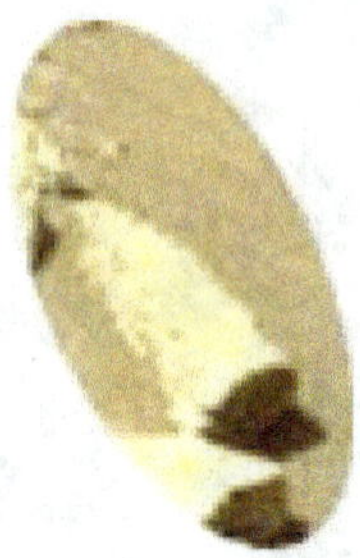

SOL 1454-B, highlighted area 10 (above): At center, a fossilized post, or perhaps a small cement pillar? Look at the square block below it. Logic says building materials.

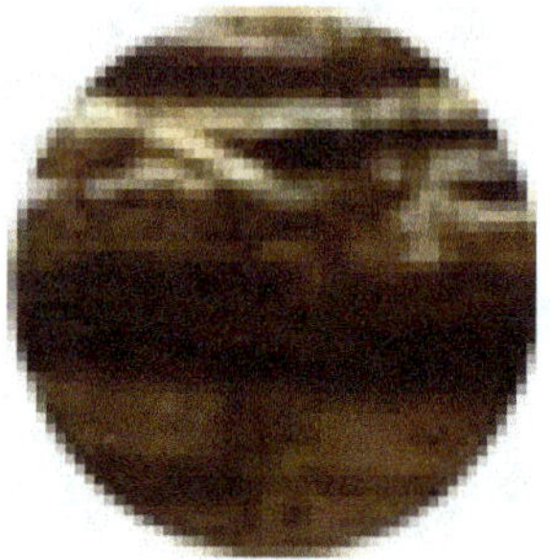

SOL 1454-B, highlighted area 11 (above): At the center of this image is what appears to be a clearly and beautifully engraved Roman numeral "2." Study it under magnification. Note that this is not the first "2" we have examined on the surface of Mars. Did ancient Martians attach special significance to this particular number?

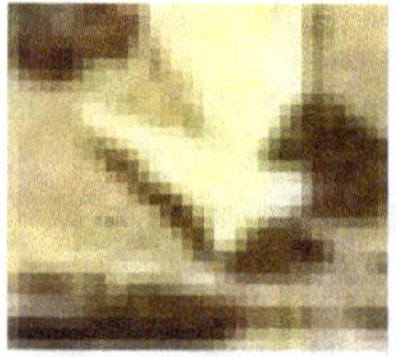

SOL 1454-B, highlighted area 12 (above): Another spear tip like object. We cannot know yet whether these are natural or artificial. It is possible that there are geologically formed ones mixed with others that were manufactured by a superior intelligence.

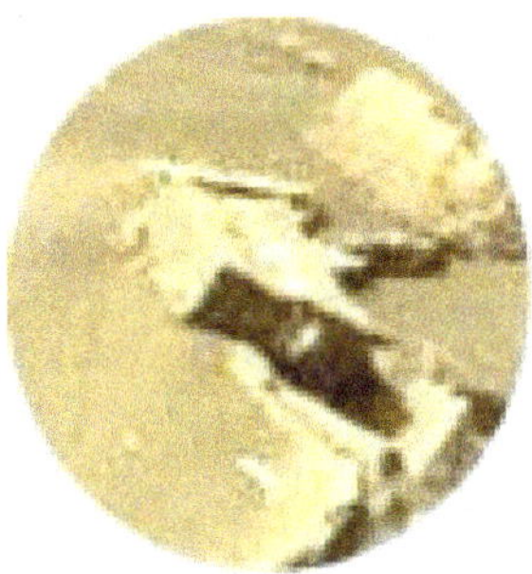

SOL 1454-B, highlighted area 13 (above): A square sedimentary rock, or an artificially layered object made from paper, plastic, metal, or an unknown material? Examine the oddly shaped debris lying around it.

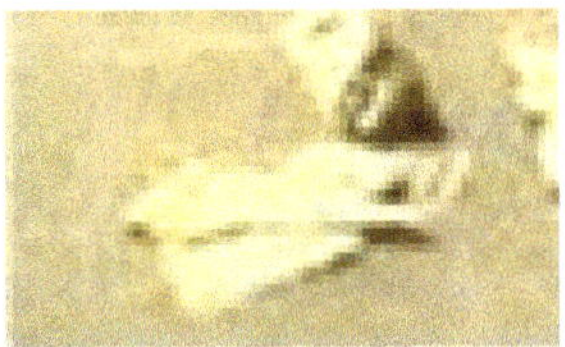

SOL 1454-B, highlighted area 14 (above): The delicate rectangular object in the center looks like a broken computer keyboard, or perhaps a damaged integrated circuit board. It is sitting on an unidentifiable flat square. Are they associated with one another? And what is the block like, porous object sitting in the sand behind them?

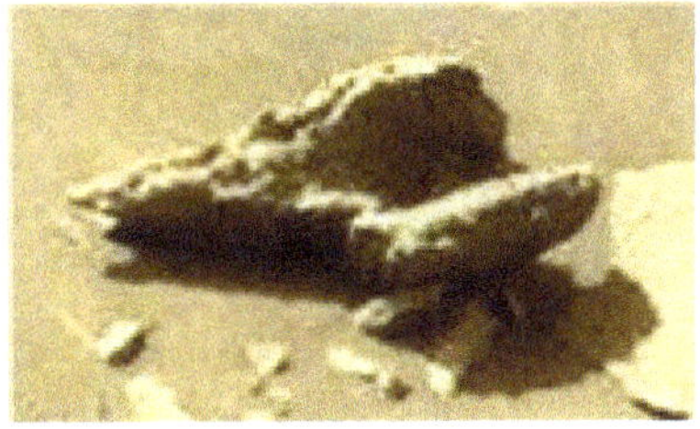

SOL 1454-B, highlighted area 15 (above): This damaged roundish piece of cement like material looks like it might have been a support for a piping system of some kind.

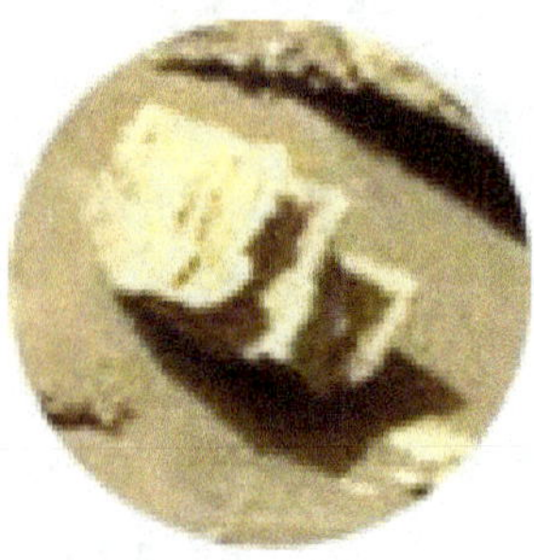

SOL 1454-B, highlighted area 16 (above): This odd damaged object is made up of flat, tray like squares. What is it?

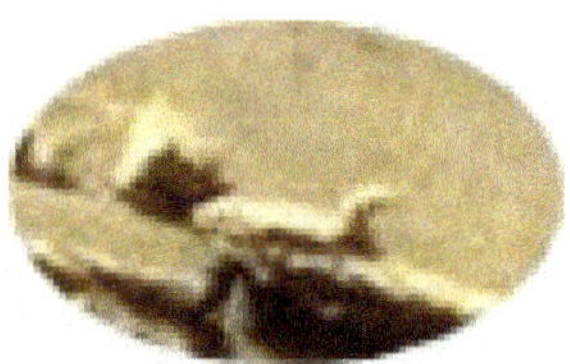

SOL 1454-B, highlighted area 17 (above): The flat object at lower center has a curved handle like appendage on its right end. To its left there is a small pyramid like object with a square base, facing uphill.

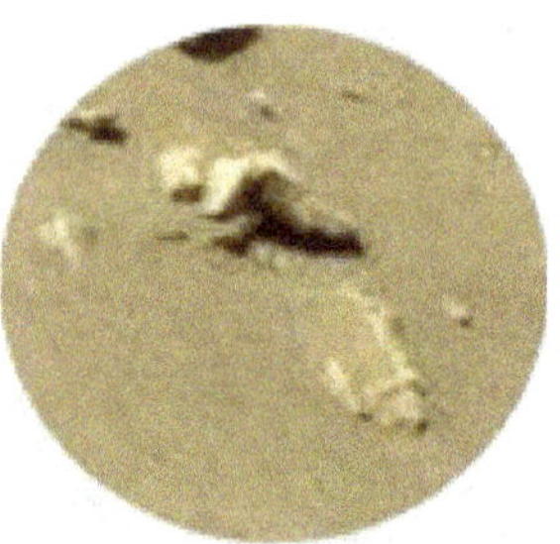

SOL 1454-B, highlighted area 18 (above): Peculiar shapes are grouped together in this little section of Martian hillside. At around 10:00 a picture frame like item barely pokes out of the sand. At about 4:30 there is a strange, rectangular, box like object with 90 degree angles. At upper center there is a pile of unknown objects that looks artificial, and thus out of place.

Chapter 28

ODDITIES

SOL 1454-C: EARTHDATE SEPTEMBER 8, 2016

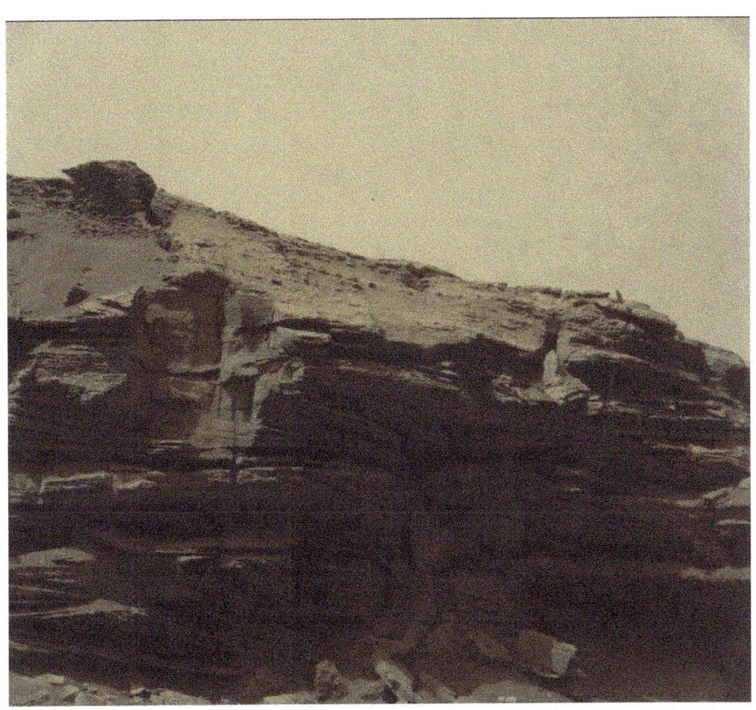

Original image. "Taken by Mast Camera (Mastcam) onboard NASA's Mars rover Curiosity on Sol 1454 (2016-09-08, 03:02:37 UTC)." Image credit: NASA/JPL-Caltech/MSSS.

ANALYSIS OF ODDITIES, SOL 1454-C

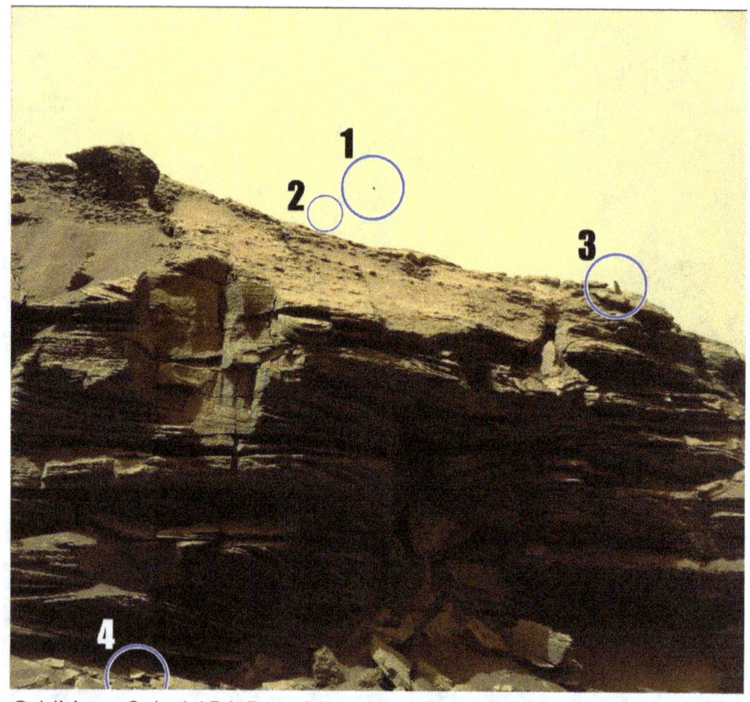

Oddities, Sol 1454-B, with anomalies circled. Image colored, brightened, and sharpened by the author to improve contrast. This does not seem to be the site of an explosion or devastation, natural or artificial. Nonetheless, a few singular items show up.

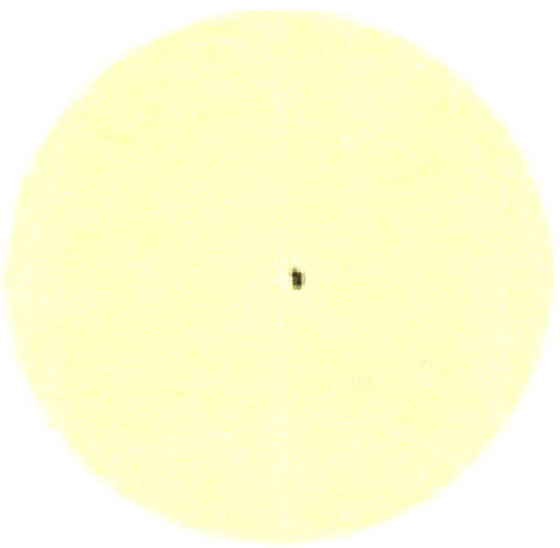

SOL 1454-C, highlighted area 1 (above): According to conventional scientists this is a digital artifact or a "dead pixel." Since it, or something similar to it, repeatedly appears in many of NASA's Mars' photos, this is probably correct. It only appears once in the full frame image of Sol 1454-C (above), however—possibly because the horizon here is not wide enough for it to repeat (as we see in some of NASA's large sweeping mosaic photos). In some cases could it possibly be what it actually looks like: a can like, alien object (with an appendage on the upper left side) flying over the crest of a large Martian hill? See closeup below.

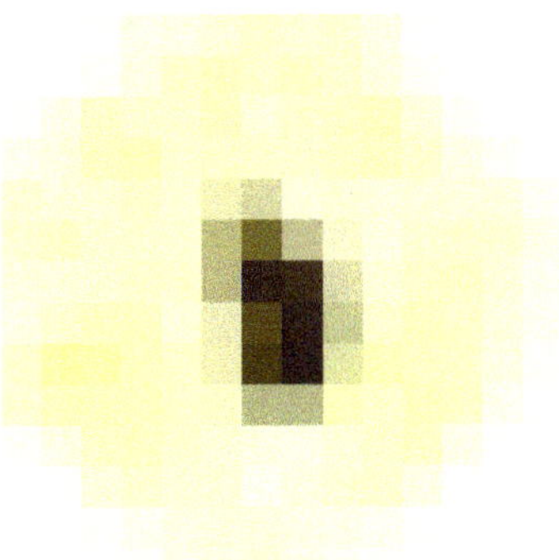

SOL 1454-C, closeup of the object described in area 1 above. Is this a "camera glitch," a "piece of dirt" on Curiosity's lens, an extraterrestrial spacecraft, or perhaps something else entirely?

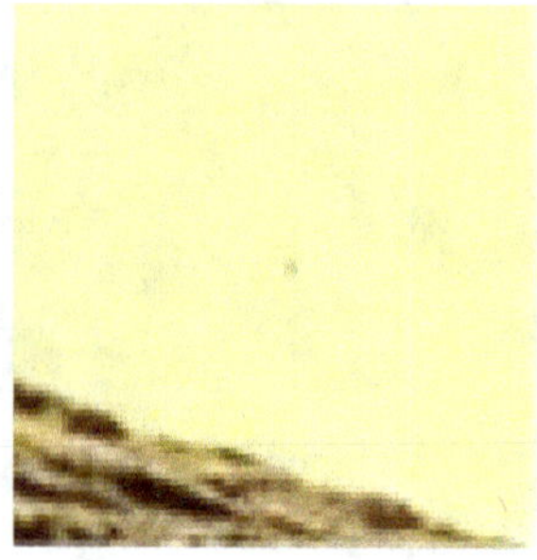

SOL 1454-C, highlighted area 2 (above): In the same photo Curiosity also captured this "dot" in the Martian sky. It is located to the lower left of the object described in area 1. See closeup below.

SOL 1454-C, closeup of the object described in area 2 above. It may be similar or identical to the aerial object described in area 1, but it is further from the camera, indicating that it is almost certainly something separate and unique. The little gray "smudges" around it could be actual digital artifacts; or, if it is a UFO, heat, electrical, or magnetic waves. Beyond pure speculation, nothing can be positively known about the object at this point. All we can do is add it to our ever growing list of Martian anomalies.

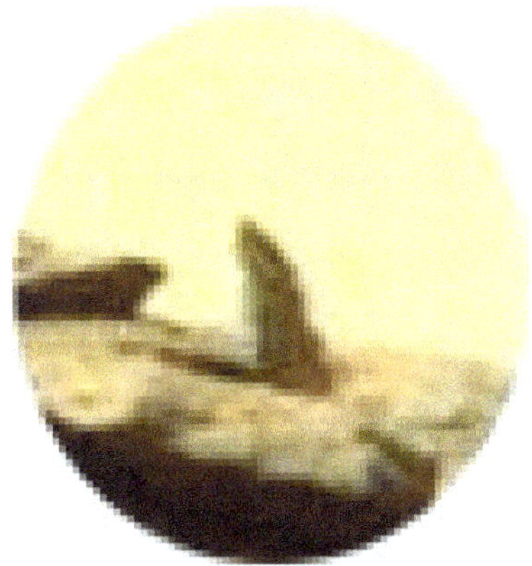

SOL 1454-C, highlighted area 3 (above): This is probably "nothing more than a rock" sticking up out of the Martian surface. Since the gravity on Mars is much less than the gravity on Earth, odd delicate structures like this one are more likely to form on the Red Planet. Thus it could be easily explained as a geological formation. However, authentic science demands that we sustain final judgement until it is thoroughly examined in situ.

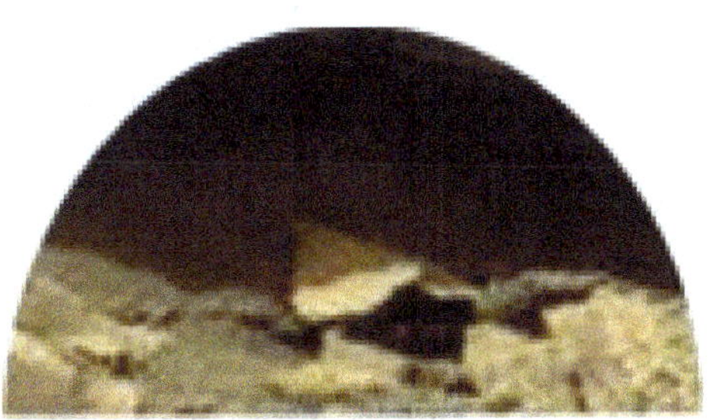

SOL 1454-C, highlighted area 4 (above): There are countless arrowhead like objects, such as this one, spread out across the surface of Mars. I acknowledge that many of them are, without doubt, natural rock formations. At the same time, I also acknowledge that some of them may be artificial, created by the hand of an advanced race of unknown, unrecognized beings.

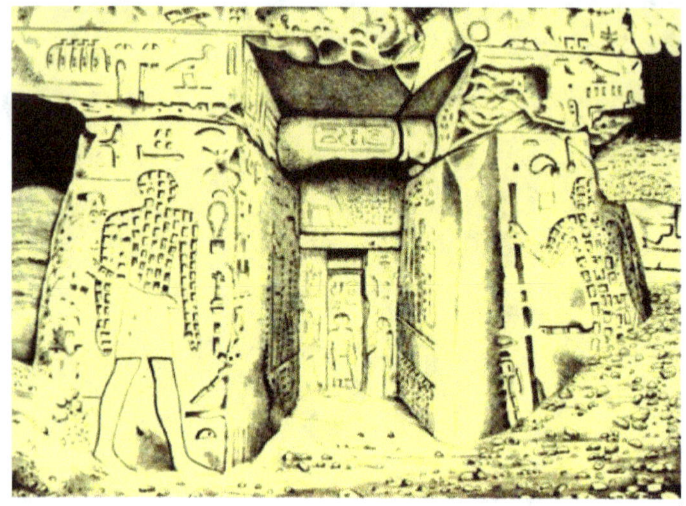

Above: This tomb of "Nofre Ma, an Egyptian of the 3rd Dynasty" (original 19th-Century caption) bears remarkable similarities to some of what I am calling "The Martian Anomalies." Are the two civilizations connected? Some believe they are—and that there is abundant evidence to prove it.

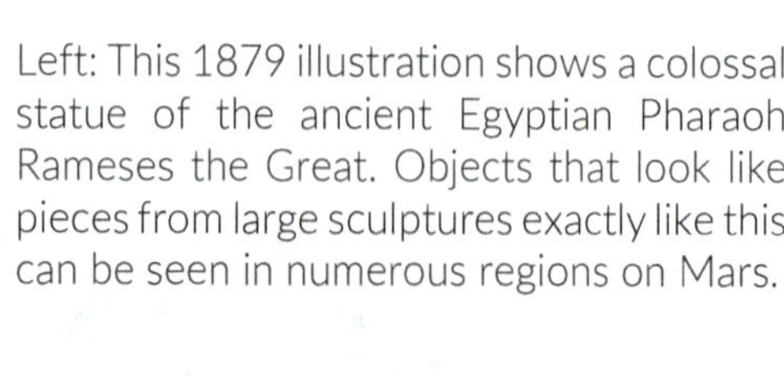

Left: This 1879 illustration shows a colossal statue of the ancient Egyptian Pharaoh Rameses the Great. Objects that look like pieces from large sculptures exactly like this can be seen in numerous regions on Mars.

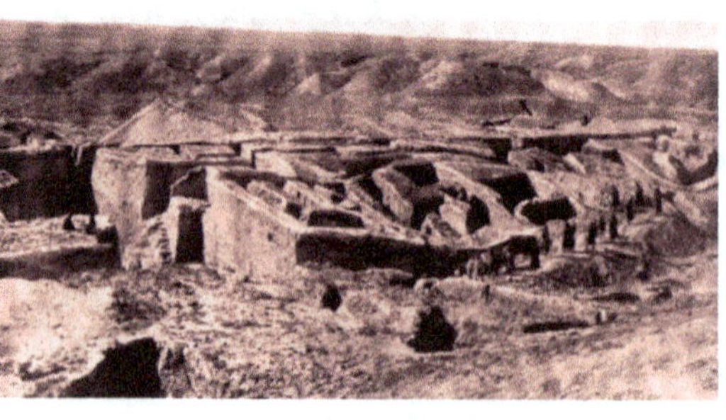

Above right: Aerial images taken from above Mars reveal countless "cityscapes" almost identical in appearance to the ruins of this 4,500 year old Babylonian town. What conclusions can be drawn from this observation?

Chapter 29

ODDS & ENDS

PIA16453: EARTHDATES OCTOBER & NOVEMBER 2012

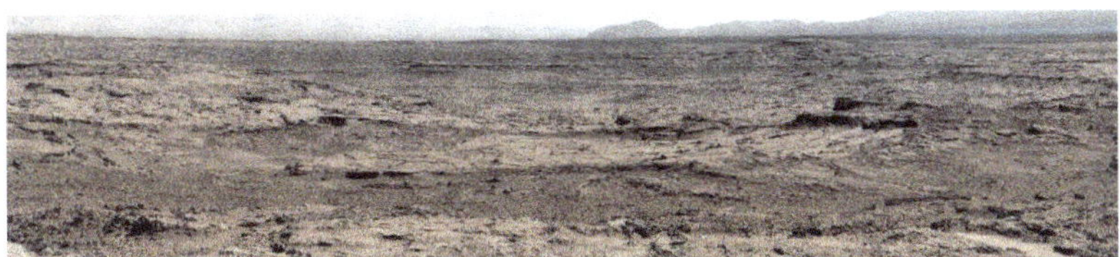

Original image. "This panorama is a mosaic of images taken by the Mast Camera (Mastcam) on the NASA Mars rover Curiosity while the rover was working at a site called 'Rocknest' in October and November 2012." Image credit: NASA/JPL-Caltech/Malin Space Science Systems.

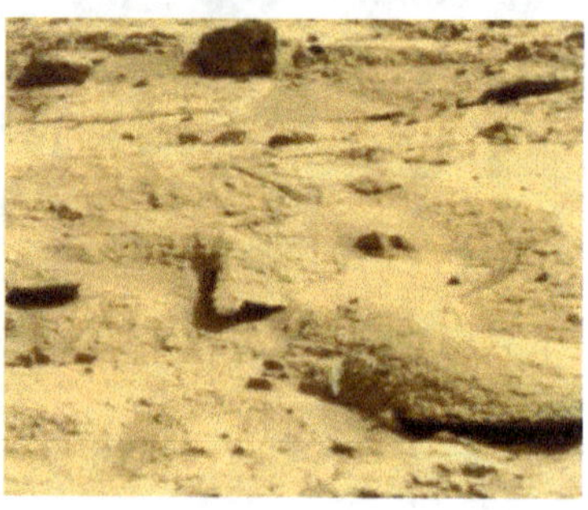

PIA16453-1 (above), colored, brightened, and sharpened by the author for contrast (as were all of the images in this chapter). There are several triangular objects in this image, and also an outline of a triangle in the sand (near center). To the lower right of the arrowhead like object a small, rounded, horseshoe shaped object sits on a rock. Under magnification numerous other anomalies are visible.

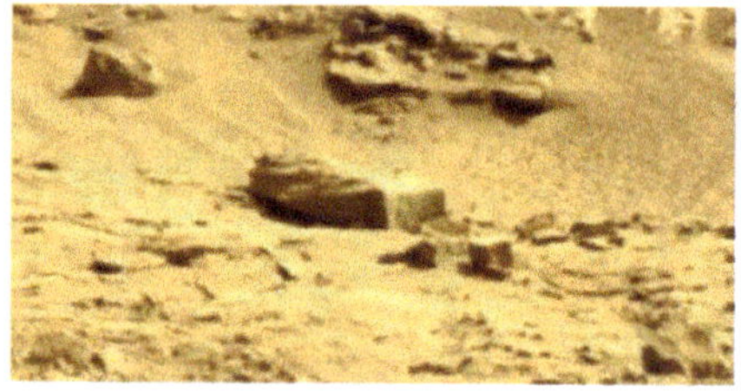

PIA16453-2 (above). A group of blocks and squarish shapes. The large rectangular object at center looks like the damaged base of a statue or monument. A myriad of other bizarre items can be seen strewn about in all directions.

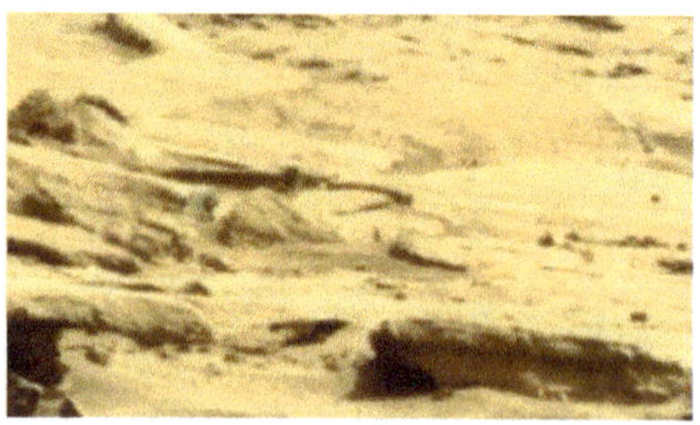

PIA16453-3 (above). At center, odd swirling "designs" in the sand. Also visible are a number of roundish and squarish shapes, as well as a "carpet" like object at around the 8:00 position.

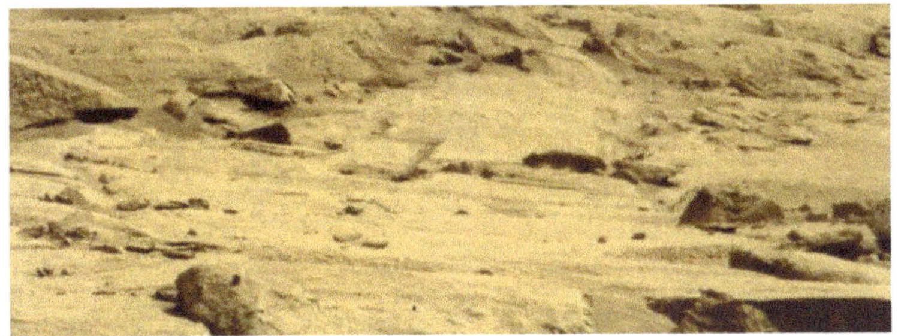

PIA16453-4 (above). Numerous artificial looking shapes and objects fill this image. The longer one examines it under magnification the more one will see, from triangles and round items to squares and rectangles. What appear to be bricks can be seen in the cleared space below the partial square that has been "etched" or "dug" into the dirt (lower left of center).

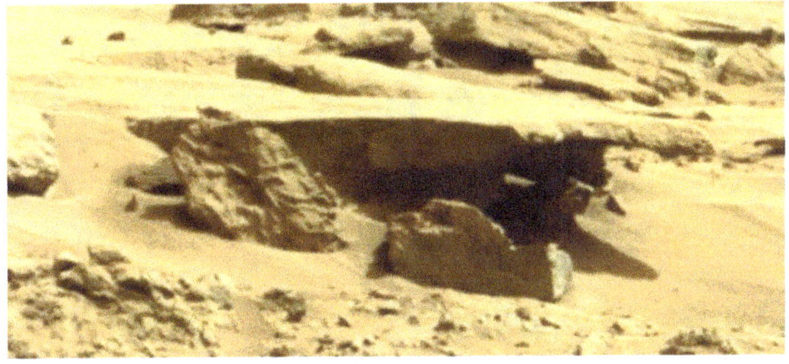

PIA16453-5 (above). Very little of this debris appears natural, and some of it, like the large broken "cornerstone" (right lower center), seems overtly unnatural. Unfortunately for truth seekers, the Martian landscape seems to encompass both geology and artificiality, which conventional science appears to use to its advantage to both confuse the public and conceal what could be photographic evidence of ancient, and possibly even current, intelligent life on Mars. Note the markings, including a "cross," on the large rock left of center.

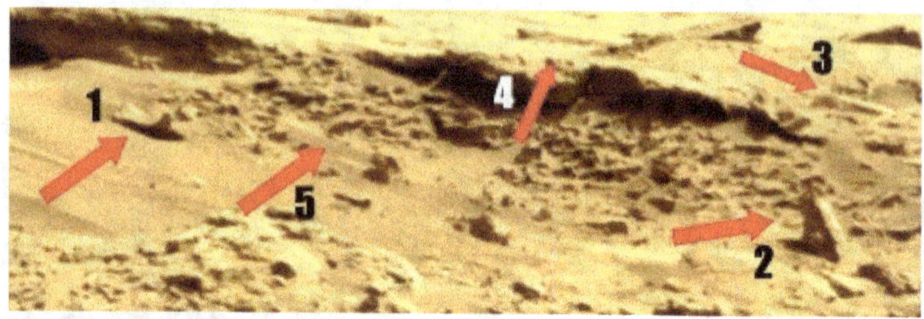

PIA16453-6 (above). 1) A broken piece of pottery? 2) An odd object leaning on a "stand"? 3) A bright white 90 degree angle and a horseshoe shaped object sit next to one another. 4) Across nearly the entire top of this image a number of bright, flat, rectangular, paving stone like objects lay side by side in the sand. 5) Peculiar bits of debris are grouped together along and below the edge of what may have once been a house floor.

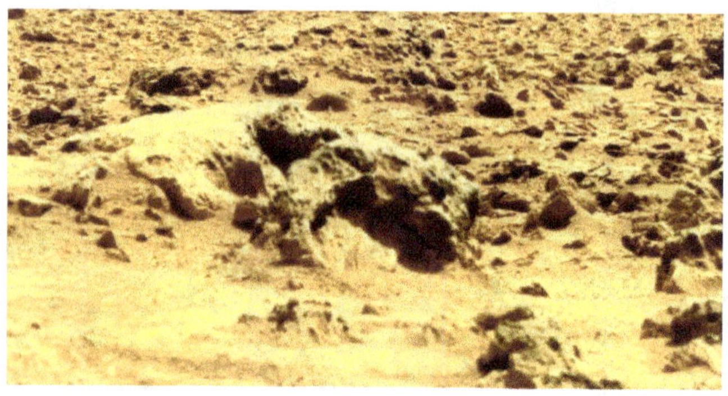

PIA16453-7 (above). The scoriated like "rock" at center has a large round hole "cut" out of it. Natural or artificial? Study the objects surrounding it.

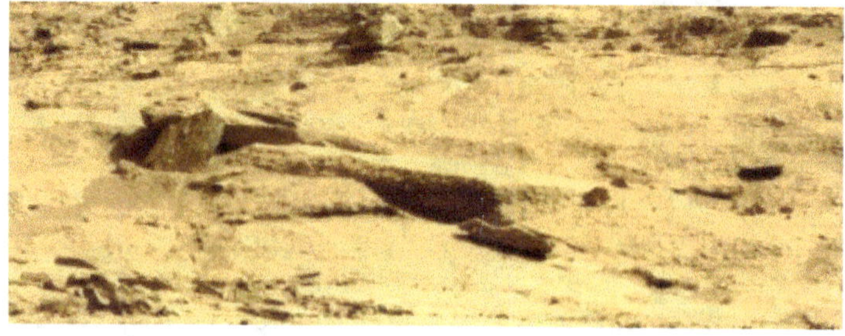

PIA16453-8 (above). A large, square, "cement" like piece sits at center. Off its lower front corner there is a serpentine like object. "Bricks," squares, triangles, 90 degree angles, and "boards" fill out the area.

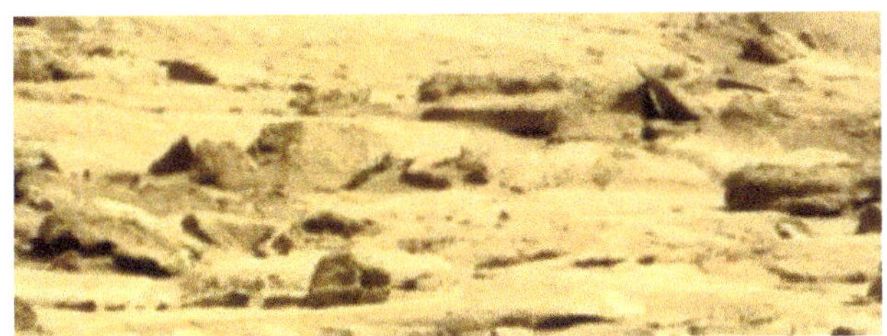

PIA16453-9 (above). Image inventory: Rectangular cement like blocks, several small pyramids, squares, triangles, straight lines, strange cavities "carved" into rocks, "boards," square railroad tie like objects, oddly "cut" rocks, trenches, a large square impression in the ground, rocks with seemingly intentionally aligned holes in them, and a host of other peculiar and unnatural looking objects. Science cannot yet prove these are natural.

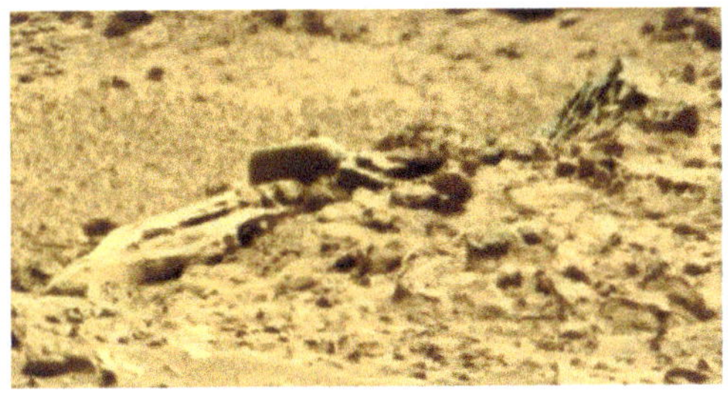

PIA16453-10 (above). Brightly colored objects mix with gun metal gray objects here, all possessing odd shapes. In particular, magnify the large, bizarre gray item at 1:00. The site of exploded ancient machinery?

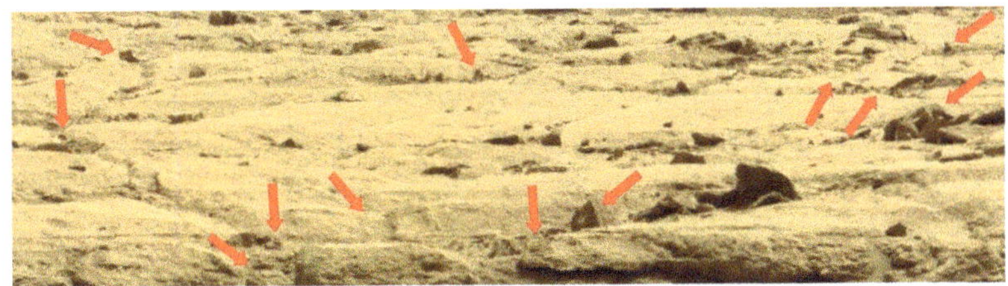

PIA16453-11 (above). This large area with "marked off" square patches (running across the entire upper center) looks like it could be the former location of a house or building. In addition, under magnification dozens of strangely shaped objects, which I have red arrowed, can be seen surrounding it. This image is worthy of detailed scrutiny.

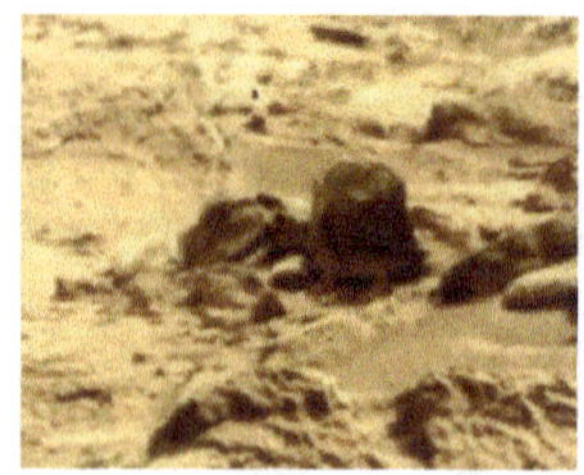

PIA16453-12 (left). This interesting little patch of Martian ground, rife with artificial looking items, appears to have been digitally airbrushed with fake sand (upper center and lower right). If "it is all natural rock," why the coverup?

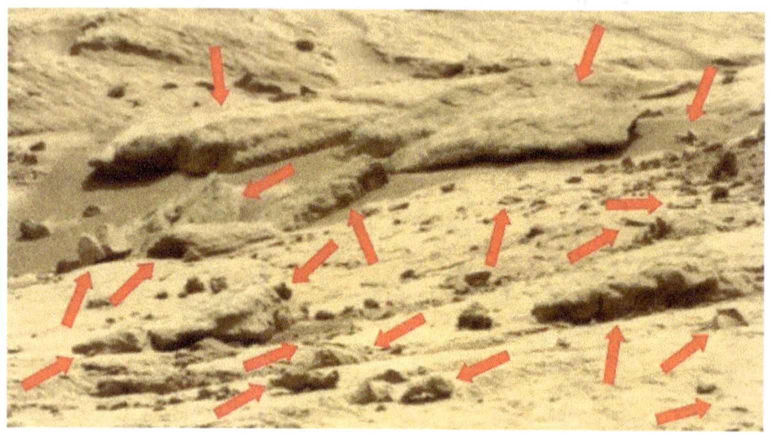

PIA16453-13 (above). This entire area appears to be the remains of a toppled temple and perhaps some of its demolished statuary. I have red arrowed some of the more obvious signs of what could be artificiality. Among the items visible under magnification are what look like a square "pillar," a round "column," triangles, "trays," "claws" or "fingers," the "Star Trek insignia," gravestone shaped objects, and a sarcophagus like object with "writing" on it half buried in the sand. Note the perfectly round holes "punched" into the Martian soil in the lower right corner. Upon close inspection the perceptive reader will find many other anomalies.

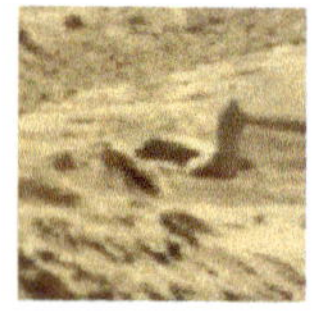

PIA16453-14 (right). A group of strange objects congregate on a Martian hillside. Though they appear artificial, at this time it cannot be irrefutably proven. But neither can establishment scientists irrefutably prove they are natural.

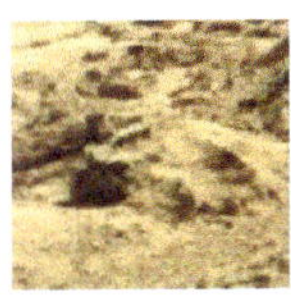

PIA16453-15 (left). At center left a rounded, "brick," well like object, surrounded by a host of other unusual shaped items.

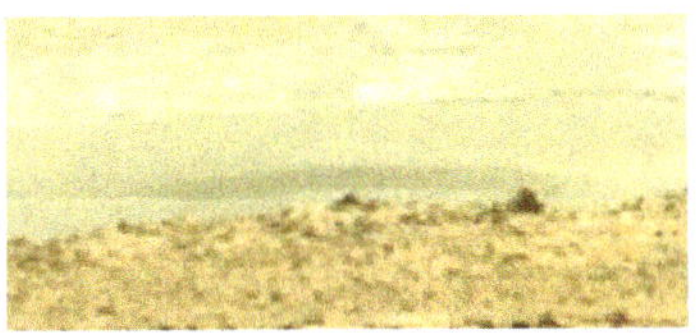

PIA16453-16 (above). Pareidolia or pyramids on the middle horizon?

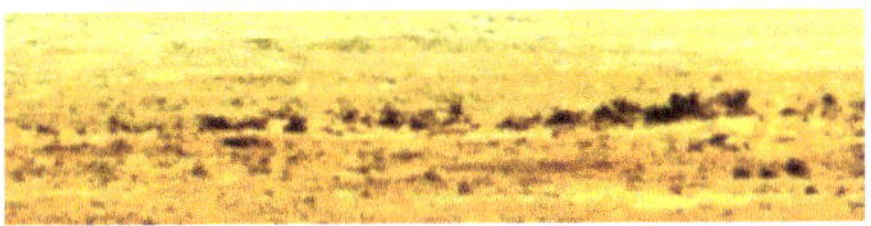

PIA16453-17 (above). A collapsed structure or natural rock formations?

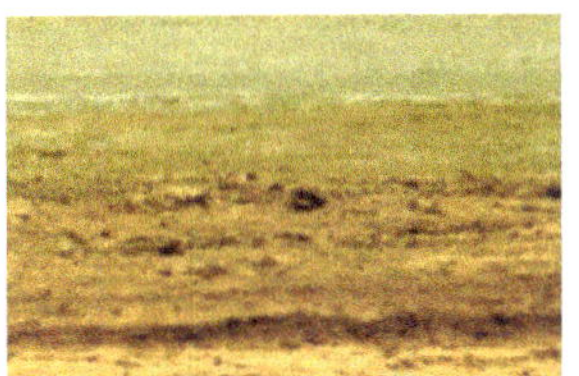

PIA16453-18 (above). What is the roundish, dark, hollowed out object in the center of this image? It has similarities to some of the hollow objects shown in PIA16453-17 above.

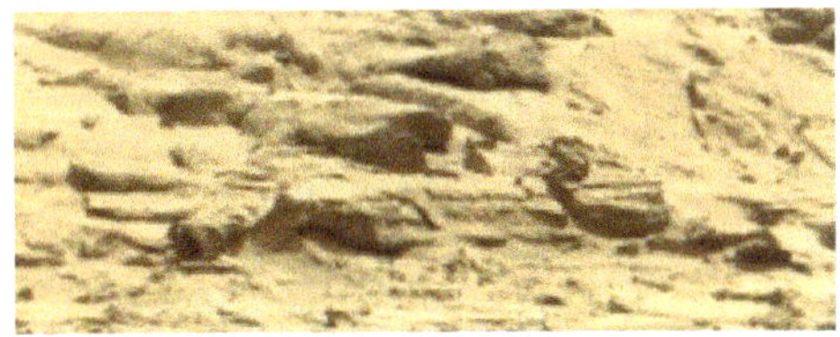

PIA16453-19 (above). Peculiar objects grouped together. Though there are undoubtedly natural rocks involved, many of these objects display characteristics of artificiality. Study this entire image carefully.

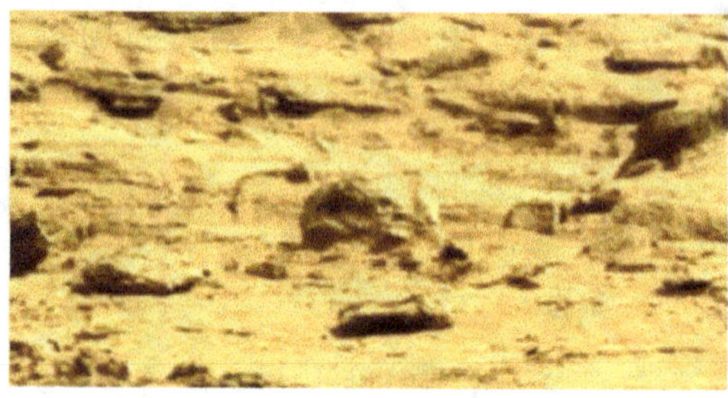

PIA16453-20 (above). More strange shapes that look out of place, including a skull like object (center), and a partially crushed box like object (to the right of the "skull"). A myriad of other curious objects lie about.

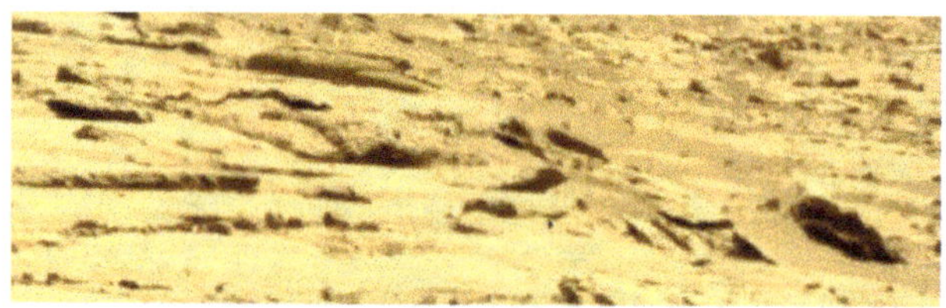

PIA16453-21 (above). To call the objects in this image bizarre would be an understatement. Note the writhing "serpent" at 10:00; a "U" shaped "magnet" at about 5:00; a "collapsed" object at 4:00; an upside down "Y" looking object just right of center; and numerous flat square objects. Were any of these made by a technological race?

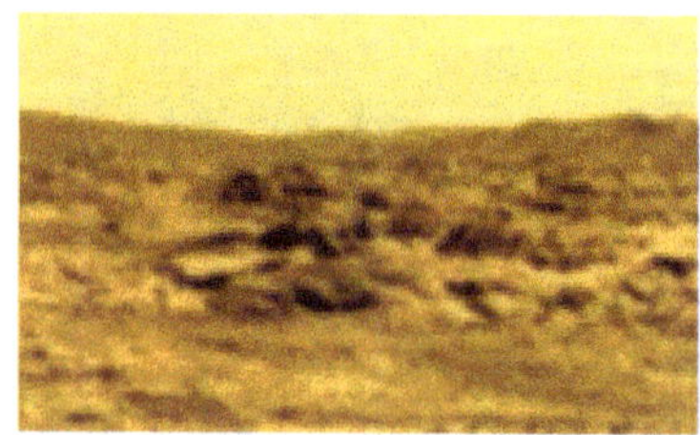

PIA16453-22 (above). Unusual shapes grouped together on a distant Martian slope. Rock rubble or artificial debris?

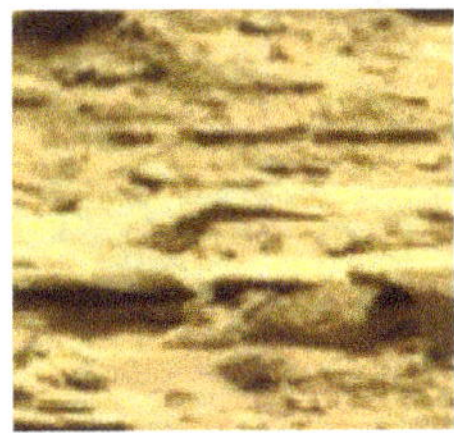

PIA16453-23 (above). Center, a strange delicate finger like object juts out above the Martian surface.

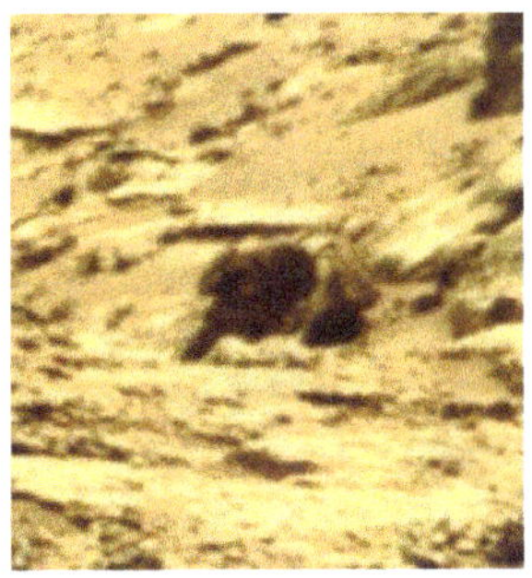

PIA16453-24 (above). This wheel like object is located on a hillside, half buried in the yellow red Martian soil. It looks like a piece of industrial machinery.

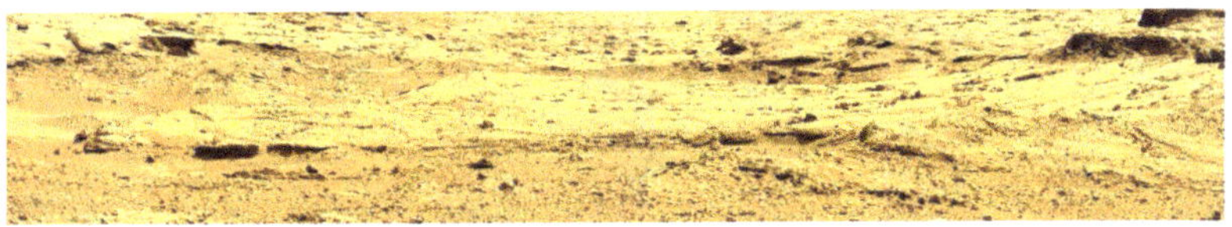

PIA16453-25 (above). Possible ancient remnants of another square "house foundation." Magnification reveals what appear to be "bricks" and other building materials in the surrounding area.

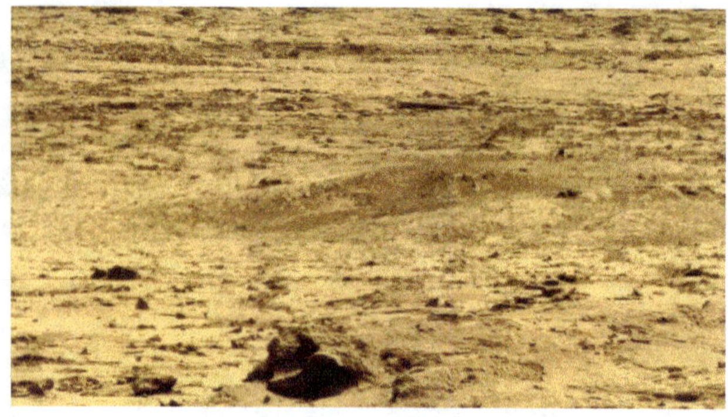

PIA16453-26 (above). Center, a rectangular cleared area "scraped" into the ground; bottom left of center, the "head" of a wrench like object, with the "profile" clearly visible, points skyward; at 4:30 a winding, car muffler like object sits on top of the Martian soil. In the upper distance, around the 12:30 position, a railroad tie like object lies half buried in the sand. As usual, a wide assortment of squares, triangles, rectangles, and circles can be seen scattered about. A million year old artifact debris field?

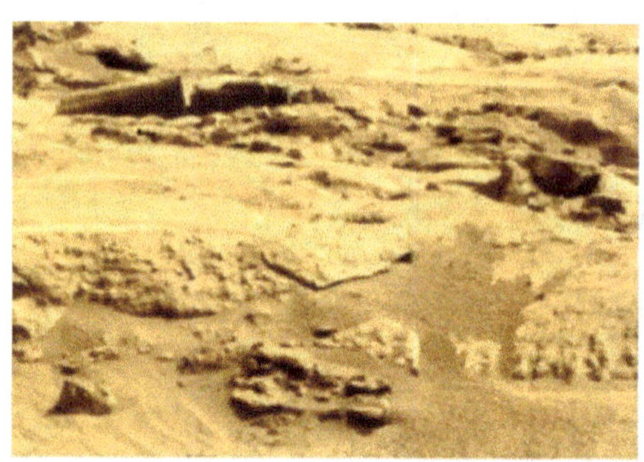

PIA16453-27 (above). Numerous artificial looking objects can be seen in this NASA image, including flat squarish "sheets," blocks, rectangles, half circles, triangles, "U" shapes, and arrowhead shapes—among others. At around the 2:30 mark there is a mysterious block like object with a dark square on its right end. Note the circular looking "column" laying on its side in the upper left. Magnification will bring many other curious items in this photo to light.

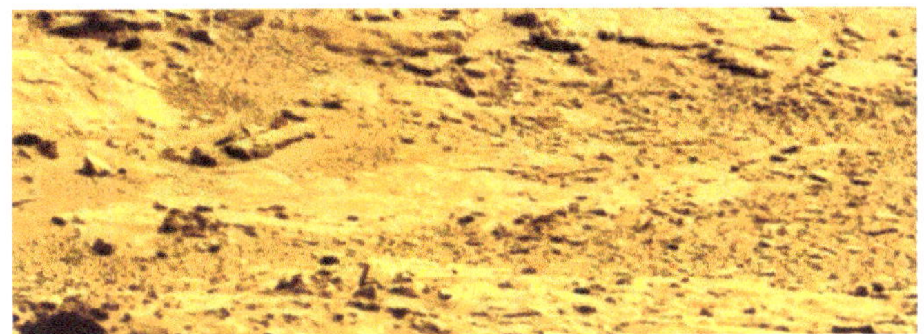

PIA16453-28 (above). Another rectangular clearing surrounded by unusual "rock" debris on the Martian surface.

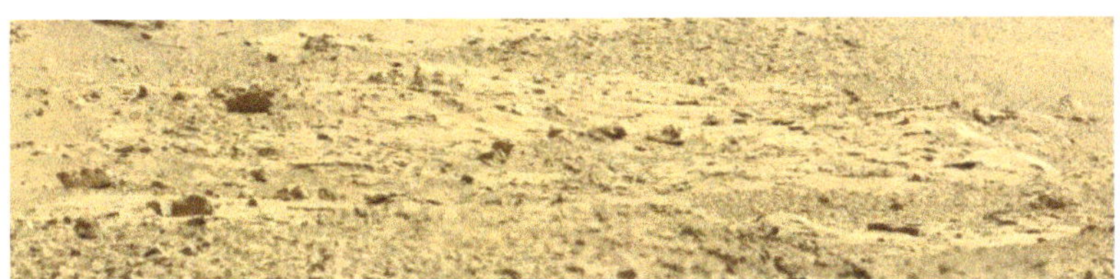

PIA16453-29 (above). What appears to be a "disrupted" area with peculiar "rock" debris in and around it. Bear in mind that this site could be hundreds of thousands of years old; i.e., time, dust, and weathering have concealed most of whatever artificial structures might have been here originally. Seen under magnification the surface detritus includes flexible "carpet" like material, squares, "V"s, triangles, rectangles, circles, horseshoe shapes, "Y"s, stick like objects, crescents, "bricks," and arrowhead like items, to name but a few.

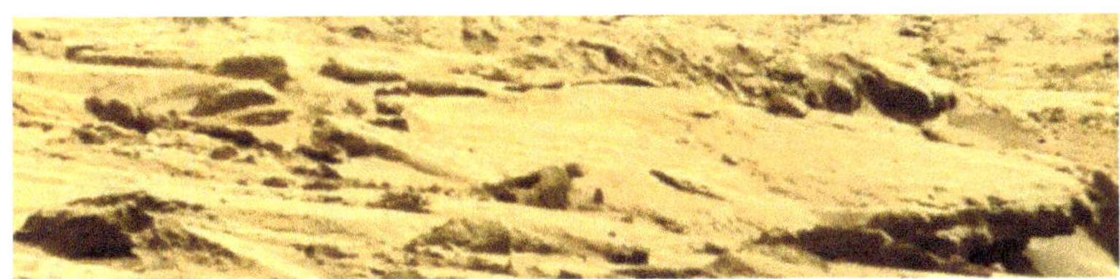

PIA16453-30 (above). This cleared square is bordered on several sides by rocks or some unknown material, and bears serpentine like "designs." In the lower left corner is an object with a square opening. At 10:00 there is a half circle on top of a rectangular "base." At 11:00 there is tray like object. "U"s, crescents, circles, triangles, and rectangles are also present.

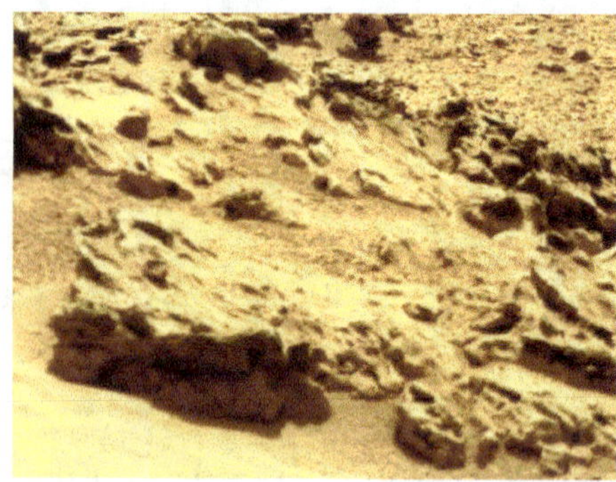

PIA16453-31 (above). When magnified this area is pockmarked with dozens of "V"s, "Y"s, squares, circles, and numerous strange shapes. Toward the lower left a tiny "broken" square box like object sits on top of a large "rock." The "rock" itself has a number of perfect triangles "impressed" into it. Natural or artificial?

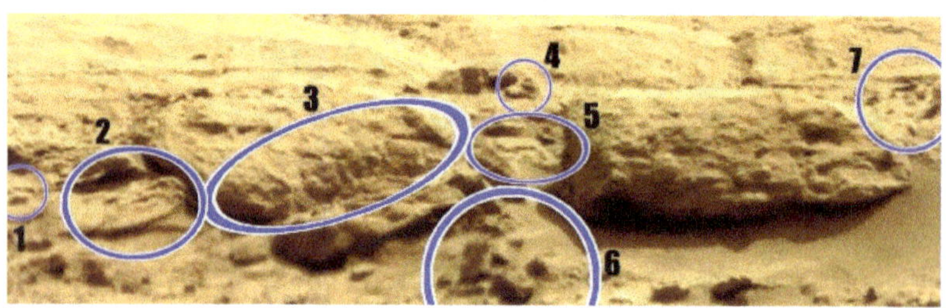

PIA16453-32 (above). Artificiality seems to abound in this small area. 1) A possible hieroglyph of some kind; 2) Remnants of an unknown object; 3) More potential lettering, including a thickly "carved" numeral 3 in relief; 4) A perfect "S" in relief; 5) The letter "C" and an upside down "L" shape on top of a damaged squarish shape; 6) This bizarre object has two triangle "eyes" and a square "mouth"—what is it? 7) More unidentifiable letters or items.

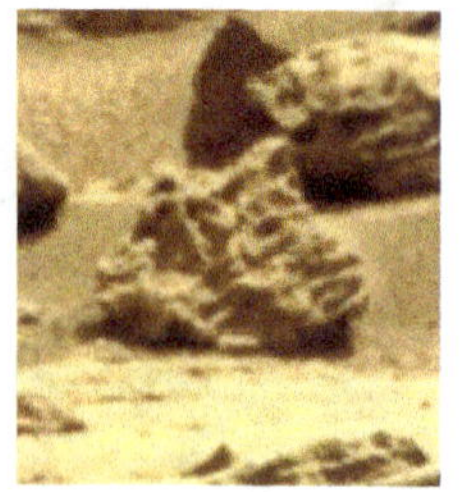

PIA16453-33 (right). Let us call this weird honey combed object what NASA calls it: a "rock." Now all that is needed is verifiable proof rather than opinion based on bad science.

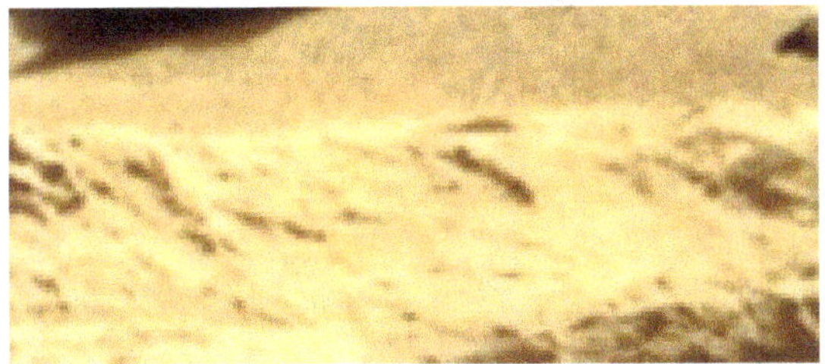

PIA16453-34 (above). There are at least three "left pointing arrows" in relief on the surface of this rock. How did they get there? Mindless geology or creative intelligence?

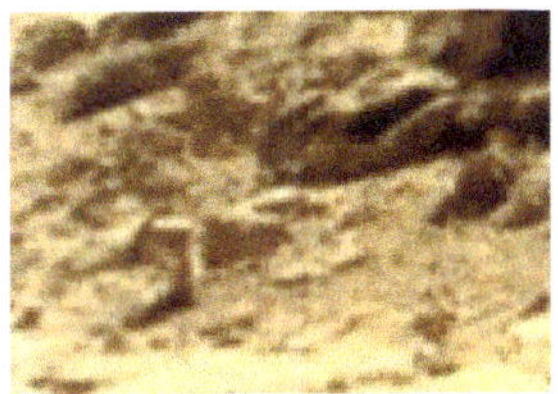

PIA16453-35 (left). A clam like object, a horseshoe like object, and a periscope like object, along with numerous rectangles and odd shapes, are grouped together here. Will we ever know what they really are?

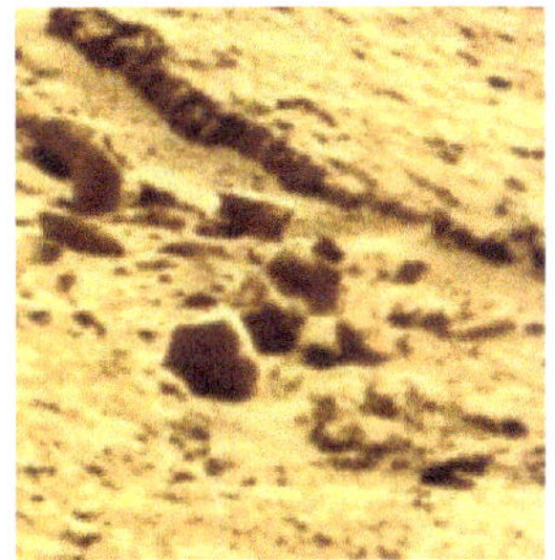

PIA16453-36 (right). This extraordinary collection of strangely shaped objects sits at the bottom of a Martian hill. Stone rubble or building debris?

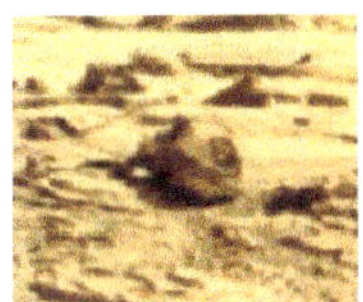

PIA16453-37 (left): Under magnification the "rock" at center has a square structure and a shield like shape "emblazoned" on its forward facing side. It sits amidst a wondrous assemblage of bizarrely shaped objects and "designs" running across the Martian surface.

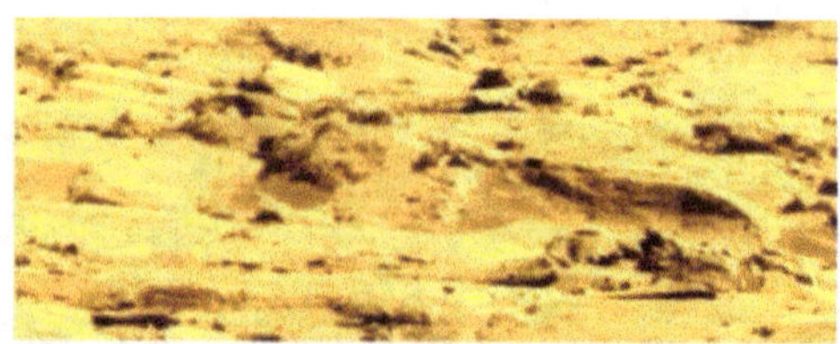

PIA16453-38 (above). Along with numerous queerly shaped objects, a myriad of pyramidal shapes dot this Martian landscape. The smooth, curving "rock" at 3:00 is interesting. Magnify everything.

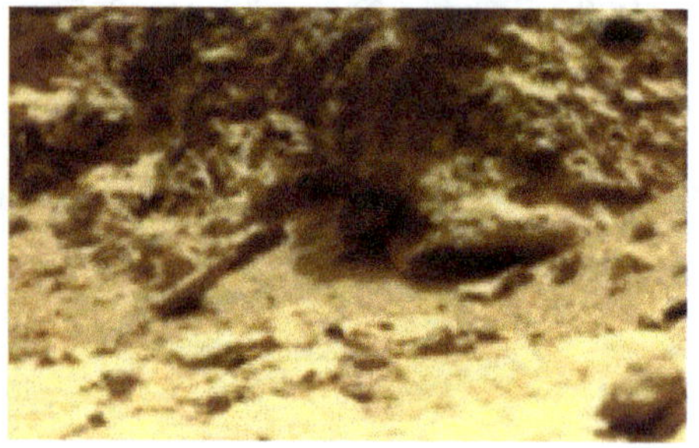

PIA16453-39 (above). Study this incredible photo closely. I am sure mainstream astronomers have a scientific explanation for what is going on here. But the truth is that they cannot be sure—and the guileless ones will admit it.

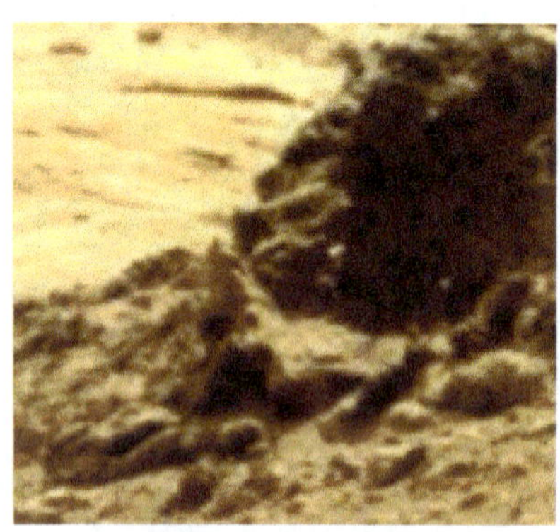

PIA16453-40 (left). Under magnification this pile of suspiciously artificial looking objects is what one imagines broken pieces of machinery might look like after being subjected to thousands of years of Mars' harsh climate.

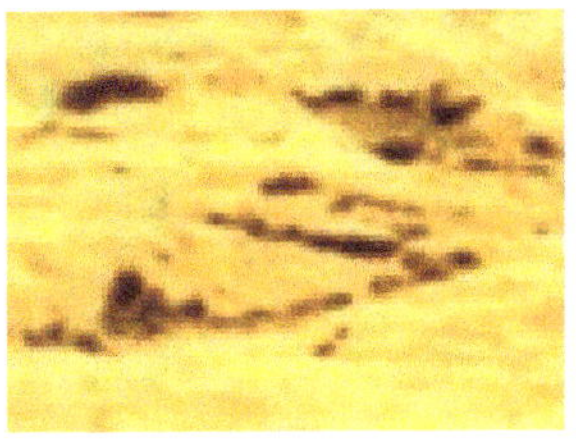

PIA16453-41 (right). Bizarre shapes, circles, and squarish "designs" on a Martian foothill.

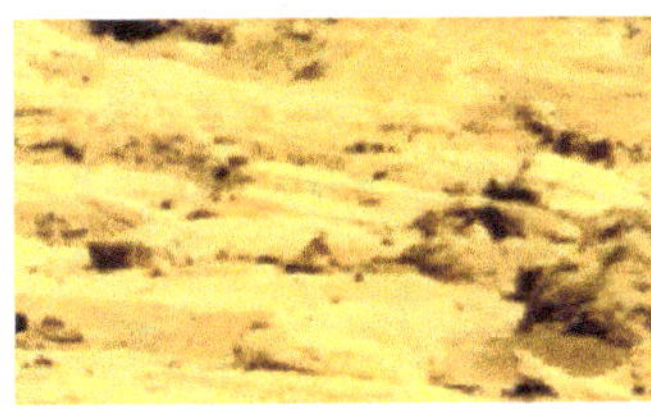

PIA16453-42 (left). Martian soil that appears to have been intelligently landscaped.

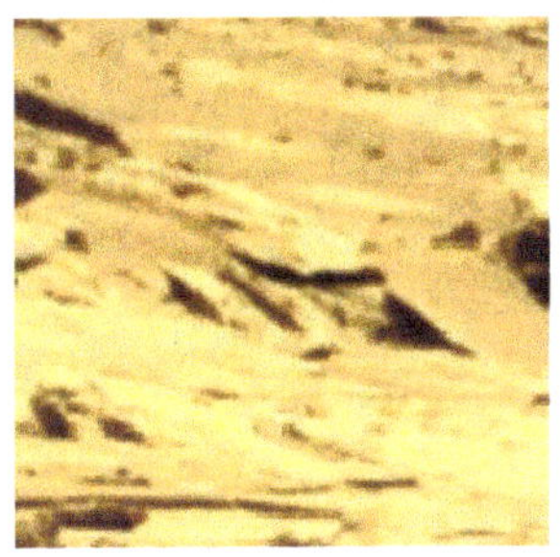

PIA16453-43 (right). More "landscaped" terrain with numerous curious objects lying about.

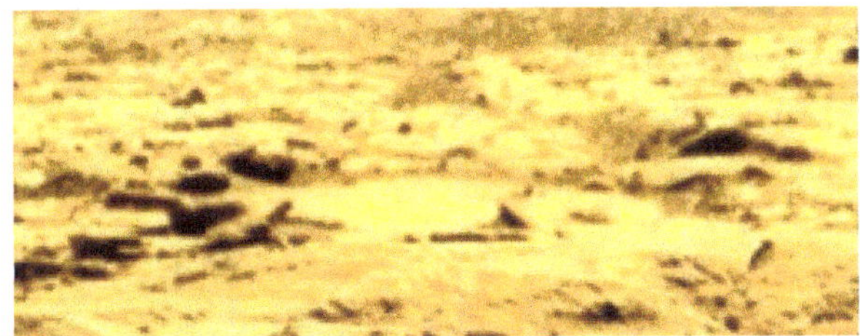

PIA16453-44 (above). A closeup of the car muffler like object (left) described in PIA16453-26. Under magnification a magnificent collection of odd objects spring up, including a square "clearing" (lower center).

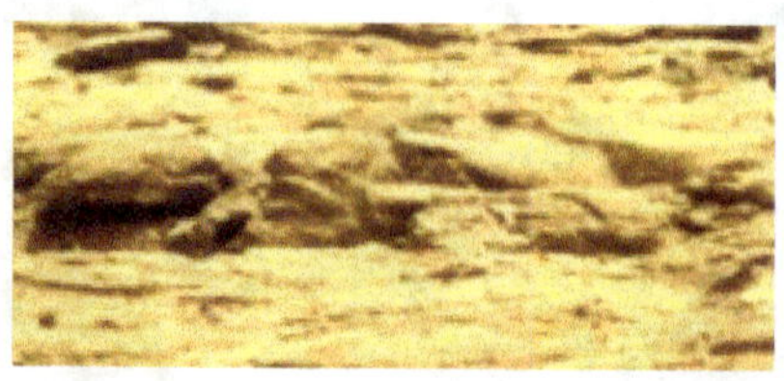

PIA16453-45 (above). Image inventory: Half circles, squiggly serpentine like "designs" and shapes, straight lines, and a "rock" with a "handle" on top at right upper center. The "rock" at right lower center has what appears to be hieroglyphs of some kind carved into it. Another photo worth studying under magnification.

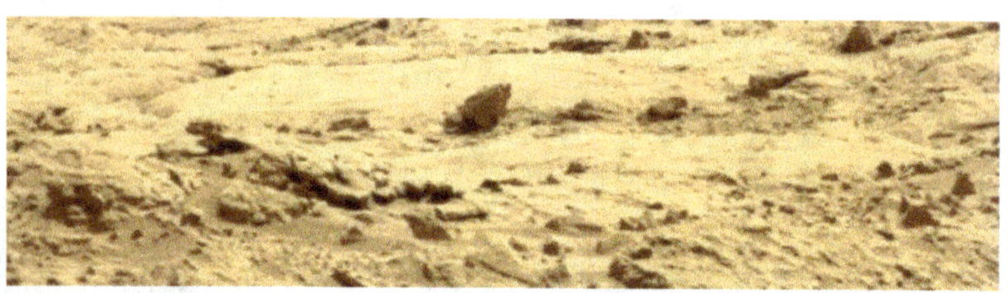

PIA16453-46 (above). More square clearings surrounded by weird objects, markings, and shapes.

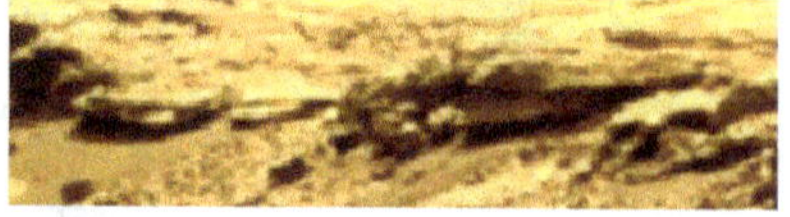

PIA16453-47 (above). I cannot be sure what the jigsaw puzzle piece like object is on the far right. And neither can anyone else. It may be a "rock." But it looks metallic.

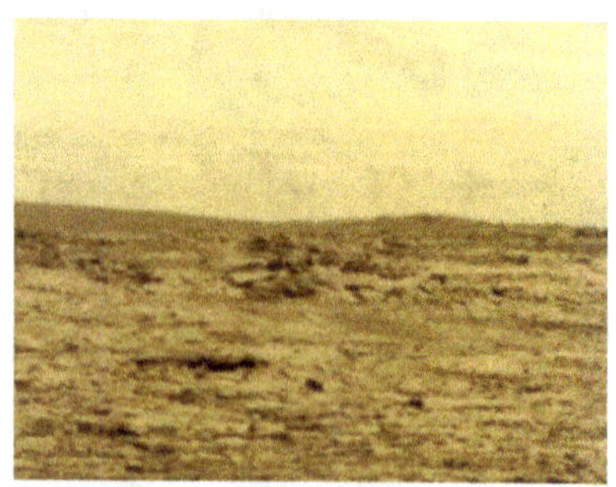

PIA16453-48 (right). There are a number of artificial looking objects in this image, but they are too far away to properly understand. Under magnification the rocky berm at center right appears to have unfamiliar symbols and text on it, including a wavy line.

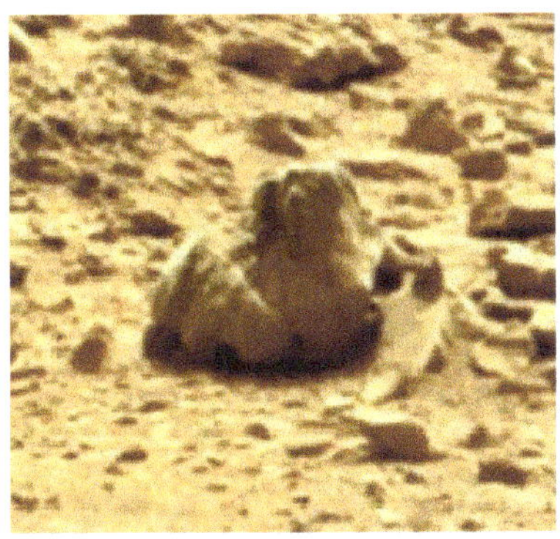

PIA16453-49 (above). This clump of "rock" looks like the shattered remains of a statue, or perhaps ornamental art. Take note of the many strange shapes and items in the surrounding area.

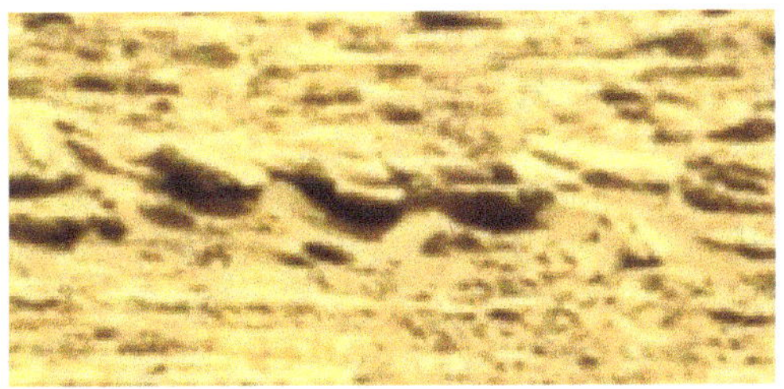

PIA16453-50 (above). Are these the ancient remains of a bathtub, a pool, or possibly something we cannot conceive of? "V"s, "U"s, rectangles, triangles, horseshoe shapes, "boards," and crescents lie spread about.

Chapter 30

ODDS & ENDS

PIA10214: EARTHDATES NOVEMBER 6-9, 2007

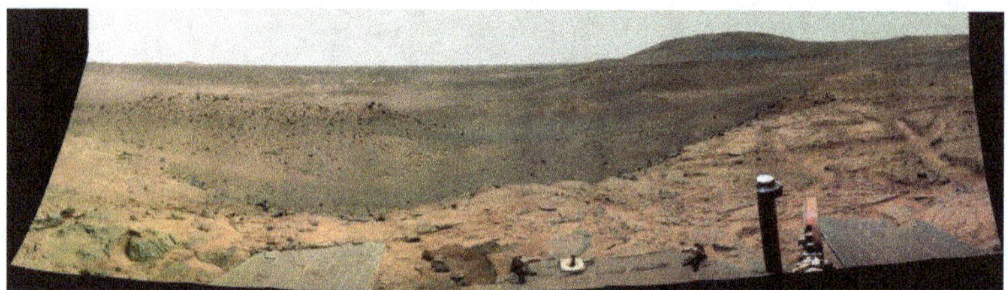

Original image. "NASA'S Mars exploration rover Spirit captured this westward view from atop a low plateau where Spirit spent the closing months of 2007. . . . After several months near the base of the plateau called 'Home Plate' in the inner basin of the Columbia Hills range inside Gusev Crater, Spirit climbed onto the eastern edge of the plateau during the rover's 1,306th Martian day, or sol (Sept. 5, 2007). It examined rocks and soils at several locations on the southern half of Home Plate during September and October. It was perched near the western edge of Home Plate when it used its panoramic camera (Pancam) to take the images used in this view on sols 1,366 through 1,369 (Nov. 6 through Nov. 9, 2007). . . . This view combines separate images taken through Pancam filters centered on wavelengths of 753 nanometers, 535 nanometers and 432 nanometers. It is presented in a false-color stretch to bring out subtle color differences in the scene." Image credit: NASA/JPL-Caltech/Cornell University.

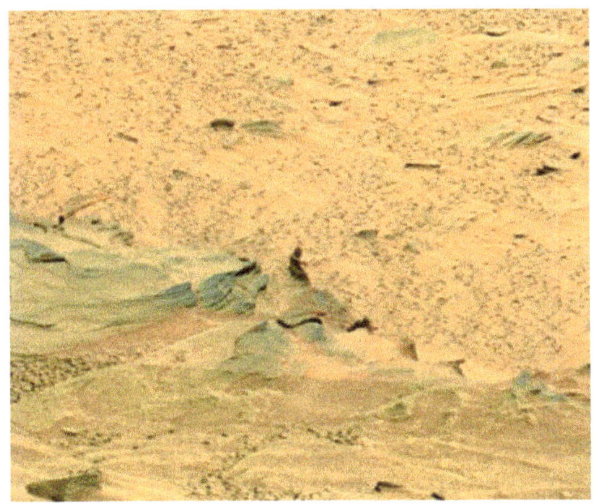

PIA10214-1 (above): A long distance view of the famous, or infamous (depending on one's perspective), "frozen woman statue," also known as the "Bigfoot statue." Is this a stunning piece of natural sculpture by Mother Nature, or an actual statue made by the hands of an intelligent and artistically sophisticated race of ancient Martians? Mainstream scientists expect that one will see a humanoid shape, as they do themselves, but quickly dismiss it as "pareidolia"—for in their minds, it cannot be anything but a "rocky outcropping." This is understandable: they have limited themselves to a natural non-extraterrestrial explanation for nearly everything. When it comes to the Martian anomalies, however, most of us do not possess what might called the arrogance of certainty. Instead, we prefer regarding this unusual looking object with a more critical eye. And this is what it tells us: We cannot possibly know what it is until one of our kind stands before it and studies it in person. See the closeup of the "frozen woman statue" below.

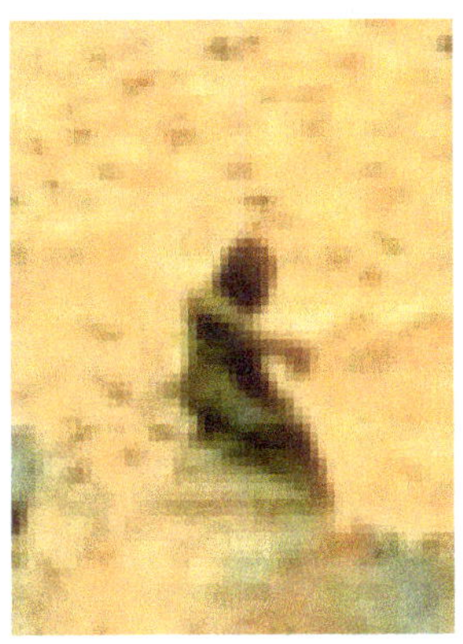

PIA10214-2 (left): Closeup of the "woman" or "Bigfoot statue" on Mars. By entertaining the extraterrestrial hypothesis, I am practicing authentic science: searching for truth without preconceived notions—and accepting whatever that truth turns out to be. Not being restricted by unscientific rules, like those held by many conventional thinkers, I cannot and do not rule out the possibility that this object is the creation of intelligent life. Only time will reveal the facts about what it is and who or what made it.

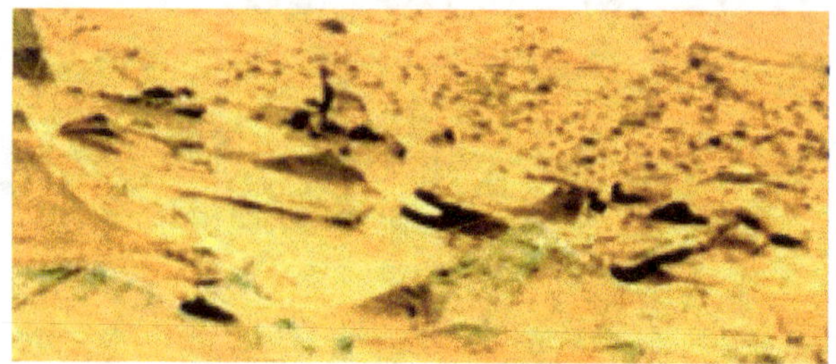

PIA10214-3 (above): Odd shapes and peculiar looking objects and debris. All "rocks."

PIA10214-4 (right): Here we have what appears to be a digitally airbrushed section of rock (the darkish square at center). If so, someone is attempting to hide something important from the public.

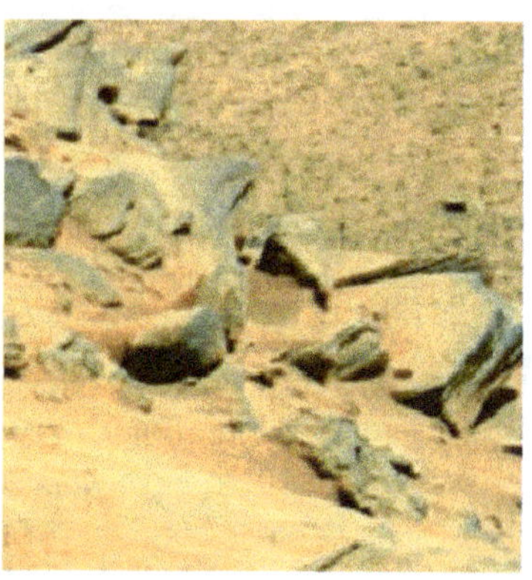

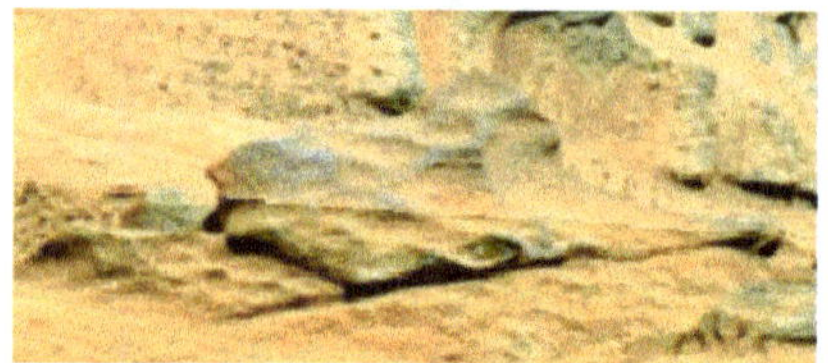

PIA10214-5 (above): Animal fossils, ancient Martian artwork, or natural geology? (The horizontal line across the middle is a digital "stitch," created when multiple images were "sewn" together to create the full frame panoramic shot. See the original image above.)

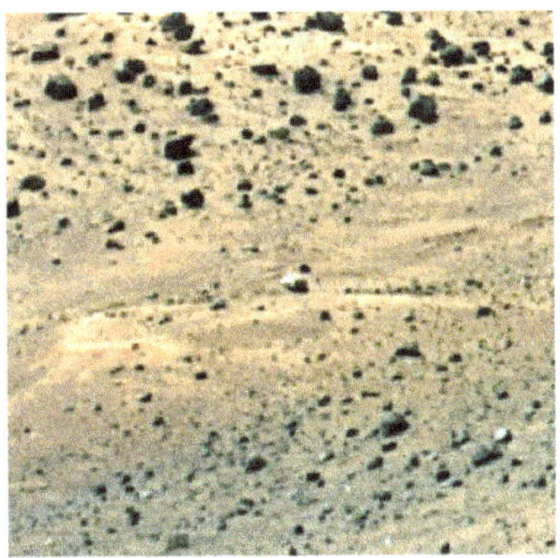

PIA10214-6 (above): There are two bright, squarish, metallic like objects in this image: one at center and one at 5:00. Both look quite dislocated in the dry reddish Martian dirt.

PIA10214-7 (right): The squarish "natural rock formation" near the top of this hill shares many traits with structural foundations found here on Earth. Of course, it could be pure geology. It could also be pure artificiality.

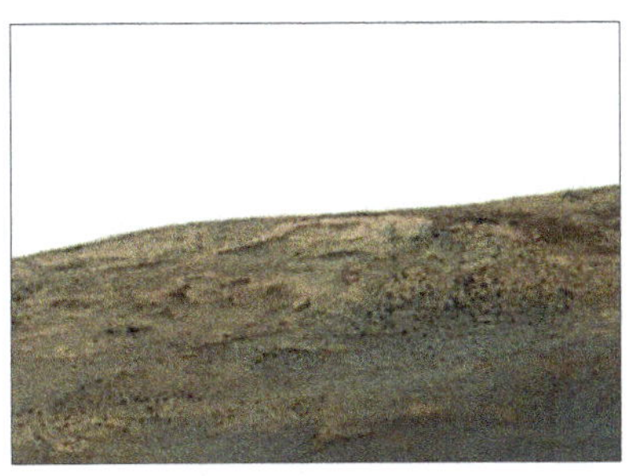

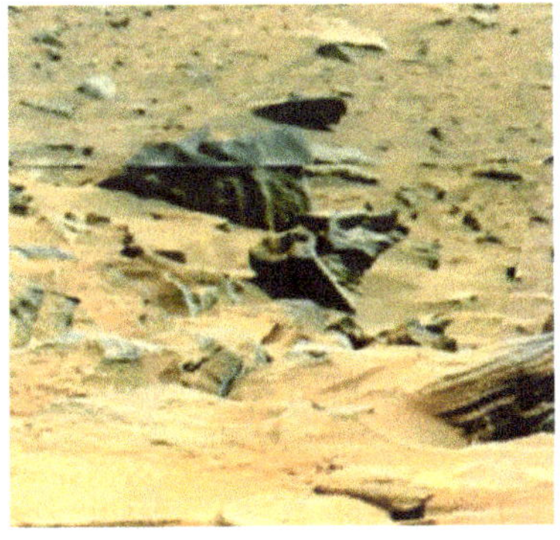

PIA10214-8 (left): There will be many bizarre objects to examine—like those that carpet this entire image—when scientists finally step foot on the Red Planet. (Note: A digital "stitch" runs through the large dark "rock" in the upper third of this photo.)

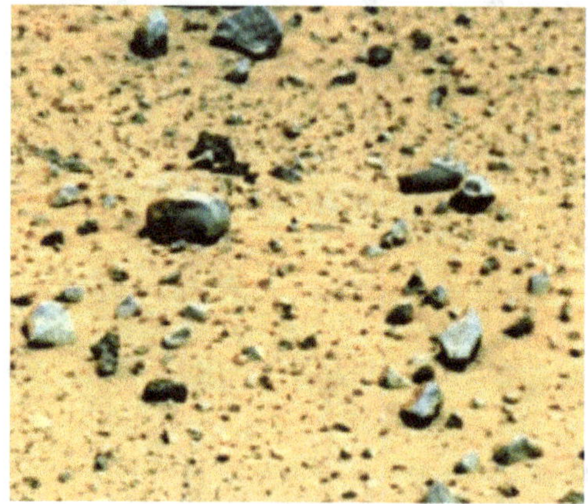

PIA10214-9 (above): At the 2:00 position there is a small, nightstand clock like object with a perfect circle "cut" into it. At the 8:00 position a small soap tray like object sits in the Martian dirt. Numerous other fantastical objects fill the frame.

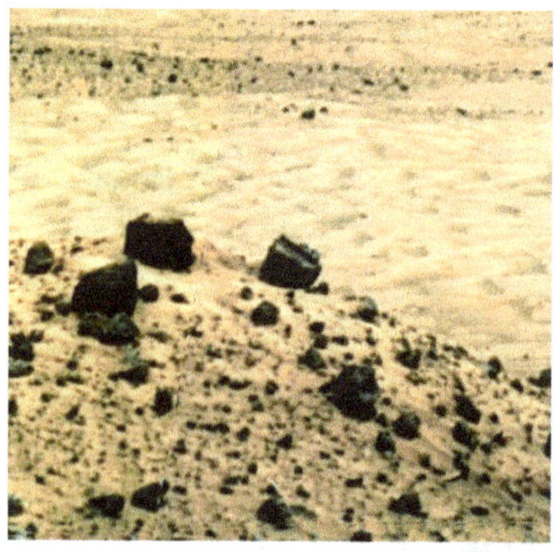

PIA10214-10 (above): The dark, block like "rock" at center has a large, evenly cut groove along its top, giving it the look of an ancient piece of machinery.

Chapter 31

SUSPICIOUS "ROCKS"

SOL 3333: EARTHDATE DECEMBER 21, 2021

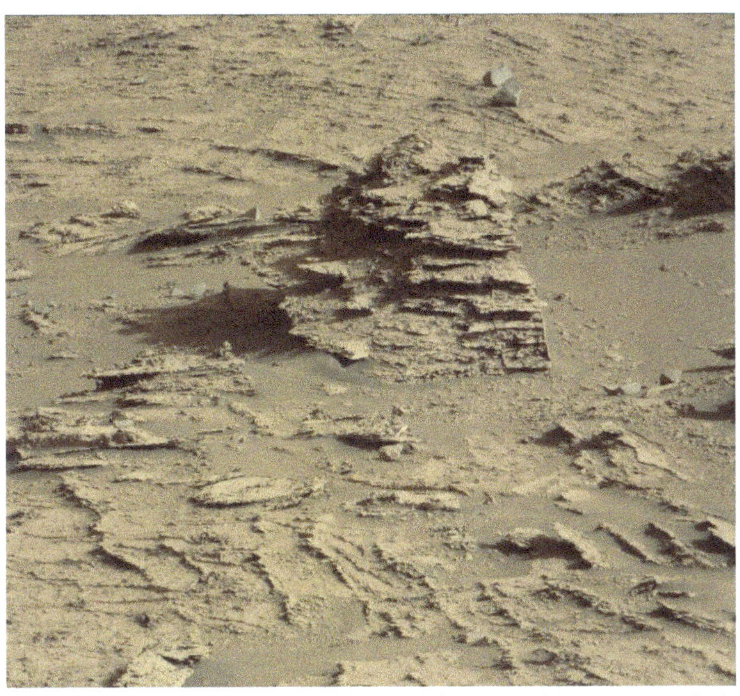

Original image. "Taken by Mast Camera (Mastcam) onboard NASA's Mars rover Curiosity on Sol 3333 (2021-12-21, 20:17:19 UTC)." Image credit: NASA/JPL-Caltech/MSSS.

ANALYSIS OF SUSPICIOUS "ROCKS," SOL 3333

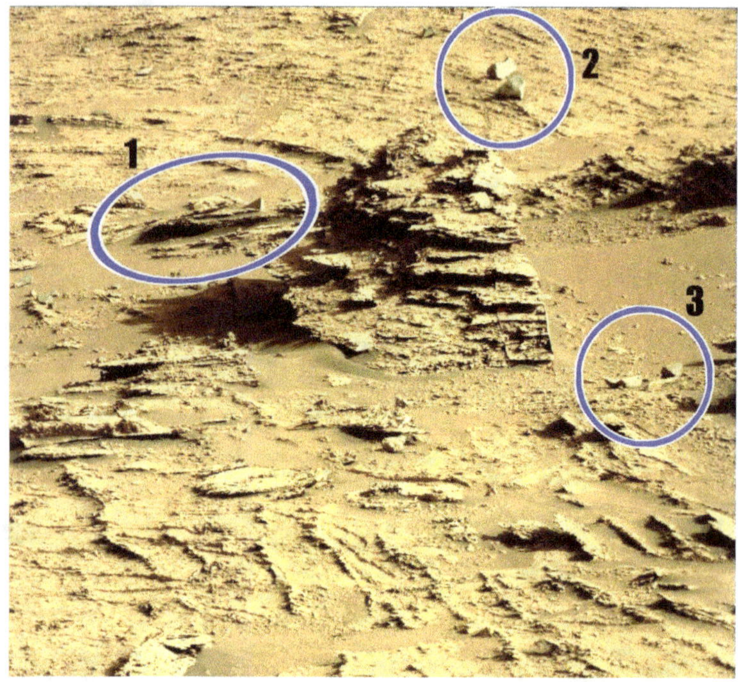

Suspicious "Rocks," Sol 3333, with anomalies circled. Image colored, brightened, and sharpened by the author to improve contrast. This does not look like one of the many major debris fields (filled with probable ancient artifacts) that we have previously examined. However, several unusual "rocks" can be seen laying on the Martian surface, each which deserves mention.

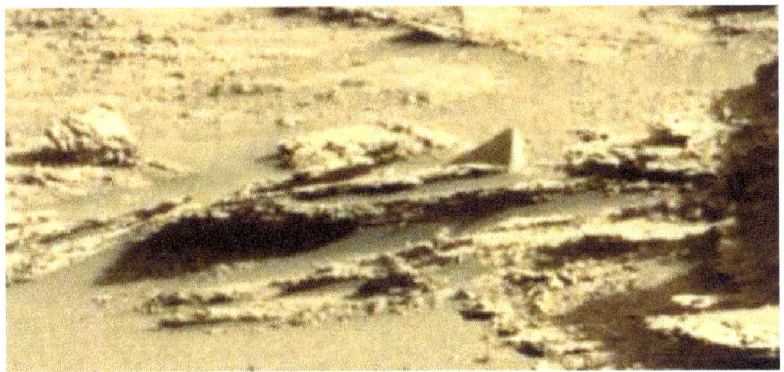

SOL 3333, highlighted area 1 (above): Center right, an object with a perfectly formed 90 degree angle lies partly submerged in the Martian sand. It looks very much like the corner of a building ledge.

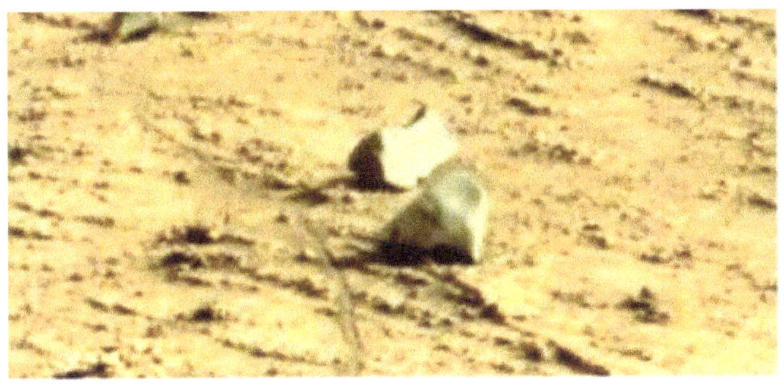

SOL 3333, highlighted area 2 (above): Odd "rocks" that appear to be sculpted pieces from a statue.

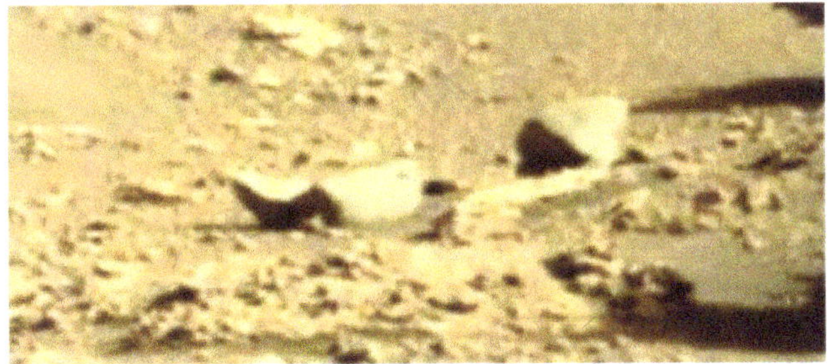

SOL 3333, highlighted area 3 (above): Pottery shards, the remains of building materials, damaged pieces of artwork, or ordinary Martian rocks? The pyramidal object at 2:00 has a square "base."

Chapter 32

ODDS & ENDS

PIA17766: EARTHDATE JANUARY 28, 2014

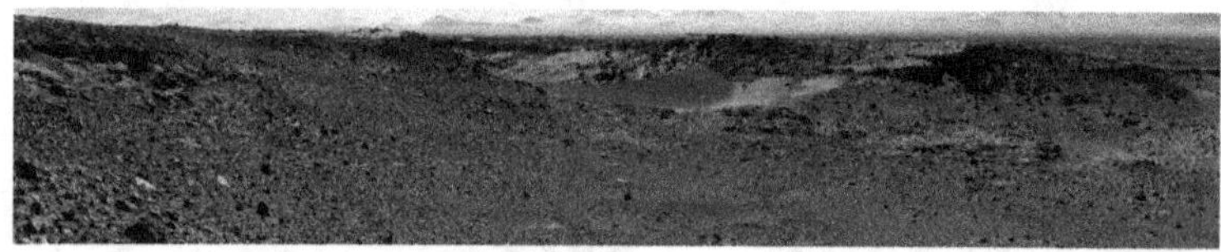

Original image. "This scene combines images taken by the left-eye camera of the Mast Camera (Mastcam) instrument on NASA's Curiosity Mars rover during the midafternoon, local Mars solar time, of the mission's 526th Martian day, or sol (Jan. 28, 2014). The sand dune in the upper center of the image spans a gap, called 'Dingo Gap,' between two short scarps. The dune is about 3 feet (1 meter) high. The nearer edge of it was about 115 feet (35 meters) away from the rover's position when the component images were taken, just after a Sol 526 drive of 49 feet (15 meters)." Image credit: NASA/JPL-Caltech/MSSS.

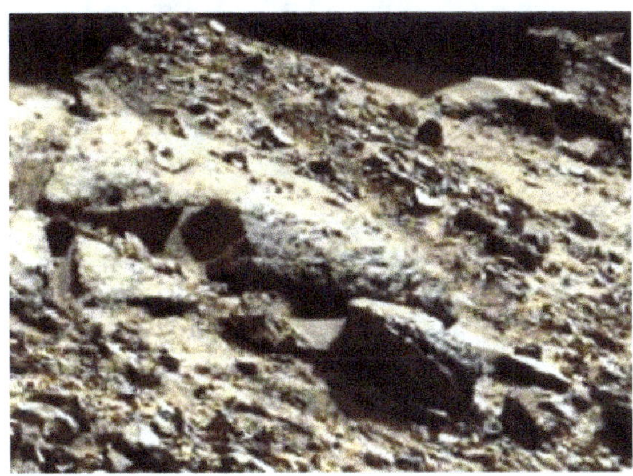

PIA17766-1 (above), colored, brightened, and sharpened by the author for contrast (as were all of the images in this chapter). The strange "rock" at center left is bomb shaped and is covered in peculiarities, including triangle and square cavities, as well as a host of what appear to be symbols and the letter "C" (see closeup below). This bomb like object is sitting amidst a myriad of other bizarre objects, items, shapes, and what appear to be hundreds of manufactured artifacts—too numerous to detail here.

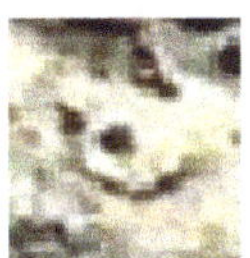

PIA17766-Low resolution closeup of the letter "C" in image 1 above.

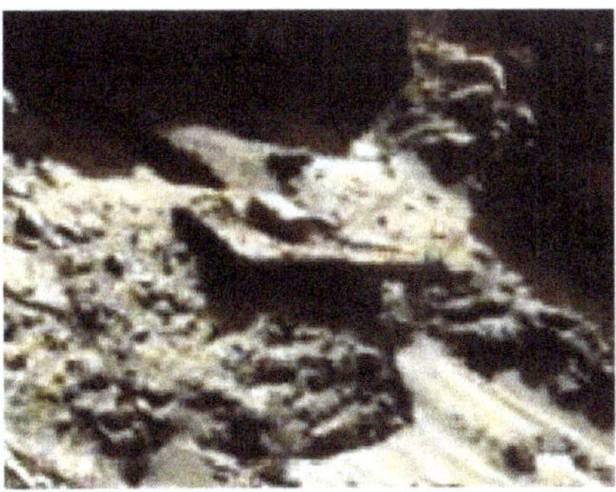

PIA17766-2 (above): What appears to be a brick sits on top of a flat, oddly shaped, metallic looking object. This looks like bomb debris.

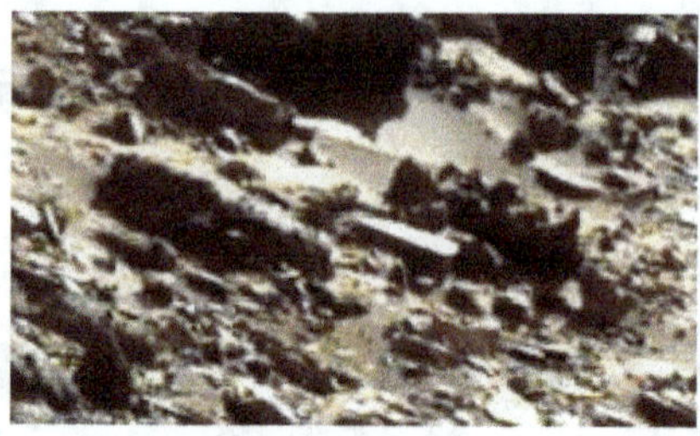

PIA17766-3 (above): Several objects of interest appear in this image. In the center is a shiny, gray, brick like object. Just above it and to its right sits a serrated object that looks like it might have been a mining drill bit. The large, dark, rectangular object to the left of the "brick" looks like a concretized piece of machinery. At about the 2:30 position there is an item with a 90 degree angle. Numerous other unidentifiable objects lie strewn about.

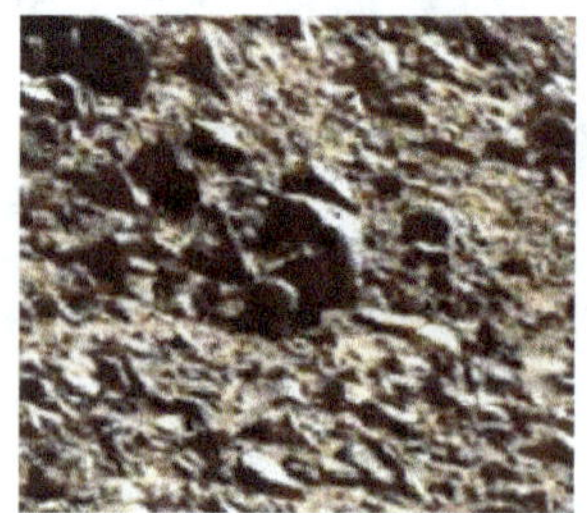

PIA17766-4 (left): The object laying on its side at center looks like a child's top or a small drill bit belonging to a power tool. It appears to have a metal (bronze?) shank coming out of the center (pointing left and downward).

PIA17766-5 (right): At center is an object with a 90 degree angle, and a rounded object above it with a serrated edge. These do not look like products of Nature—and neither do the hundreds of odd items surrounding them.

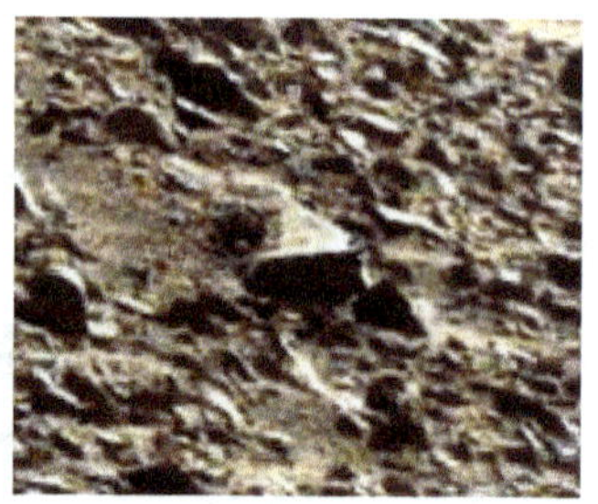

PIA17766-6 (left): This plaque like object looks quite ordinary and could easily pass for a common Martian rock—except for the fact that we have already discovered and examined a number of objects with this *exact* shape. This suggests artificiality.

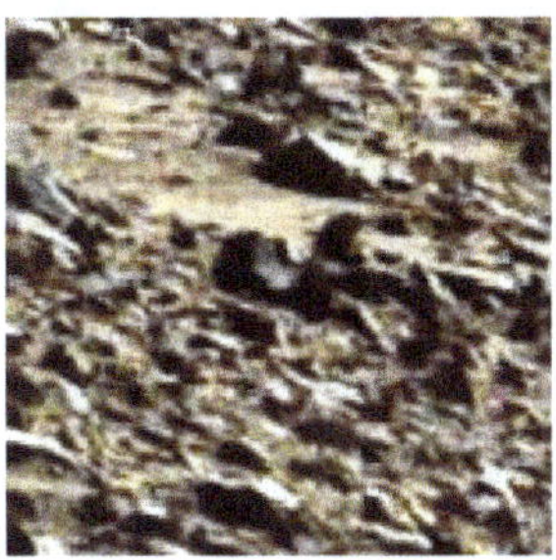

PIA17766-7 (above): The broken, black object at center looks like a bracket of some kind; perhaps one that held a pipe.

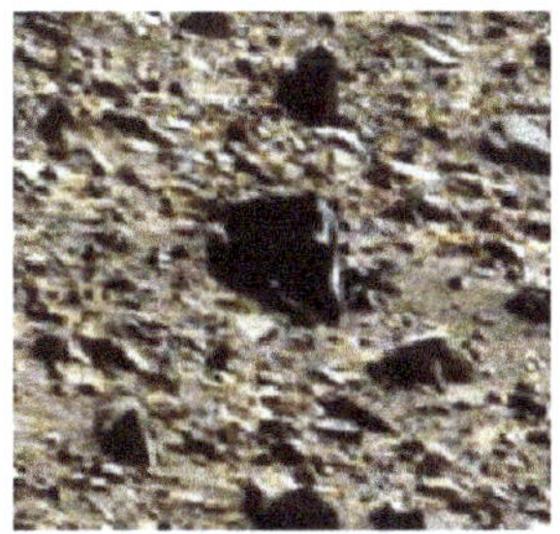

PIA17766-8 (above): The dark bluish object at center appears to have a symbol or an "n" like letter stamped on its right side (barely visible due to the angle of the camera).

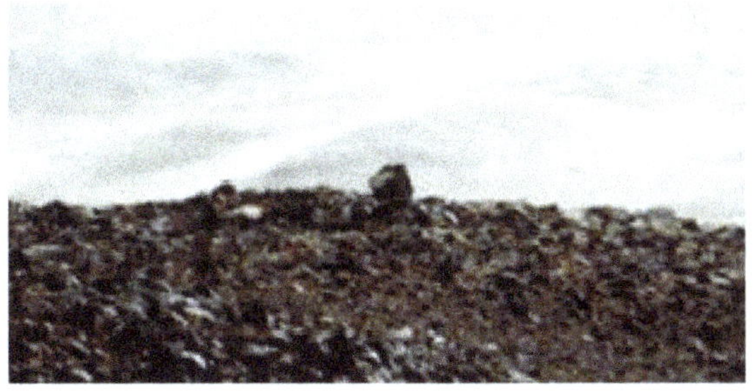

PIA17766-9 (above): A strange cube like object sits on one of its corners atop a rocky mountain ridge.

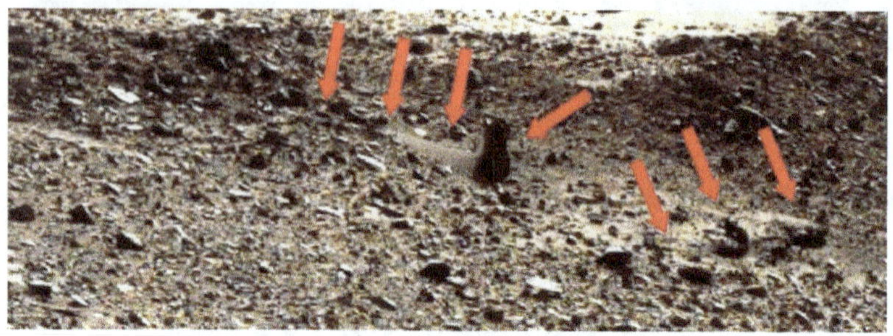

PIA17766-10 (above): What appears to be a cluster of rusted pipes poke out of the side of a Martian hill. Were these once part of an ancient Martian pipeline that was used to move fresh water, sewage, oil, gas, or electricity?

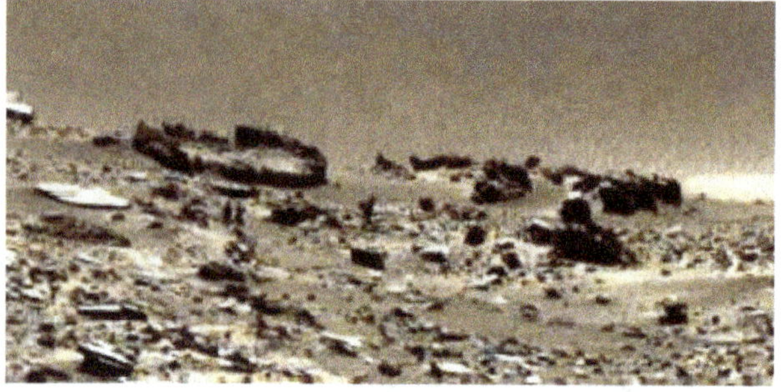

PIA17766-11 (above): At left of center the remains of a circular object sit on the Martian surface. It is surrounded by other objects and shapes of unknown origin and purpose, including blocks, squares, rectangles, "boards," "sheet metal," "Y"s, "V"s, and "bricks."

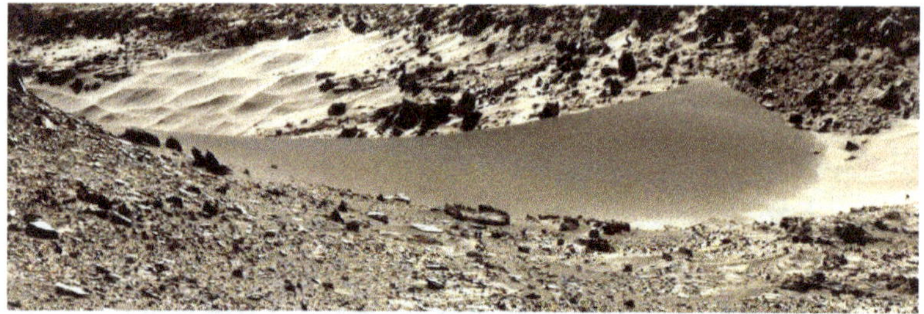

PIA17766-12 (above): What is the great brown patch swooping across the center of this image? It appears to be digital manipulation. If so, it is one of the most obvious examples of pseudo sand we have yet seen. Judging from the many artificial looking objects encircling this probable fake dune, there must be something quite shocking underneath it.

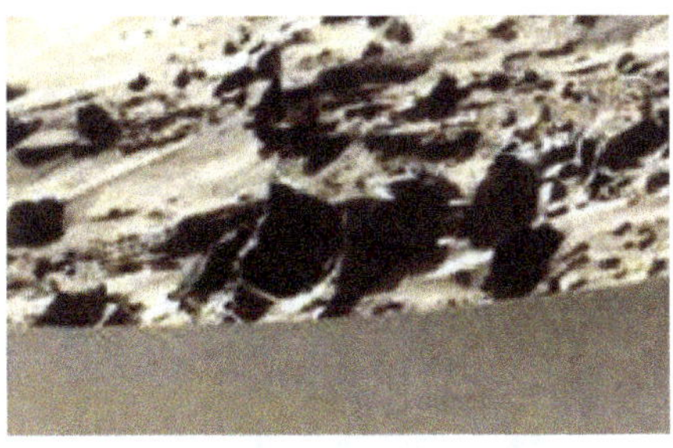

PIA17766-13 (above): Weird looking objects on the other side of what looks like a digitally created sand dune, described in PIA17766-12 above.

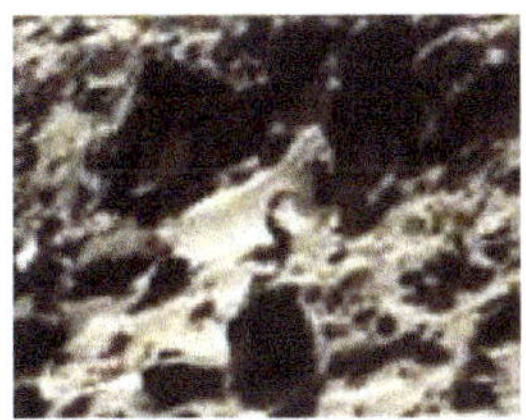

PIA17766-14 (above): Is there anything in this image that looks natural? If the "S" like object at center is natural, how was it made? If it is artificial what is it? The same questions apply to the circular objects and the "X" in the lower left.

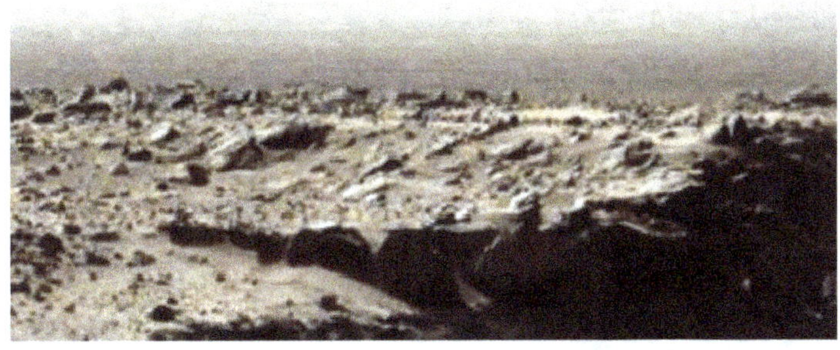

PIA17766-15 (above): This Martian slope is covered in strange shapes reminiscent of Roman lettering and numerals. Among others there is a clear "S," an "e," a "c," an "O," and the number "4." Mainstream scientists like to say that not every anomaly on Mars is necessarily artificial. I like to say that not every anomaly on Mars is necessarily pareidolia.

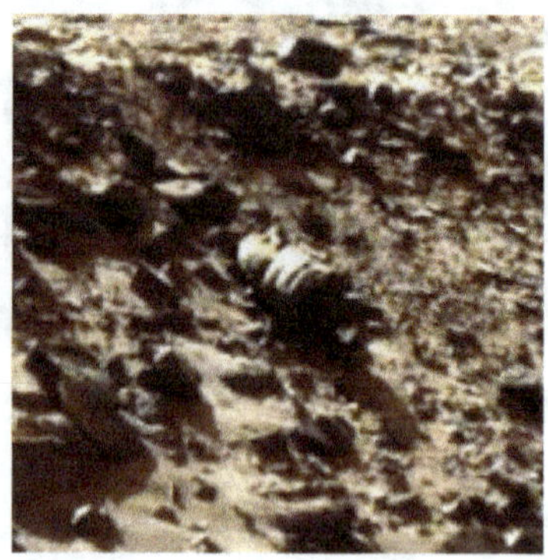

PIA17766-16 (above): At center is what would probably be described by most thinking people as a small damaged robot, a broken down engine, or a shattered metal or plastic piece of electronic equipment of some kind. The artificially bright object seems to have ribbed panels, a "head," "arms," and "feet," giving it some resemblance to the famed *Star Wars* droid R2-D2. Strangely, there seems to be a large appendage attached to it, for at its "bottom" end it is casting a long shadow (which is pointing toward to the bottom right corner of the image) down the hillside.

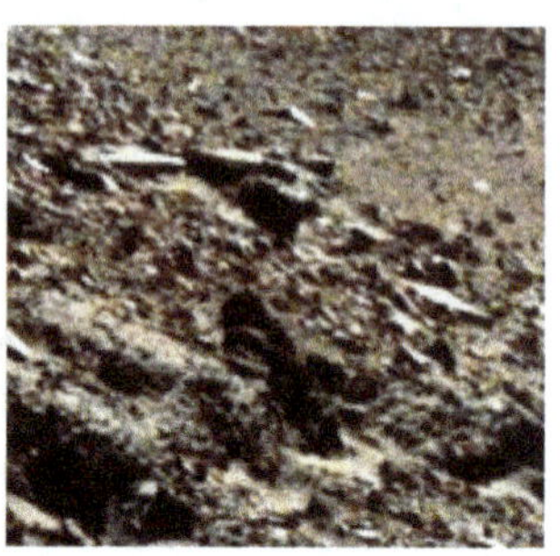

PIA17766-17 (right): A number of unfamiliar objects litter this image. Mainstream science is quite certain it can identify them. I am not certain at all. Study them closely under magnification and decide for yourself.

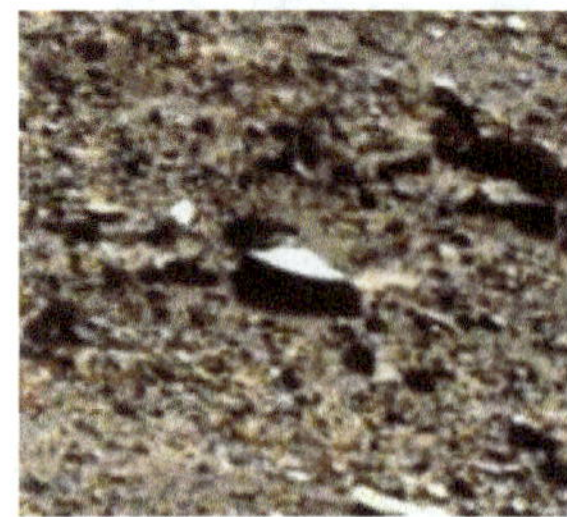

PIA17766-18 (left): At center sits a thick, shiny, gray, out-of-place, triangular shape. What is it? Despite the strident proclamations of skeptics, the question remains open. It must remain open if we wish to practice authentic science.

Chapter 33
CONSTRUCTION SITE

SOL 1039: EARTHDATE JULY 9, 2015

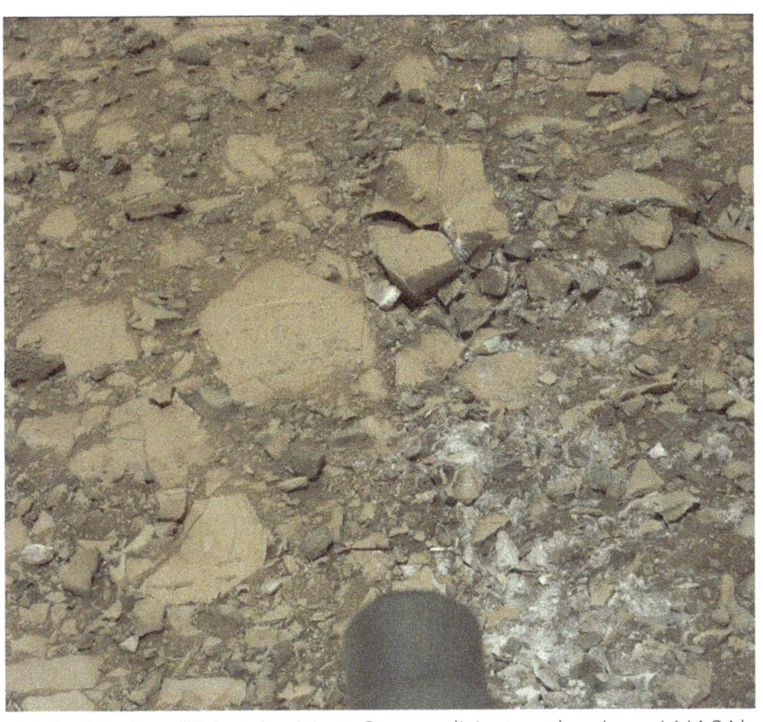

Original image. "Taken by Mast Camera (Mastcam) onboard NASA's Mars rover Curiosity on Sol 1039 (2015-07-09, 16:47:55 UTC)." Image credit: NASA/JPL-Caltech/MSSS.

ANALYSIS OF CONSTRUCTION SITE, SOL 1039

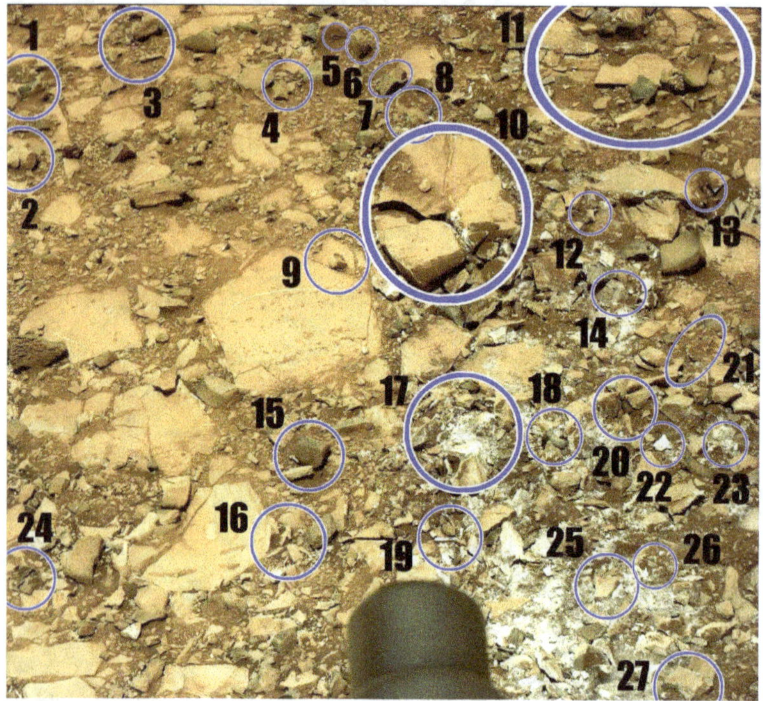

Construction Site, Sol 1039, with anomalies circled. Image colored, brightened, and sharpened by the author to improve contrast. With its countless objects that look like they were collected from a building zone, few NASA photos demonstrate more clearly the probability that an intelligent race once inhabited the Red Planet than this one. After carefully studying it for many hours, I have concluded that there are literally hundreds of probable artifacts visible in this one small area alone, far too many to examine in an introductory book like *The Martian Anomalies*. Thus, I have chosen what I believe are the most interesting 27 anomalous sections for analysis. It is startling to realize that in this image we may be looking at the devastated remains of a Martian family home. A magnifying glass is recommended.

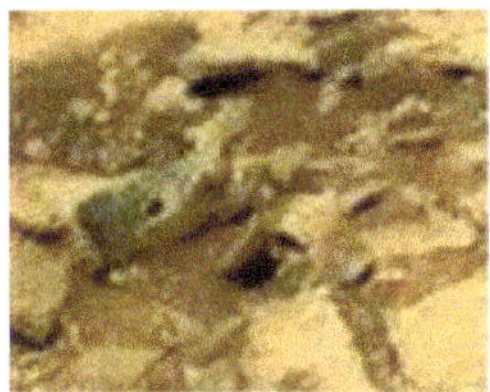

SOL 1039, highlighted area 1 (above): Left of center a gray rectangular item with a hole in its middle. To its lower right, a roundish, silver, spring like object (with ribbing or lines running around it) peeks out of the Martian soil.

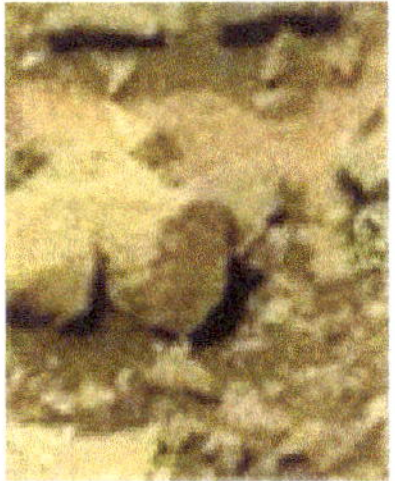

SOL 1039, highlighted area 2 (above): The strange round object at center appears to be made of an artificial cement like material.

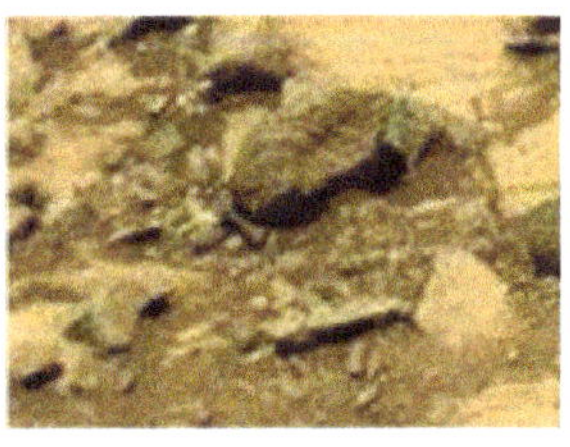

SOL 1039, highlighted area 3 (right): A host of tiny artificial looking objects fill this image. Two examples: Just left of center is what looks like a nail. Below it sits a long, narrow, whitish item with squared sides and a small circular protrusion on one end. Other unidentifiable building debris lies scattered about.

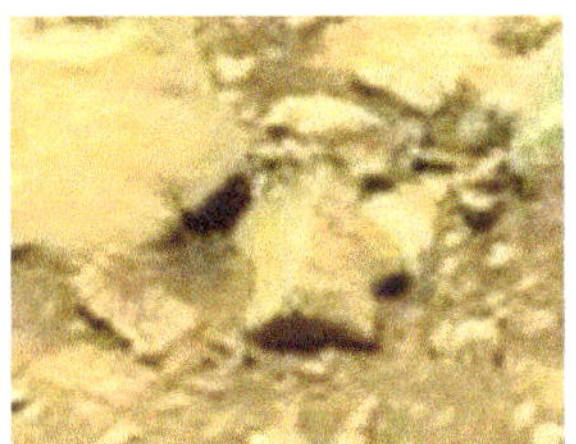

SOL 1039, highlighted area 4 (left): At 2:00 we see another nail like object and also a bracket like object sitting next to one another. At center a broken off "corner piece." At 8:00 an artificial looking rectangular piece of "woven" material. To its lower right a "belt buckle." To its lower left, a small brick like object embedded in the soil.

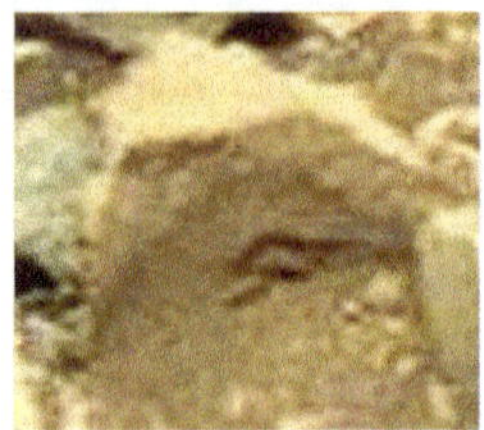

SOL 1039, highlighted area 5 (left): This smooth hexagram shaped "rock" has a diamond shaped Eye of Horus like design stamped or carved into its facing end, just one of thousands of ancient Egyptian like items on Mars.

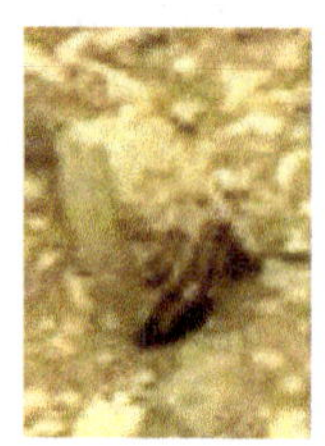

SOL 1039, highlighted area 6 (right): This bizarre crushed looking "rock" has a slit in one end and what appears to be a little letter "O" carved into it.

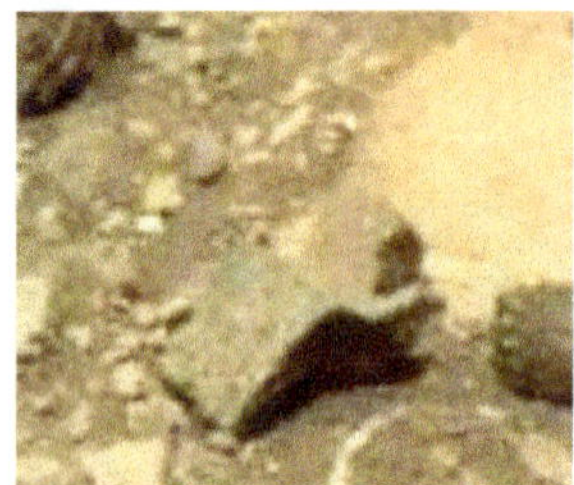

SOL 1039, highlighted area 7 (left): Whatever this object is, it is almost impossible that it is natural.

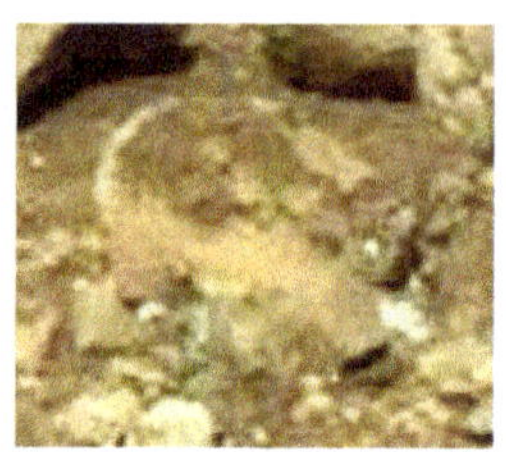

SOL 1039, highlighted area 8 (right): A fish like object pokes out of Mars' red sand. (Its "mouth" is at 5:00; its "tail" sweeps upward and over to the right.) Artificial or natural?

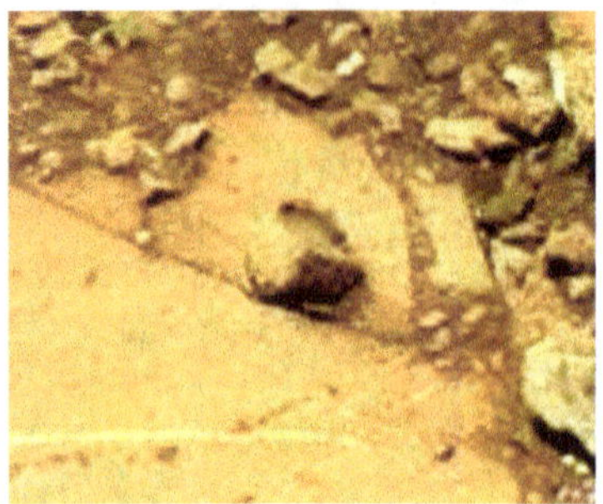

SOL 1039, highlighted area 9 (left): At center a small square object with a perfect circle impressed or carved into one end. It is surrounded by hundreds of items that look just like our building materials here on Earth.

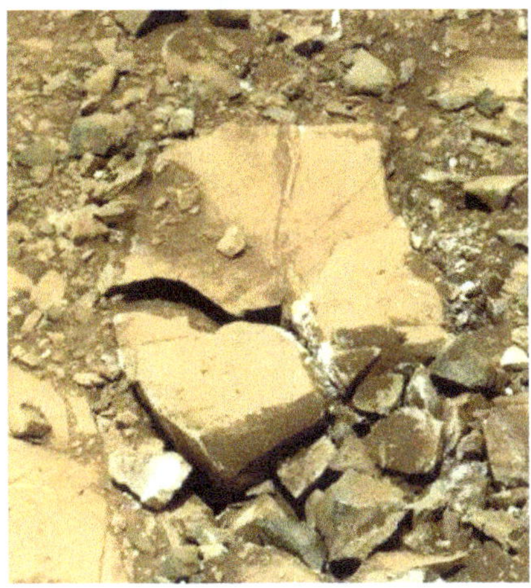

SOL 1039, highlighted area 10 (above): Though at first glance this seems like an average Martian rock, it bears various traits of possible artificiality: smooth sides and edges, a paving stone like appearance, and its location in the middle of what look like other types of manufactured building materials.

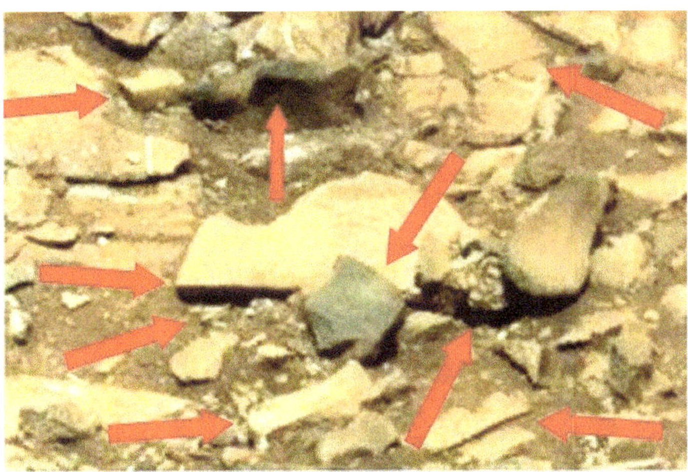

SOL 1039, highlighted area 11 (above): Under magnification this tiny space, probably only inches in width, contains a wealth of artificial looking items, a few which I have red arrowed: "bricks," shattered "paving stones" with mathematically accurate 90 degree angles, asphalt like "rocks," "nails," "cornerstones," "arrowheads," and a "rock" with peculiar "designs" or "patterns" of some kind emblazoned on one side.

SOL 1039, highlighted area 12 (right): The light, oddly shaped object, with a 90 degree angle, at center is surrounded by square brick like objects and smooth but broken paving stone like objects. What looks like a lock or buckle lies just above the center.

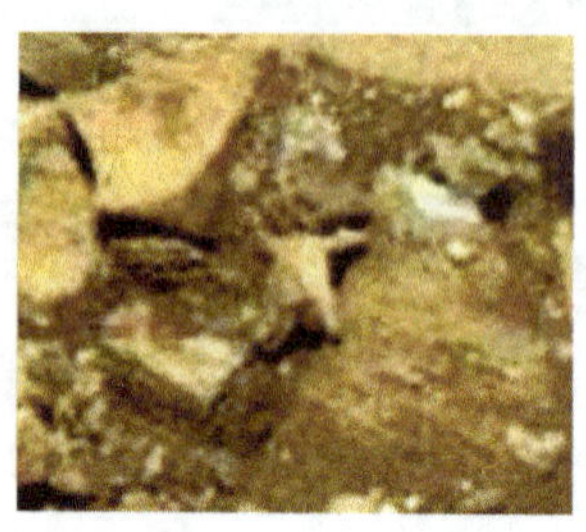

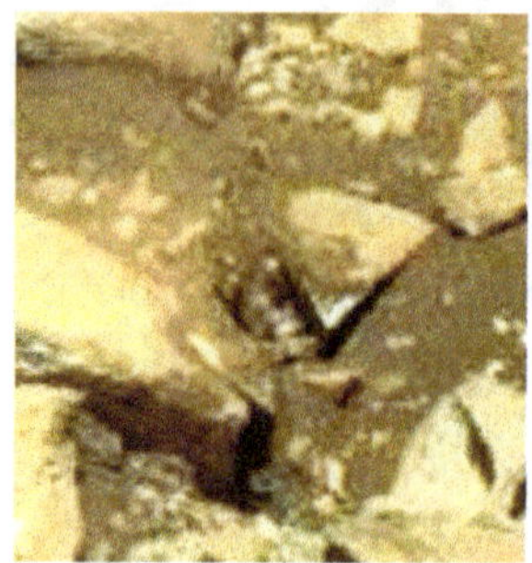

SOL 1039, highlighted area 13 (left): At the center of this image, to the left of the tan triangular "rock" with a curious silver tip, is a very small but interesting item that looks similar to a coiled piece of metal. Most of it is buried in the sand. One shiny side of it is exposed, however—perhaps revealing what we were never meant to see.

SOL 1039, highlighted area 14 (right): At the center of this low resolution image there appears to be a perfectly carved Roman letter "O" on the side of a pyramidal like "rock."

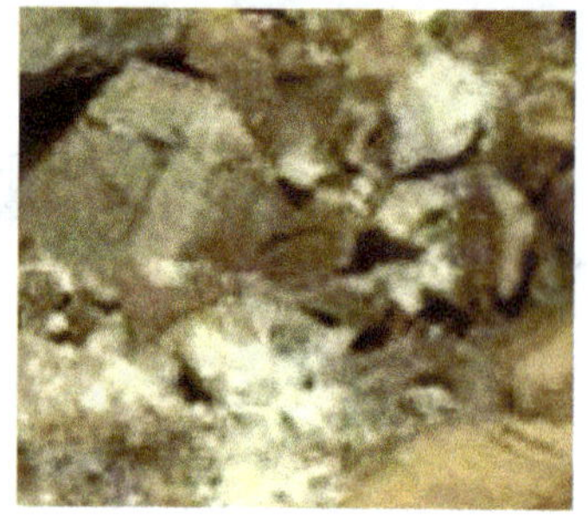

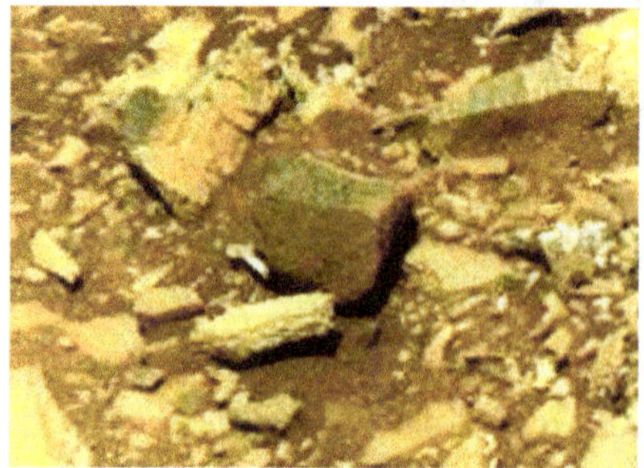

SOL 1039, highlighted area 15 (left): At center is a thick, grayish, horse's hoof shaped object that appears layered. To the left of it is a shiny metallic key like object laying on top of the rubble. Closely study the debris around them.

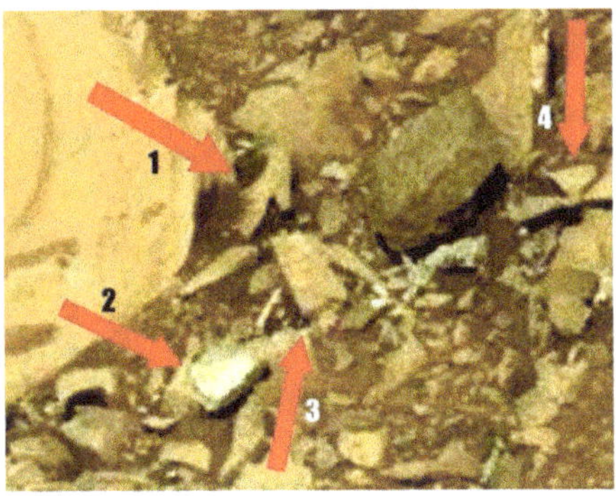

SOL 1039, highlighted area 16 (above): Bizarre litter fills this entire section: 1) This leaf like item looks out of place; 2) A white pottery shard, or perhaps a piece of drywall or some other type of building material? 3) This little rectangular object looks like a piece of wood, complete with "saw cuts" along its upper side and "splinters" sticking out of its bottom end; 4) A set of small fan blades? Among the rest of the mostly unidentifiable "building" debris there are also what appear to be "nails" and "screws," as well as "arrowhead" and "spearhead" like items.

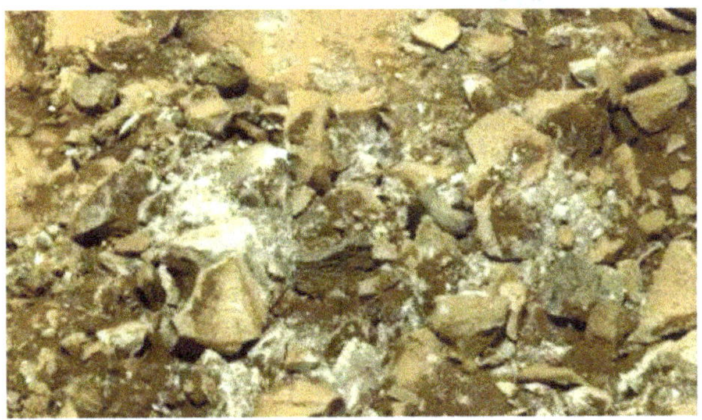

SOL 1039, highlighted area 17 (above): This is a small section I cut from NASA's full frame photo (above) in order to show a closeup of the white, powdery like substance that covers a large swath of the Martian soil here. Is it a naturally occurring Martian mineral, like silica, or something artificial, such as crumbling plaster? If these are the remains of an ancient building site, it could very well be the latter.

SOL 1039, highlighted area 18 (right): At center is a bent, blackish gray object that appears to be hollow. It looks like a rubber piece of fuel line hose used in engines. (See area 20 below.)

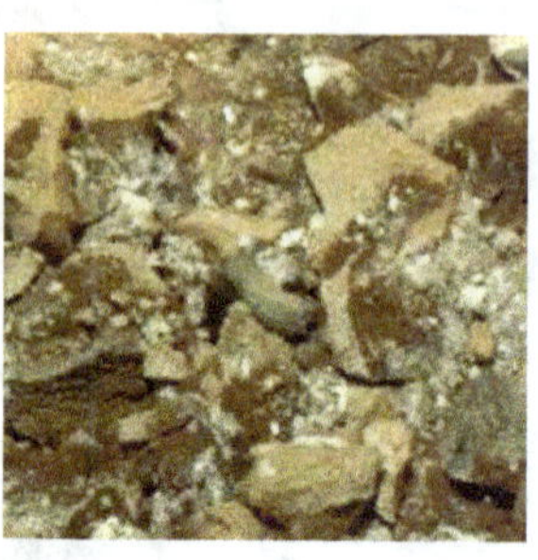

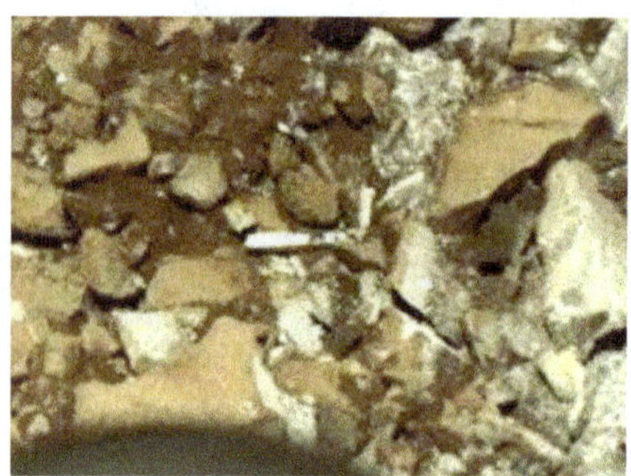

SOL 1039, highlighted area 19 (left): Is the bright white "L" shaped object at center made of wood, plastic, metal, or an unknown Martian material? Or is it simply a "rock," as conventional scientists label it? Look at what is around it.

SOL 1039, highlighted area 20 (right): At center another black, rubber, fuel line hose like object. Is it from the same source as the object described in area 18?

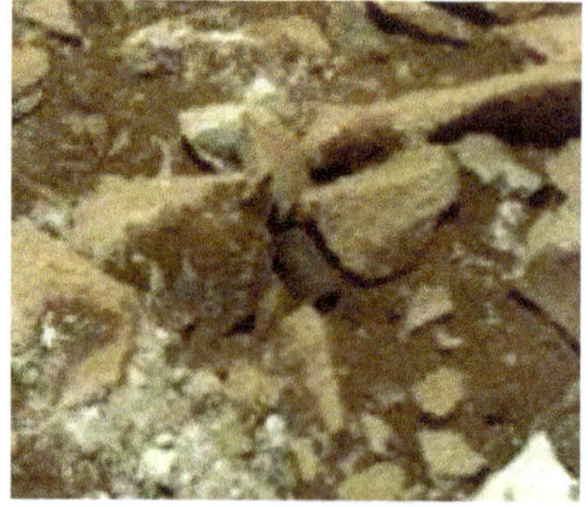

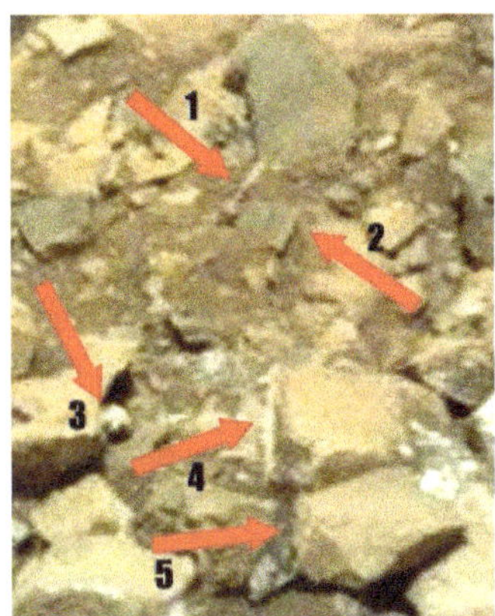

SOL 1039, highlighted area 21 (above): More evidence of probable artificiality. 1 and 2): Two parallel pieces of what look like metal or wire poke out from the sand; 3): A small, peculiar white object that appears to be some type of household item; 4 and 5): Damaged "bricks" lying in the rubble of what the evidence suggests is an ancient construction site.

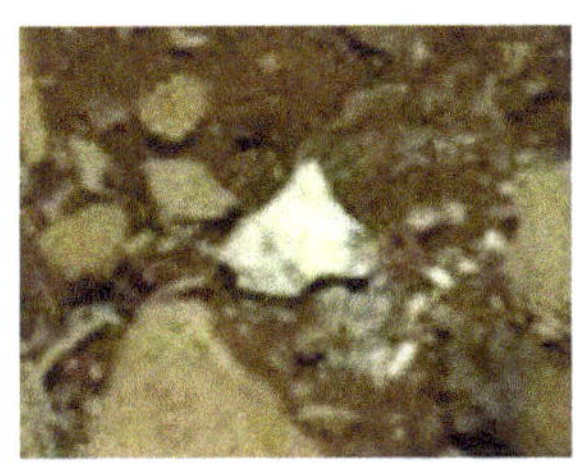

SOL 1039, highlighted area 22 (left): Pottery shard, plaster chip, or natural rock?

SOL 1039, highlighted area 23 (right): 1) What appears to be a stylized "X" in relief carved into a "rock"; 2) What looks like a large silver "0" embedded in a "rock"; 3) A small, silvery, curved, tube like object protrudes out from behind a "rock." Is this a piece of metal from a machine? Why are all of these unusual items clumped together in this specific location?

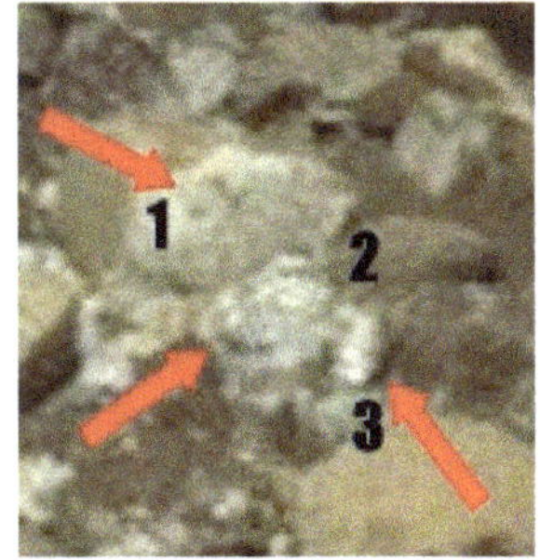

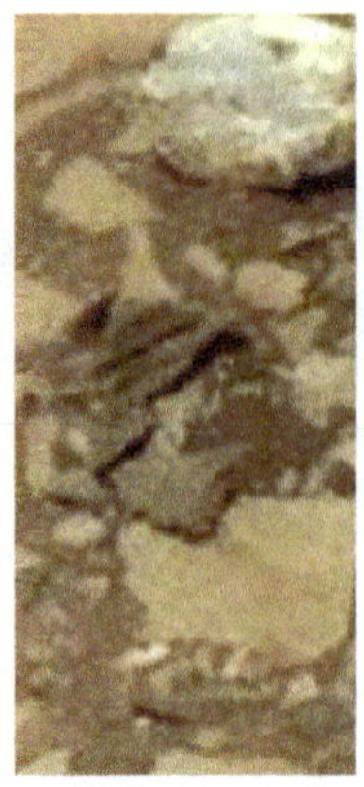

SOL 1039, highlighted area 24 (left): Left of center: More thin, fragile, leaf like refuse resting on the Martian soil. Are these items damaged roof shingles, or perhaps something wholly unknown to us? They appear similar to the delicate "leaf litter" described in area 16.

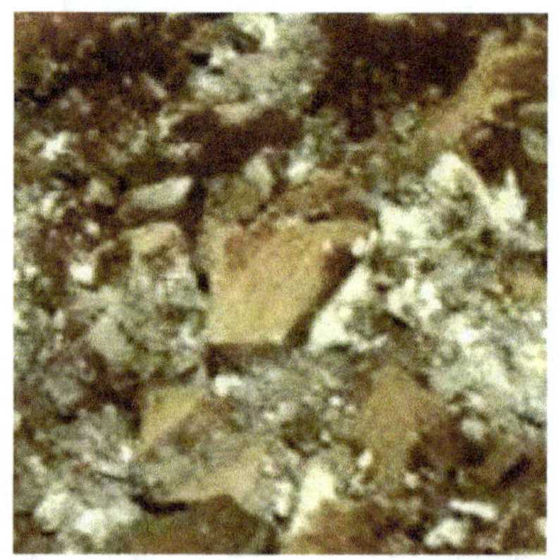

SOL 1039, highlighted area 25 (above): The "brick" at center sits amidst a pile of broken "bricks" and debris, perhaps the result of an explosion, or some kind of catastrophic astronomical or atmospheric event. The U.S. army psychic (remote viewer) discussed in my introduction said that in the distant past Mars seems to have experienced a planet wide "environmental disturbance" after passing through a comet's tail. Is what we are looking at here the result of that disaster?

SOL 1039, highlighted area 26 (right): More of what appears to be clear evidence of artificiality on Mars. At center a Roman capital letter "O" or possibly a lower case "d." Above it is a peculiar, damaged square object. Both are surrounded by rubble, "plaster dust," Martian "junk," and other types of seemingly structural debris.

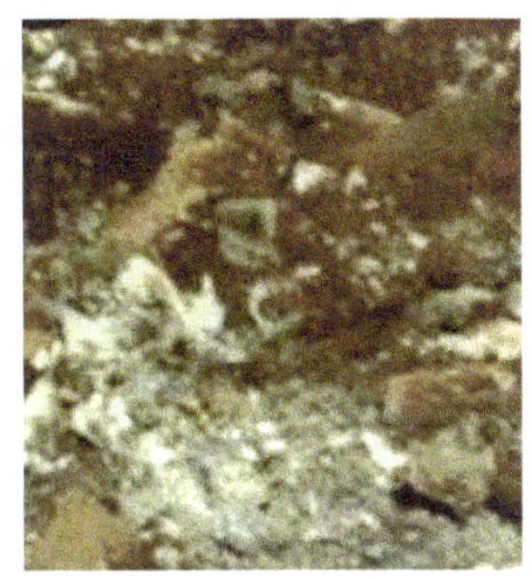

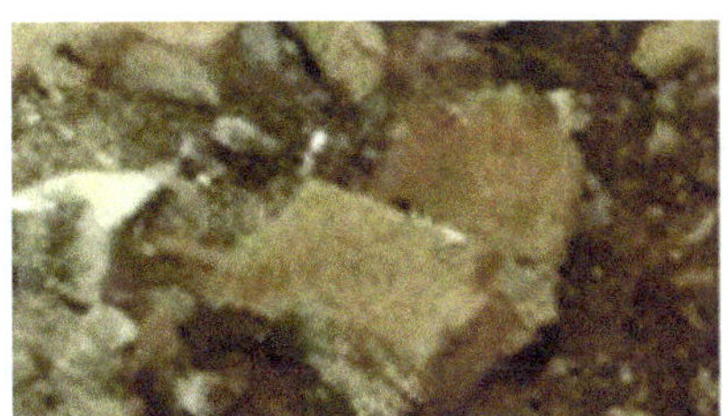

SOL 1039, highlighted area 27 (above): Smashed "bricks." To their left is a bright white object with what appears to be a rusty piece of metal reinforcing bar (rebar) sticking out of its middle.

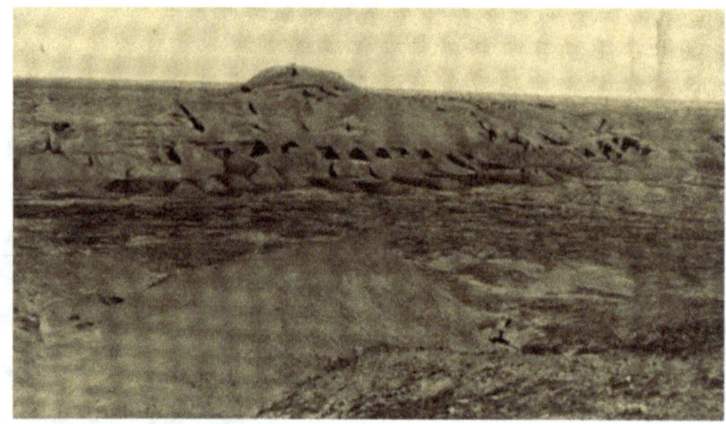

After carefully studying the Martian anomalies in this book, one might be forgiven for thinking that this is a photo of the Red Planet. After all, there are locations on Mars with nearly identical features. However, this is a photo of an ancient area on our planet, one known as "Temple Hill," at Nippur, Babylonia (modern Iraq). The mound on the horizon at the top left represents the remains of a 4,500 year old ziggurat, which shares traits with a Martian anomaly known as the "Tholus" in the Cydonia region.

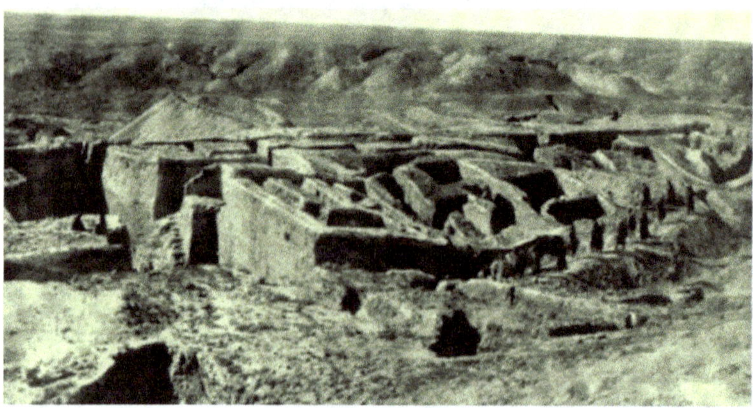

A view from Temple Hill (see image above) toward the ancient Babylonian village below it. These ruins are less than 5,000 years old. What would similar ruins look like if they had been sitting on the surface of Mars for 1 million years? In my opinion, they would look very much like many of the images in this book.

Chapter 34

TREE STUMP

SOL 1647: EARTHDATE MARCH 25, 2017

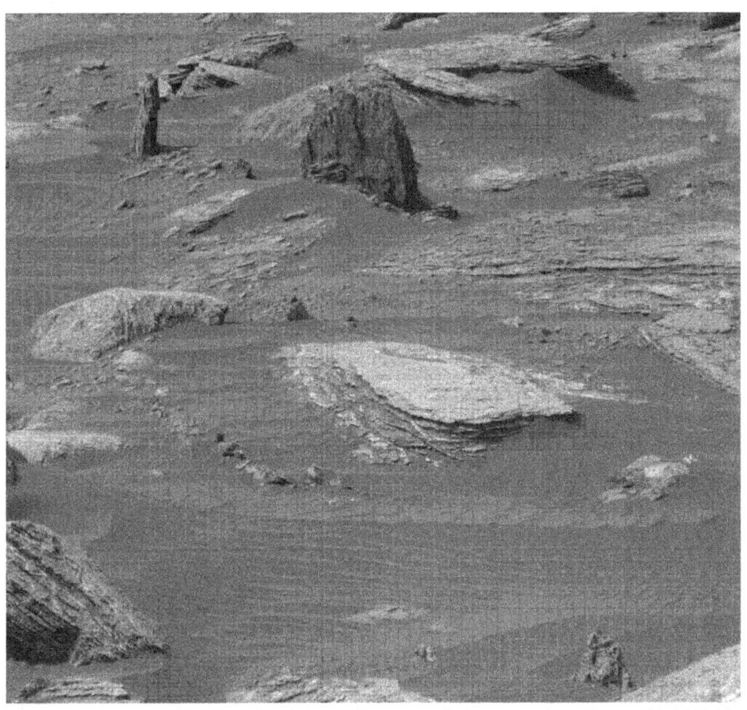

Original image. "Taken by Mast Camera (Mastcam) onboard NASA's Mars rover Curiosity on Sol 1647 (2017-03-25, 07:19:49 UTC)." Image credit: NASA/JPL-Caltech/MSSS.

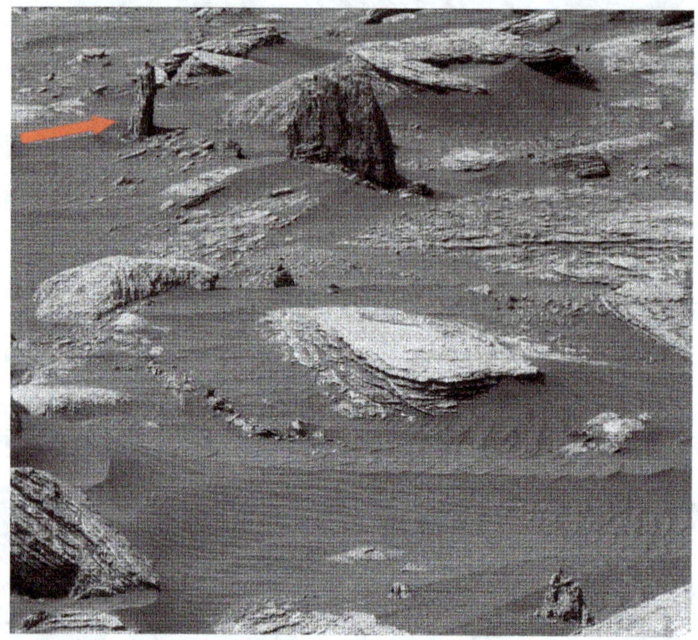

Above: Sol 1647, brightened and sharpened by the author for contrast. Only one object of anomalous interest appears in this image, which I have red arrowed in the upper left.

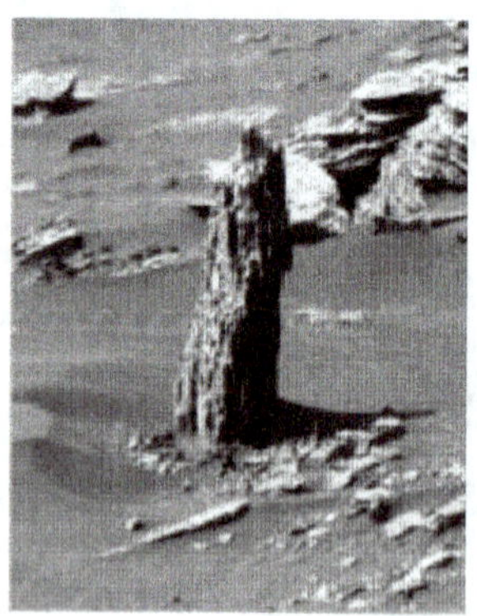

Above: Sol 1647, closeup of highlighted area above. Believers and anomalists maintain that this is a tree stump, and that it proves the existence of ancient vegetation on Mars. Mainstream scientists assert that it is almost certain trees never grew on Mars and that this is nothing but an odd rock formation. The truth is that it will remain unidentifiable until it is examined in person.

Chapter 35

ARTIFICIAL LIGHT

SOL 589: EARTHDATE APRIL 3, 2014

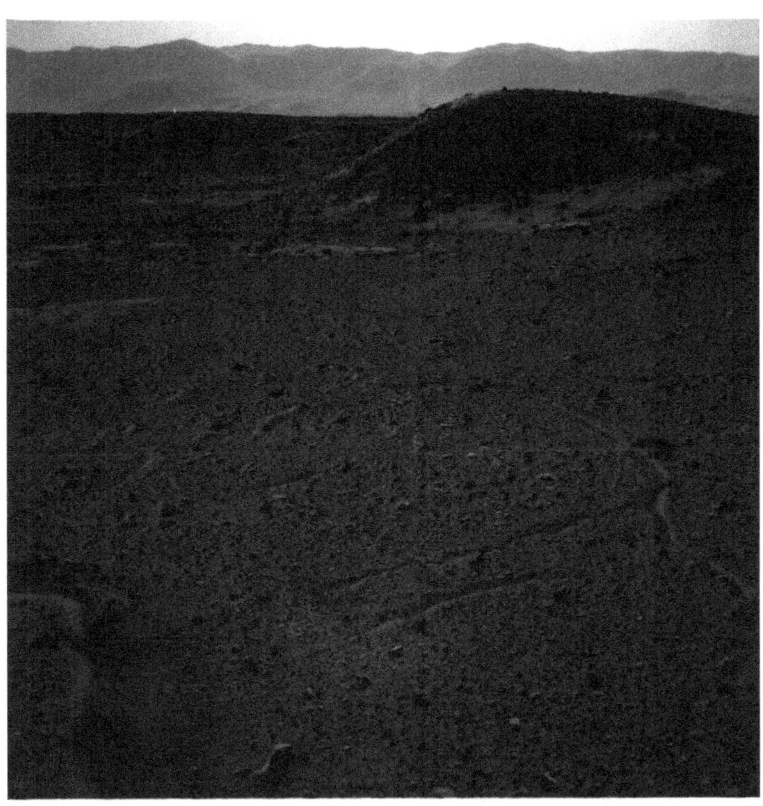

Original image. "Taken by Right Navigation Camera onboard NASA's Mars rover Curiosity on Sol 589 (2014-04-03, 10:00:03 UTC)." Image credit: NASA/JPL-Caltech.

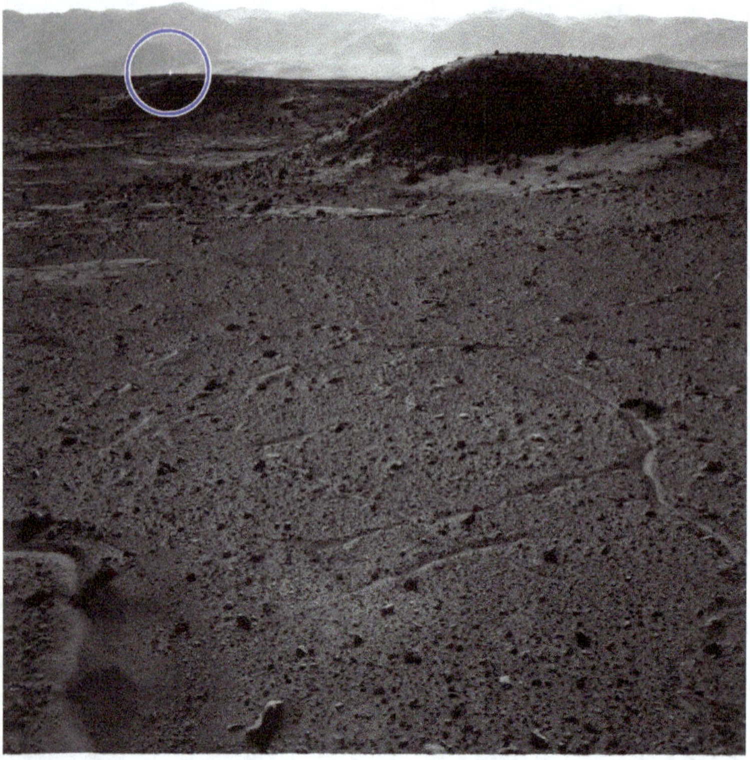

Above: Artificial Light, Sol 589, brightened and sharpened by the author for contrast. I have highlighted only one object in this image: upper left.

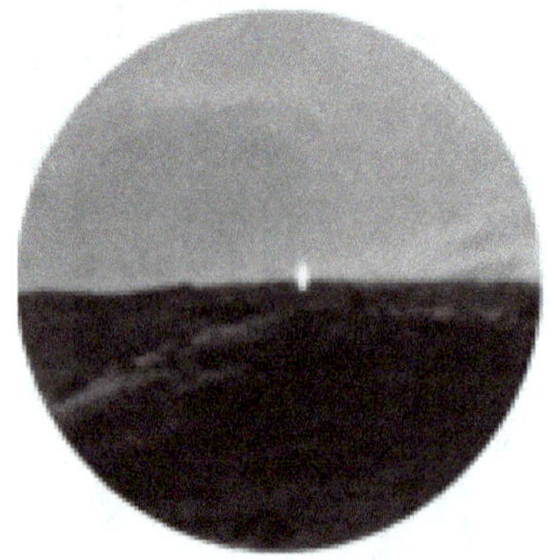

Left: Sol 589, closeup of highlighted area above. What is this bright cross like light shining from atop a Martian hill? Conspiracy theorists claim it is a sign that Mars currently hosts intelligent life. Orthodox scientists say it is most likely a "shiny rock," a "vent hole leak" in the camera, or a flash of "cosmic rays." Interestingly, the light disappeared immediately after this photo was taken.

Chapter 36

MARS RAT

SOL 52: EARTHDATE SEPTEMBER 28, 2012

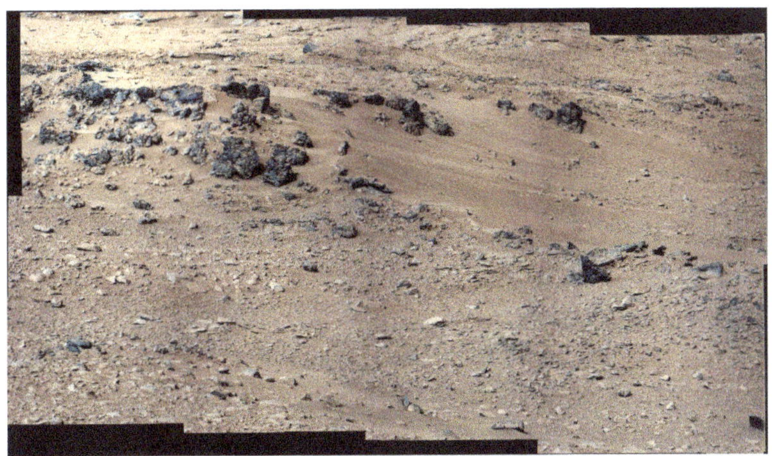

Original image. "This view is a mosaic of images taken by the telephoto right-eye camera of the Mast Camera (Mastcam) during the 52nd Martian day, or sol, of the mission (Sept. 28, 2012), four sols before the rover arrived at Rocknest. The Rocknest patch is about 8 feet by 16 feet (1.5 meters by 5 meters)." Image credit: NASA/JPL-Caltech/MSSS.

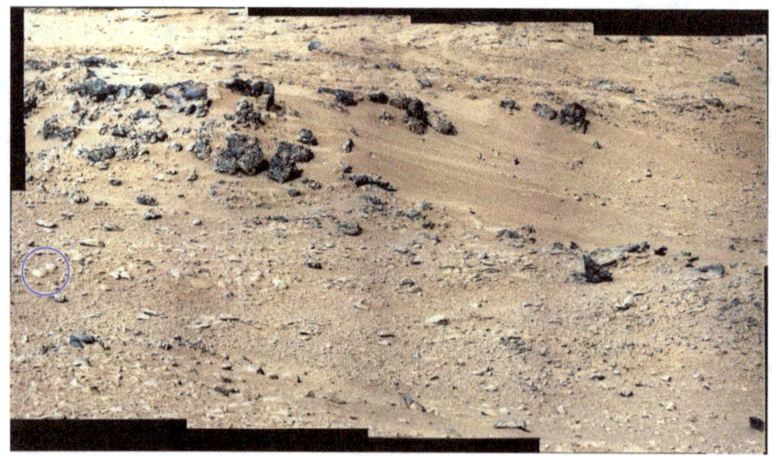

Above: Mars Rat, Sol 52, brightened and sharpened by the author for contrast. I have highlighted only one object in this image: lower left.

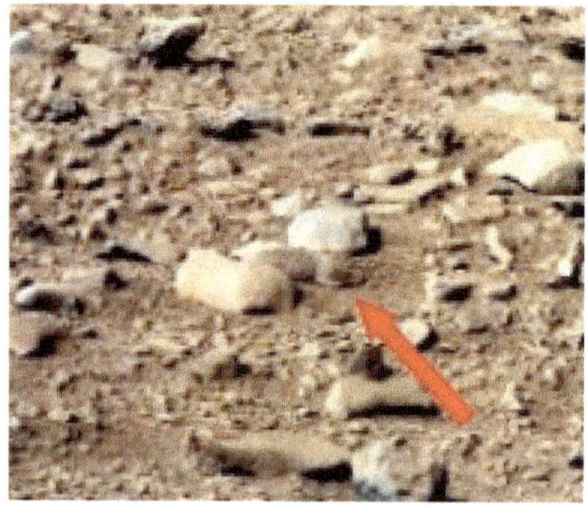

Above: Sol 52, closeup of the "Mars rat." A rodent like object appears to be sitting between two stones at center.

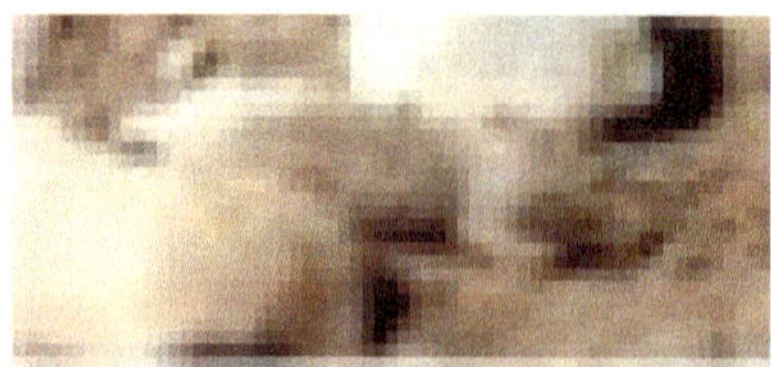

Left: Extreme closeup of the "Mars rat." Inanimate rock or living creature?

Chapter 37

ARTIFICIAL SPHEROID

SOL 746: EARTHDATE SEPTEMBER 11, 2014

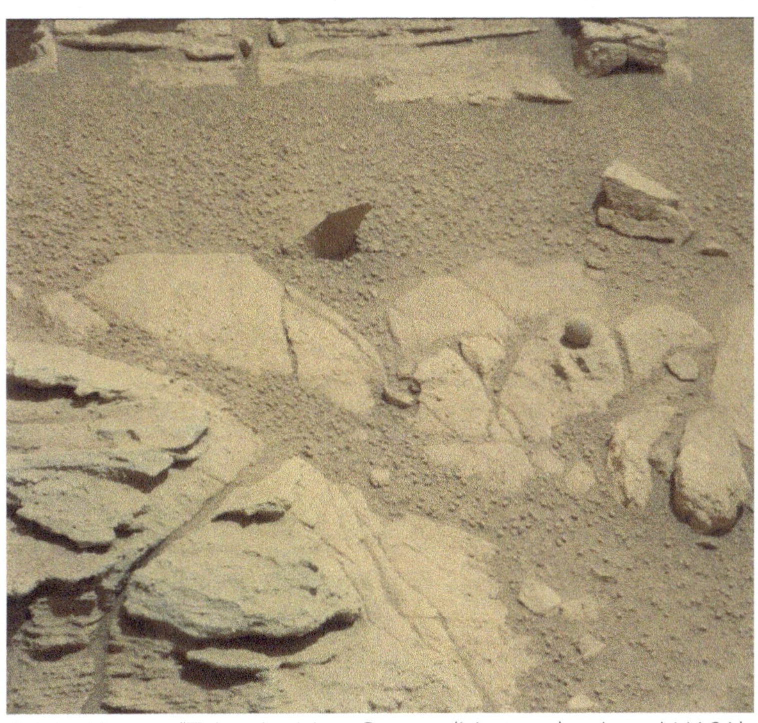

Original image. "Taken by Mast Camera (Mastcam) onboard NASA's Mars rover Curiosity on Sol 746 (2014-09-11, 14:46:58 UTC)." Image credit: NASA/JPL-Caltech/MSSS.

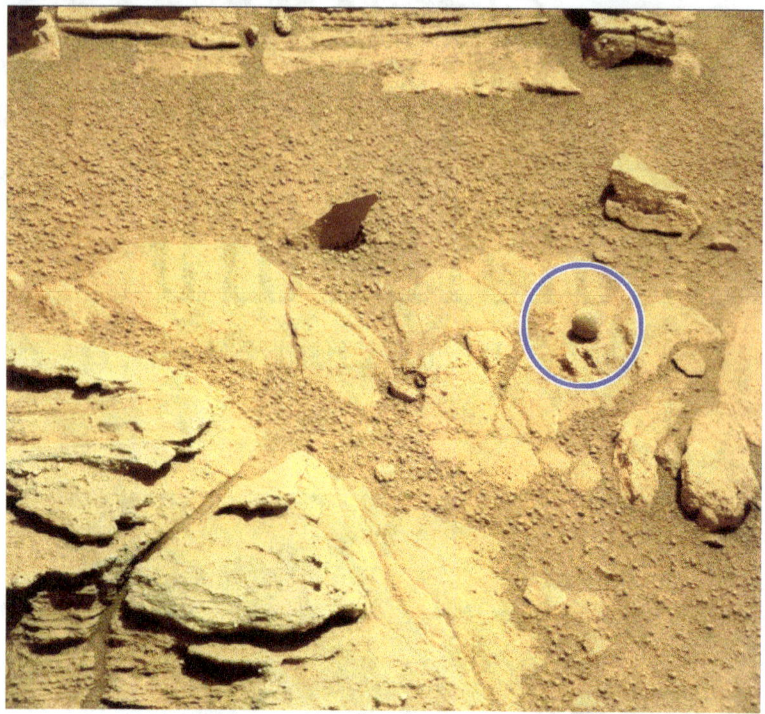

Above: Sol 746, brightened and sharpened by the author for contrast. I have highlighted only one object in this image: center right.

Right: This near perfectly round cannon ball like sphere perched on a rock on the surface of Mars would be exciting news for anomalists were it not for a simple fact: Martian wind and water can create mineral packed balls like this one through a process called concretion. Despite this, many believers in the Martian hypothesis maintain that this little grayish orb was made by an advanced race.

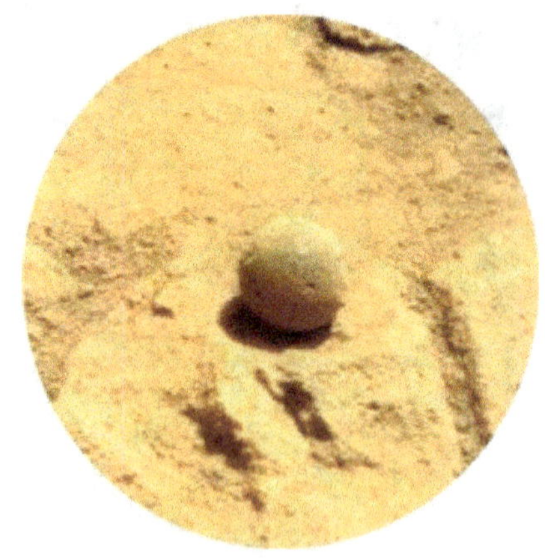

Chapter 38

TELESCOPE DISH

SOL 644: EARTHDATE MAY 29, 2014

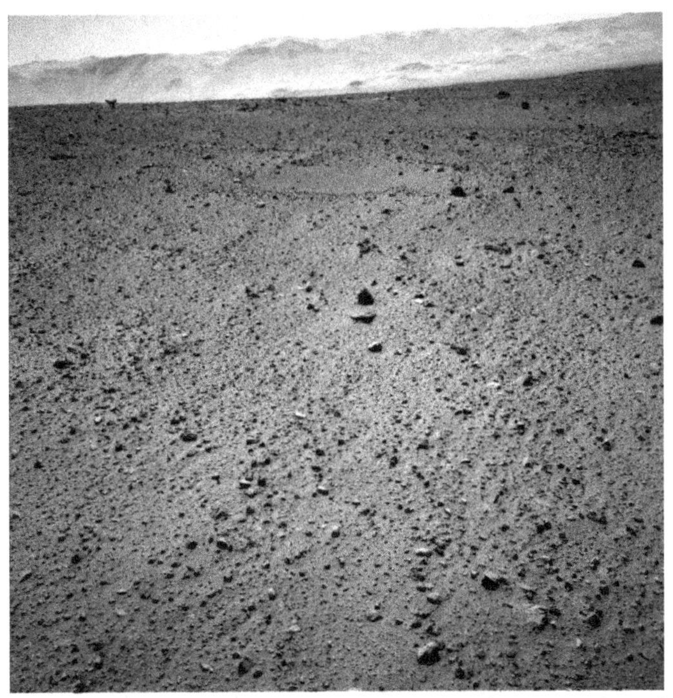

Original image. "Taken by Right Navigation Camera onboard NASA's Mars rover Curiosity on Sol 644 (2014-05-29, 20:05:24 UTC)." Image credit: NASA/JPL-Caltech.

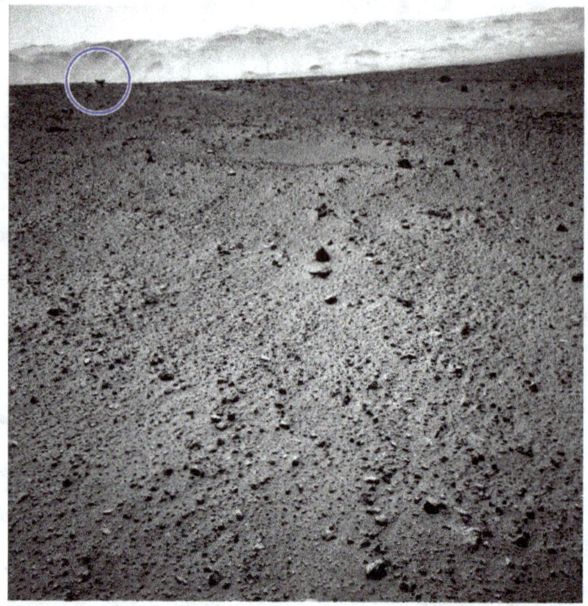

Above: Telescope Dish, Sol 644, brightened and sharpened by the author for contrast, with one highlighted area.

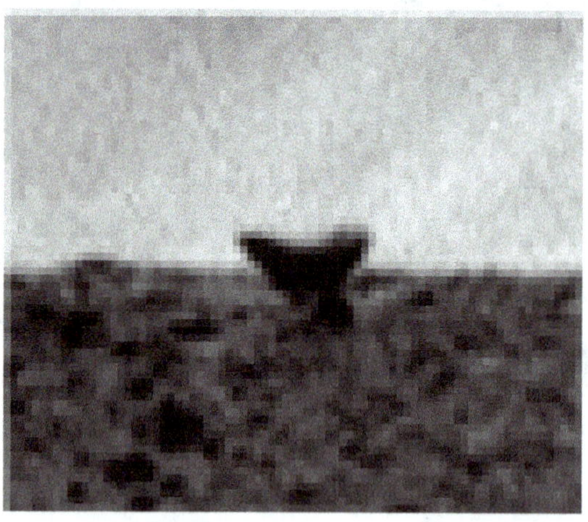

Above: Sol 644, extreme closeup of highlighted area above. Believers maintain that this highly unusual, seemingly hollowed out object is the dish of a radio telescope, or perhaps a scuttled spacecraft. Conventional scientists call it a "rock." Either way, how did it end up in this barren location?

Chapter 39

ODDS & ENDS

PIA23042: EARTHDATES DECEMBER 19, 2018

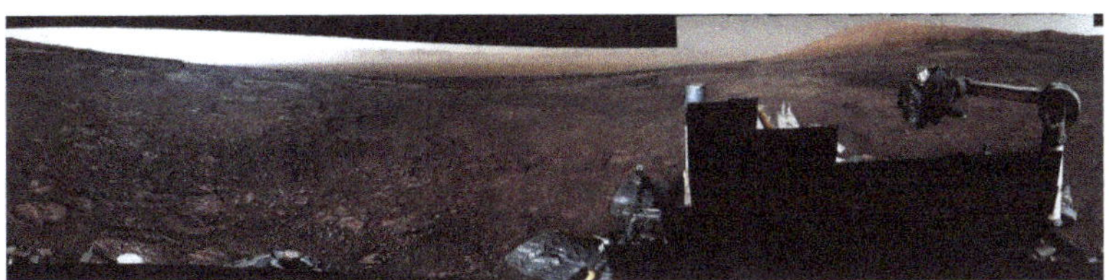

Original image. "This panorama was taken on Dec. 19, 2018 (Sol 2265), by the Mast Camera (Mastcam) on NASA's Curiosity Mars rover. The rover's last drill location on Vera Rubin Ridge is visible, as well as the clay region it will spend the next year exploring. The scene combines 122 images taken with Mastcam's left-eye camera." Image credit: NASA/JPL-Caltech.

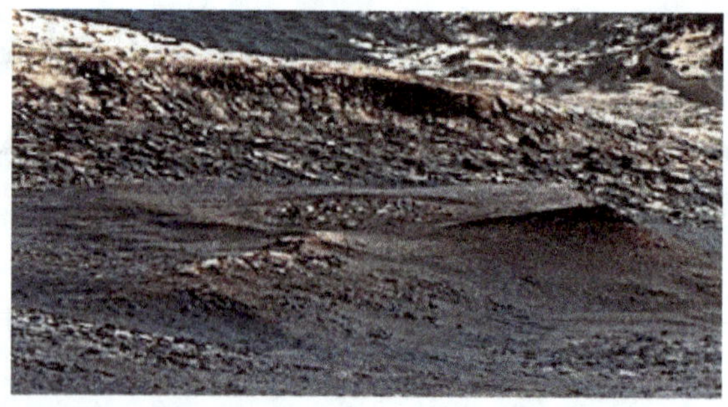

PIA23042-1 (above), colored, brightened, and sharpened by the author for contrast (as were all of the images in this chapter). A circle impression in a square rise.

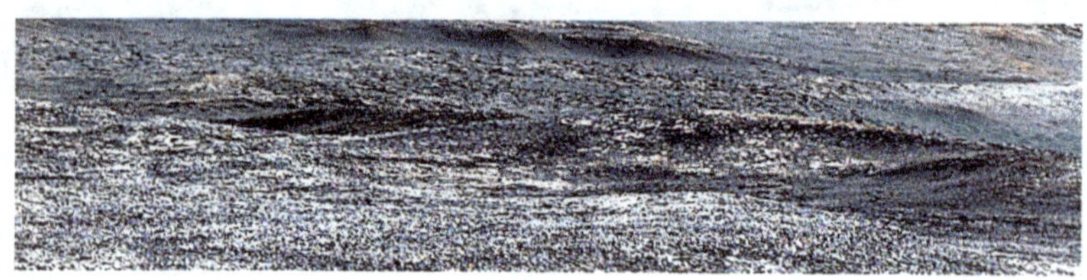

PIA23042-2 (above): These "cleared," square impressions in the Martian soil stand out against a backdrop of small, haphazardly scattered stone and gravel fields. The squares appear to have small rock "borders" around them.

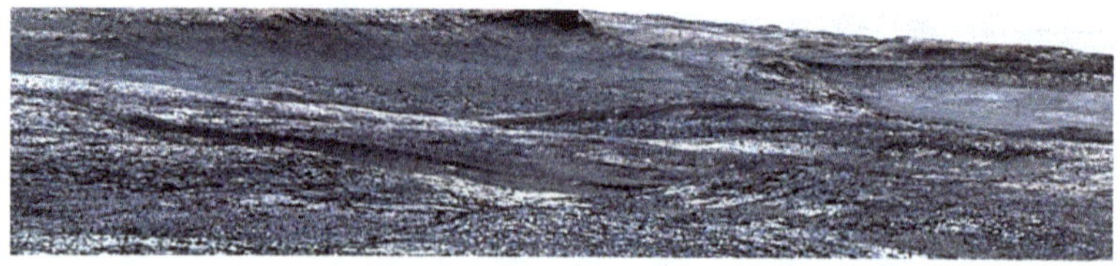

PIA23042-3 (above): More "cleared" square impressions.

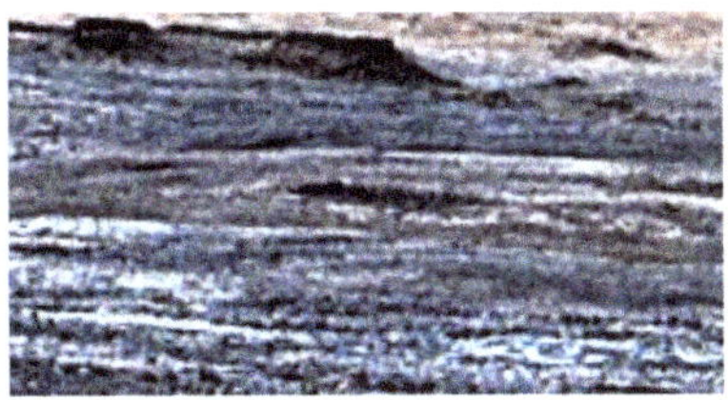

PIA23042-4 (above): This rectangular impression in the surface is further away from Curiosity's camera than those described in images 2 and 3. Note the odd shapes and markings at 9:00 and 3:00, and also in the lower right corner. All of this could be the result of natural astrogeological processes. It could also be the result of something else.

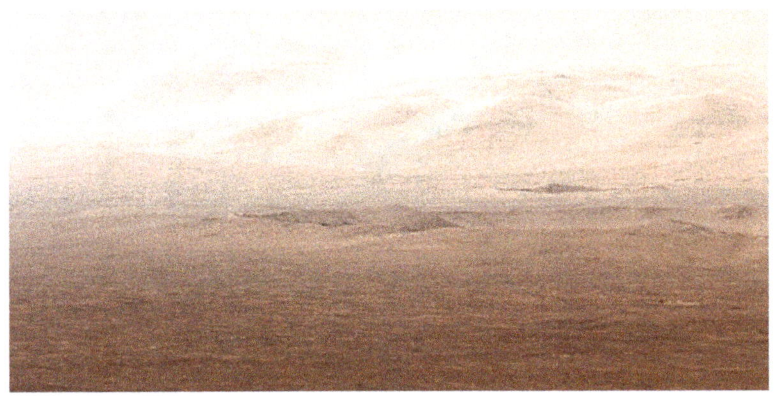

PIA23042-5 (above): Several massive square and rectangular holes or recesses can be seen in this long distance image of a Martian plain. How were they created? If by intelligent life, what purposes did they serve?

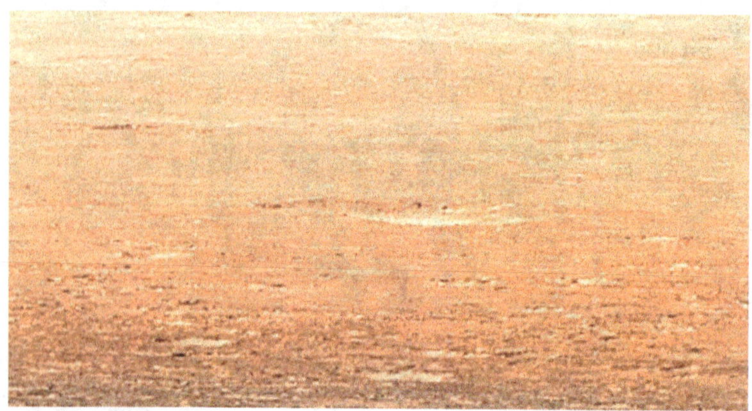

PIA23042-6 (above): Another large rectangle depression in the Martian dirt. Natural formations or artificial structures?

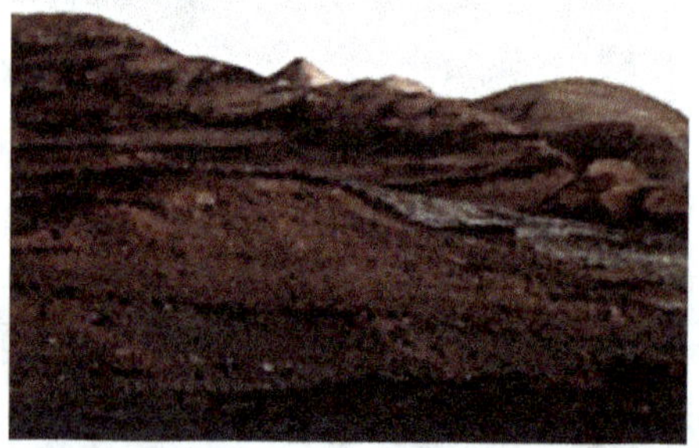

PIA23042-7 (above): In this image Curiosity appears to have captured the top of a pyramid on the crest of a distant Martian mountain. To its immediate right part of another mysterious structure can be seen just over the ridge. Both are the same color and stand out from their surroundings. What are they? Note the bizarre shapes and "designs" on the forward facing slopes. Some look like building foundations, others like lettering, symbols, or possibly artwork. On the far right a large circular shape with a "U" cut out of it can be seen laying sideways in the dirt.

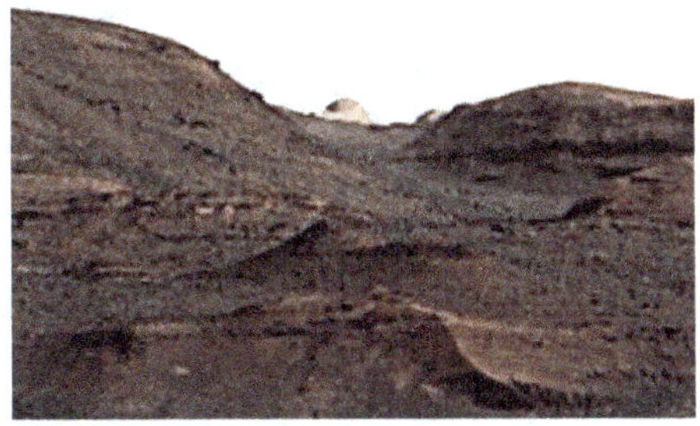

PIA23042-8 (above): Just to the left of the "pyramid" described in image 7, Curiosity came across this dome like structure on the same mountaintop. Again, to its right another strange structure looms just out of sight. The darkish slope on the right contains numerous odd markings, while at the lower right what looks like the remnants of a building foundation can be seen.

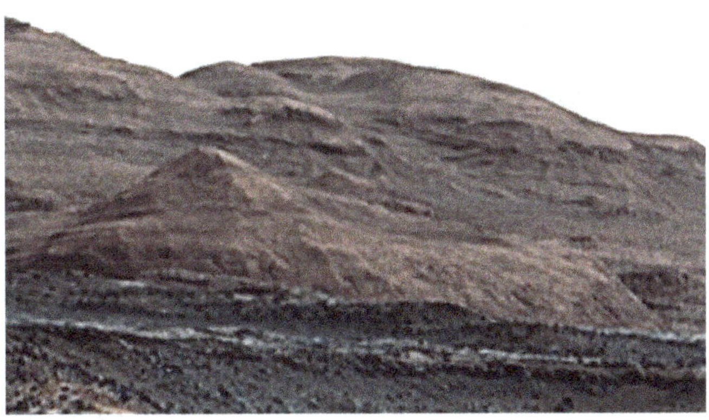

PIA23042-9 (above): Left of center is what seems to be the remnants of an ancient pyramid, identical in appearance to some of the oldest pyramids on our planet. On the slopes above it are what look like tumbled logs or structural beams half buried in the Martian soil. Other odd "designs" and objects abound. At the center of the mountaintop we can see what appears to be one of Mars' ubiquitous pointing "arrows" etched into the soil. Just below it and slightly left a "rock" with three large holes sits perched above the valley. This could all be pareidolia—or not.

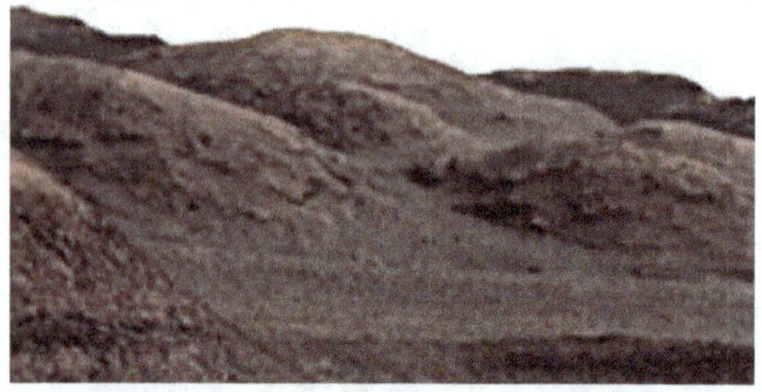

PIA23042-10 (above): These slopes brim with peculiar shapes, objects, and possibly symbols. Ancient Martian writing or art might look exactly like this after hundreds of thousands of years.

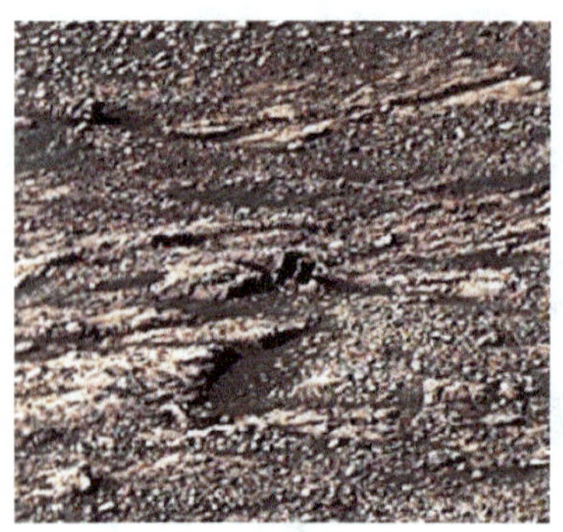

PIA23042-11 (left): What is the small box like object at the center of this image? It is missing a side and a top. We know what conventional scientists would call it. Look at it under magnification. What would you call it?

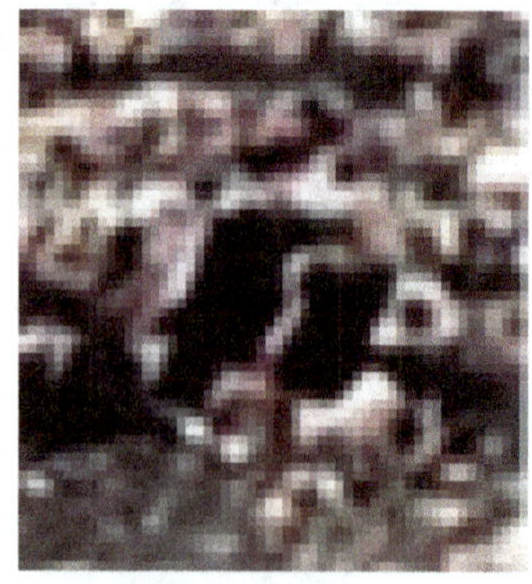

PIA23042-11 (right): Closeup of object described in the previous image. Zooming in on it only emphasizes its high strangeness—along with the many artificial looking oddments around it.

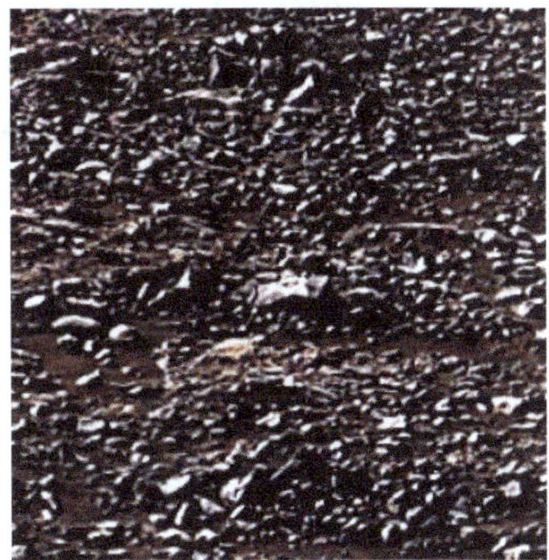

PIA23042-12 (right): At the center of this image there is a starry, sheriff's badge like object with a perfect gold circular object embedded in its center. What is it, and what is it doing in a Martian scree field?

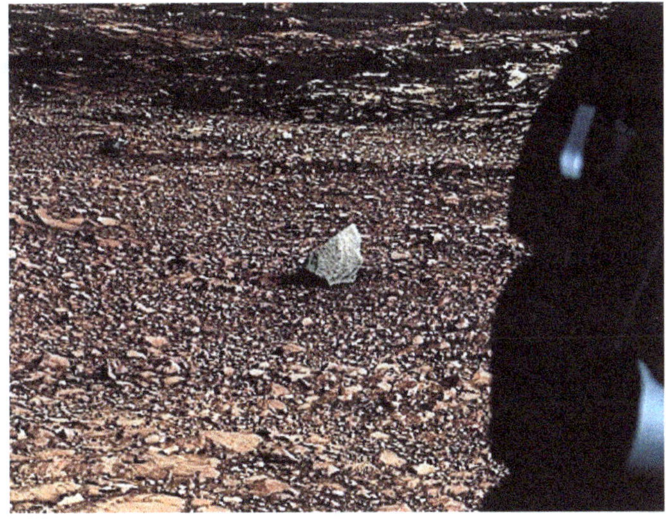

PIA23042-13 (above): At far right is part of the Curiosity rover. At the center of this image, however, sitting all alone in the middle of a talus slope, is a peculiar greenish and blackish object bearing "designs" and a jagged "outline" along one edge. It looks like nothing else in the full frame panorama photo. To NASA it is a "rock." Could it be a broken piece of Martian pottery or artwork? Whether it is natural or artificial, how did it end up in this specific location?

Chapter 40

ODDITIES

PIA17083: EARTHDATE JULY 24, 2013

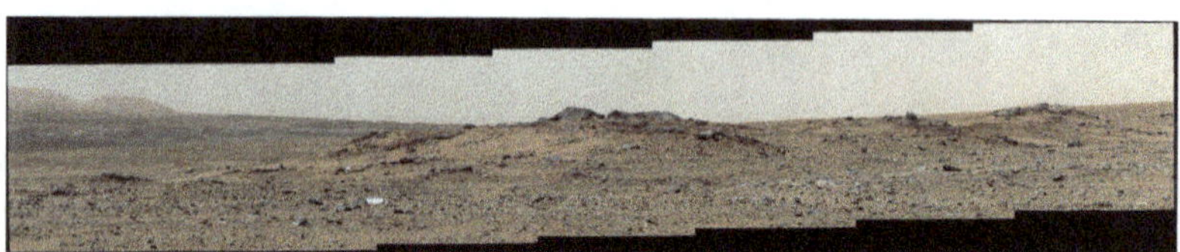

Original image. "This scene combines seven images from the telephoto lens camera on the right side of the Mast Camera (Mastcam) instrument on NASA's Mars rover Curiosity. The component images were taken between 11:39 and 11:43 a.m., local solar time, on 343rd Martian day, or sol, of the rover's work on Mars (July 24, 2013). A rise topped by two gray rocks near the center of the scene is informally named 'Twin Cairns Island.' It is about 100 feet (30 meters) from Curiosity's position. The two gray rocks, combined, are about 10 feet (3 meters) wide, as seen from this angle." Image credit: NASA/JPL-Caltech/MSSS.

ANALYSIS OF ODDITIES, PIA17083

Divided into Three Sections Due to Large Image Size

Section 1

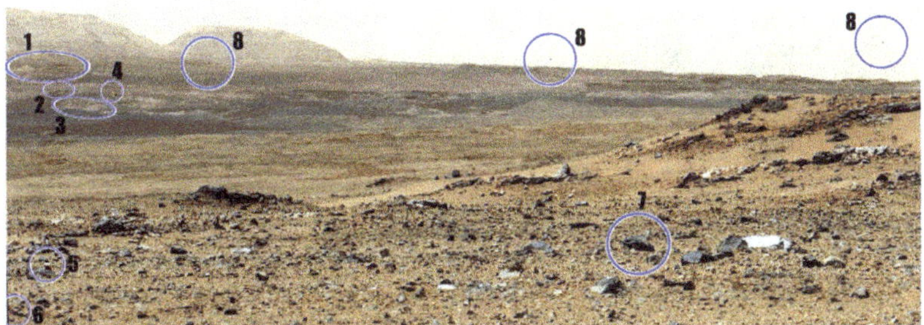

PIA17083-SECTION 1, brightened, colored, and sharpened by the author for contrast. This section, along with the following two sections, taken from the full frame NASA image of the Twin Cairns Island region (above) does not appear to be a "bomb site," but rather an area with random debris strewn about, as if dropped by hikers along a walking trail. In this particular image, which I have named Section 1 (left third of full frame photo), we find at least eight anomalies of interest. There are many more. But these will suffice to demonstrate the theory that a superior intelligence may have once lived—and may still live—on Mars.

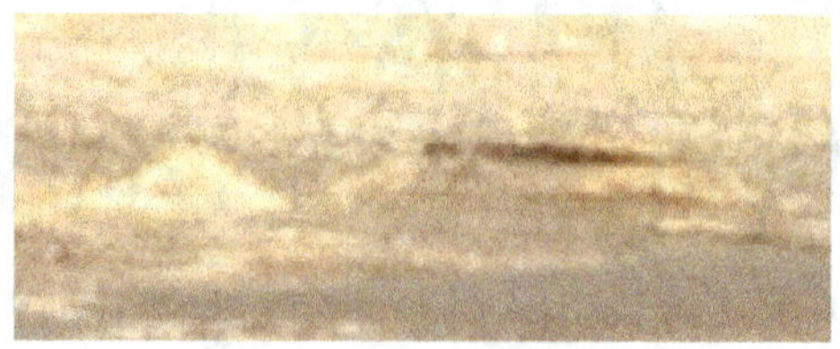

PIA17083-Section 1, area 1 (above): In this blurry long distance view of a Martian hillside, a large white triangle (on the left) and a cleared rectangular area (on the right) are plainly visible. Who or what made them? If something intelligent, what was or is their purpose?

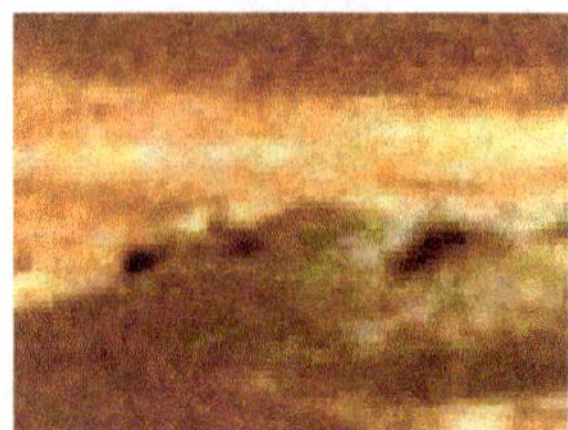

PIA17083-Section 1, area 2 (left): This odd "U" shaped structure sits just below the two structures described in area 1 above. Though pixelated in this closeup, one can just make out several small upright structures on the left side of the "U."

PIA17083-Section 1, area 3 (above): Just below the object described in area 2 above, we find this large whitish rectangle in the soil, a common sight on Mars.

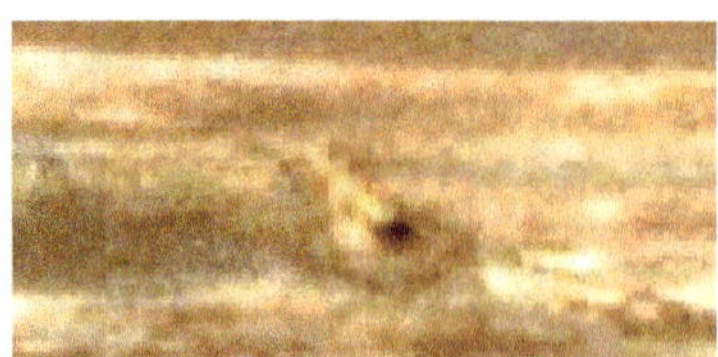

PIA17083-Section 1, area 4 (left): We cannot be sure what this log like or pipe like object is poking out from the side of a Martian hill. We can only say that it does not look natural. The fact that is in the same little area as the objects described in areas 1, 2, and 3, engenders curiosity.

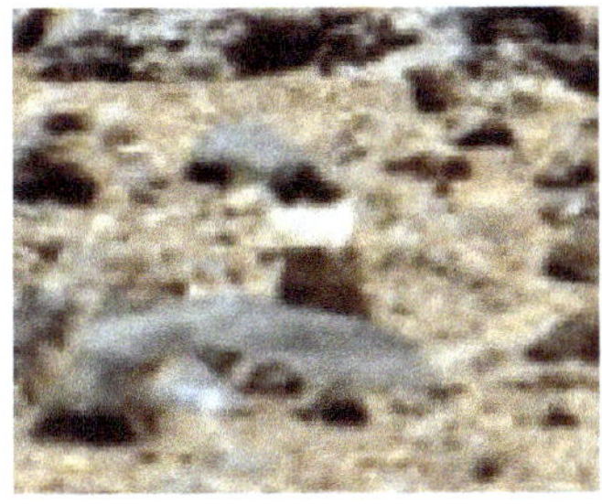

PIA17083-Section 1, area 5 (left): What is the extraordinary, white, box like object at center? It looks like a small, broken, child's playhouse with a covered porch and damaged supporting posts. "Windows" and perhaps a "door" can also be seen.

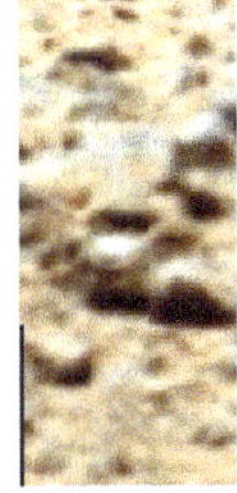

PIA17083-Section 1, area 6 (left): The hollow triangular object at center appears to be filled with something. Ordinary sediment deposit or an artificial material?

PIA17083-Section 1, area 7 (right): This bizarre object has the appearance of fossilized tree roots. NASA, however, has declared that trees *probably* never existed on the Red Planet.

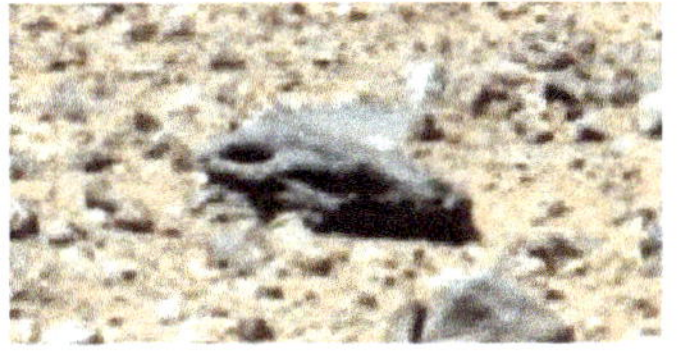

PIA17083-Section 1, area 8 (below left): I have given the number 8 to all of the tiny UFO like objects in the full frame shot. Scientists label them "digital artifacts," or in some cases "dead pixels." Since they appear numerous times in many of the images they are captured in, this may be true. However, do digital artifacts have "appendages," or produce pink, green, and yellow "magnetic fields" around themselves—as can be clearly seen in this photo? We must also consider the possibility that a digital artist intentionally copied and pasted one *real* UFO multiple times in some NASA photos as a form of disinformation: data meant to conceal the truth, influence public opinion, or distract, manipulate, and mislead the populace.

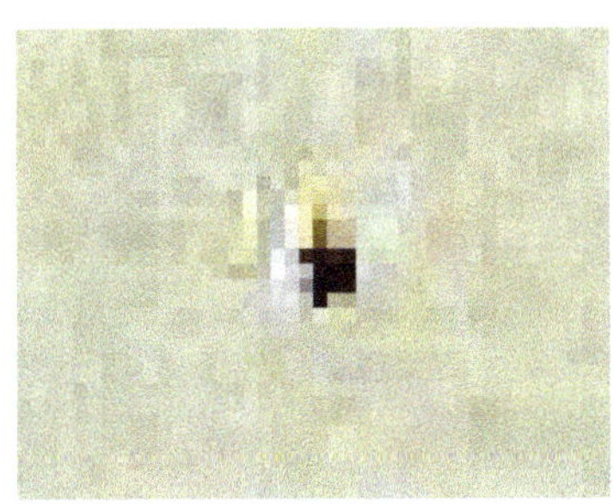

Section 2

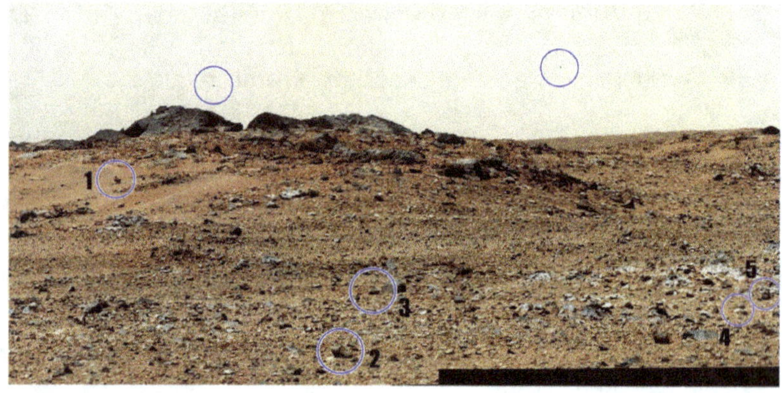

PIA17083-SECTION 2, brightened, colored, and sharpened by the author for contrast. This section (middle third of the full frame photo) contains five anomalies of interest in the Twin Cairns Island region. (Note: For information on the two circled aerial objects, see Section 1, number 8, above.)

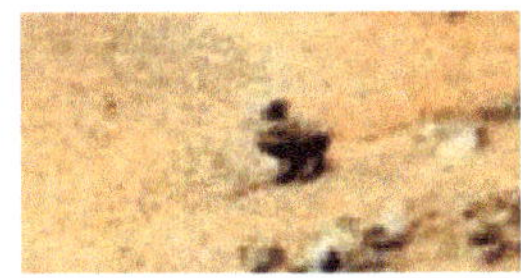

PIA17083-Section 2, area 1 (above): I have named this unusual object the "Mars Duck," due to its remarkable resemblance to the common waterfowl. It seems to have a "head," a "body" with "folded wings," a "tail," and two "webbed feet." If it is not a statue or effigy of an avian species of some type, it is an exceptional rock formation indeed.

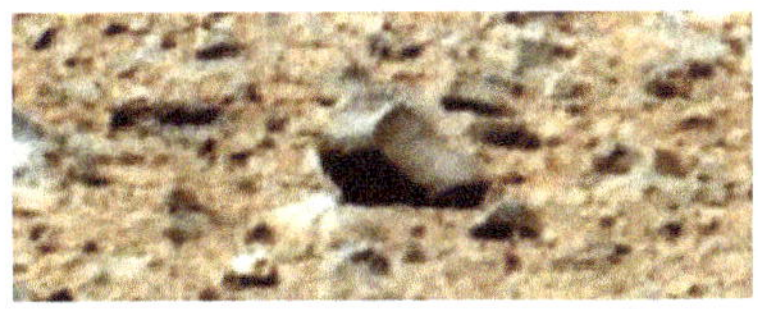

PIA17083-Section 2, area 2 (above): I have named the odd object at center the "Loaf of Bread." We can be confident that this is not what it is. However, until future astrogeologists examine it in situ, it is a handy moniker.

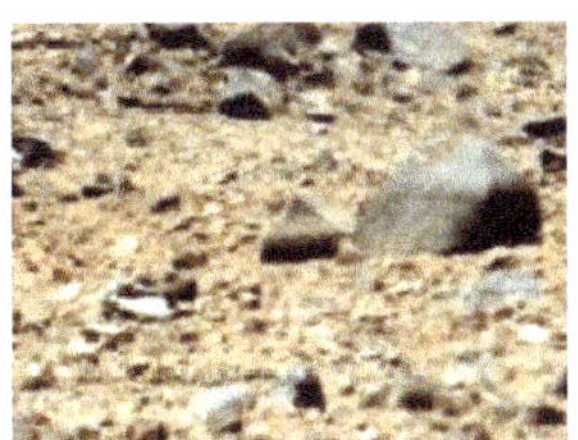

PIA17083-Section 2, area 3 (above): At the center of this photo there is an out-of-place triangular object. It appears artificial and has a ridge running along all three facing sides. To its lower left (about 8:00), there is a weird "U" shaped item sitting next to a brilliant white "shard" of some kind. These are located in the midst of a host of squares, circles, straight lines, and other unusual shapes.

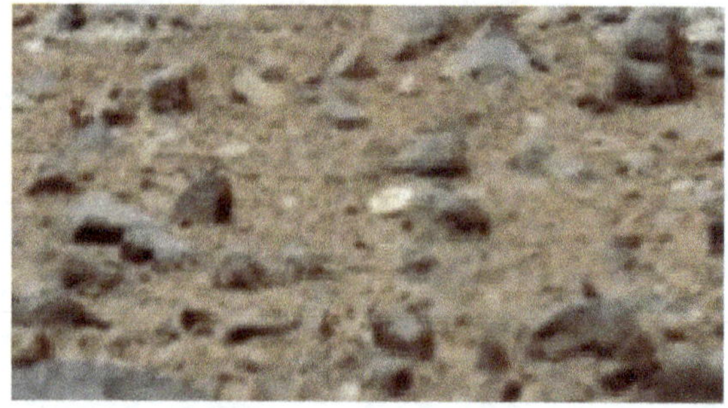

PIA17083-Section 2, area 4 (above): I have named the bright, shiny, whitish object at the center of this image the "Toy Flying Saucer." Though "damaged," it has the shape and appearance of a classic hat shaped flying saucer, complete with a "rim" and a "dome." Is this an ancient Martian child's plaything or, as establishment scientists claim, an ordinary rock?

PIA17083-Section 2, area 4 (left): An extreme closeup of the "Toy Flying Saucer" described in area 3 above.

PIA17083-Section 2, area 5 (right): What are the strange objects in this photo? Perhaps we are not supposed to know.

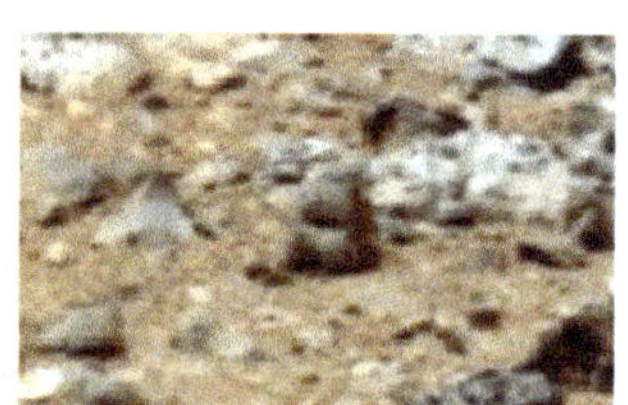

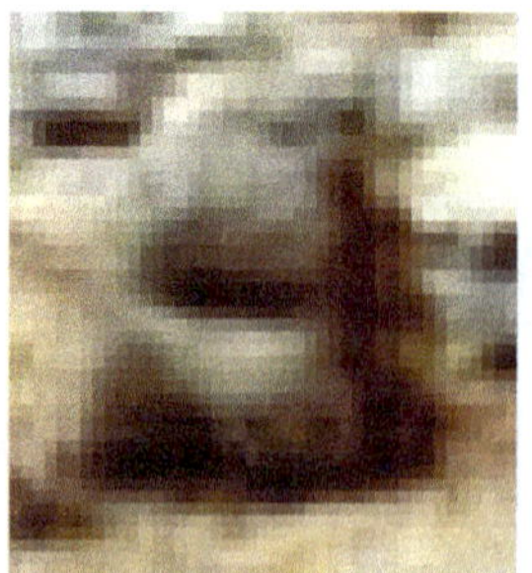

PIA17083-Section 2, area 5 (left): An extreme closeup of the weird object at center in area 5 above. Could Nature create such a thing? Possibly. But so could a technological life form.

Section 3

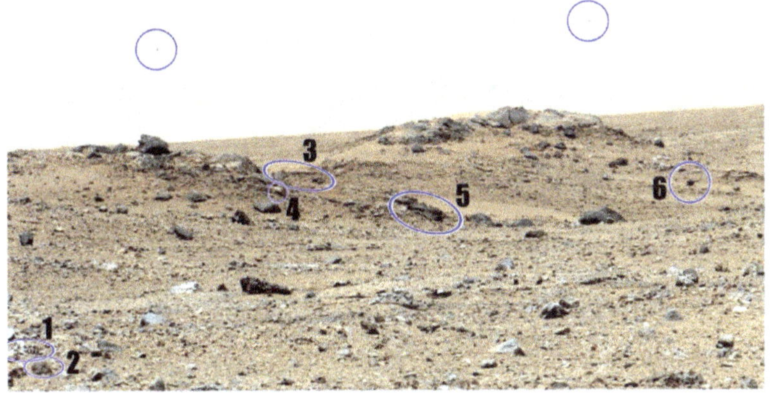

PIA17083-SECTION 3, brightened, colored, and sharpened by the author for contrast. This section (right third of the full frame photo) contains six anomalies of interest, again located in the Twin Cairns Island region. (Note: For information on the two circled aerial objects, see Section 1, number 8, above.)

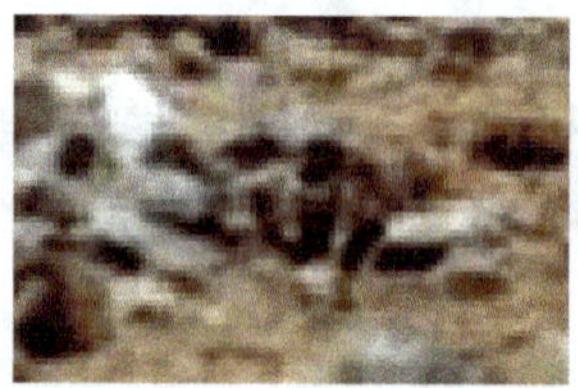

PIA17083-Section 3, area 1 (above): This primitive skull like object peers out from the reddish brown Martian soil. A host of other strange artificial looking items surround it.

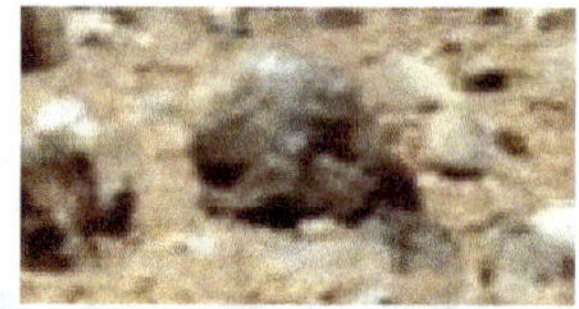

PIA17083-Section 3, area 2 (above): Under magnification this bizarre, spiraled, roundish object appears to have a Celtic cross like hole or cavity "cut" into its right side. While this is almost certainly pareidolia, it is scientifically important to conjecture.

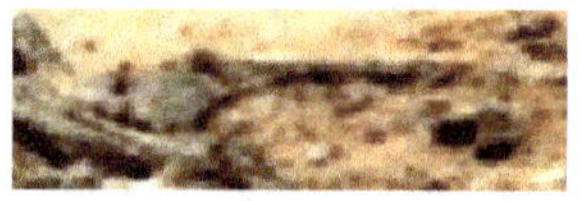

PIA17083-Section 3, area 3 (above): Under magnification at least two unusual artificial looking "formations" occur in this pixelated image: a circular shape surrounding a cleared patch of ground (center), and a pie like object with a triangular "piece" cut out of it (lower right). Many other oddities can be seen, but are too numerous to discuss here.

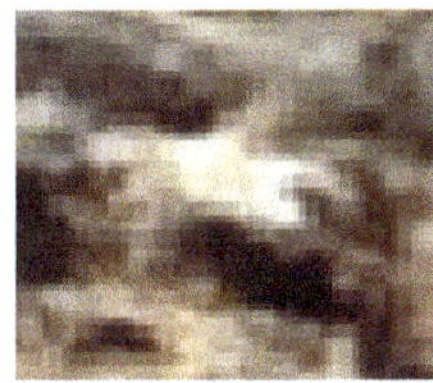

PIA17083-Section 3, area 4 (left): Heavy pixelation does not prevent one from seeing potential artificiality in the odd white object at center.

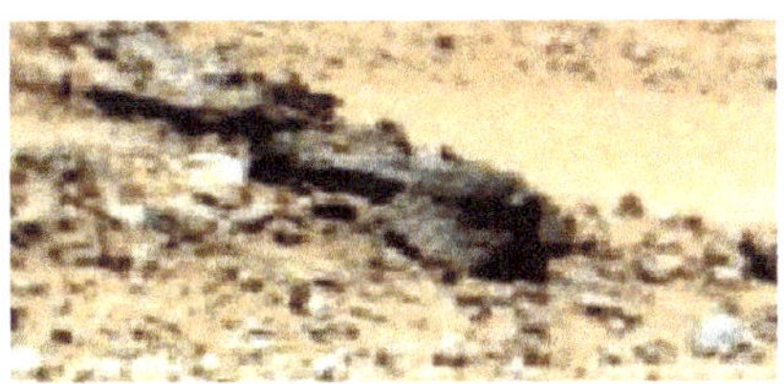

PIA17083-Section 3, area 5 (above): This object resembles the remains of an aquatic animal of some kind, perhaps something similar to our skates and rays (Rajiformes) here on Earth. As we have discussed previously, it has been shown that Mars almost certainly once had surface rivers, lakes, and oceans—and probably still hosts aquifers and groundwater.

PIA17083-Section 3, area 6 (right): The tallish black object at center is round and has a squarish hollow center, suggesting it is an artifact.

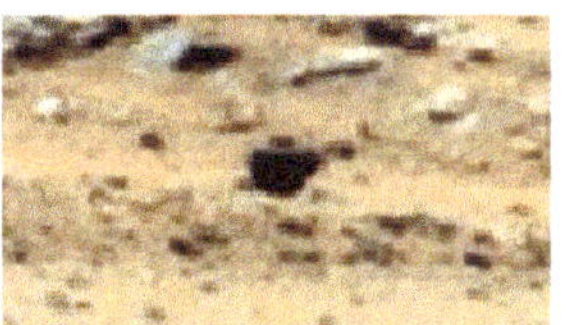

PIA17083-Section 3, area 6 (left): An extreme closeup of the object described in area 6 above. Could this be a piece of ancient Martian pottery, or perhaps a part from a machine of some kind? And what is the reddish color to the lower right of the object? The answers to these questions, and thousands of others, await a fascinated world.

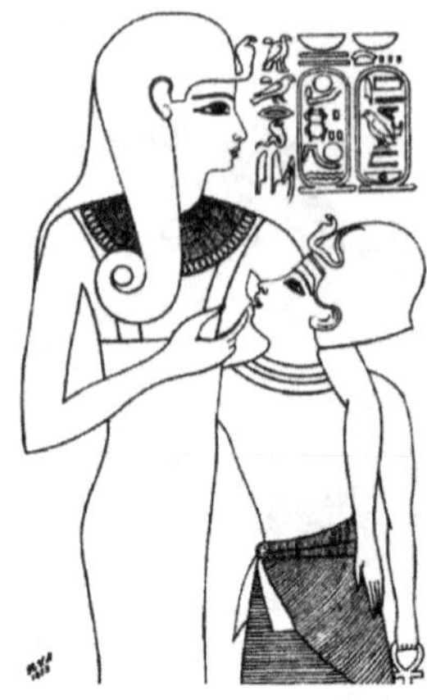

Above: As history reveals, ancient Egyptian images of the virgin mother goddess Isis and her savior son/sun Horus provided an artistic, biographical, and religious template for later emerging mother goddesses and their savior sons—in particular the Virgin Mary and Jesus. In turn, Martian anomalies host a myriad of ancient Egyptian like writing, art, statuary, buildings, and religious structures. If there was an ancient Martian civilization, is it somehow connected to Egypt, and if so, how?

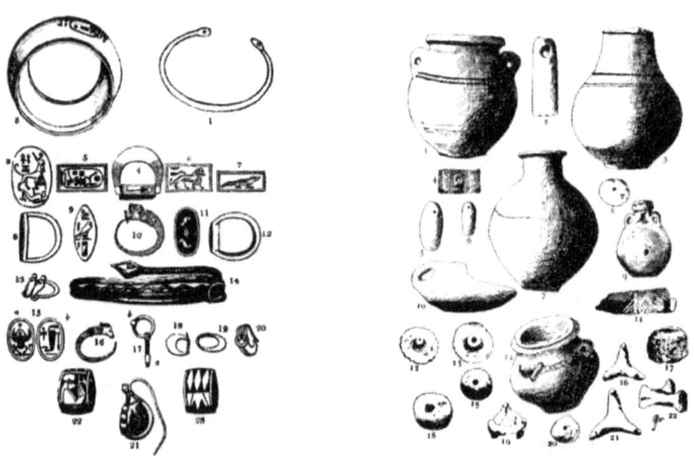

Above left: A collection of ancient Egyptian jewelry (rings, signets, bracelets, and earrings). Above right: Examples of Babylonian household goods (clay jars, spindle weights, jeweler's furnaces, bottles, lamps, sieves, and pottery rests). Thousands of items that look very much like these 4,000 year old Earth artifacts continue to turn up in Martian photos.

Section Two

Images From Above the Surface of Mars

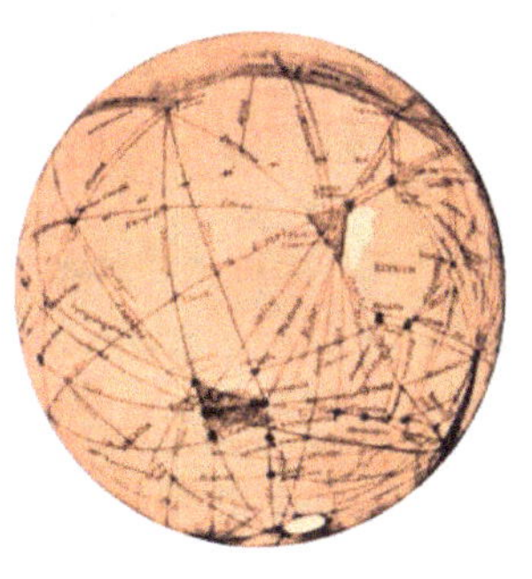

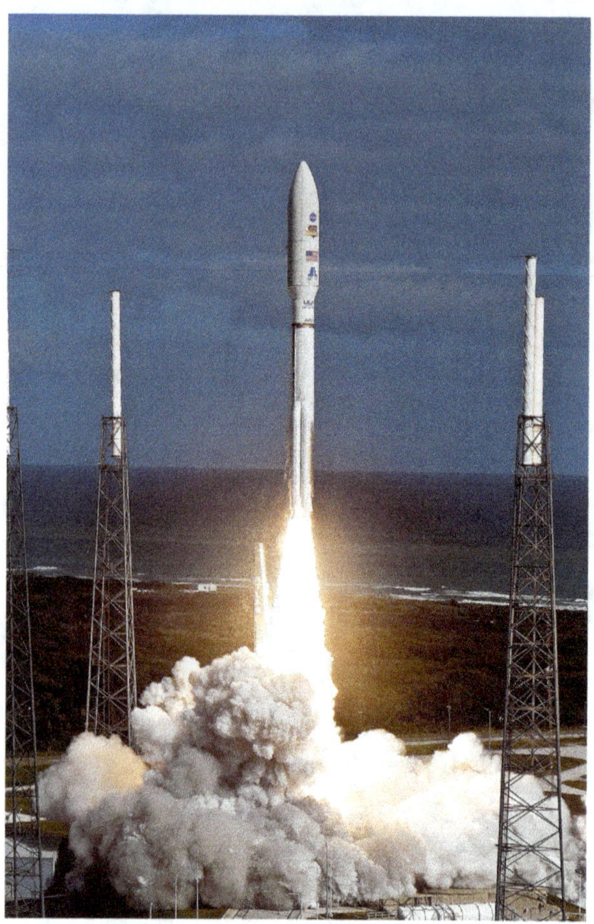

November 26, 2011: NASA's Mars Science Laboratory (the Curiosity rover) lifts off from Cape Canaveral Air Force Station, Florida, USA. A major facet of Curiosity's mission is to, in NASA's words, hunt for "potential biosignatures" on Mars. Why not actual life itself? Because conventional science *a priori* rejects the extraterrestrial theory. This means that mainstream astronomers are, to some extent, conducting their search for life on the Red Planet deductively rather than inductively (that is, by way of *a posteriori* analysis), the latter a methodology vital to authentic science. I have discussed the reasons for this approach in my introduction. Image credit: NASA/JPL-Caltech.

Chapter 41

PYRAMIDS, FACE, & CITY

CYDONIA REGION: EARTHDATES 1976-2001

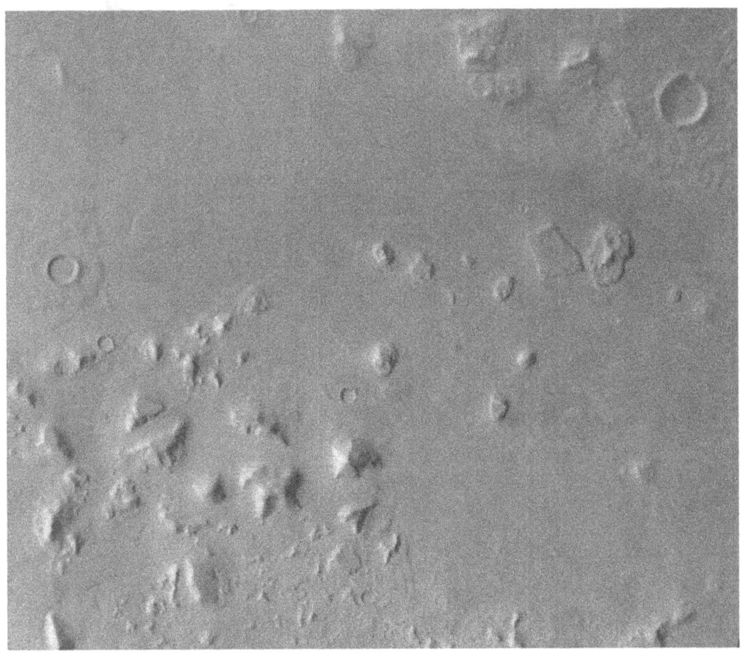

Original image. "Taken on April 14, 1998, by the Mars Global Surveyor, at 14:02:17 UT (10:02:17 a.m. EDT), on its 239th orbit around Mars." Image credit for this photo, and all other photos in this chapter: NASA/JPL/Malin Space Science Systems.

ANALYSIS OF CYDONIA REGION

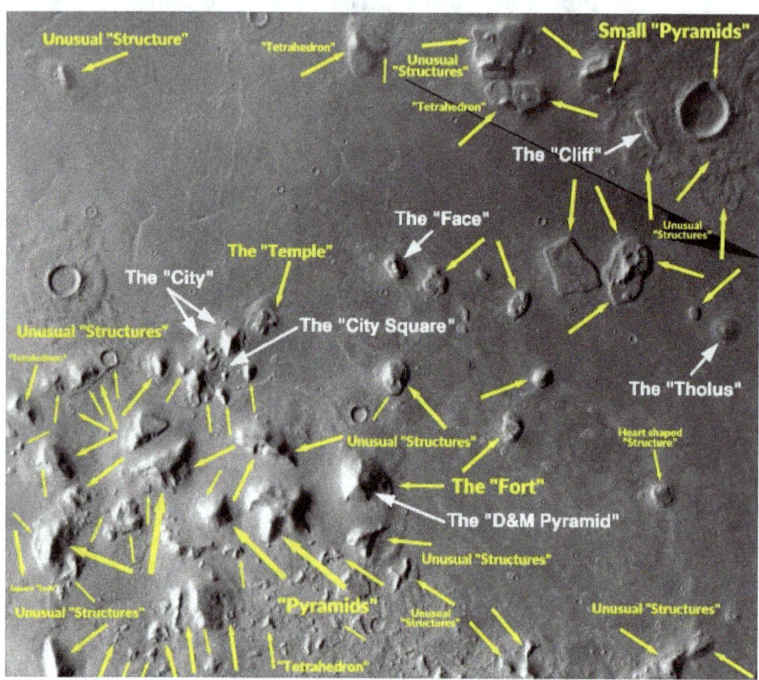

Pyramids, Face, and City: Cydonia region. This 1998 image has been brightened and sharpened by the author for contrast, with anomalies highlighted. Text and arrows in white (as well as the black arrow, upper right) are by NASA. Text and arrows in yellow are by the author (L.S.). As is apparent, I see much more in this image than NASA does—or perhaps is willing to admit publicly.

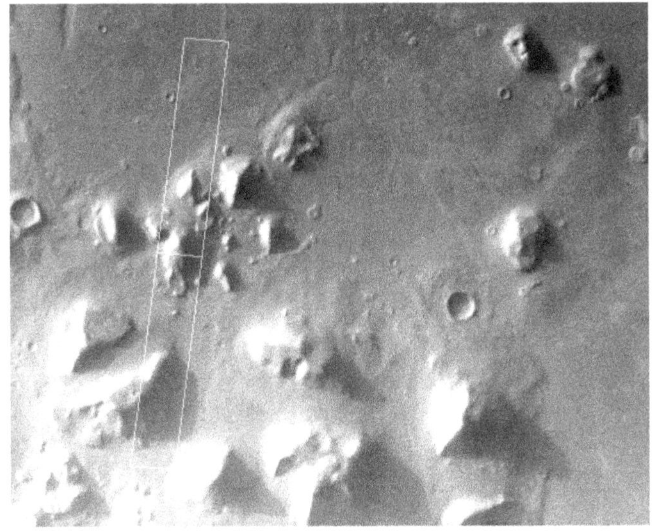

Cydonia region (above): Another look, this time zoomed in. (Note: The white box is one of NASA's image coordinates).

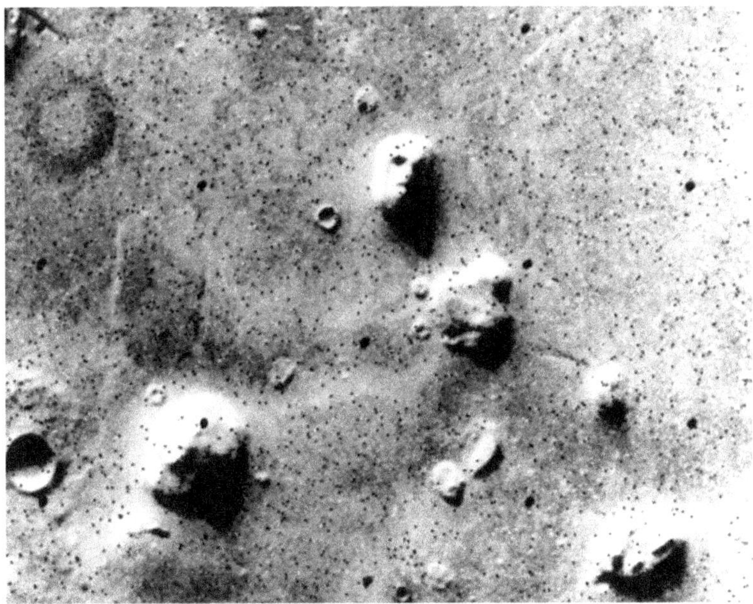

Cydonia region, "Face on Mars" (above): This famous image of the "face" (near upper center) was snapped by Viking 1 in 1976. Image credit: NASA/JPL.

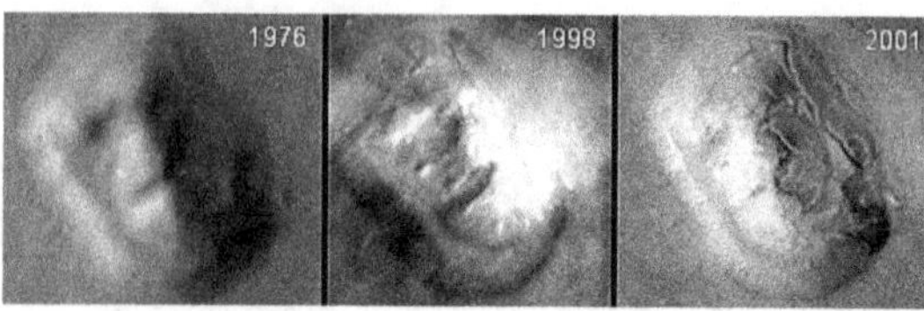

"Face on Mars" (above): Though originally the "Face on Mars" excited much wonder and curiosity as a possible sign of an intelligent Martian race, high resolution images taken during NASA's later Mars missions allegedly debunked the theory, as shown in this photo sequence spanning the years 1976 to 2001. The public's 1976 "misidentification" was chalked up to "low resolution imagery" and "shadows," which scientists now say concealed the "face's" true identity: a Martian mesa, similar to the buttes that are common to the American West on Earth. Images credit: NASA/JPL/Malin Space Science Systems.

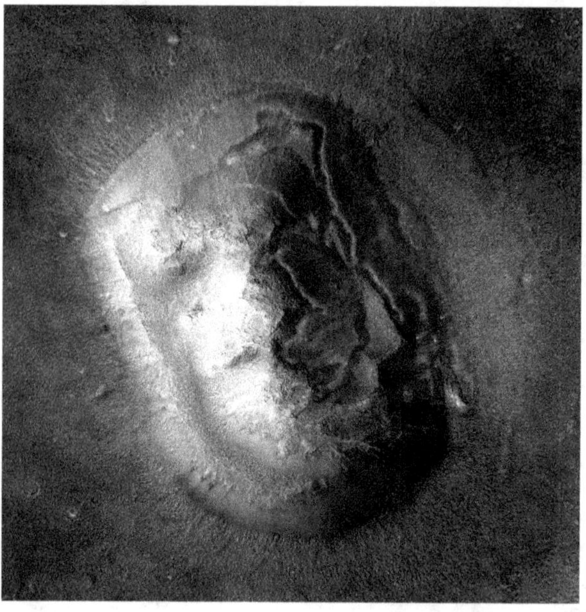

"Face on Mars" (above): A high resolution closeup of the 2001 photo of the famous head, which many believed could be a pharaonic like Martian monument. Despite being presumably "debunked" by orthodox astronomers long ago, a sizeable percentage of the public continues to assume that there has been a coverup, and that the full truth about the Face on Mars has yet to be revealed. Image credit: NASA/JPL/Malin Space Science Systems.

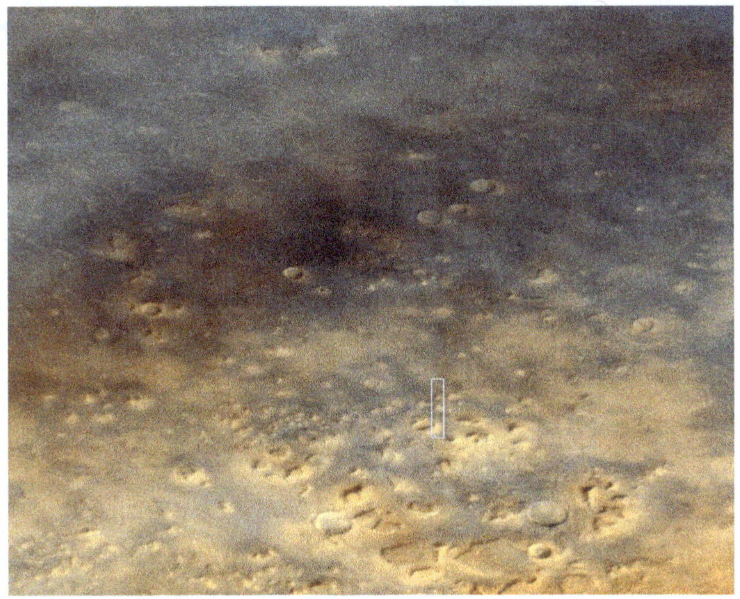

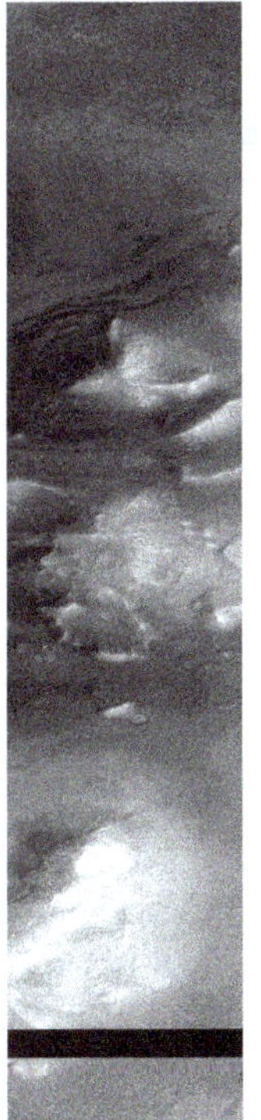

Cydonia region (above) from a great height over the surface of Mars. This low resolution wide angle view shows (in the small white rectangle) the location of the narrow vertical angle image in the lower left below. Image credit: NASA/JPL/MSSS.

Cydonia region (left), showing the section highlighted in the white rectangle in the wide angle image above. (Note: The black stripe near the bottom is part of NASA's original image.) Image credit: NASA/JPL/MSSS.

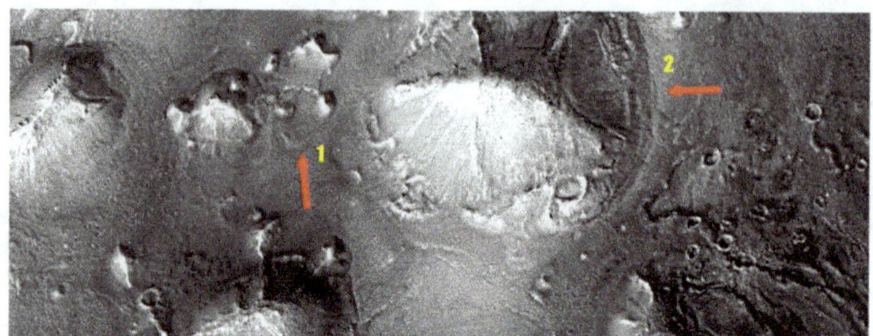

Cydonia region (above), showing 1) the "City Square" area; and 2) one of the region's massive "pyramids."

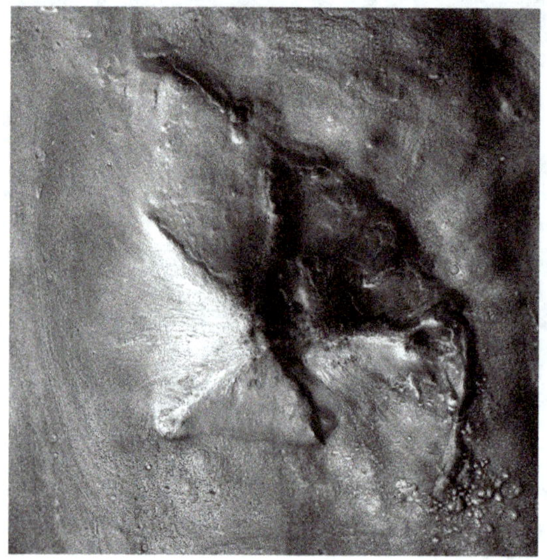

Cydonia region (above): Closeup of the "D&M Pyramid" at Cydonia (named after its discoverers, NASA computer engineers Vincent DiPietro and Gregory Molenaar), with what look like "reinforced" corners at the base. It is located near 40.7°N, 9.6°W. NASA describes it as a natural "landform . . . one of thousands of massifs, buttes, mesas, knobs, and blocks that mark the transition from the far northwestern Arabia Terra cratered highlands [on Mars] down to the northeastern Acidalia Planitia lowlands. Each block, whether shaped like a face, a pyramid, or simply a mesa, massif, or knob, is a remnant of the bedrock of northeastern Arabia [on Mars] that was left behind as erosion slowly degraded the terrain along this zone between the highlands and the lowlands." I, along with many others, would be more apt to call NASA's natural "landform" a collapsed ancient pyramid.

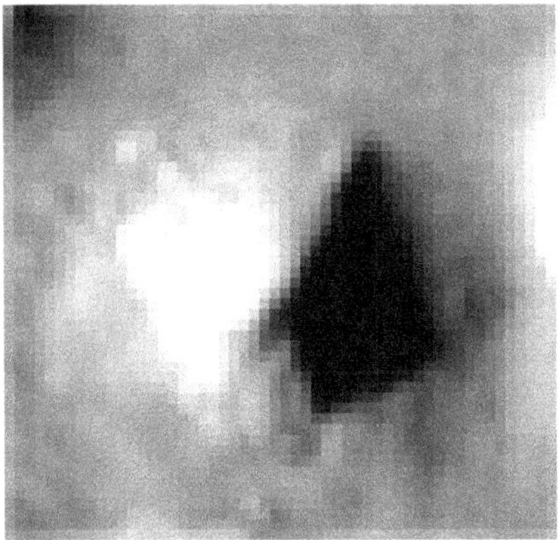

Cydonia region (above), closeup of a second Cydonian "pyramid." Note the clear triangular shadow falling to the right.

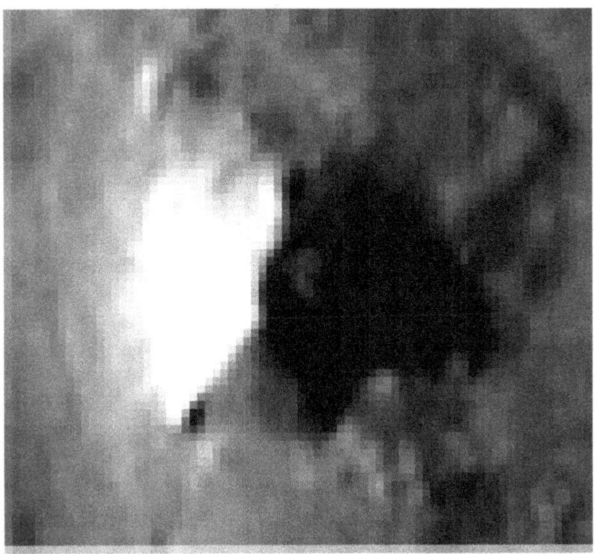

Cydonia region (above), closeup of a third Cydonian "pyramid." Note the triangular shadow (right) and the perfect triangle on the facing side (bottom).

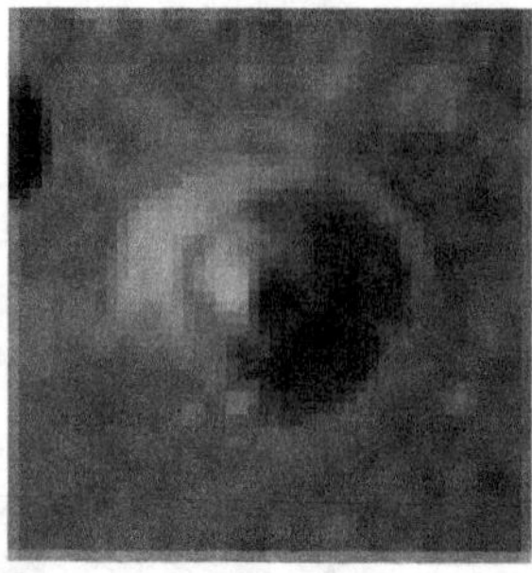

Cydonia region, closeup of the "Tholus" (above): This artificial looking structure has what appears to be a spiral staircase that wraps around it from bottom to top, making it similar in design (and possibly function) to the Mesopotamian ziggurat, a pyramidal stepped temple tower used across the ancient Middle East, often for religious purposes.

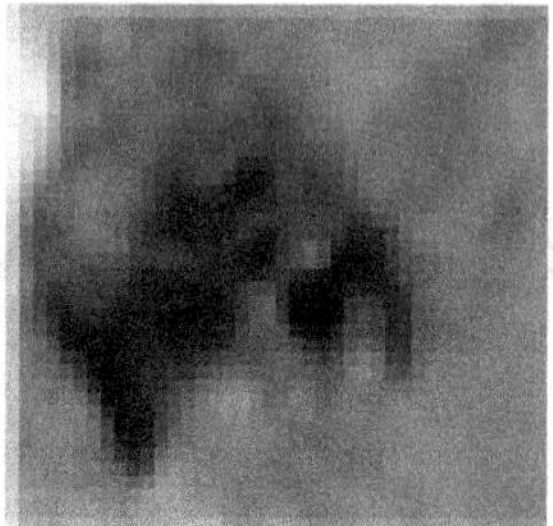

Cydonia region, closeup of the "Fort" (above): To the right of the "D&M Pyramid" sits another artificial looking feature, this one containing what I believe could be "fortified walls," "sally ports," and possibly reinforced "doorways" and protected "walkways." For these reasons it has been dubbed the "Fort"—though NASA considers it just another landform.

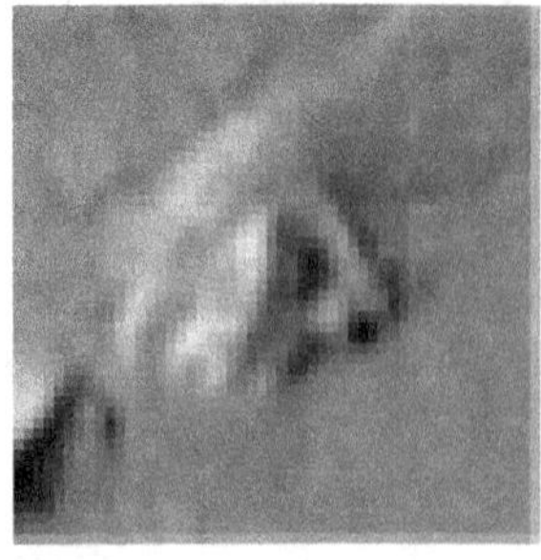

Cydonia region, closeup of the "Temple" (left): Under magnification there appears to be several small pyramids at the center, surrounded by walls, pillars, and a large plaza or courtyard. (This is my name for this structure.)

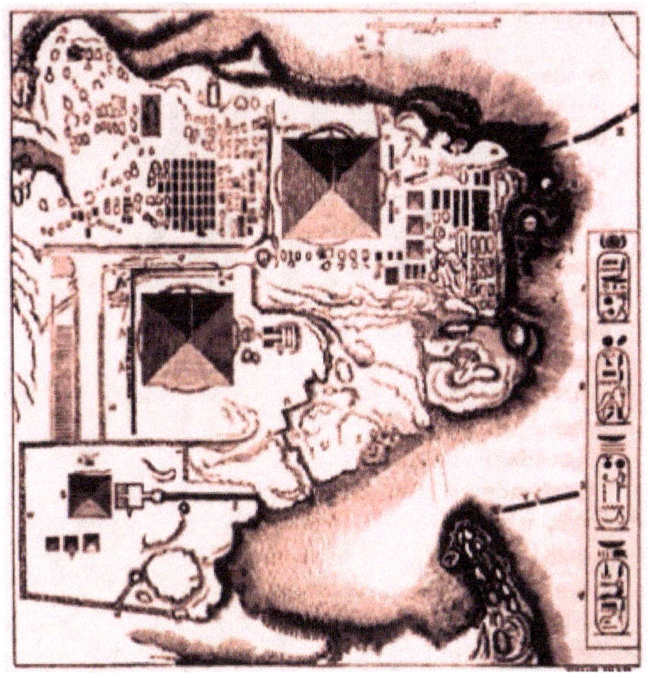

Above: This early map showing the ground plan of the famous Pyramids of Giza in Egypt reveals remarkable parallels to the "Pyramids of Cydonia" on Mars.

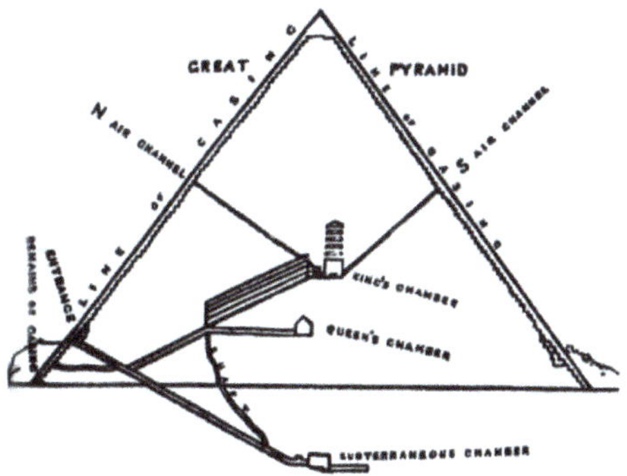

Above: A schematic of the interior of the 4,500 year old Great Pyramid at Giza, near Cairo, Egypt. When humans are finally able to explore the Martian "pyramids" at Cydonia, will they find similar entrances, staircases, channels, and chambers? The U.S. government remote viewer discussed in my introduction said yes.

Chapter 42

TREES & SHRUBS

MOC2-166B: EARTHDATE JULY 21, 1999

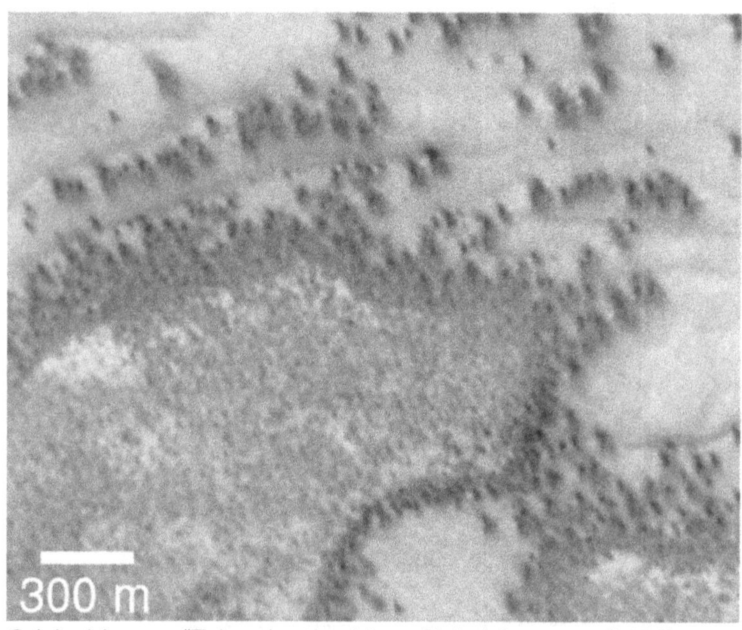

Original image. "Taken by the Mars Global Surveyor (MGS) Mars Orbiter Camera (MOC) on July 21, 1999. The dunes are located in the south polar region and were expected to be completely defrosted by November or December 1999. North is approximately up, and sunlight illuminates the scene from the upper left. The 300 meter scale is 328 yards." Image credit: NASA/MSSS.

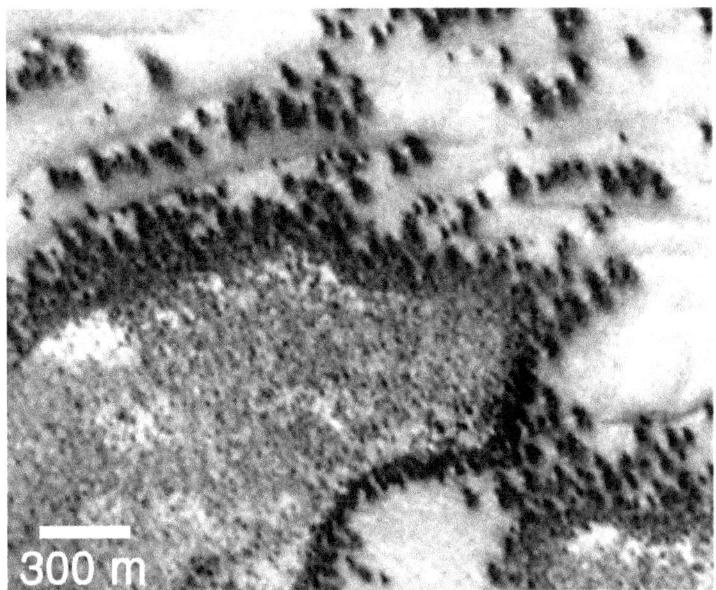

Trees and Shrubs, MOC2-166B, brightened and sharpened by the author for contrast. Conspiracy theorists assert that what we are seeing here are "trees and shrubs" growing on "sand dunes"—more proof that "there is living vegetation on the Red Planet." Like most of the other anomalies in this book, in fact, NASA does not truly know what these objects are. The following somewhat imperious reply to the above claim is what I would call a "best guess" by one of the space agency's scientists:

> "That's what almost everyone says when they see the dark features found in pictures taken of sand dunes in the polar regions as they are beginning to defrost after a long, cold winter. It is hard to escape the fact that, at first glance, these images . . . over both polar regions during the spring and summer seasons, do indeed resemble aerial photographs of sand dune fields on Earth—complete with vegetation growing on and around them! Of course, this is not what the features are. . . . The bright, smooth surfaces that are dotted with occasional, nearly triangular dark spots are sand dunes covered by winter frost.
>
> "Because the Martian air pressure is very low—100 times lower than at Sea Level on Earth—ice on Mars does not melt and become liquid when it warms up. Instead, ice sublimes—that is, it changes directly from solid to gas, just as 'dry ice' does on Earth. As polar dunes emerge from the months-long winter night, and first become exposed to sunlight, the bright winter frost and snow begins to sublime. This process is not uniform everywhere on a dune, but begins in small spots and then over several months it spreads until the entire dune is spotted like a leopard. . . ."

Chapter 43

GLASS TUBES

M0400291: EARTHDATE AUGUST 11, 1999

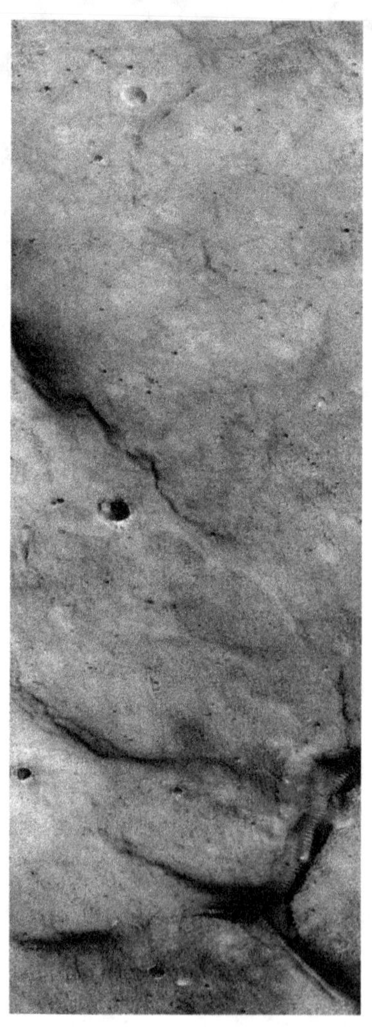

Original image. "Taken by the Mars Global Surveyor (MGS) Mars Orbiter Camera (MOC) on August 11, 1999." Image credit: NASA/MSSS.

ANALYSIS OF GLASS TUBES, M0400291

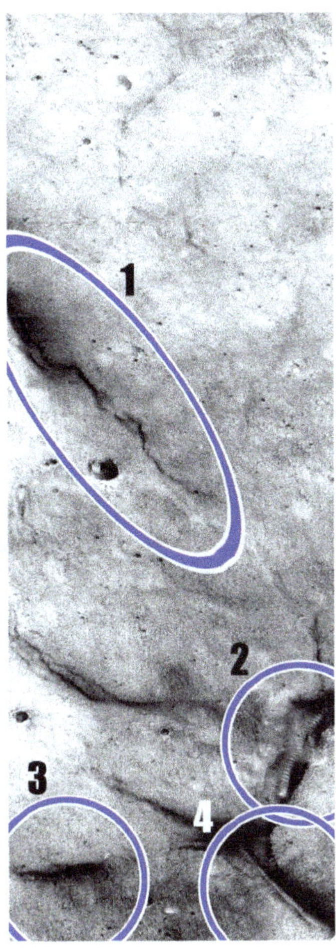

Glass Tubes, MO400291, brightened and sharpened by the author for contrast. Four massive anomalies can be seen in this image. Popularly known as "glass tubes" or "glass worms," believers generally claim that they are a type of underground transport system or possibly aqueducts. Establishment scientists affirm that they are rows of sand dunes known as "dune trains." Unfortunately, we cannot yet determine which theory is correct—of if both are incorrect.

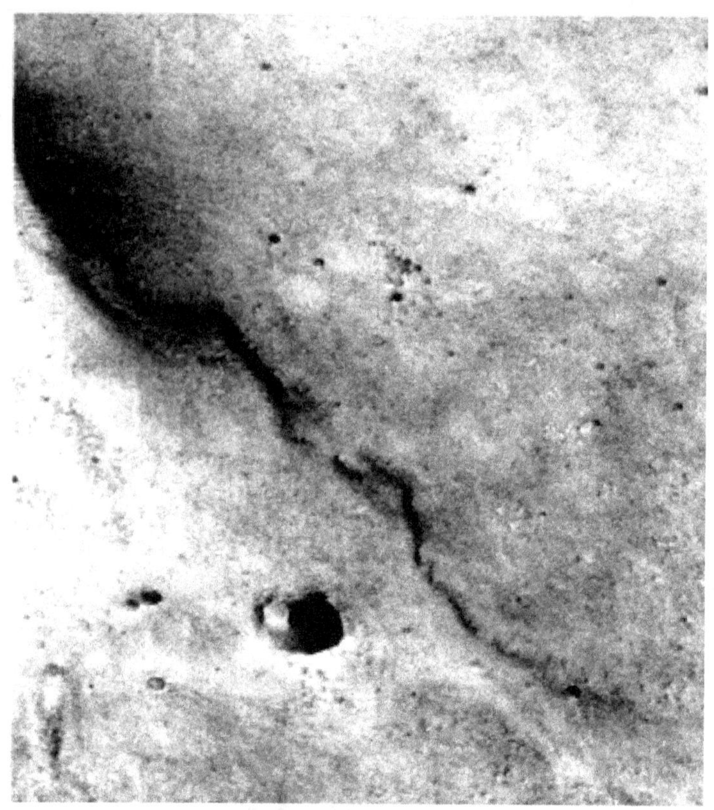

MO400291, highlighted area 1: A closeup.

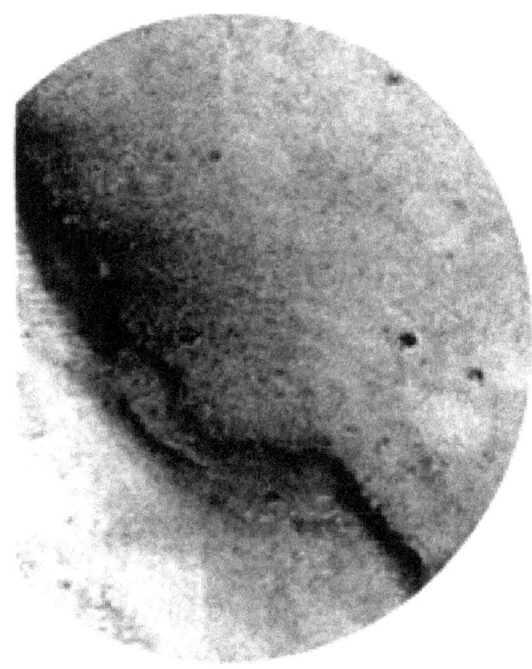

MO400291, highlighted area 1 (left): Extreme closeup of point of interest in area 1 above, showing some of Mars' alleged translucent underground "tunnels."

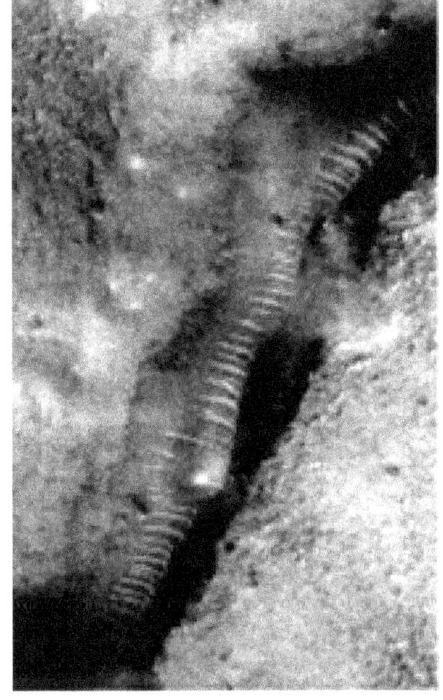

MO400291, highlighted area 2 (right): Extreme closeup.

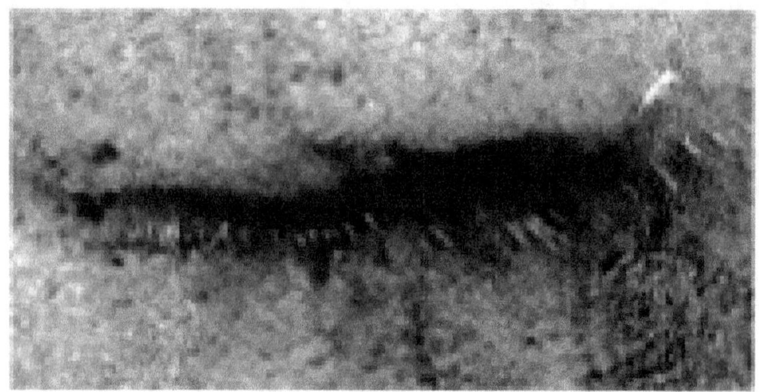

MO400291, highlighted area 3 (above): Extreme closeup.

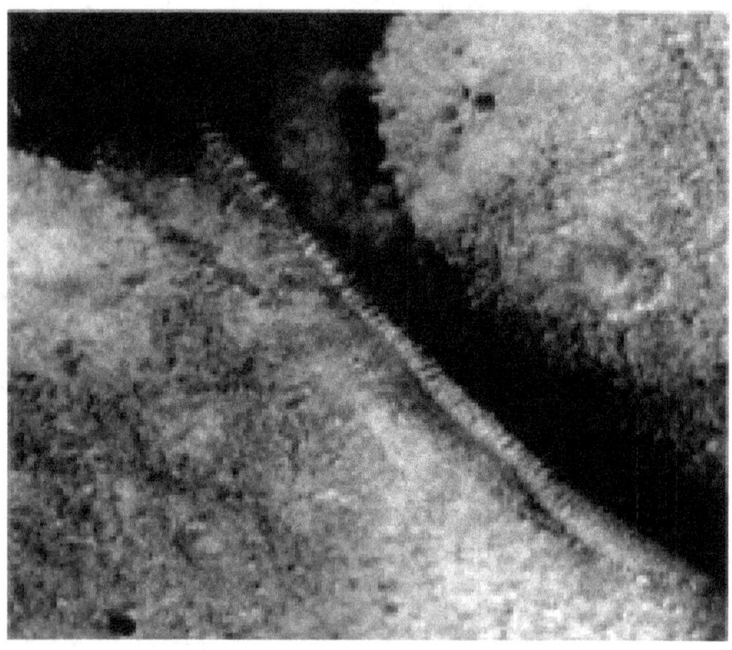

MO400291, highlighted area 4 (above): Extreme closeup of part of Mars' purported transparent, subterranean transportation system. I invite my readers to do a thorough photographic study of actual Martian sand "dune trains" (some examples of which can be found in this book). I have determined, as have others, that they look almost nothing like the "glass tubes" pictured in the images in this chapter.

Chapter 44

MARS MONOLITHS

PSP-009342-1725: EARTHDATE JULY 24, 2008

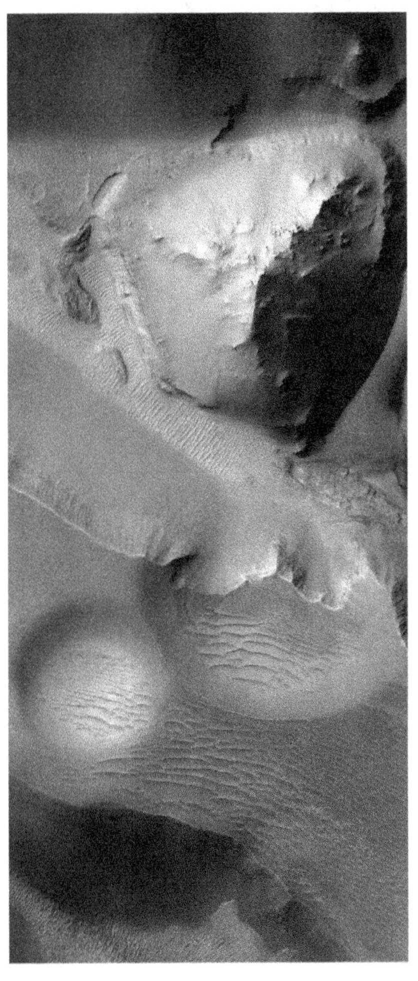

Original image. “Taken by the Mars Reconnaissance Orbiter, July 24, 2008.” Image credit: NASA/JPL/UArizona.

ANALYSIS OF MARS MONOLITHS, PSP-009342-1725

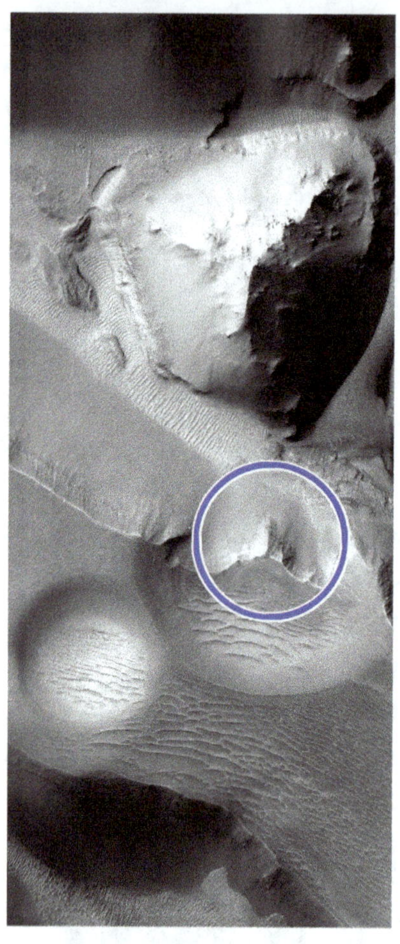

Mars Monoliths, PSP-009342-1725, brightened and sharpened by the author for contrast, with point of interest highlighted. The Mars Reconnaissance Orbiter snapped this enormous 3.1 mile wide image from an altitude of 164.1 miles above the surface of Mars. What it did not immediately detect from this great height were a group of monolith like objects inside the blue circle I have added to this, the full frame image.

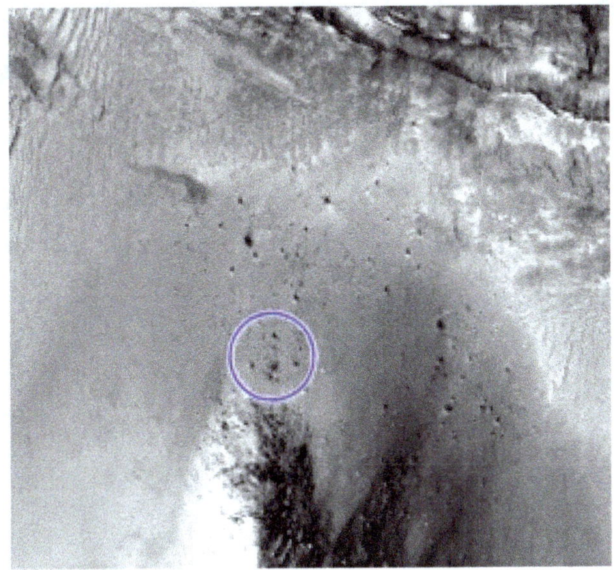

PSP-009342-1725 (above): Zoomed in for a closer view of the highlighted area.

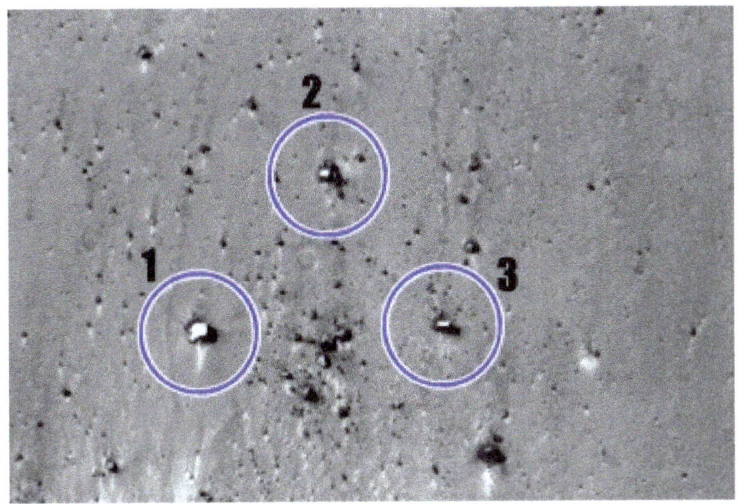

PSP-009342-1725 (above): As we move in even closer on the site of interest, three artificial looking objects begin to appear.

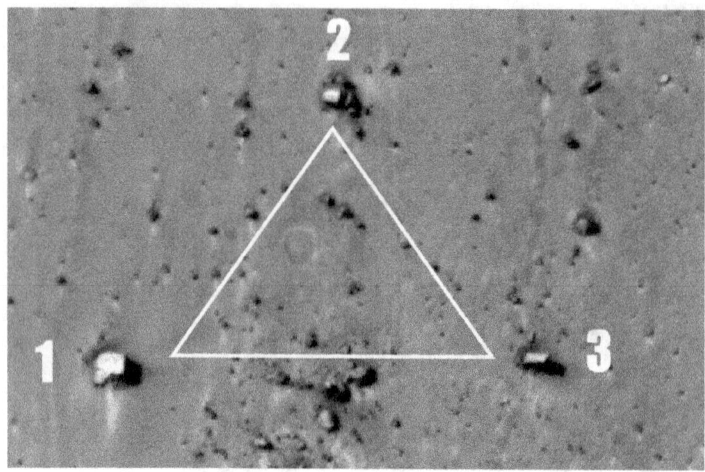

PSP-009342-1725 (above): At this zoom level all three individual objects, 1, 2, and 3, appear to bear squarish or rectangular outlines. Additionally, they are situated so as to form a slightly skewed but near perfect triangle (one that may have been knocked out of mathematical alignment by crustal movement). Is this triangulation accidental or intentional?

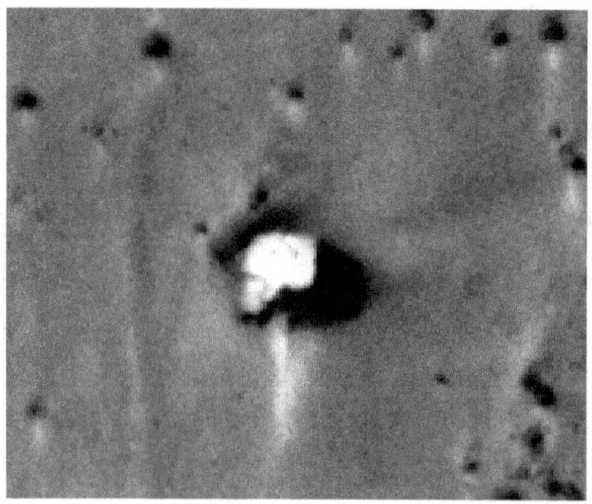

PSP-009342-1725 (above): A closeup of object 1: Though pixelated at this zoom level, it is clear that this anomaly has a square shape and is quite large and tall.

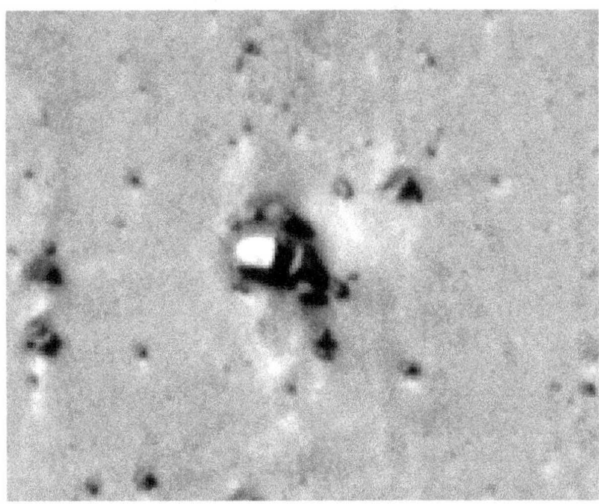

PSP-009342-1725 (above): Closeup of object 2: This shiny anomaly also possesses a square shape. To its right (or behind it) is what appears to be a second peculiar object with 90 degree angles.

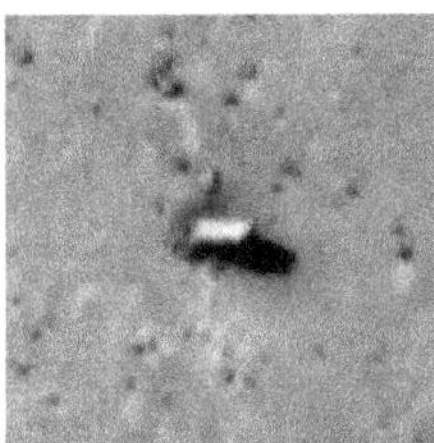

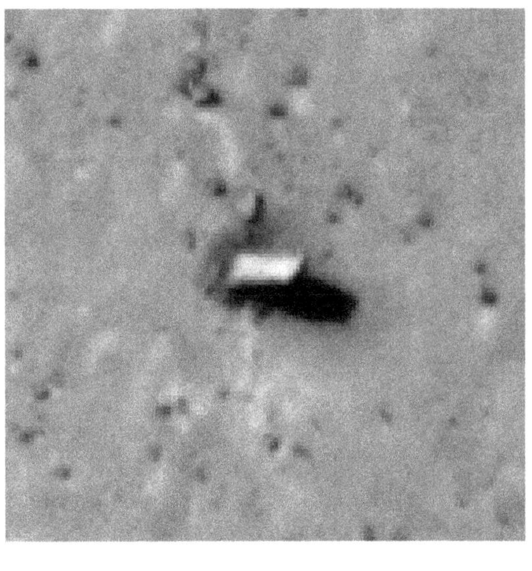

PSP-009342-1725 (above and right): Two closeups of object 3. Known as the "Mars monolith," this renowned object has long been compared to the black monolith in the 1968 Stanley Kubrick film, *2001: A Space Odyssey.* NASA dismisses all extraterrestrial claims by describing these objects as common "rectangular boulders" that have "weathered out and separated from the bedrock." In 1999, a "monolith" was also found on Phobos, one of Mars' two moons, an object that even caught the attention of American hero, former U.S. astronaut Buzz Aldrin.

Chapter 45
STAR CITY

PIA00092: EARTHDATE 1980

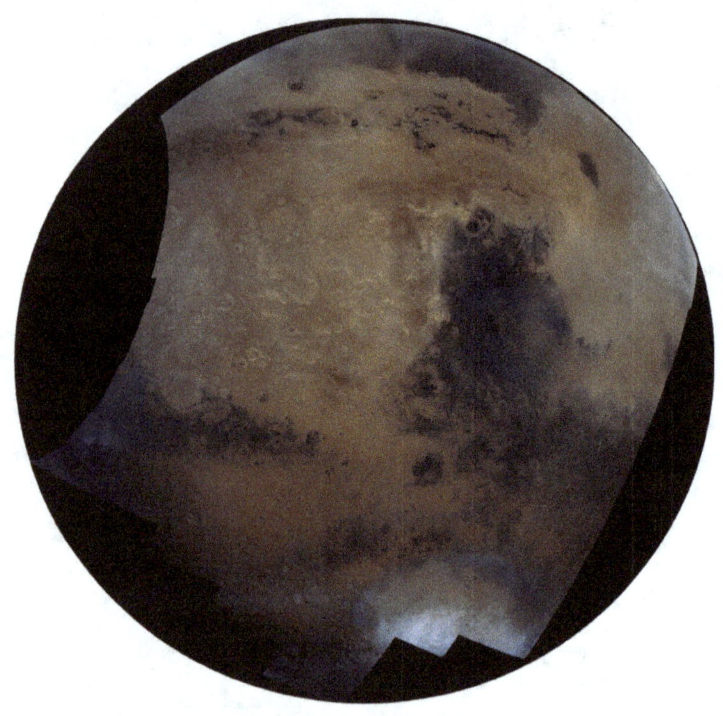

Original image (artificially colored by NASA). "This mosaic of the Syrtis Major Hemisphere is composed of about 100 red and violet filter Viking Orbiter 1 images, digitally mosaiced in an orthographic projection at a scale of 1 km/pixel. The images were acquired in 1980 during early northern summer on Mars. The large circular area with a bright yellow color (in this rendition) located in the upper left area of the image is known as Arabia. The dark blue area to the right of Arabia is called Syrtis Major Planum." Image credit: NASA/JPL/USGS.

ANALYSIS OF STAR CITY, PIA00092

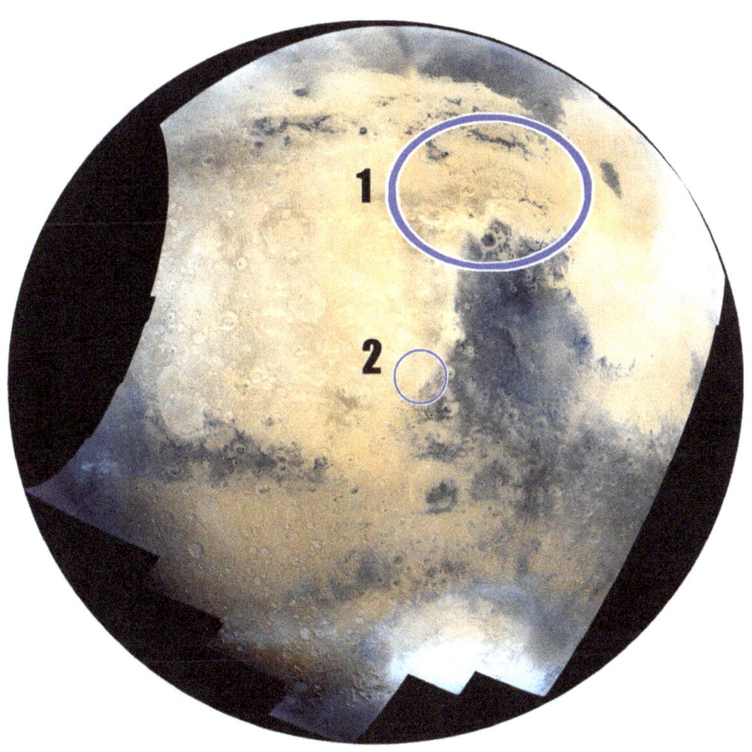

Star City, PIA00092, brightened, colored, and sharpened by the author for contrast, with two anomalies highlighted. According to NASA, the bright white areas to the south (including the Hellas impact basin at lower right) is carbon dioxide frost. Our main focus of interest, however, are the two blue circled areas, which many believe contain overt signs of an ancient intelligent race.

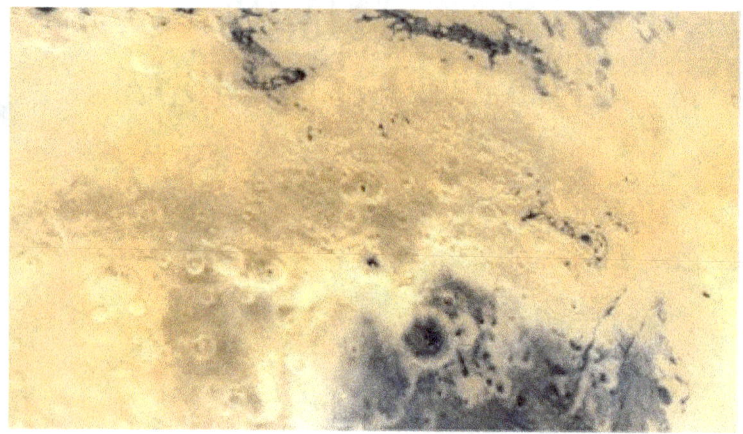

PIA00092 (above), a zoom in on highlighted area 1, the "Star City" region: From a lower altitude we can begin to discern a plethora of unusual shapes and structures.

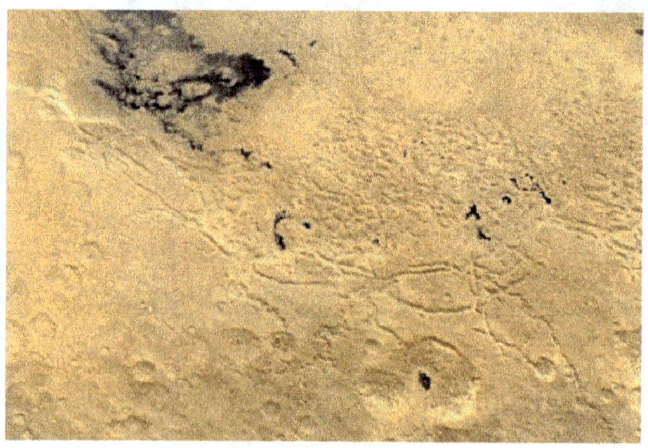

PIA00092 (above), closeup of the left side of area 1, showing what some think are the remnants of an early Martian metropolis, which would presumably include residential neighborhoods, roadways, canals, lakes, rivers, parks, industrial areas, transportation hubs, shopping areas, and a variety of infrastructure.

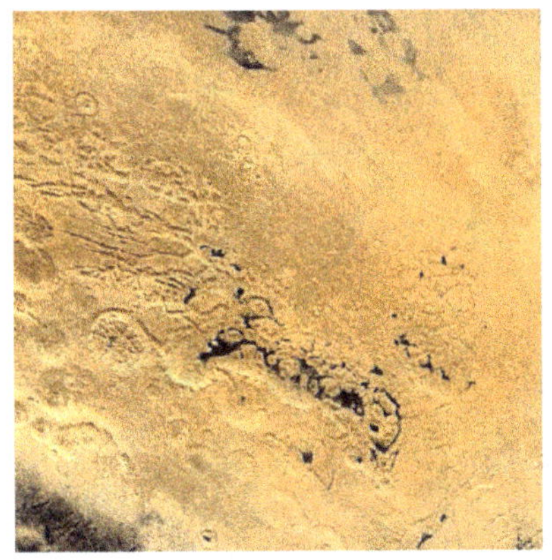

PIA00092 (left), closeup of the right side of area 1, showing the urban center popularly known as "Star City."

PIA00092 (right), extreme closeup of "Star City" (area 1), showing perhaps 1 million year old "planned" neighborhoods, subdivisions, roads, lakes, ponds, and more. NASA explains all of this away as "a low-relief volcanic shield of probable basaltic composition." In reality, as this area has never been explored by us in person, the jury is still out.

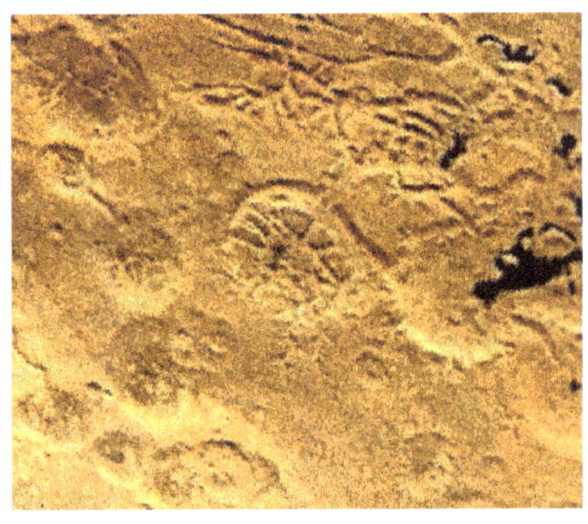

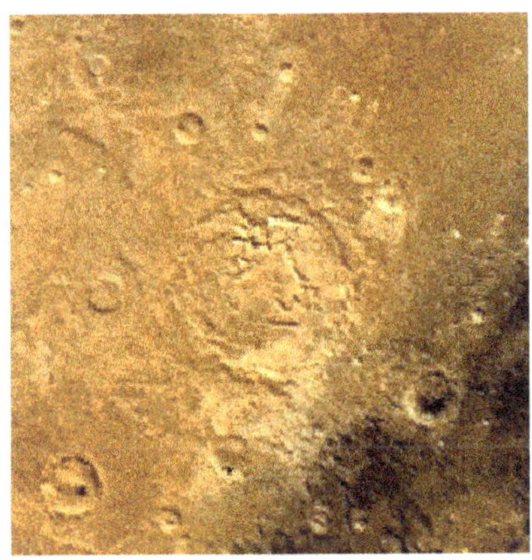

PIA00092 (left), extreme closeup of area 2, showing, under magnification, numerous artificial looking shapes, objects, and anomalies; from "roadways," "walls," and "enclosures" (all center), to perfectly rectangular, building foundation like impressions in the ground (4:30) and odd wavy structures running parallel to one another (7:30).

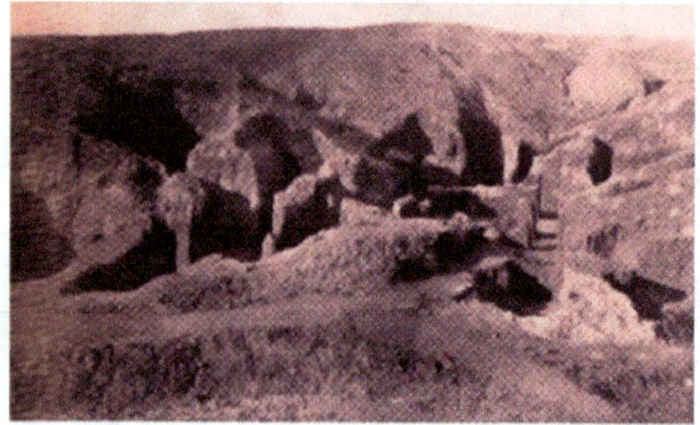

Above: Recently excavated Babylonian ruins, showing their similarity to some of the city like ruins on Mars.

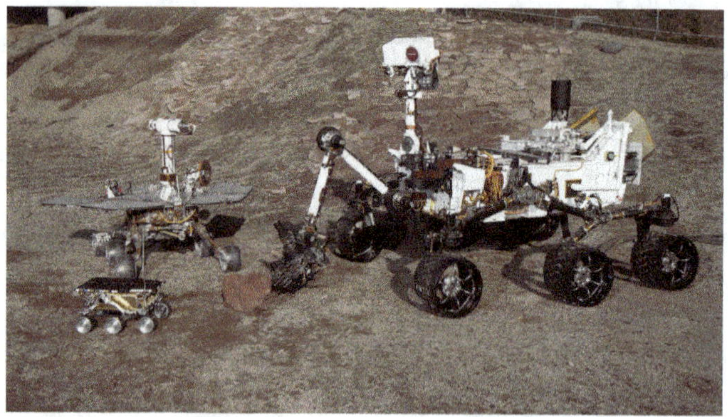

Above: Three generations of Mars rovers. The countless bizarre anomalies they have photographically recorded are largely being ignored by mainstream science. Image credit: NASA/JPL-Caltech.

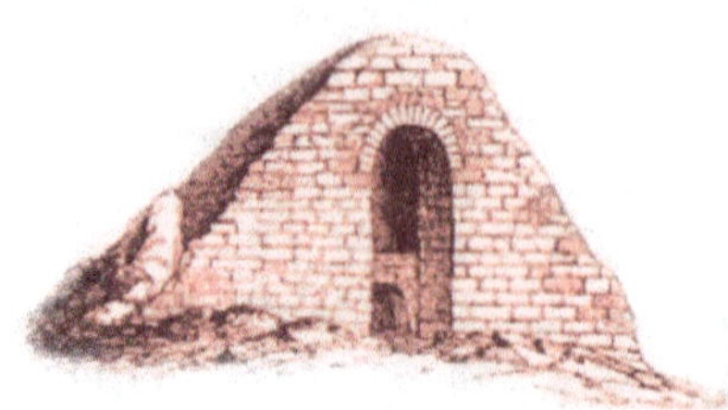

Above: As observant readers will note, with its vaulted doorway and rooms, this crude, ancient brick pyramid at Thebes shares many traits with some of the anomalies found on the Red Planet.

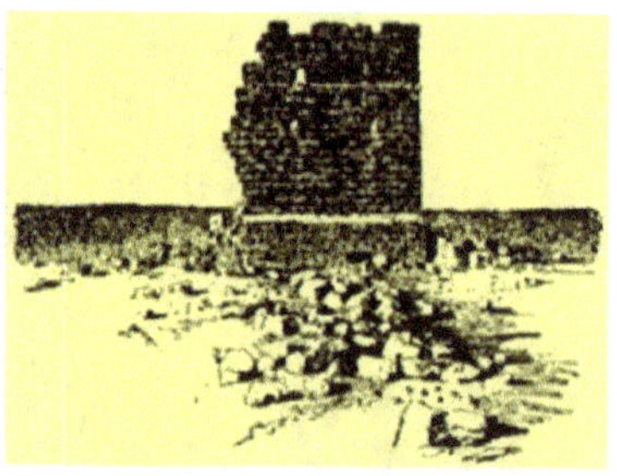

Above: These tower ruins at Kasr-el-Hair, the ancient Heliaramia in Mesopotamia, are nearly identical to anomalies I and others have discovered on the surface of Mars.

Chapter 46

THREE OTHER CYDONIA LIKE AREAS ON MARS

THE MITHRAS, MINERVA, & ARTEMIS REGIONS

Note 1): The following assemblage of cropped photos are small images that I have cut from much larger photos that were created by NASA and its affiliates over the past six decades or so. Thus, though I have not noted so in each individual photo, NASA retains image credit for this entire photographic collection.

Note 2): After cropping them from their parent images, I brightened, colored, and sharpened all of the photos in this chapter. By adding dynamic contrast, the reader will find it easier to examine the anomalies I have selected. Outside these few minor adjustments, the images are untouched and remain just as NASA originally photographed them.

Note 3): Let us acknowledge that many if not most of the following images may be or are the products of natural Martian geology. As I have painstakingly pointed out, however, there is as yet no ironclad proof of this. At the same time, it is true that there is also no proof that they are artificial.

Whichever side the reader falls on, a glaring fact remains. In my Mars research I found few objects and structures in isolation: most of the anomalies I have come across are nearly always located near or in the midst of dozens, sometimes hundreds, of other anomalies—*just as one finds with both ancient and present human habitations here on Earth*.

While conventional science will label this particular phenomenon "natural geology" as well, in my opinion it speaks favorably of the possibility that an advanced civilization once occupied the Red Planet.

The Mithras Region

During my research I discovered this area with numerous Cydonia like objects and structures. For my own personal use and research, I have unofficially named it the Mithras Region (which is in no way meant to compete with any official names it has been or will be given).

Approximate Coordinates
Latitude: 19.00818 degrees/Longitude: 177.85537 degrees

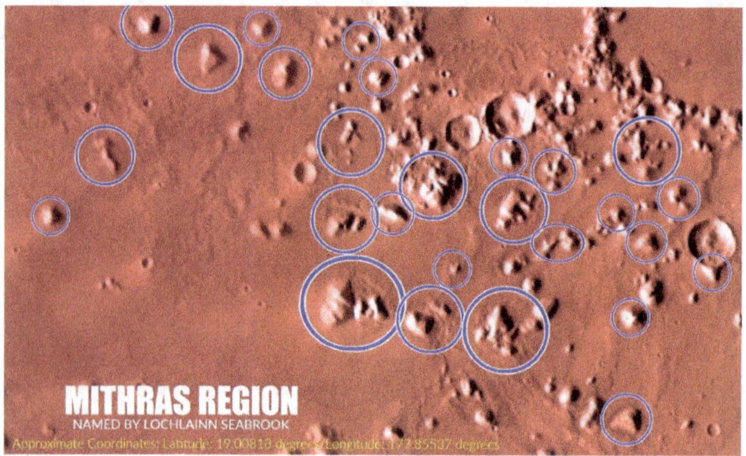

Mithras Region. Wide angle landscape, from high above the Martian surface (exact height unknown). Colored, brightened, and sharpened by the author for contrast, with some of the anomalies highlighted. Bear in mind that if any of these anomalies are indeed the remains of an ancient Martian civilization, they have been sitting in the soil of Mars for millennia and have experienced intense weathering, corrosion, erosion, and massive deterioration. Destroyed and abandoned, half buried in the sand, and covered in layers of concretion and dust, one would expect such artifacts to be almost wholly unrecognizable after tens of thousands, hundreds of thousands, or even millions of years. Hence, my recommendation is to study the following images with an inquisitive mind and a magnifying glass. Image credit for this photo (and those cropped from it): NASA/JPL-Caltech.

CLOSEUPS I HAVE CREATED OF SOME OF THE OBJECTS AND STRUCTURES IN THE MITHRAS REGION

Pay particular attention to shapes, bright areas, and shadows.

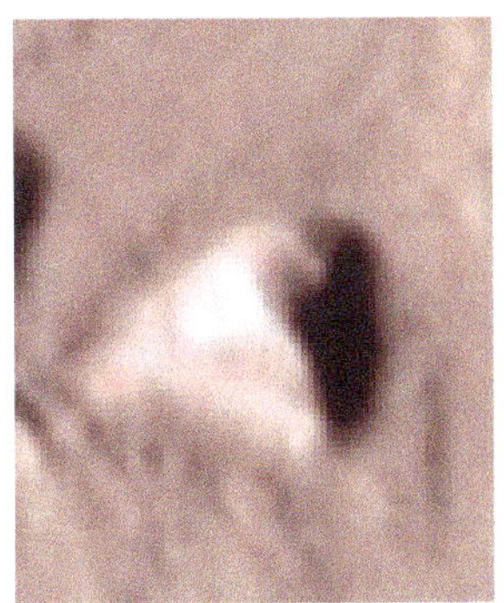

Mithras Region, Image 1.

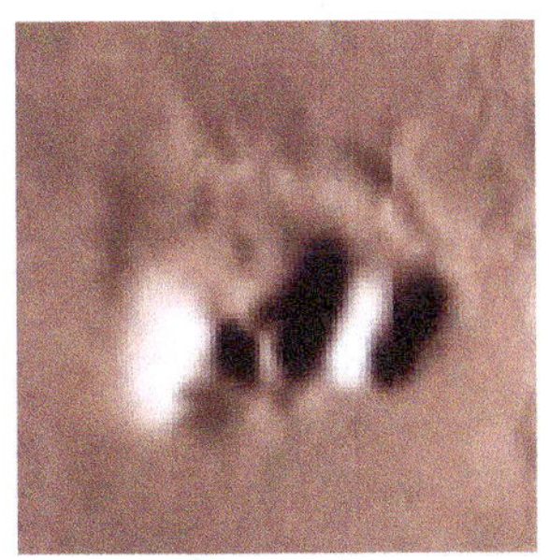

Mithras Region, Image 2.

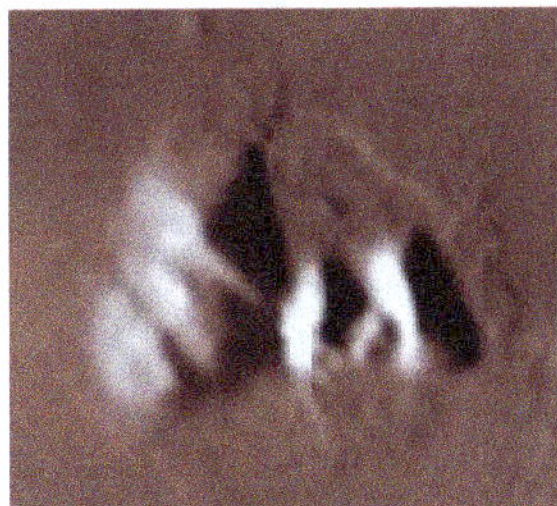

Mithras Region, Image 3.

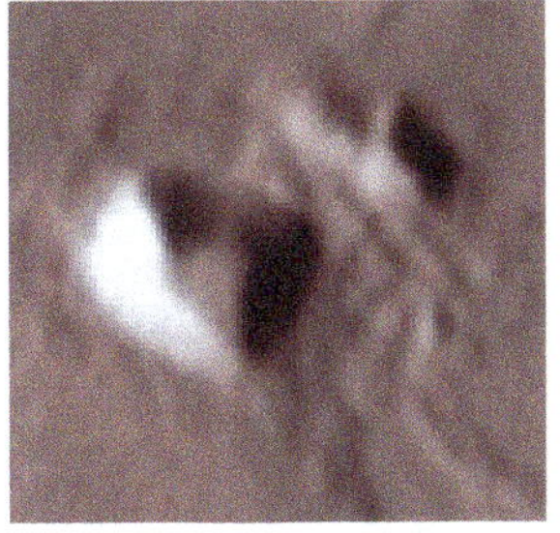

Mithras Region, Image 4.

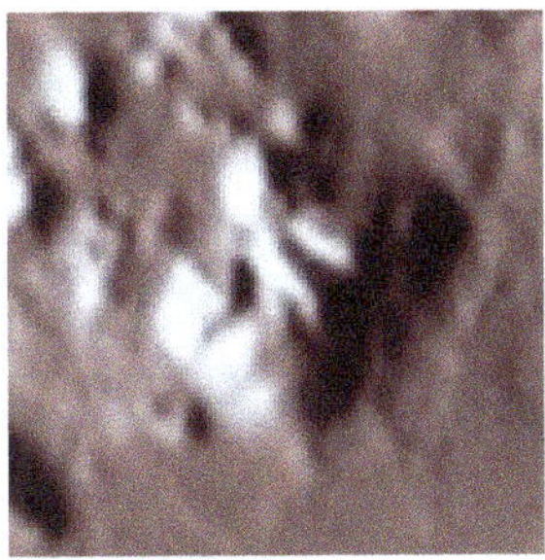

Mithras Region, Image 5.

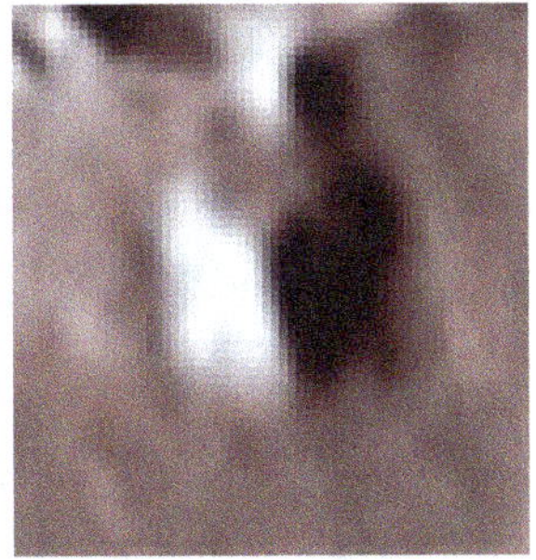

Mithras Region, Image 6.

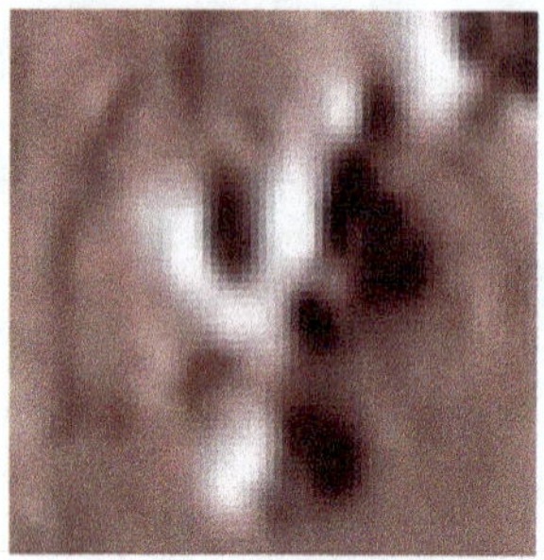

Mithras Region, Image 7.

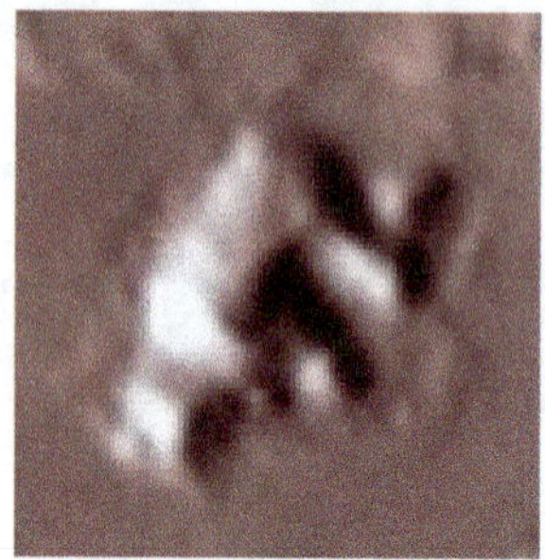

Mithras Region, Image 8.

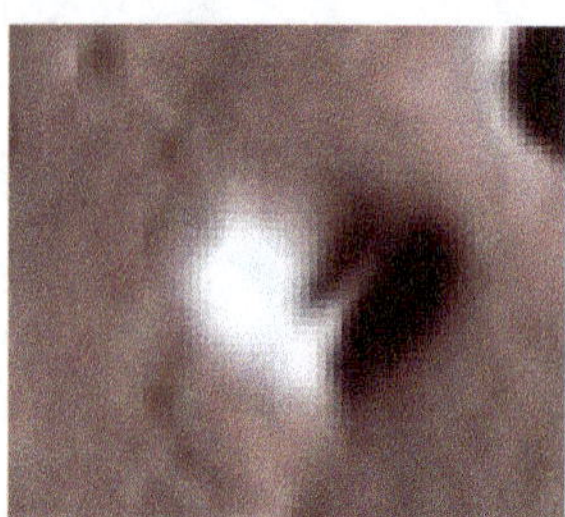

Mithras Region, Image 9.

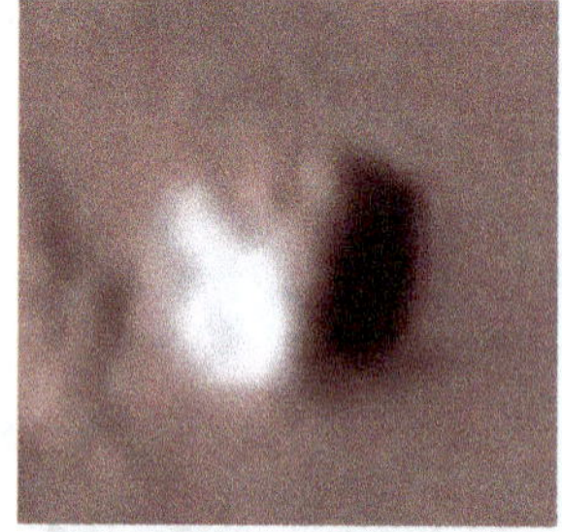

Mithras Region, Image 10.

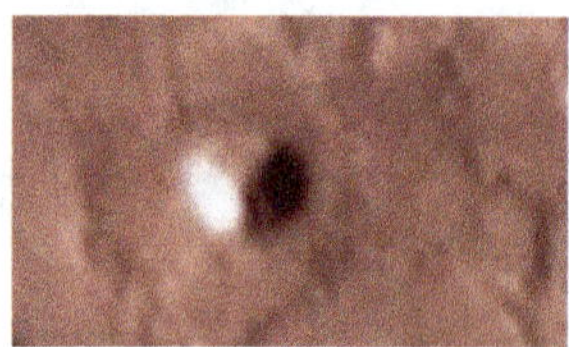

Mithras Region, Image 11.

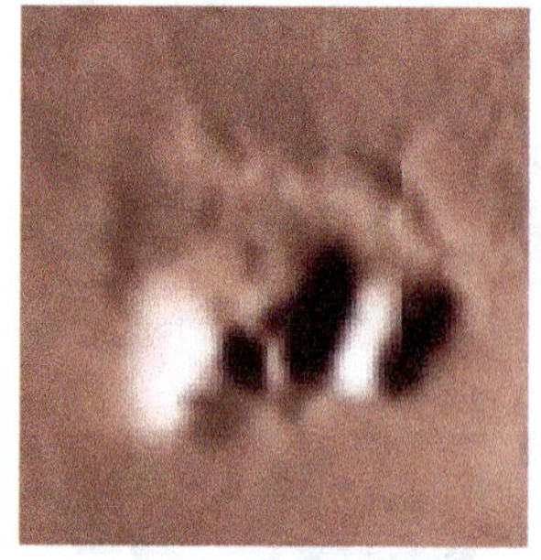

Mithras Region, Image 12.

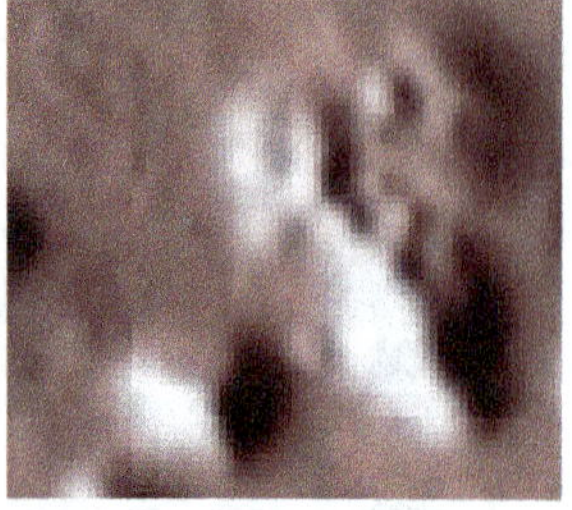

Mithras Region, Image 13.

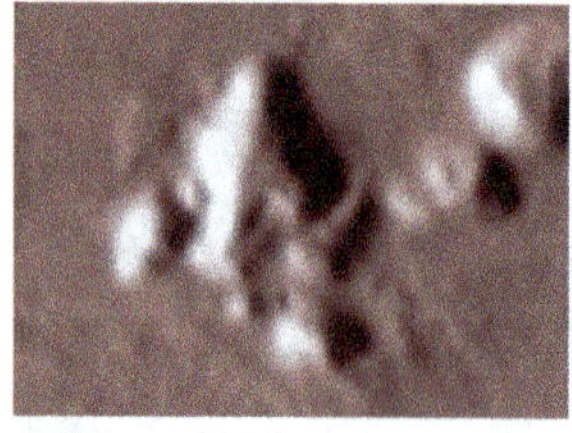

Mithras Region, Image 14.

The Minerva Region

Another area with Cydonia like objects and structures. For my own use and research I have unofficially named it the Minerva Region.

Approximate Coordinates
Latitude: 24.23492 degrees/Longitude: 169.89302 degrees

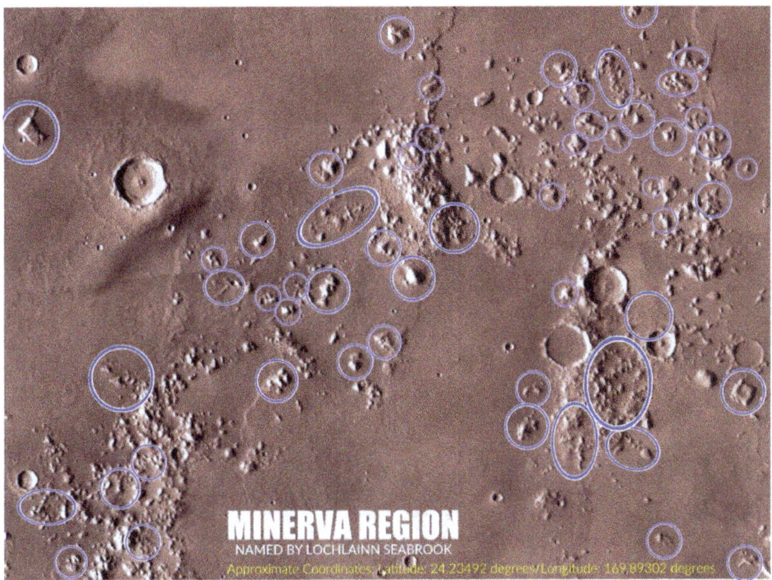

Minerva Region. Wide angle landscape, from high above the Martian surface (exact height unknown). Colored, brightened, and sharpened by the author for contrast, with some of the anomalies highlighted. Image credit for this photo and those cropped from it: NASA/JPL-Caltech.

CLOSEUPS I HAVE CREATED OF SOME OF THE OBJECTS AND STRUCTURES IN THE MINERVA REGION

Pay particular attention to shapes, bright areas, and shadows.

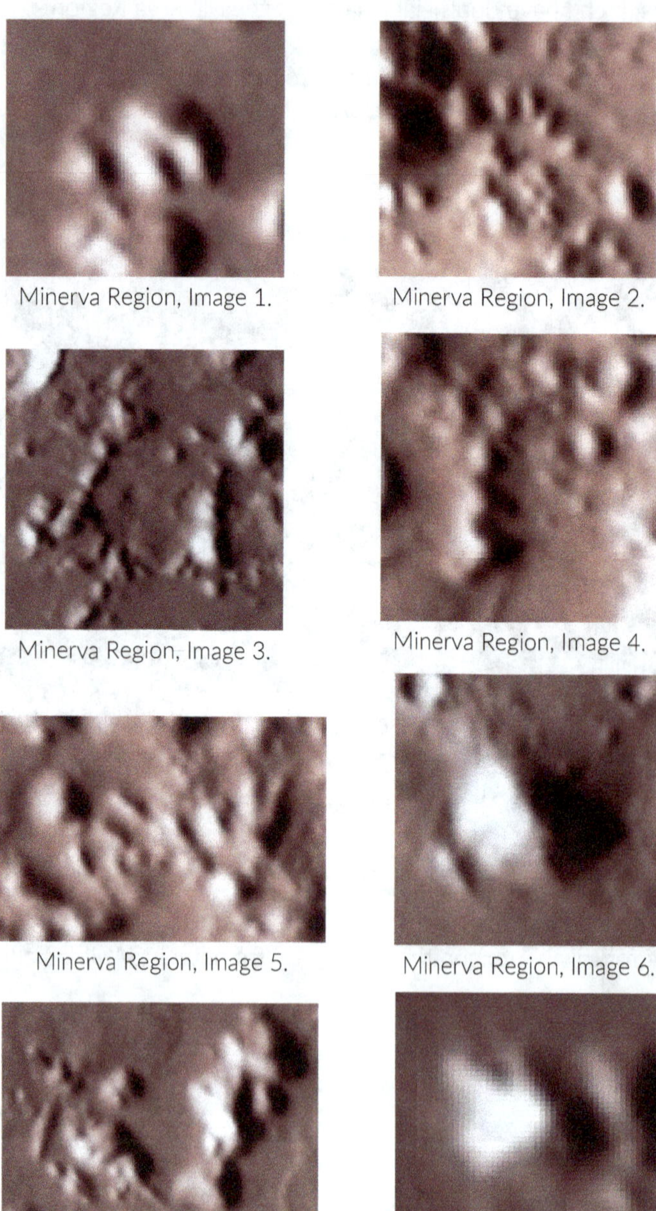

Minerva Region, Image 1.

Minerva Region, Image 2.

Minerva Region, Image 3.

Minerva Region, Image 4.

Minerva Region, Image 5.

Minerva Region, Image 6.

Minerva Region, Image 7.

Minerva Region, Image 8.

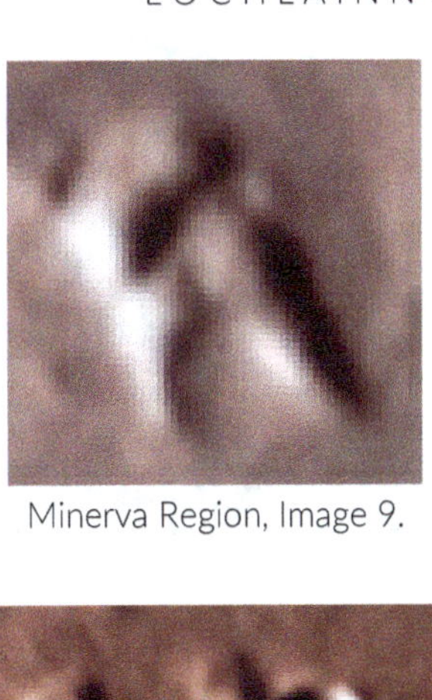
Minerva Region, Image 9.

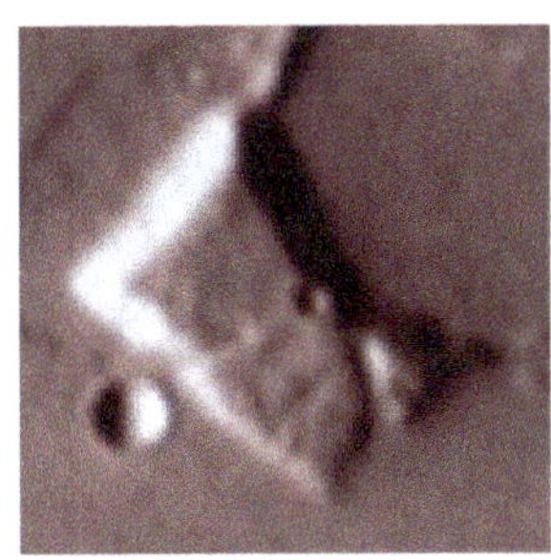
Minerva Region, Image 10.

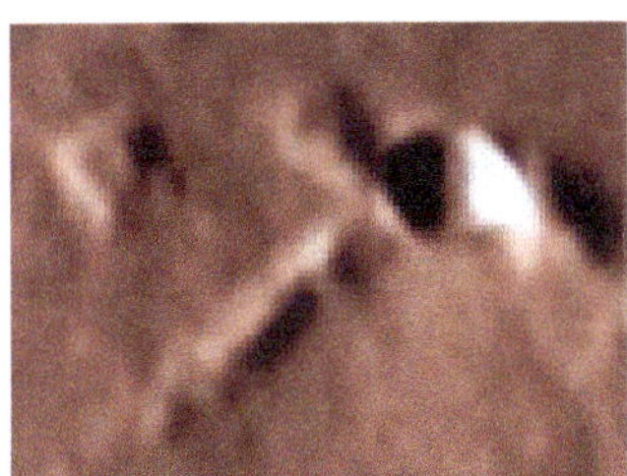
Minerva Region, Image 11.

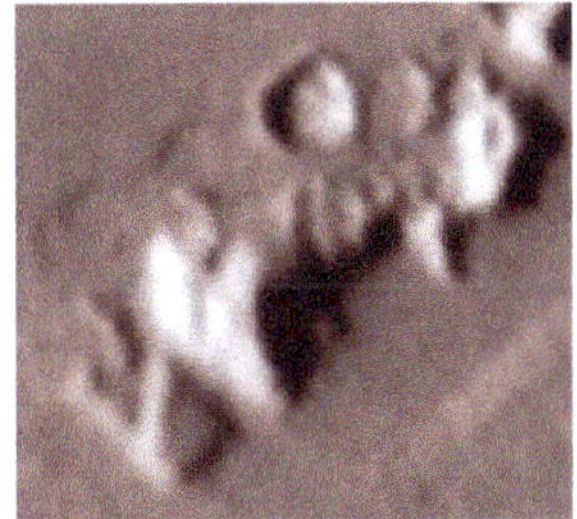
Minerva Region, Image 12.

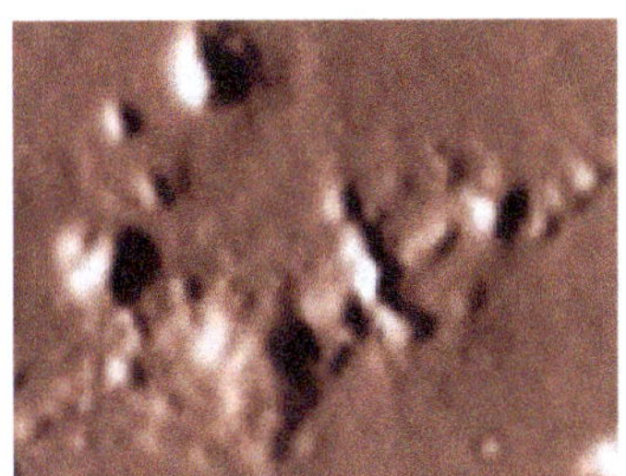
Minerva Region, Image 13.

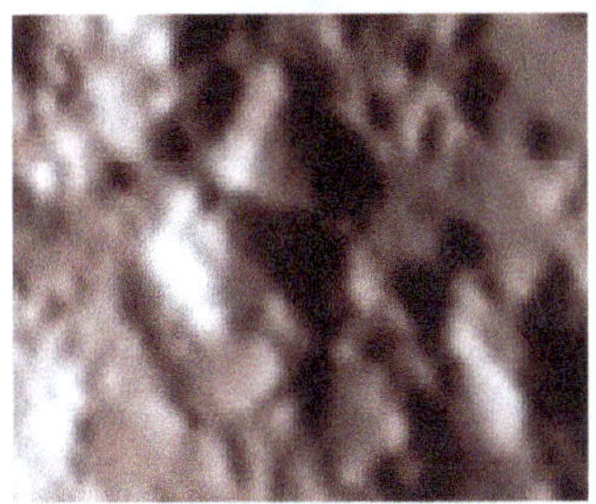
Minerva Region, Image 14.

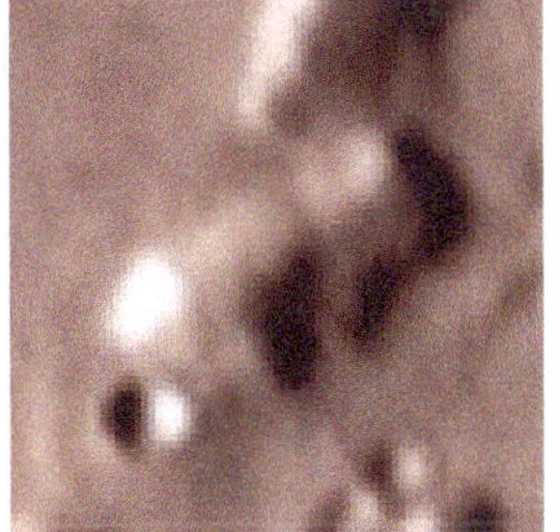
Minerva Region, Image 15.

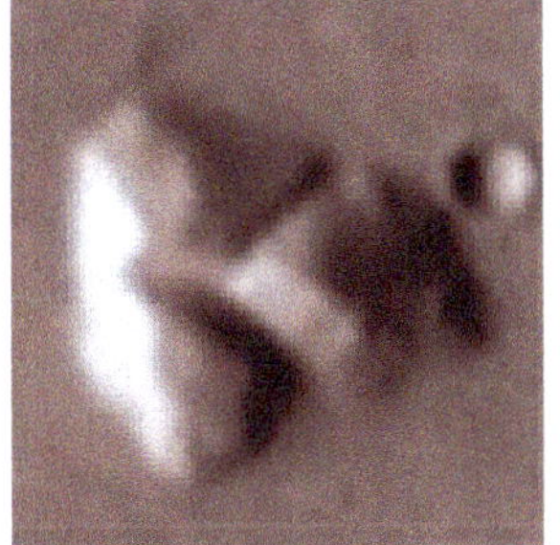
Minerva Region, Image 16.

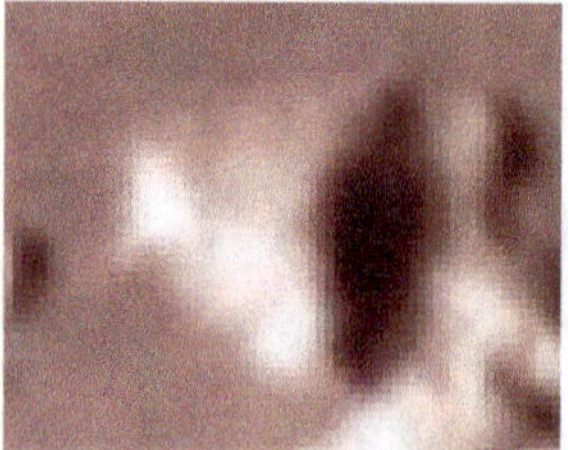
Minerva Region, Image 17.

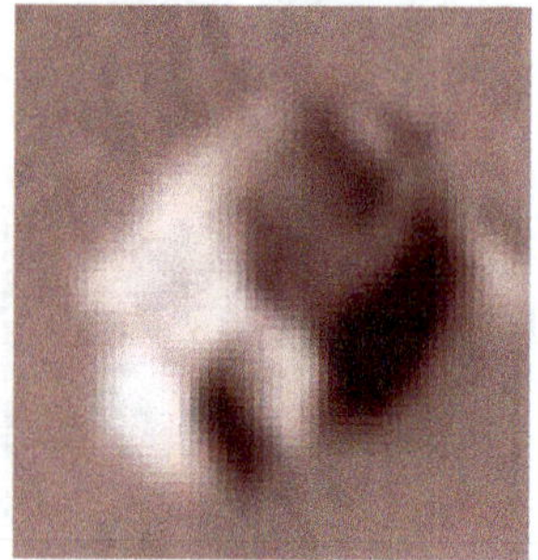
Minerva Region, Image 18.

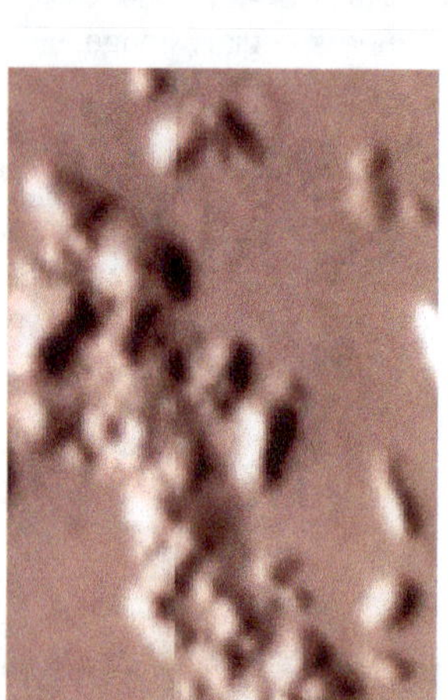
Minerva Region, Image 19.

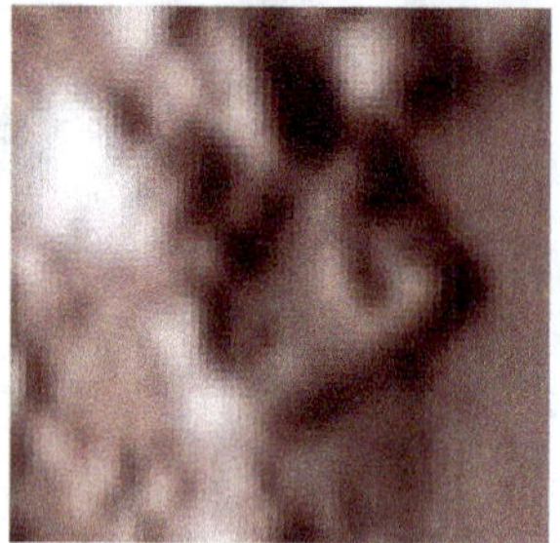
Minerva Region, Image 20.

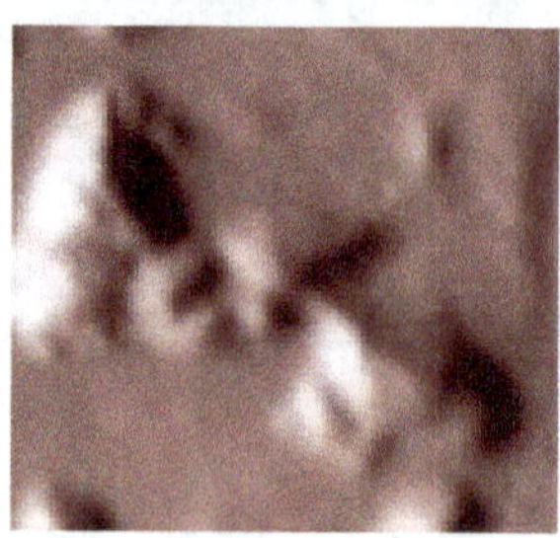
Minerva Region, Image 21.

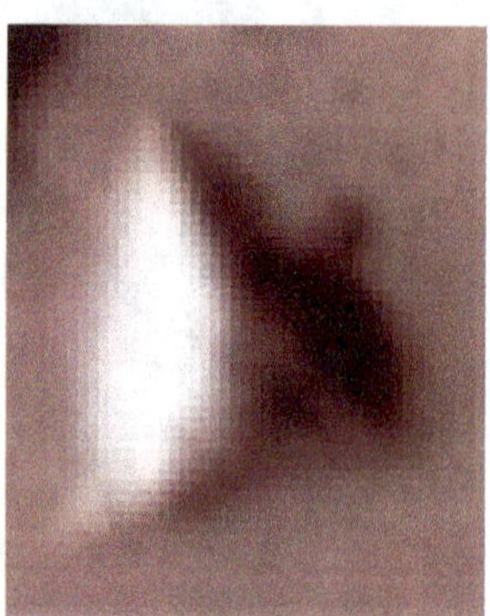
Minerva Region, Image 22.

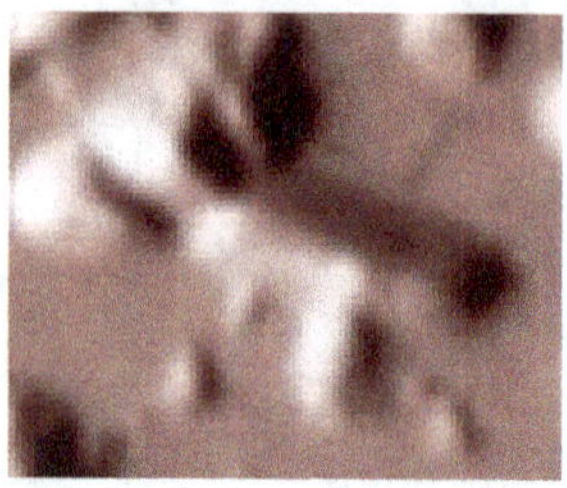
Minerva Region, Image 23.

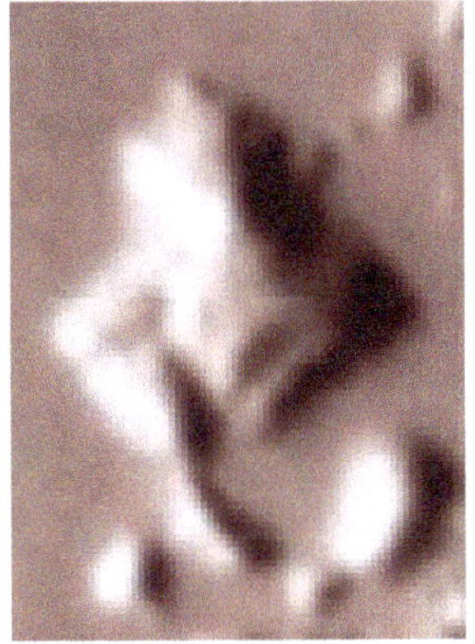
Minerva Region, Image 24.

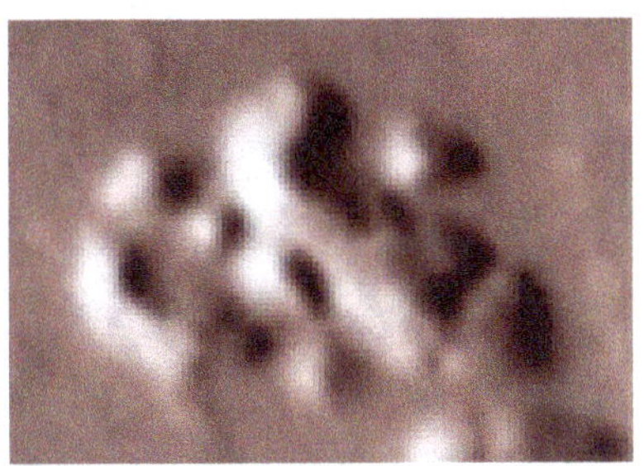
Minerva Region, Image 25.

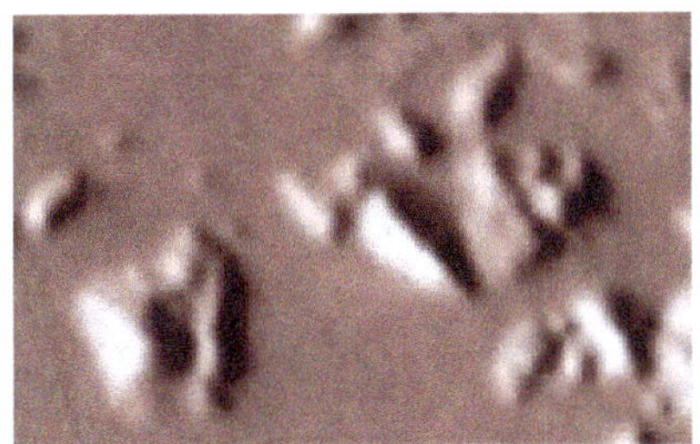
Minerva Region, Image 26.

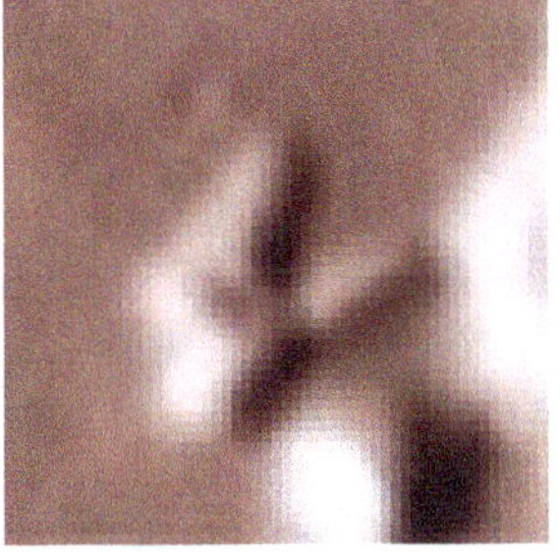
Minerva Region, Image 27.

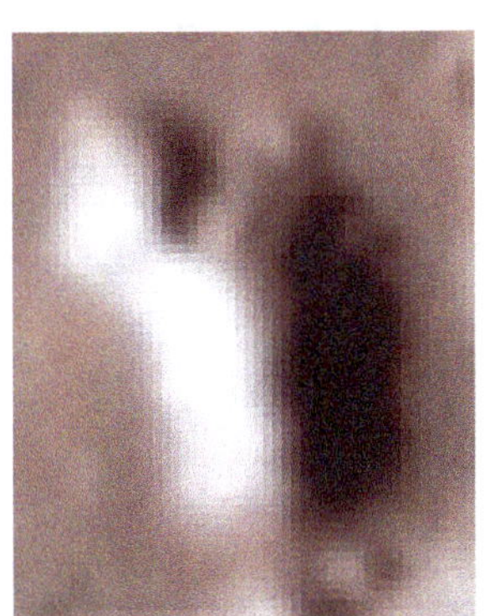
Minerva Region, Image 28.

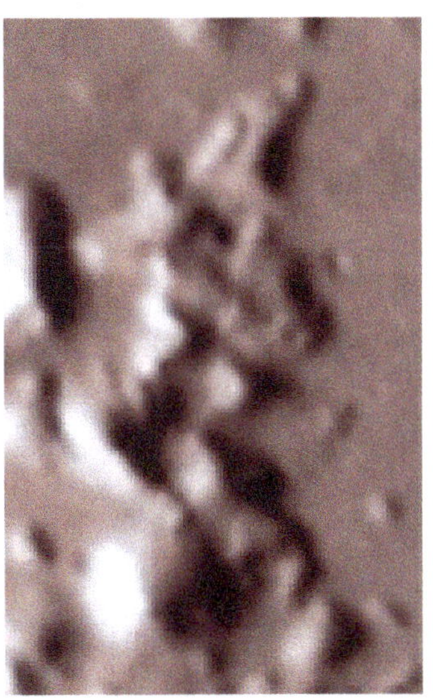
Minerva Region, Image 29.

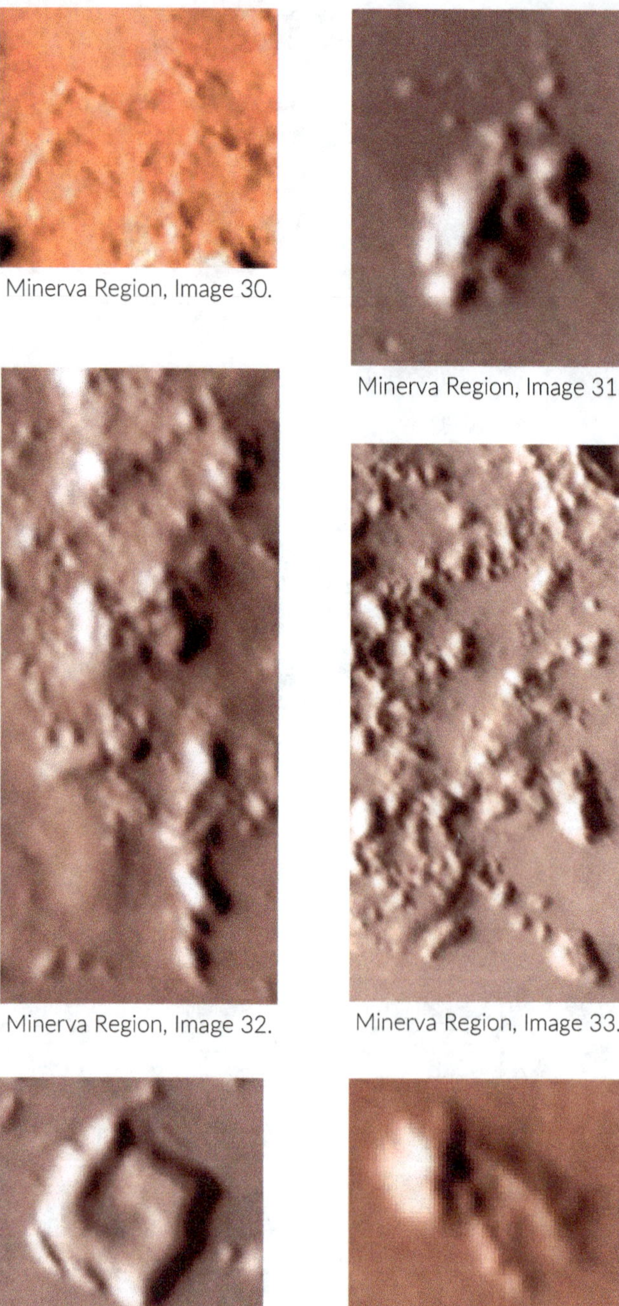

Minerva Region, Image 30.

Minerva Region, Image 31.

Minerva Region, Image 32.

Minerva Region, Image 33.

Minerva Region, Image 34.

Minerva Region, Image 35.

The Artemis Region

Another area with Cydonia like objects and structures. For my own use and research I have unofficially named it the Artemis Region.

Approximate Coordinates
Latitude: 38.4374 degrees/Longitude: 162.93976 degrees

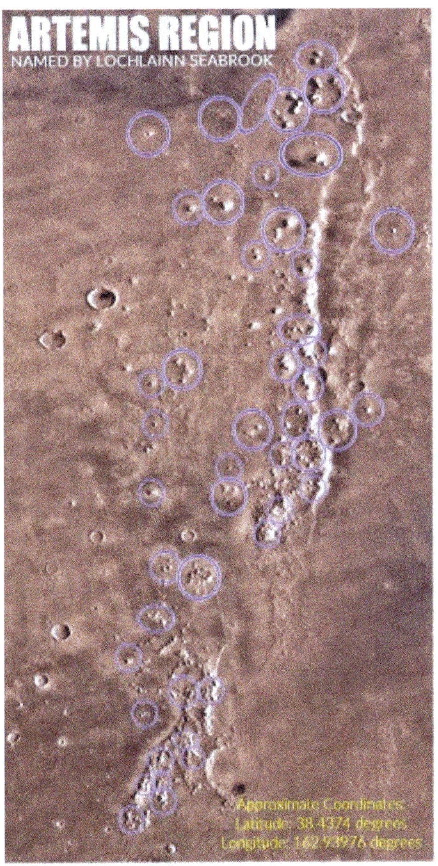

Artemis Region. Wide angle landscape, from high above the Martian surface (exact height unknown). Colored, brightened, and sharpened by the author for contrast, with some of the anomalies highlighted. Image credit for this photo and those cropped from it: NASA/JPL-Caltech.

CLOSEUPS I HAVE CREATED OF SOME OF THE OBJECTS AND STRUCTURES IN THE ARTEMIS REGION

Again, pay particular attention to shapes, bright areas, and shadows.

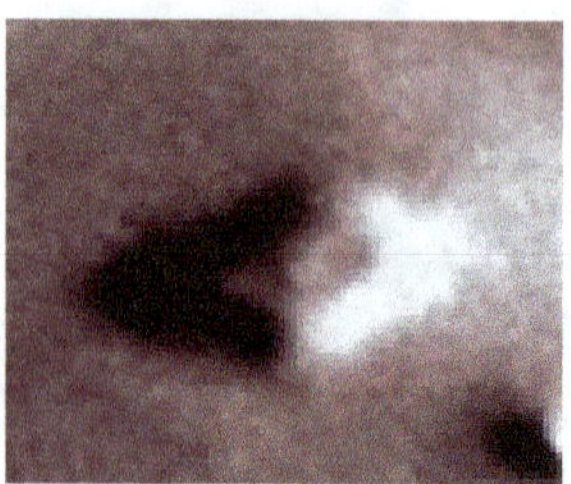

Artemis Region, Image 1.

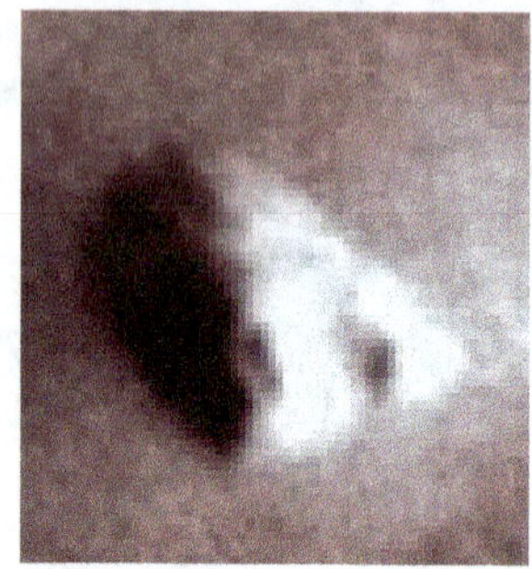

Artemis Region, Image 2.

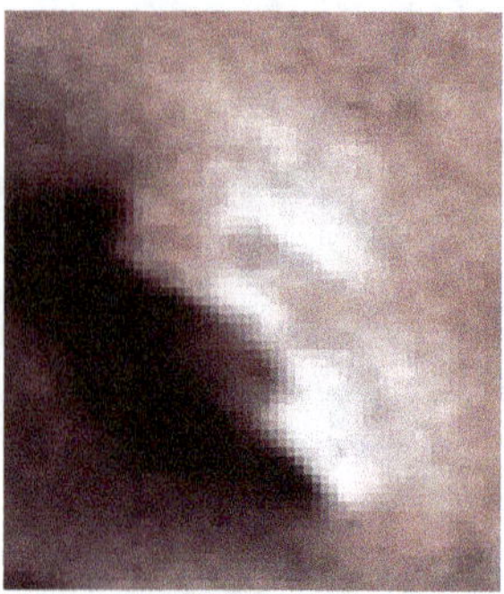

Artemis Region, Image 3.

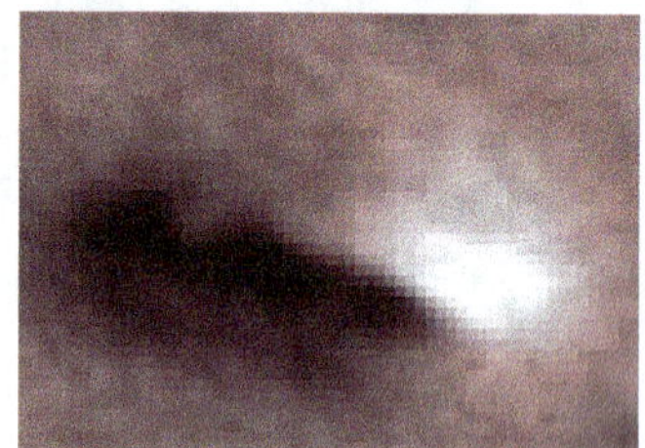

Artemis Region, Image 4.

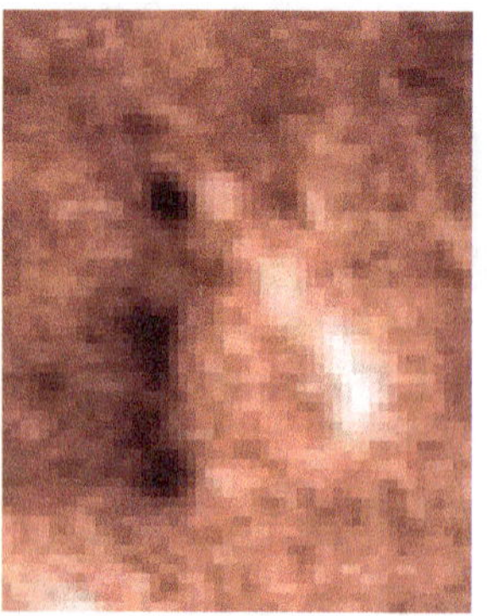

Artemis Region, Image 5.

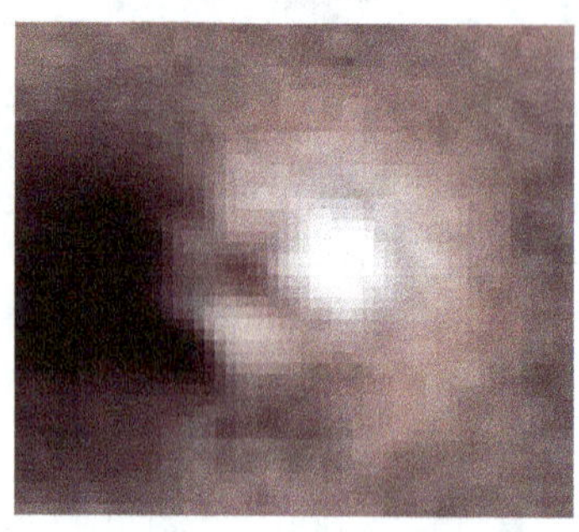

Artemis Region, Image 6.

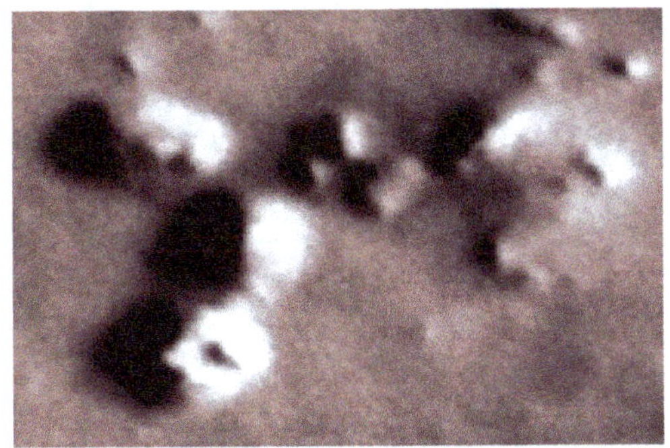

Artemis Region, Image 7.

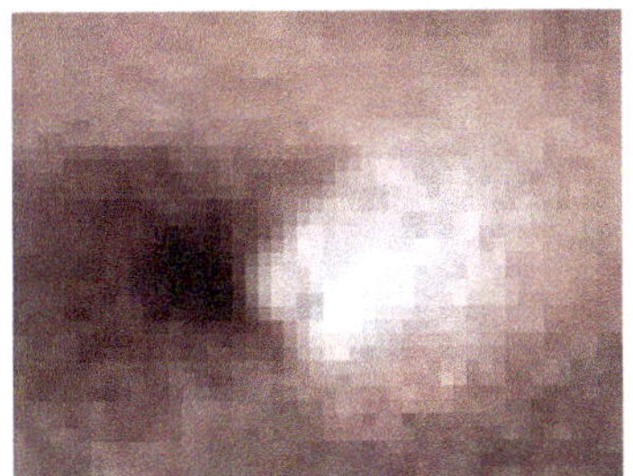

Artemis Region, Image 8.

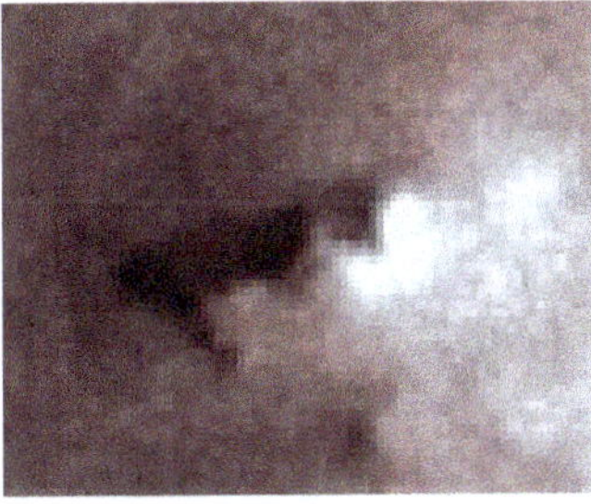

Artemis Region, Image 9.

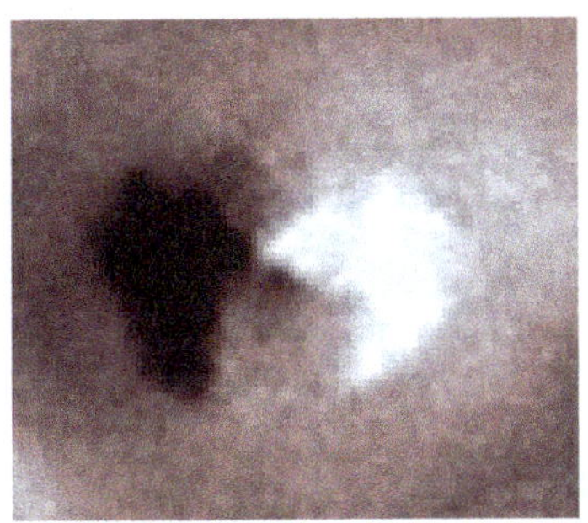

Artemis Region, Image 10.

Artemis Region, Image 11.

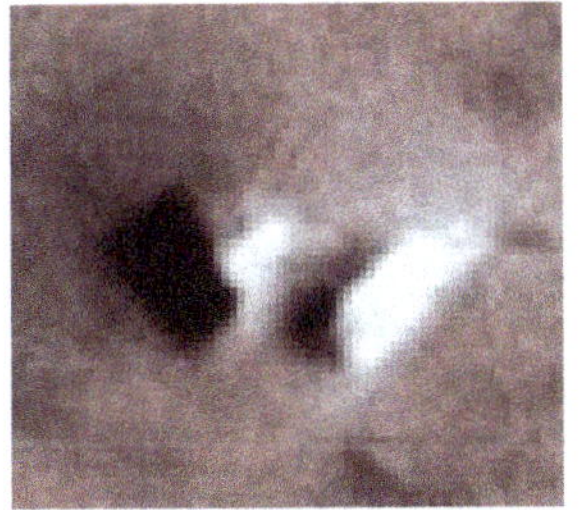

Artemis Region, Image 12.

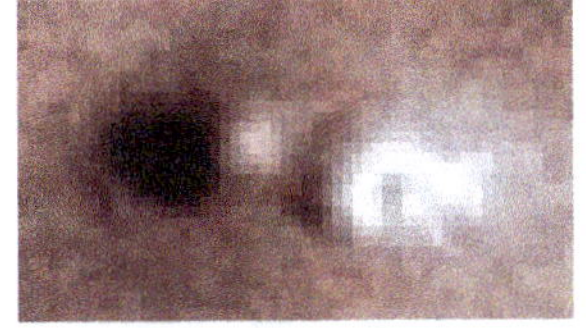

Artemis Region, Image 13.

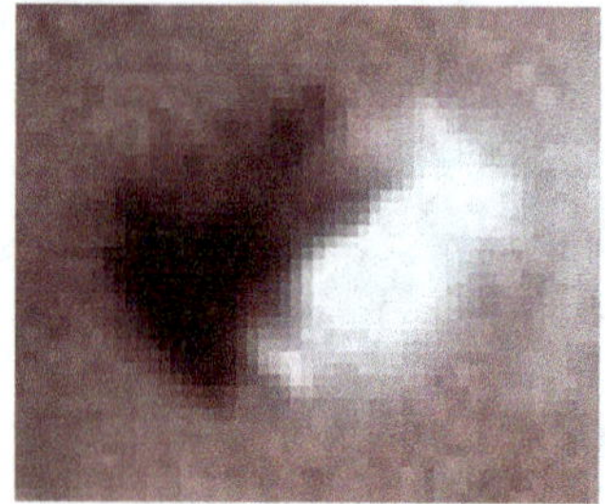

Artemis Region, Image 14.

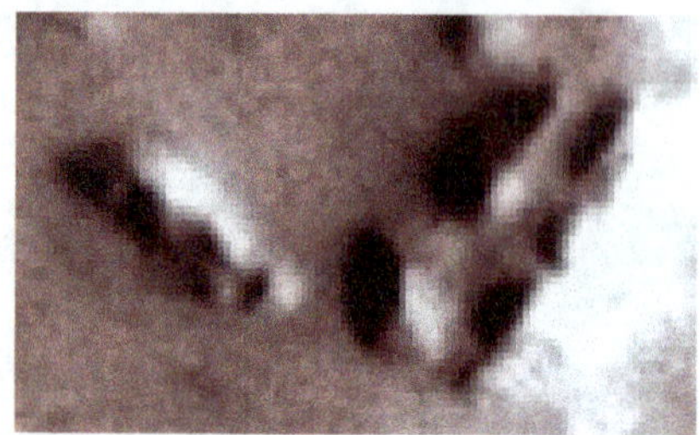

Artemis Region, Image 15.

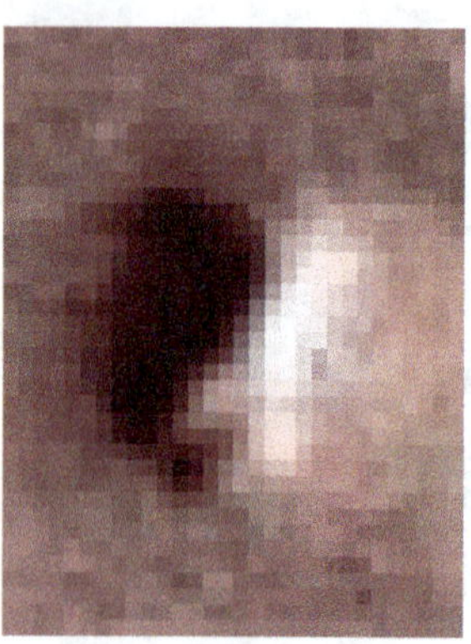

Artemis Region, Image 16.

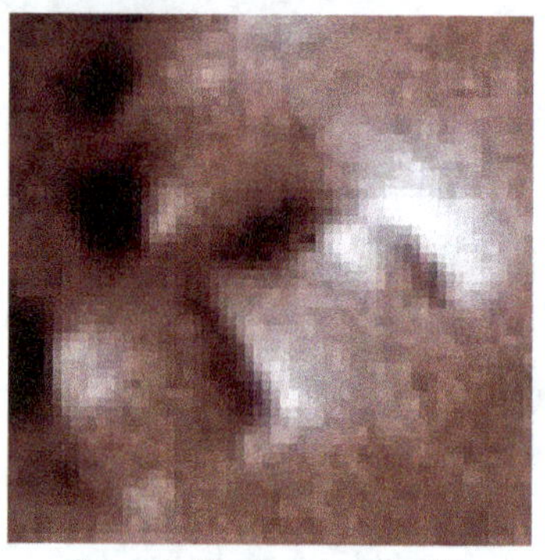

Artemis Region, Image 17.

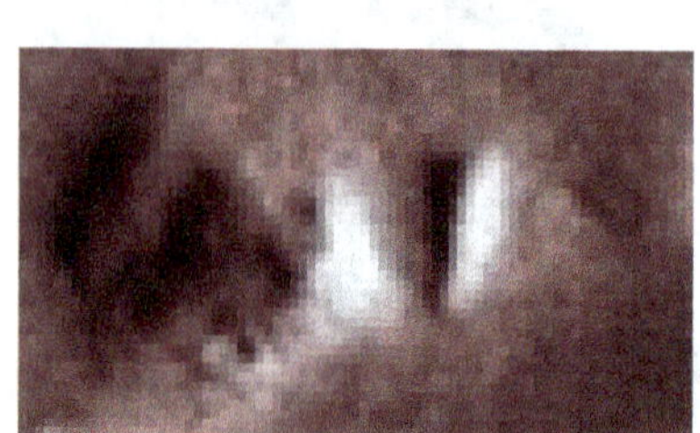

Artemis Region, Image 18.

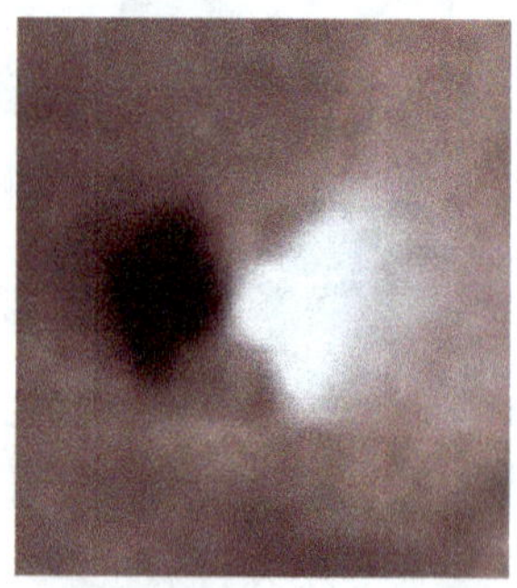

Artemis Region, Image 19.

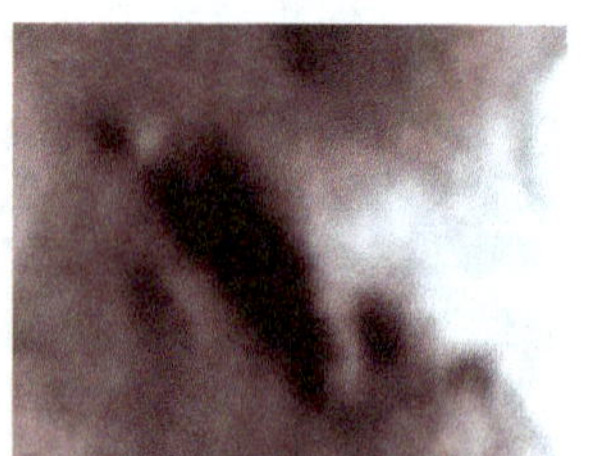

Artemis Region, Image 20.

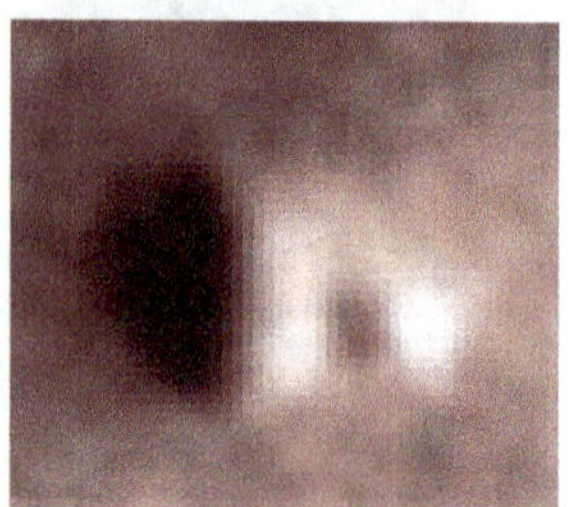

Artemis Region, Image 21.

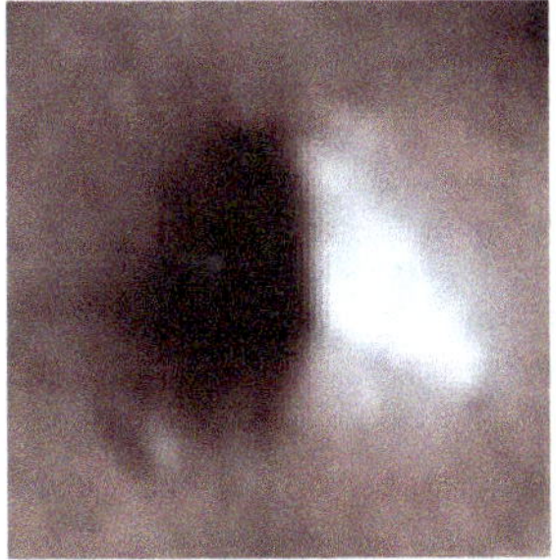
Artemis Region, Image 22.

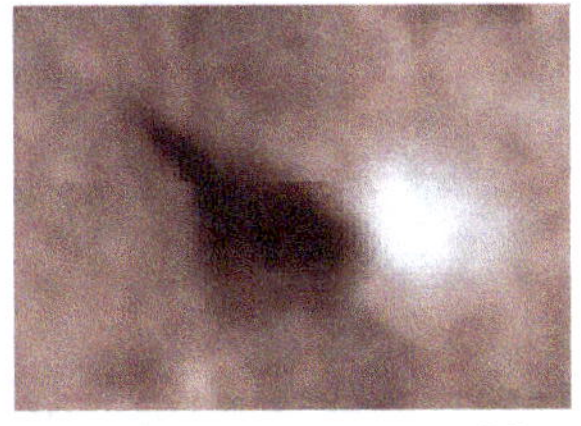
Artemis Region, Image 23.

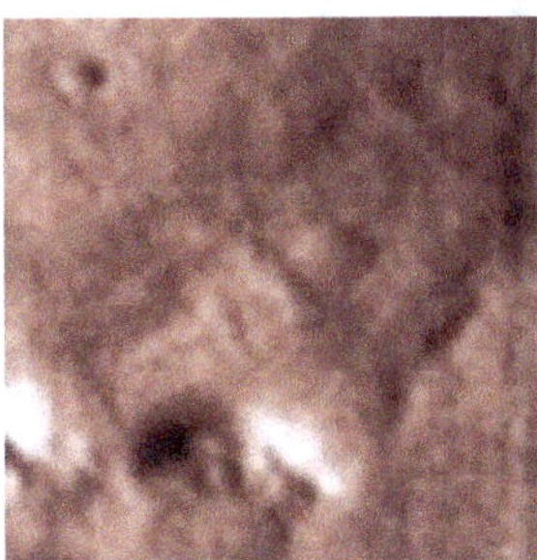
Artemis Region, Image 24.

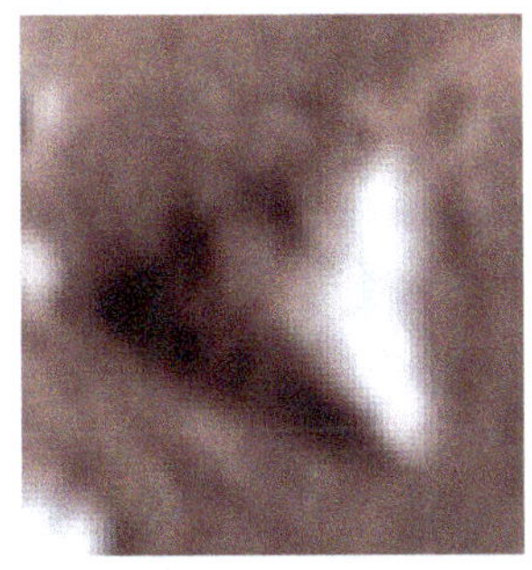
Artemis Region, Image 25.

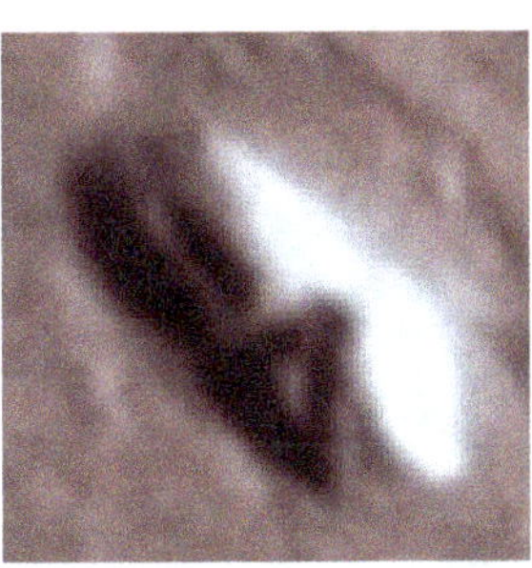
Artemis Region, Image 26.

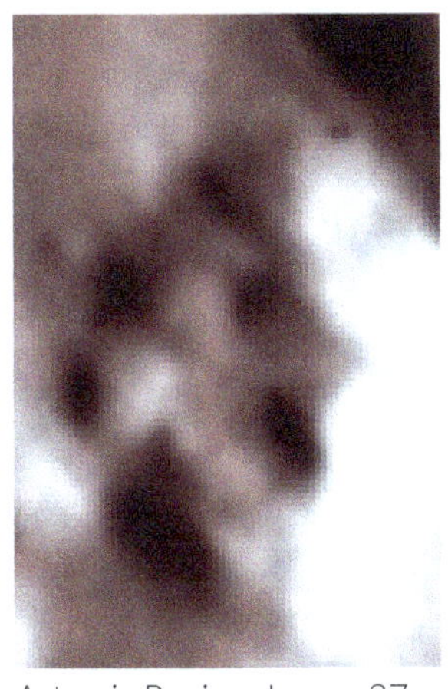
Artemis Region, Image 27.

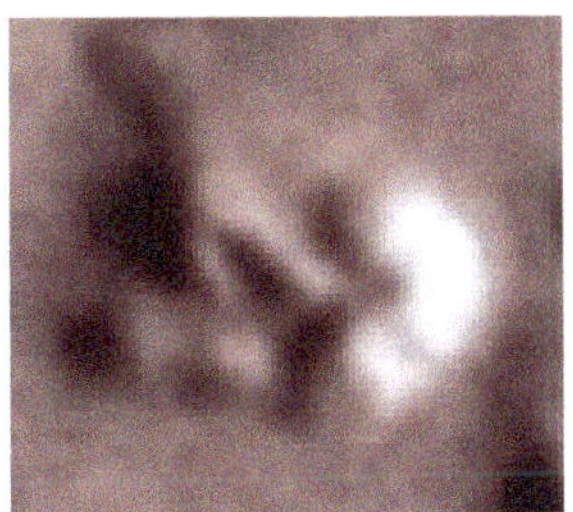
Artemis Region, Image 28.

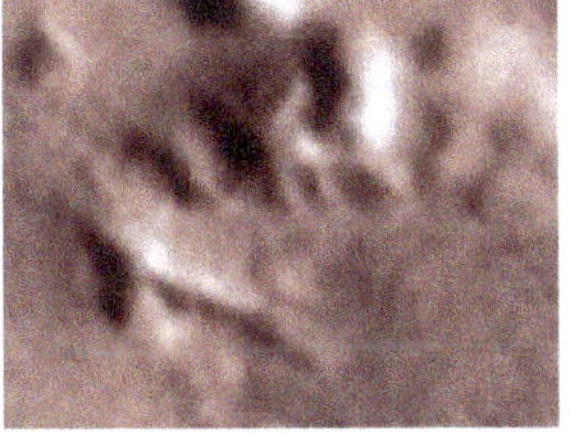
Artemis Region, Image 29.

Chapter 47

200 RANDOM MARTIAN ANOMALIES

CLOSEUPS OF RANDOM STRANGE AND UNUSUAL OBJECTS AND STRUCTURES I DISCOVERED ON THE SURFACE OF MARS

The following collection is comprised of photos I have cropped from NASA's larger parent images. I have colored, brightened, and sharpened them for added contrast and viewability. Nearly all of these photos were taken from high above the Martian surface (exact height unknown). Images credit: NASA/JPL-Caltech/MSSS/UA/USGS.

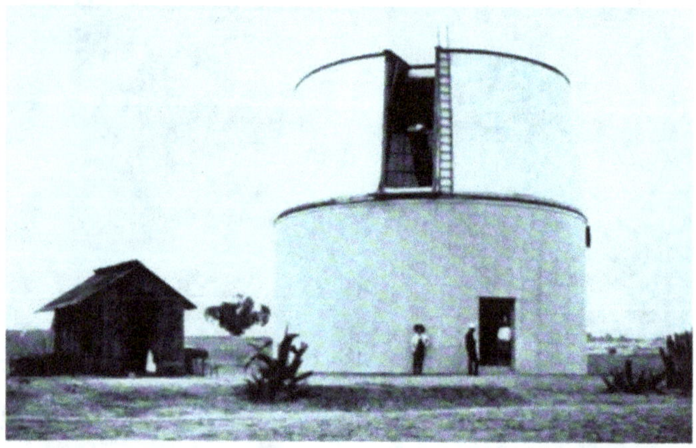

Dome of the Lowell Observatory in Tacubaya, Mexico, where Mars was observed during the winter of 1896-1897.

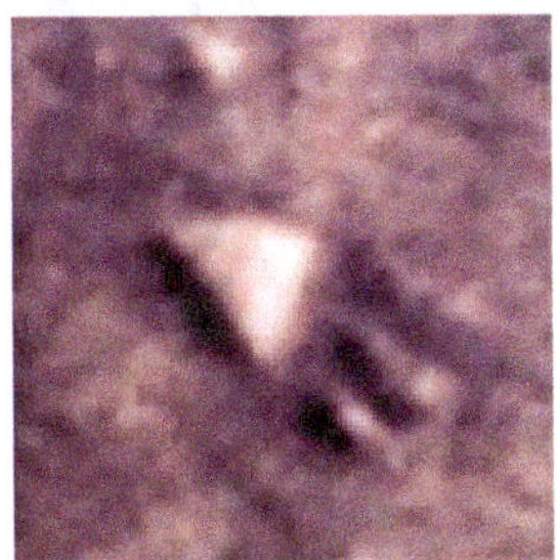
Random objects, Image 1.

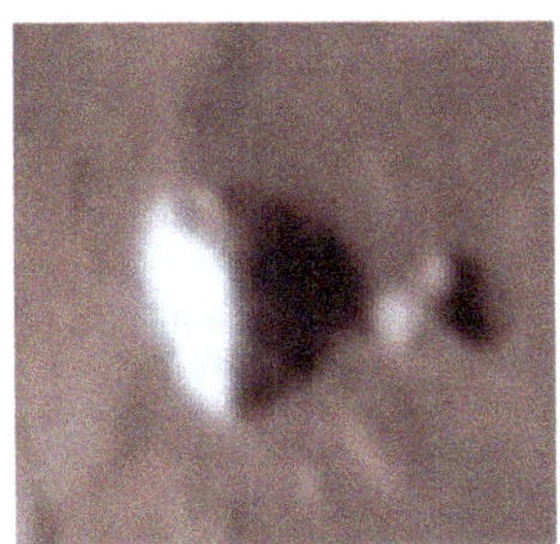
Random objects, Image 2.

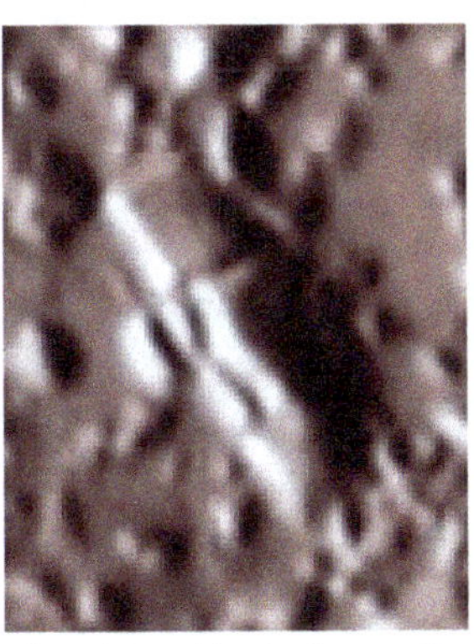
Random objects, Image 3.

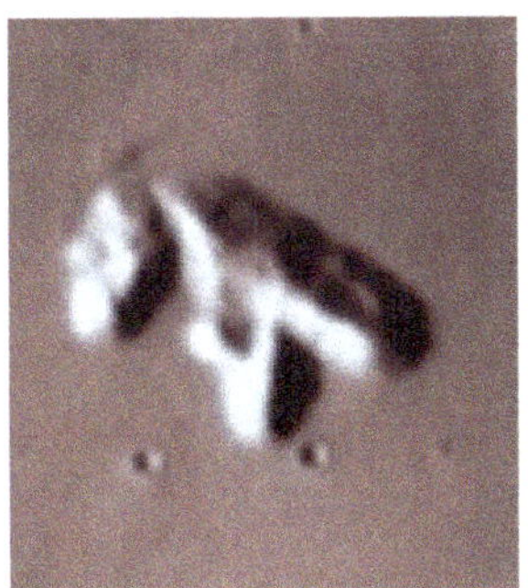
Random objects, Image 4.

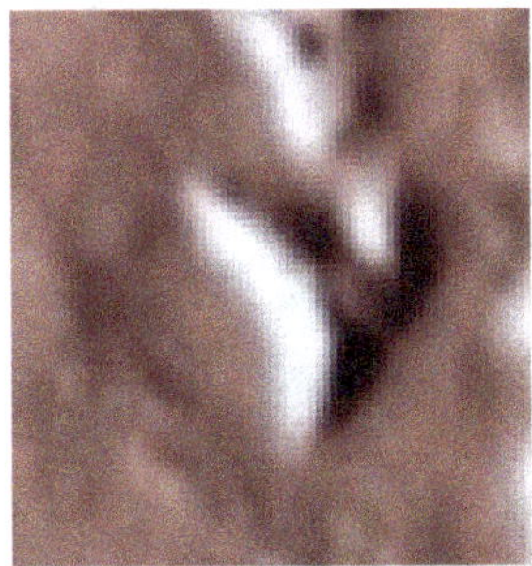
Random objects, Image 5.

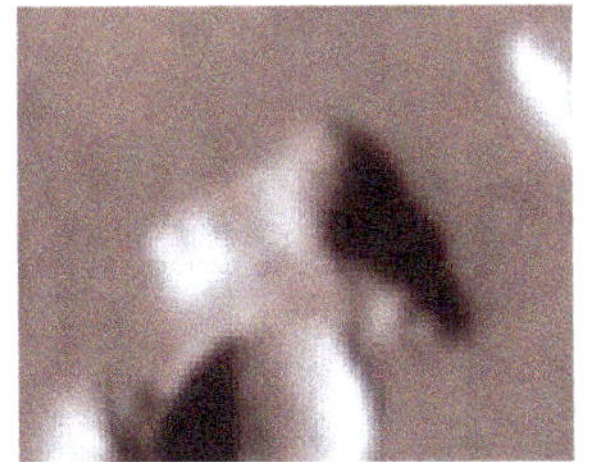
Random objects, Image 6.

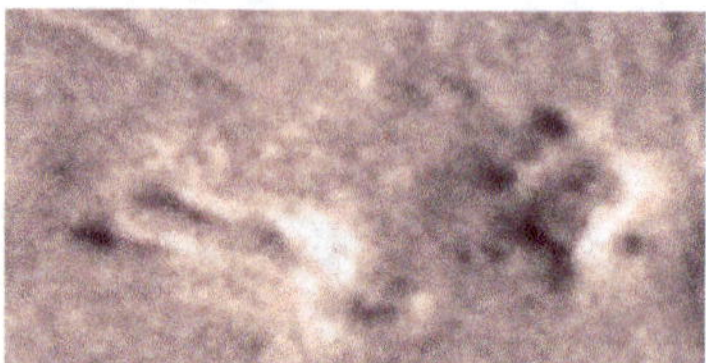
Random objects, Image 7.

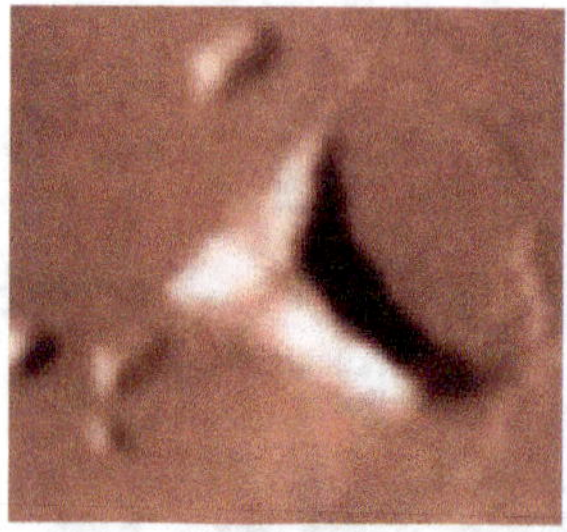
Random objects, Image 8.

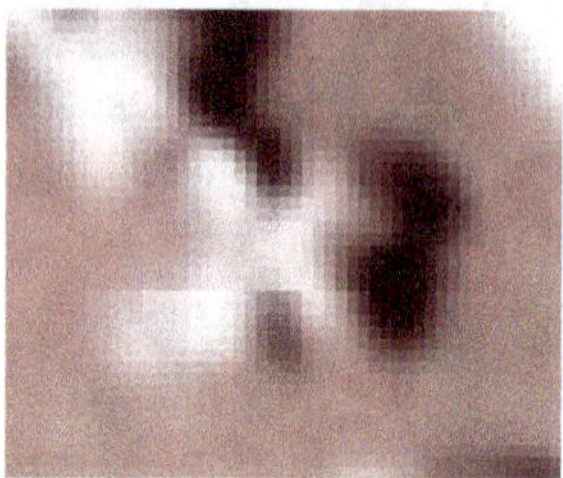
Random objects, Image 9.

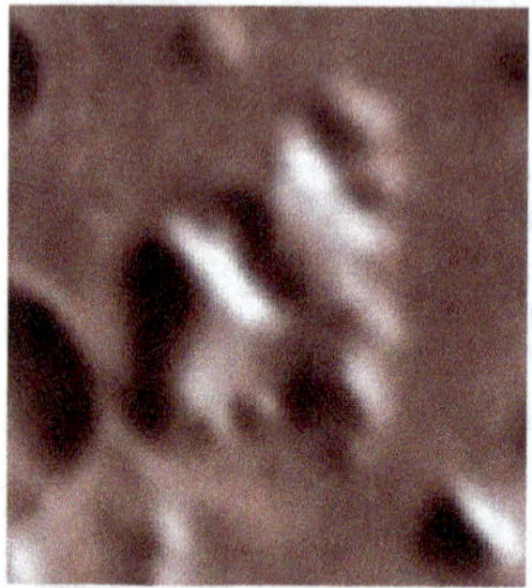
Random objects, Image 10.

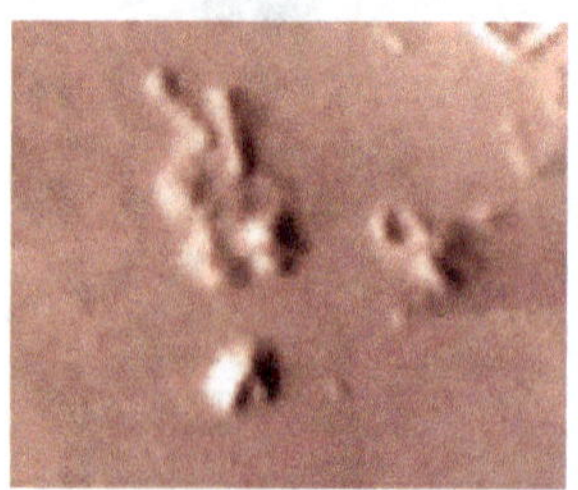
Random objects, Image 11.

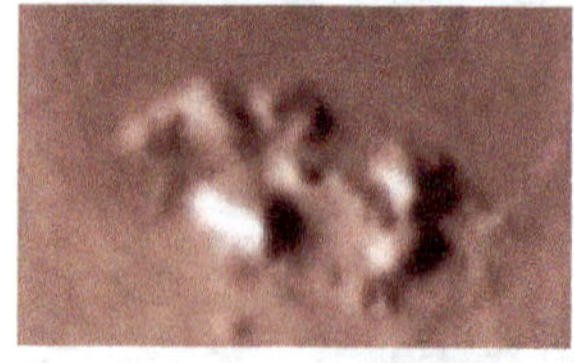
Random objects, Image 12.

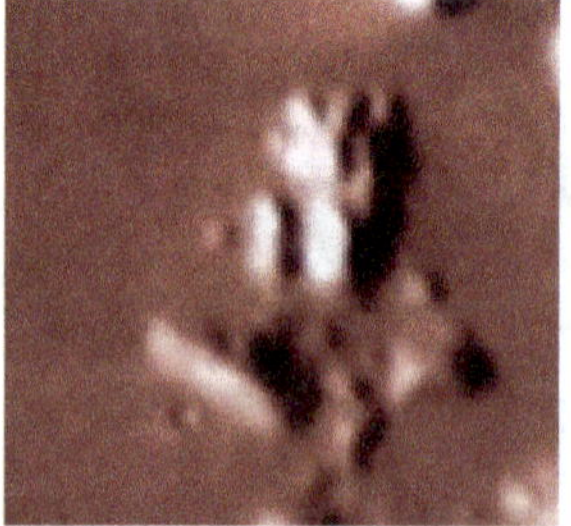
Random objects, Image 13.

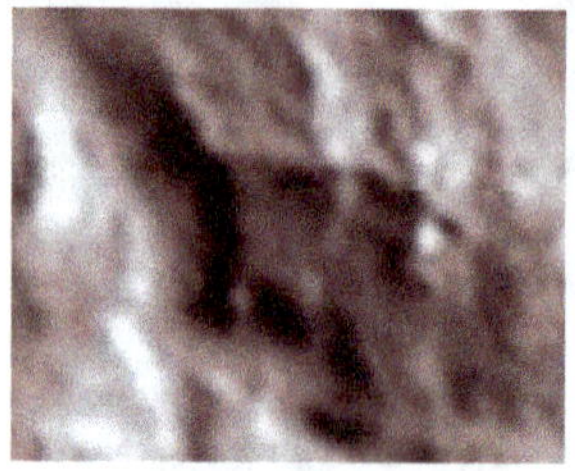
Random objects, Image 14.

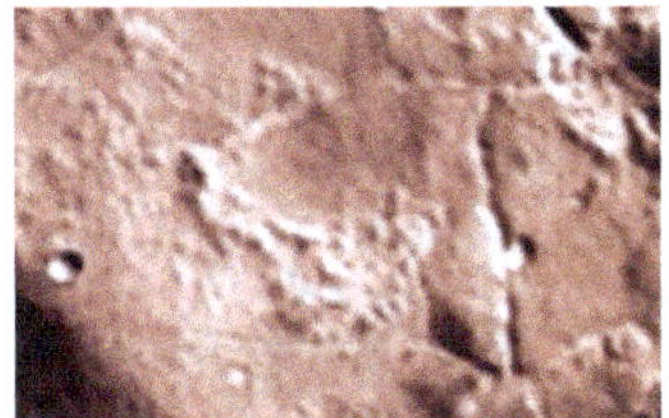

Random objects, Image 15.

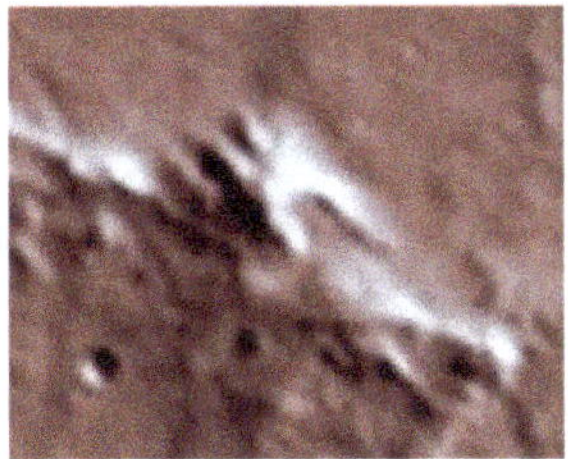

Random objects, Image 16.

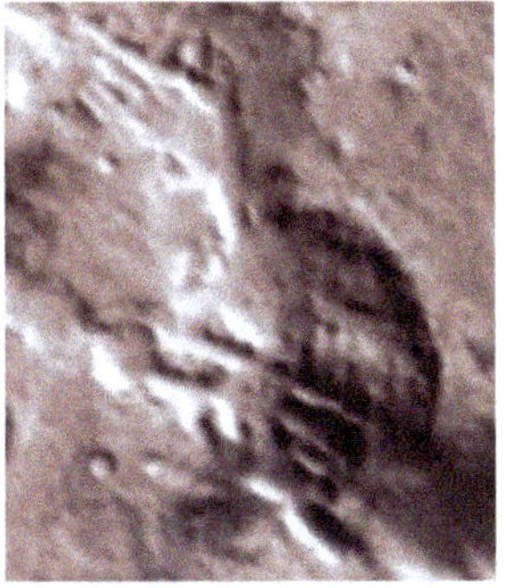

Random objects, Image 17.

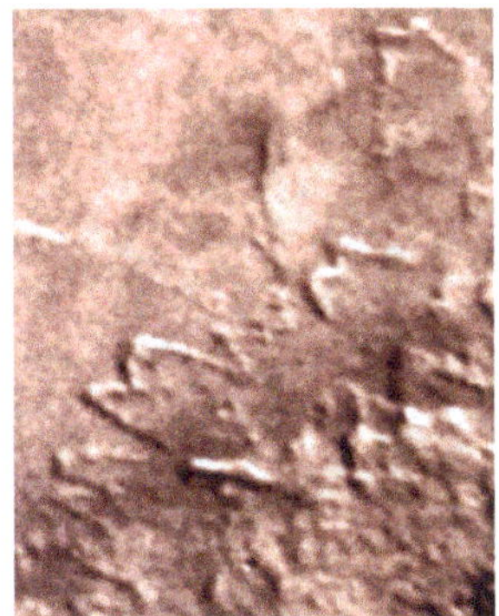

Random objects, Image 18.

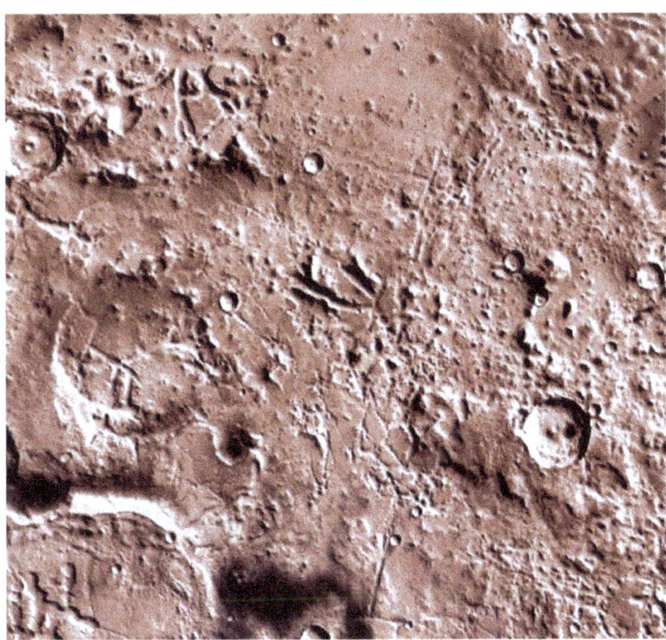

Random objects, Image 19.

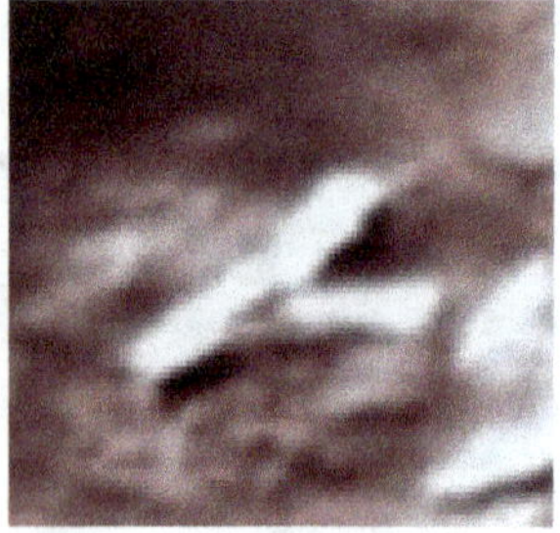

Random objects, Image 20.

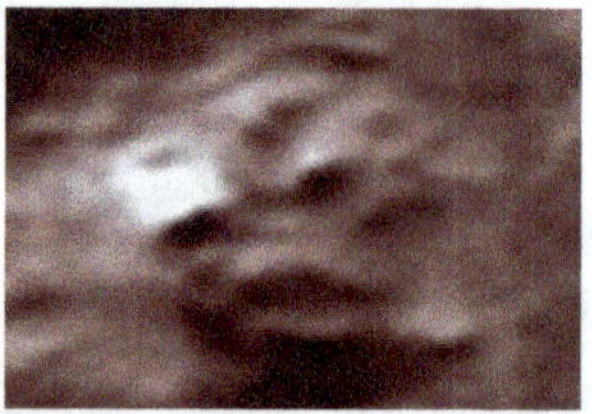

Random objects, Image 21.

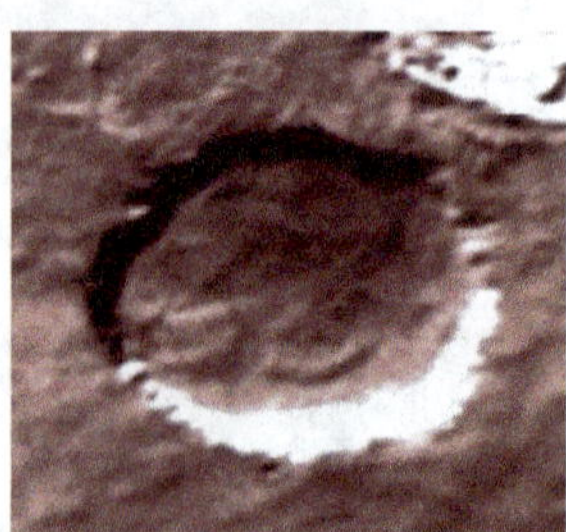

Random objects, Image 22.

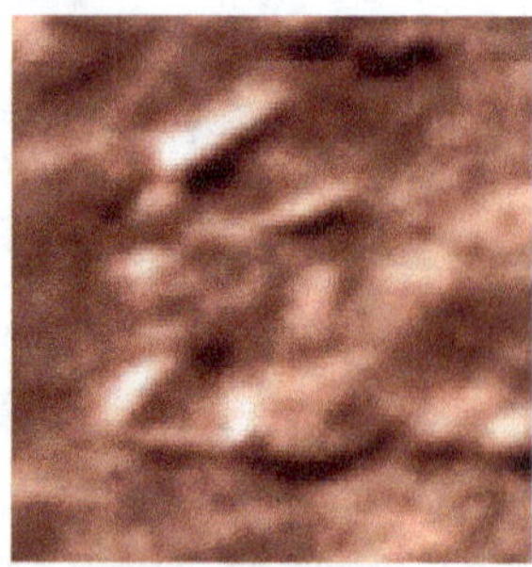

Random objects, Image 23.

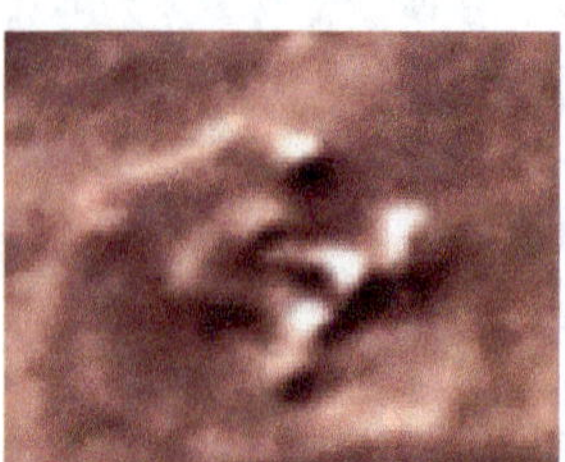

Random objects, Image 24.

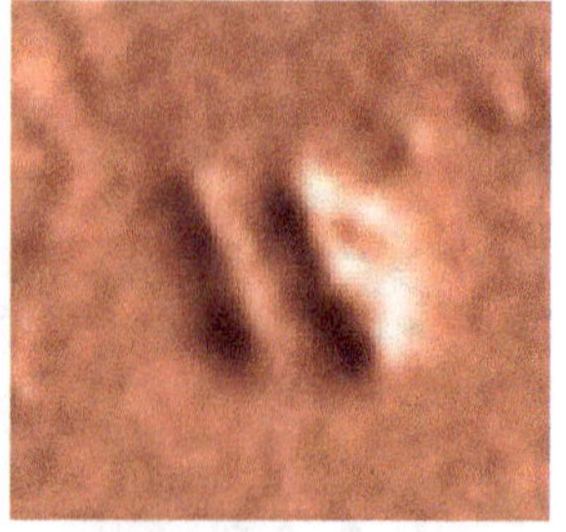

Random objects, Image 25.

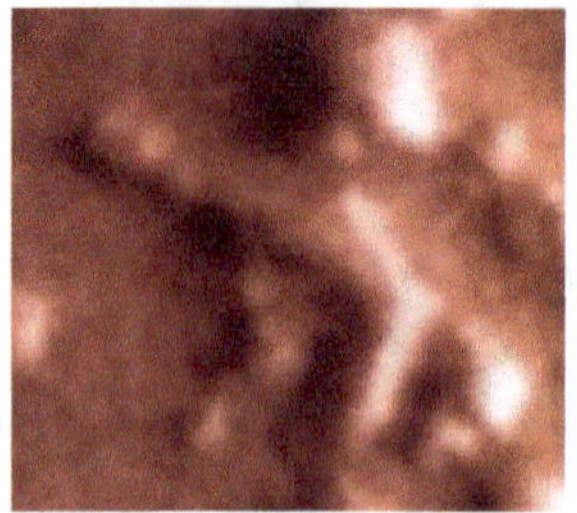

Random objects, Image 26.

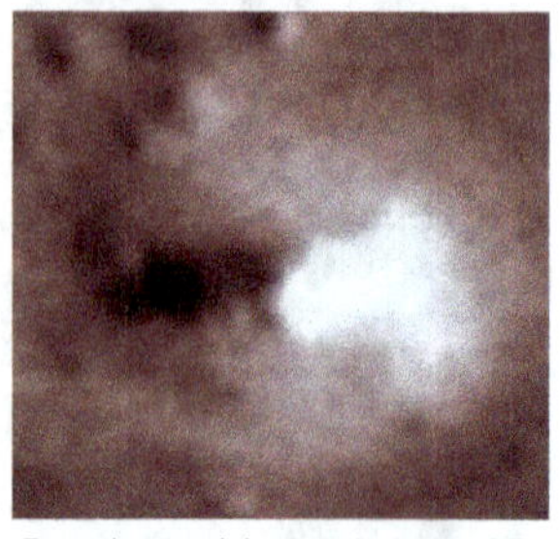

Random objects, Image 27.

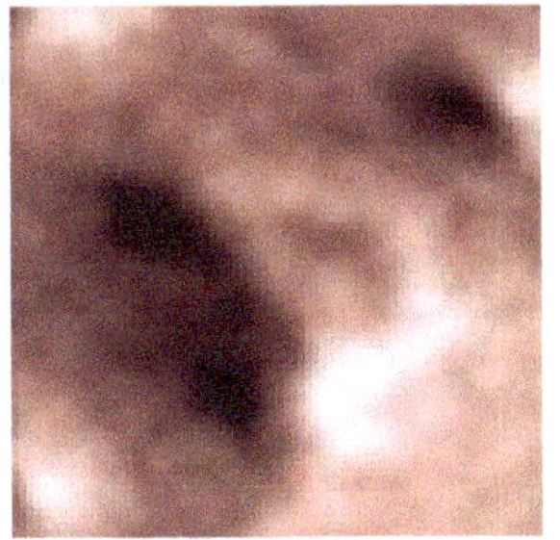
Random objects, Image 28.

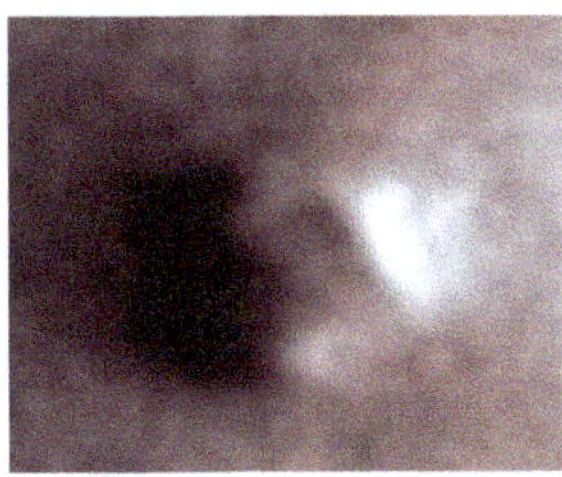
Random object, Image 29.

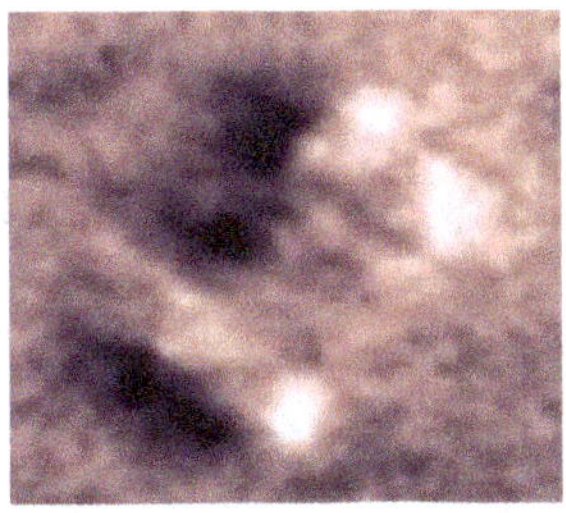
Random objects, Image 30.

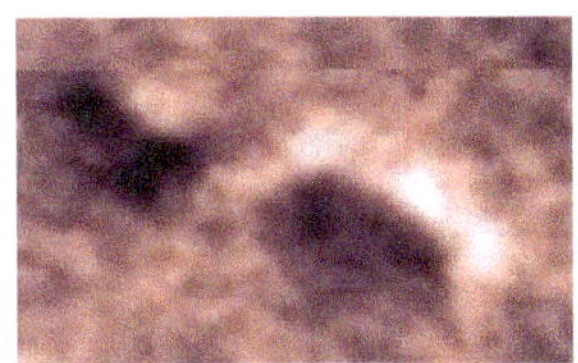
Random objects, Image 31.

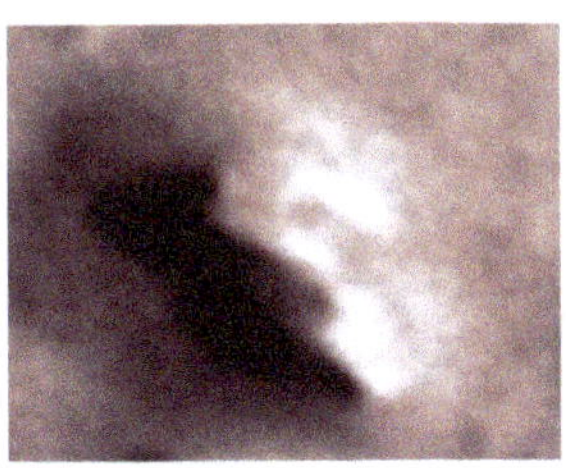
Random object, Image 32.

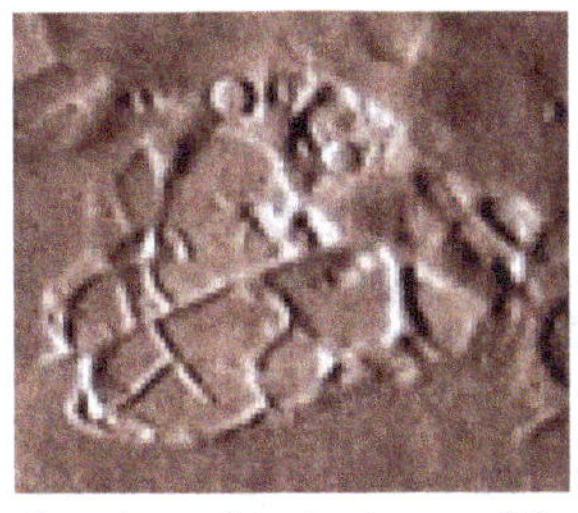
Random objects, Image 33.

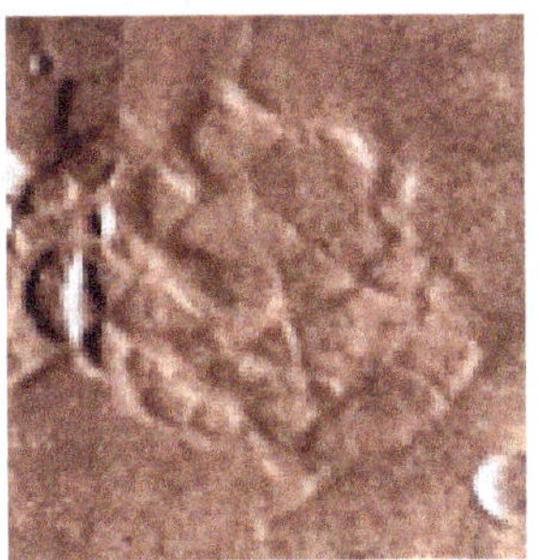
Random objects, Image 34.

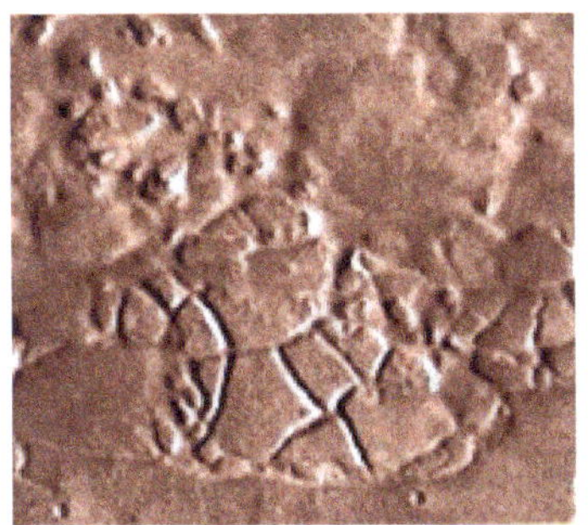
Random objects, Image 35.

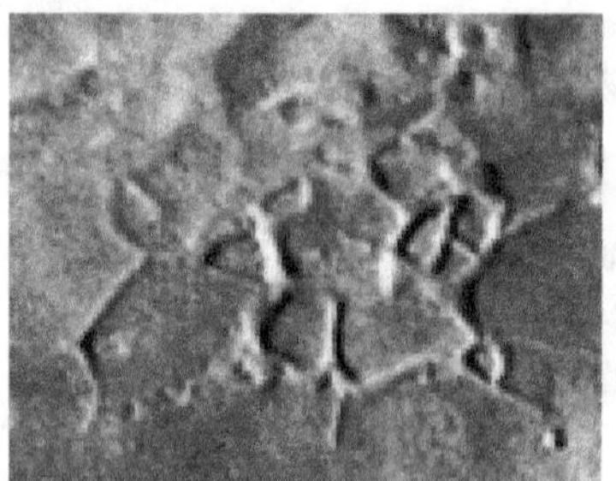
Random objects, Image 36.

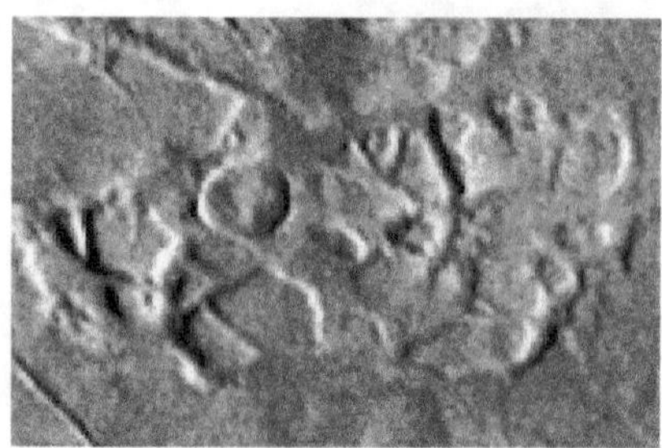
Random objects, Image 37.

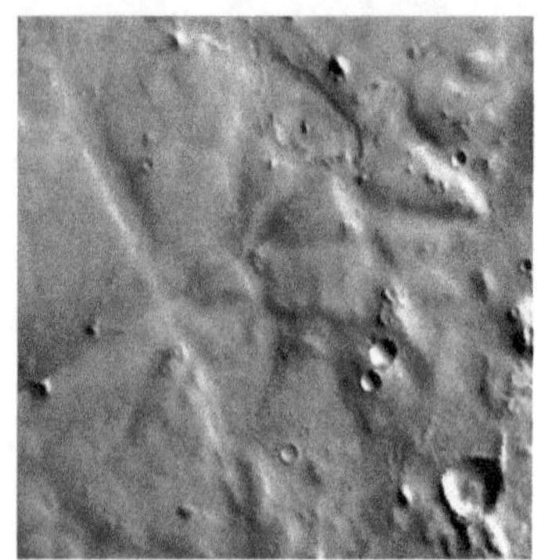
Random objects, Image 38.

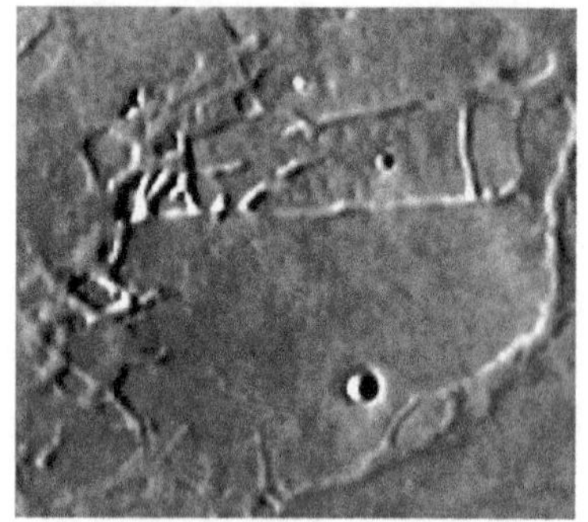
Random objects, Image 39.

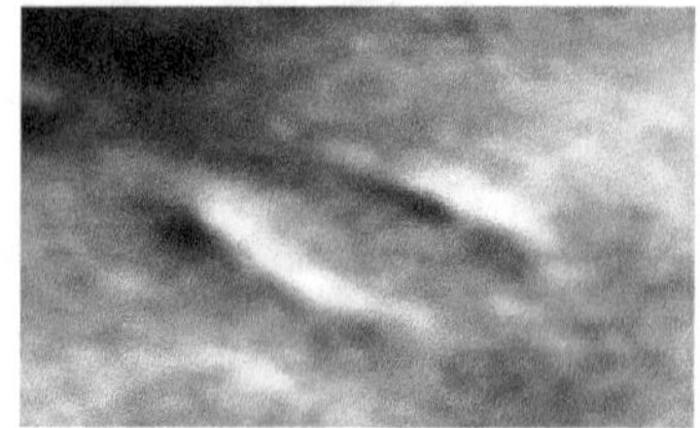
Random objects, Image 40.

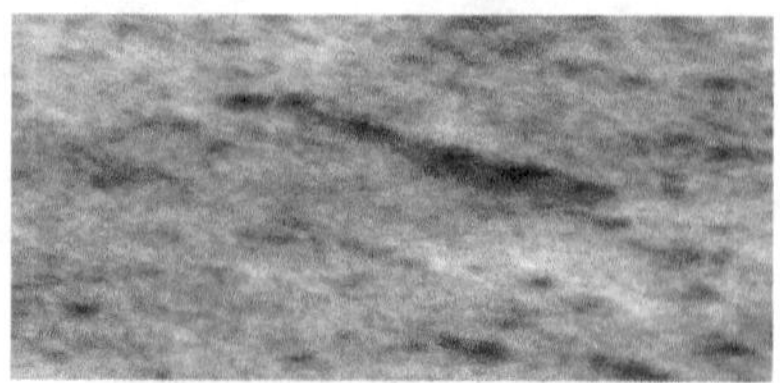
Random object, Image 41.

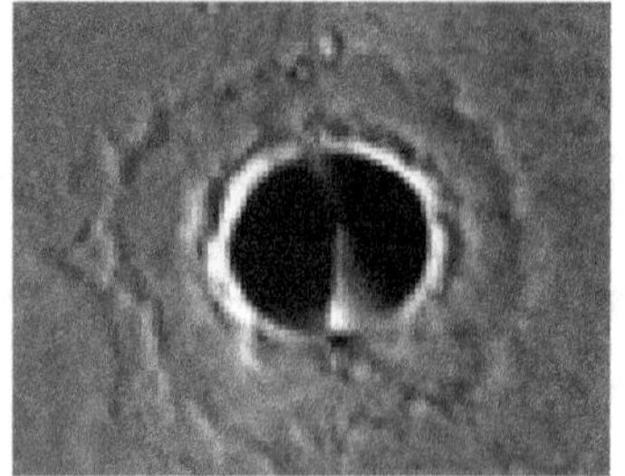
Random objects, Image 42.

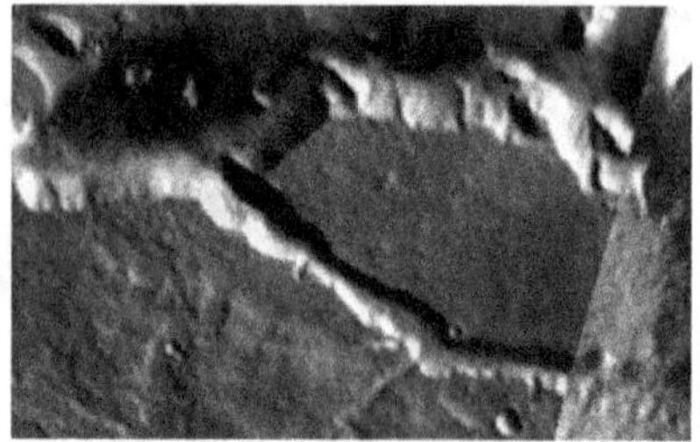
Random objects, Image 43.

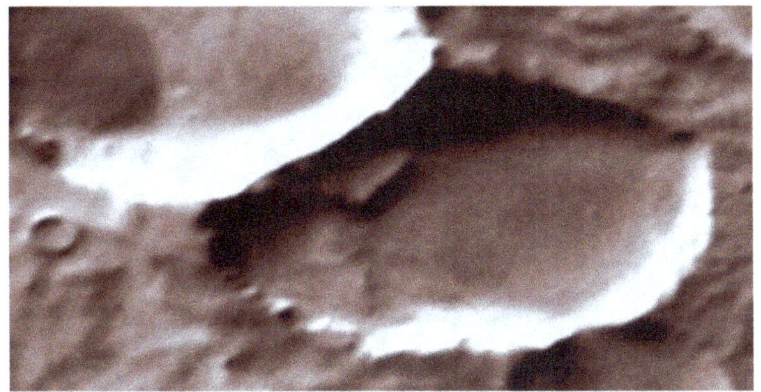

Random objects, Image 44.

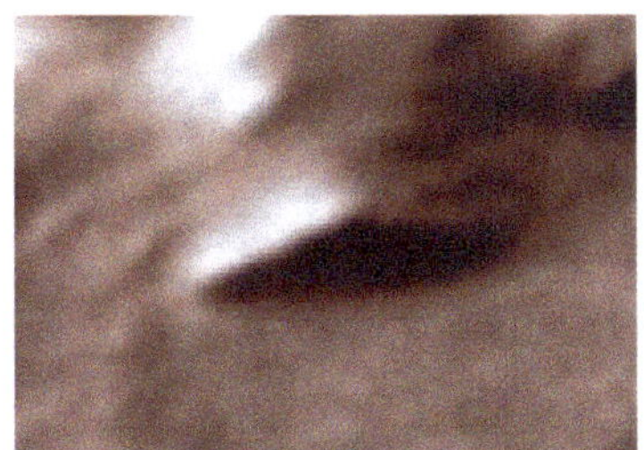

Random objects, Image 45.

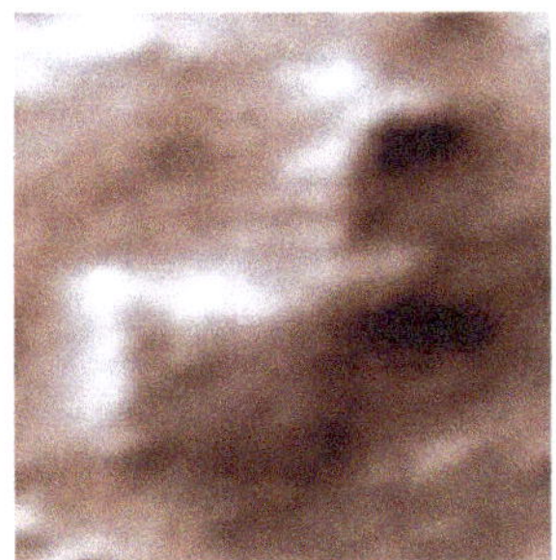

Random objects, Image 46.

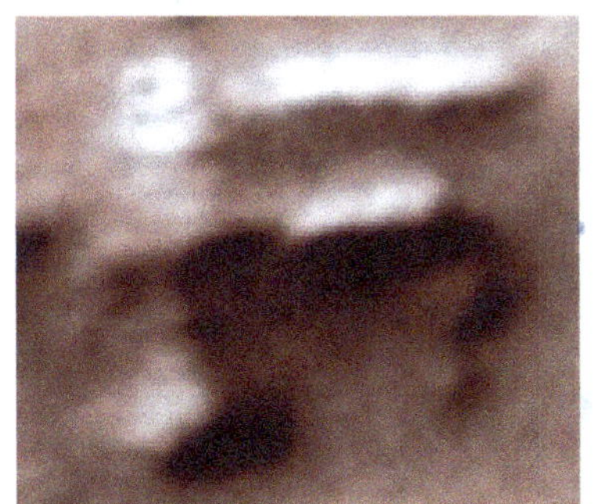

Random objects, Image 47.

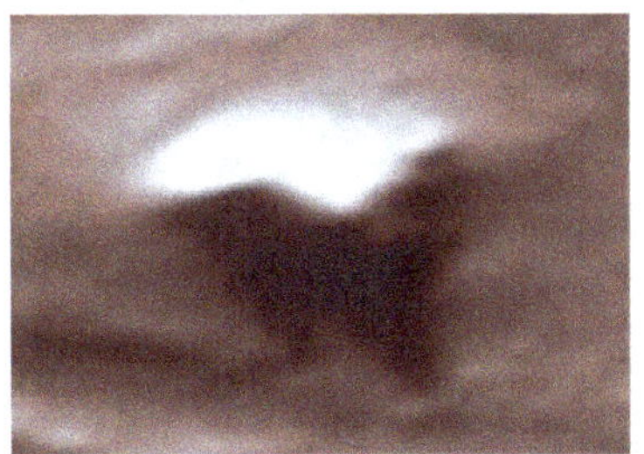

Random object, Image 48.

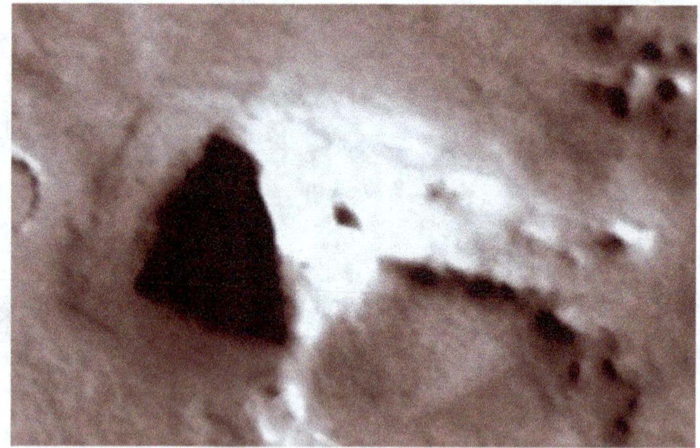
Random objects, Image 49.

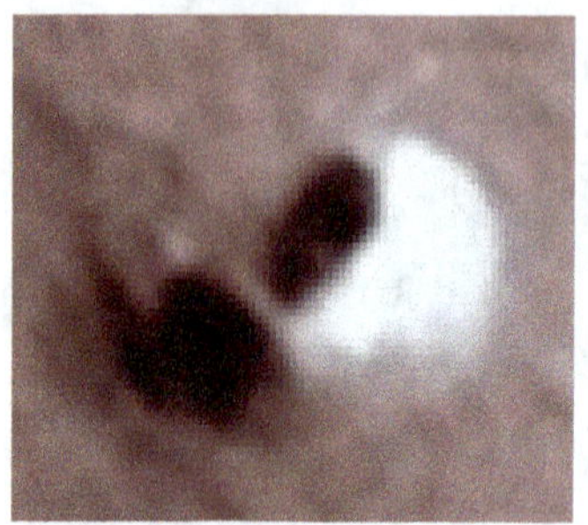
Random object, Image 50.

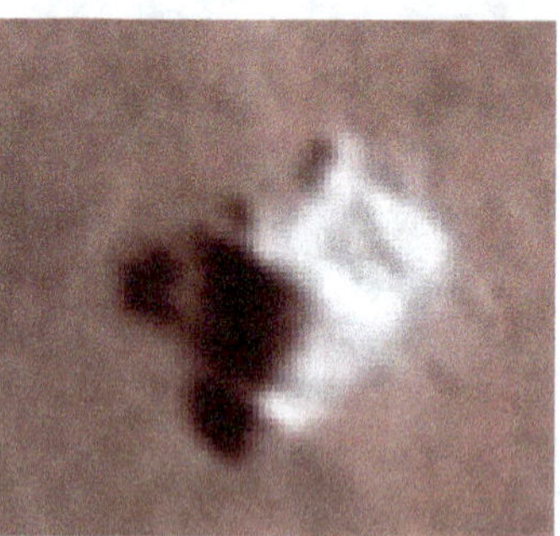
Random object, Image 51.

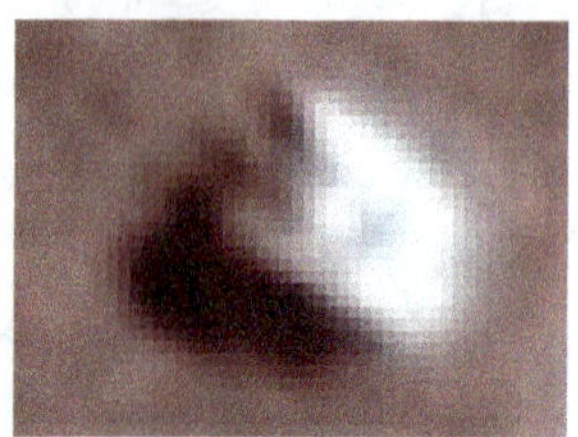
Random object, Image 52.

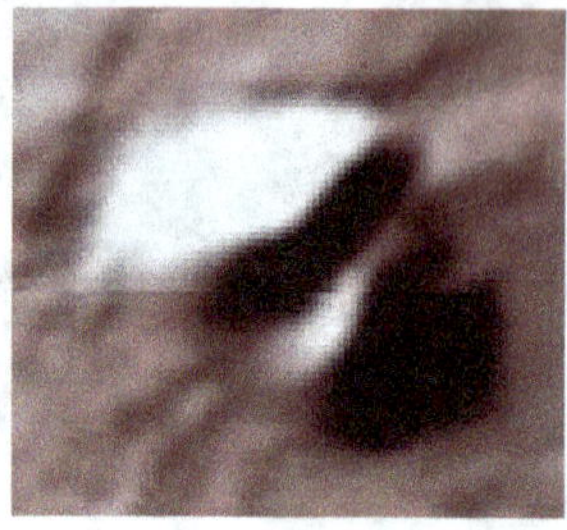
Random objects, Image 53.

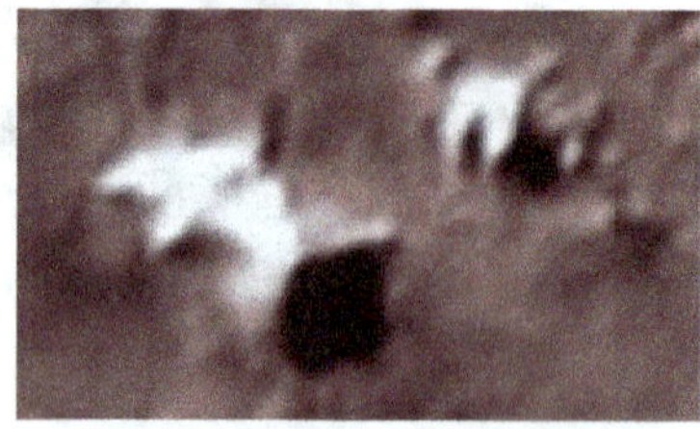
Random objects, Image 54.

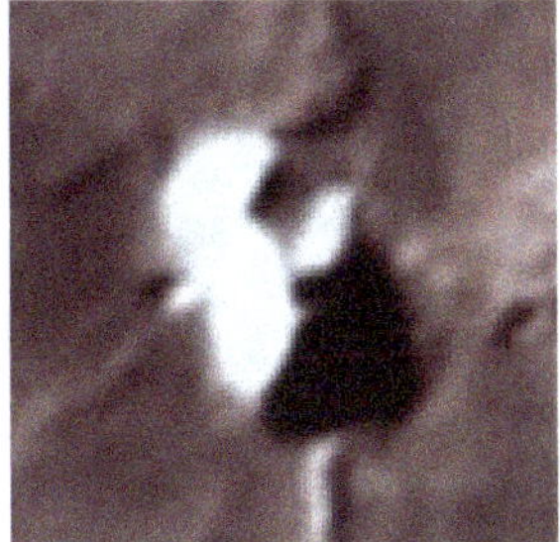
Random objects, Image 55.

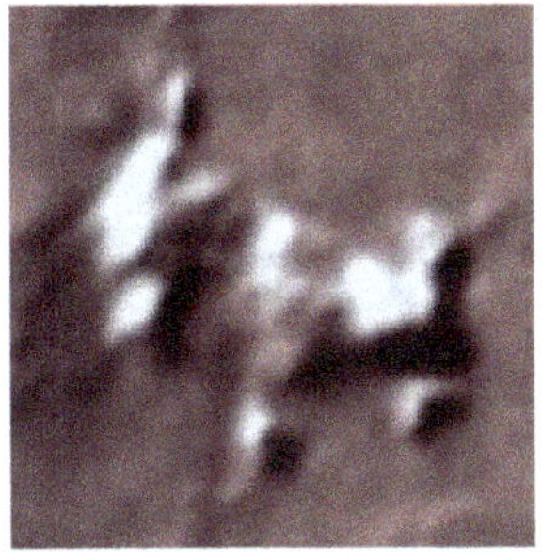
Random objects, Image 56.

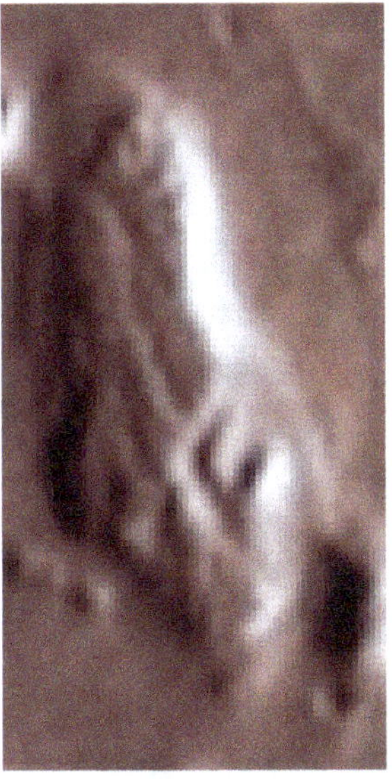
Random objects, Image 57.

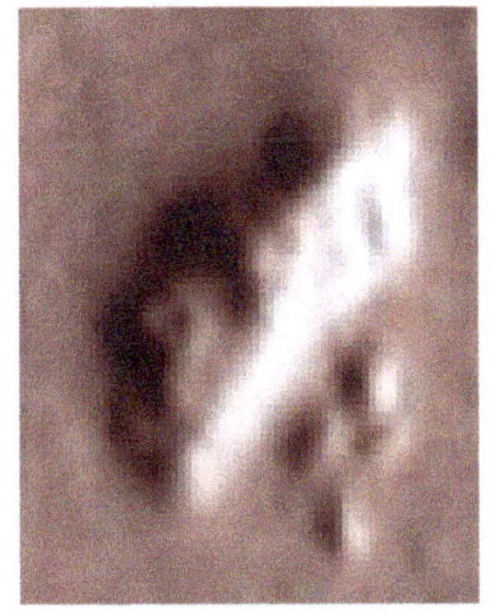
Random objects, Image 58.

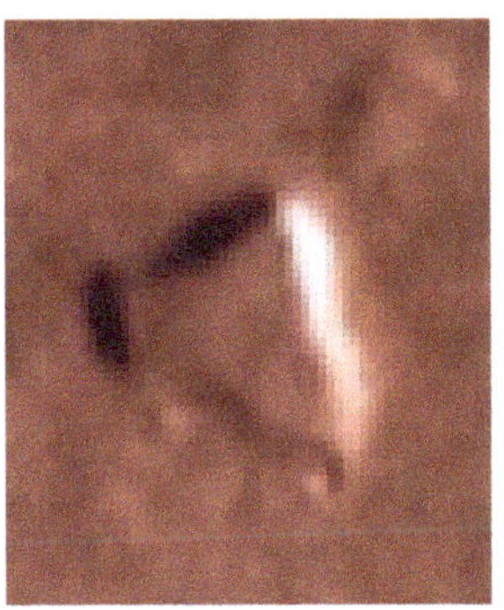
Random objects, Image 60.

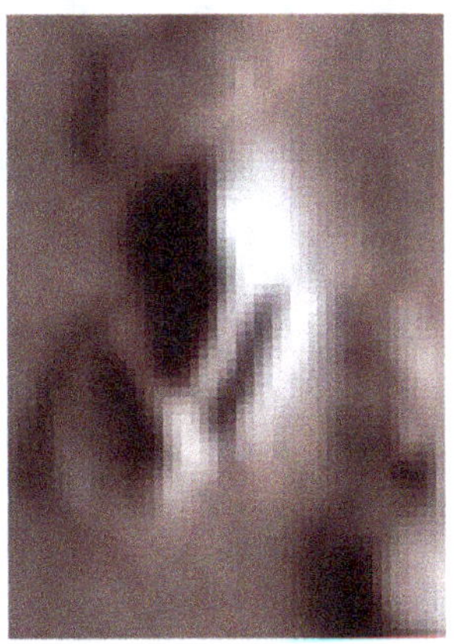
Random objects, Image 59.

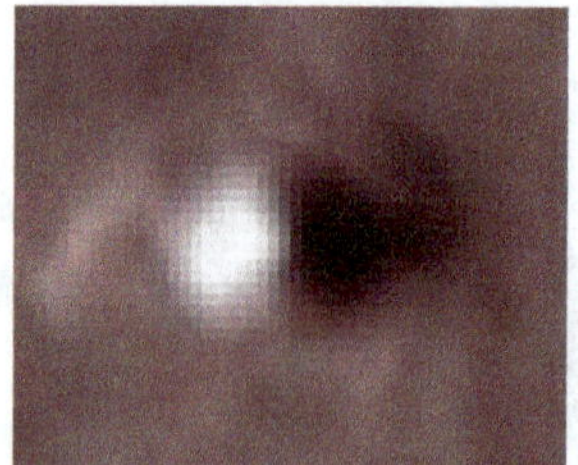

Random object, Image 61.

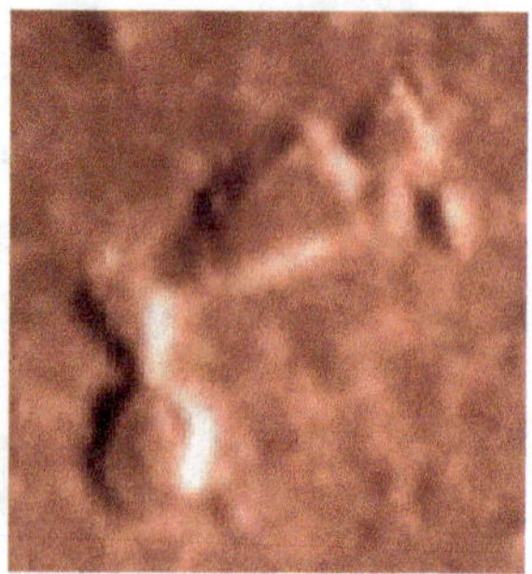

Random objects, Image 62.

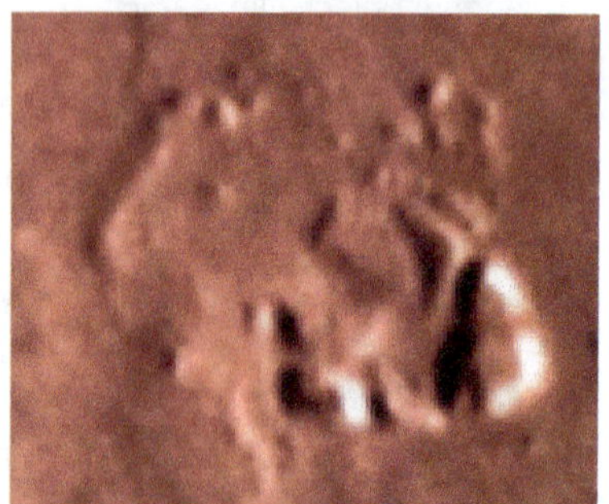

Random objects, Image 63.

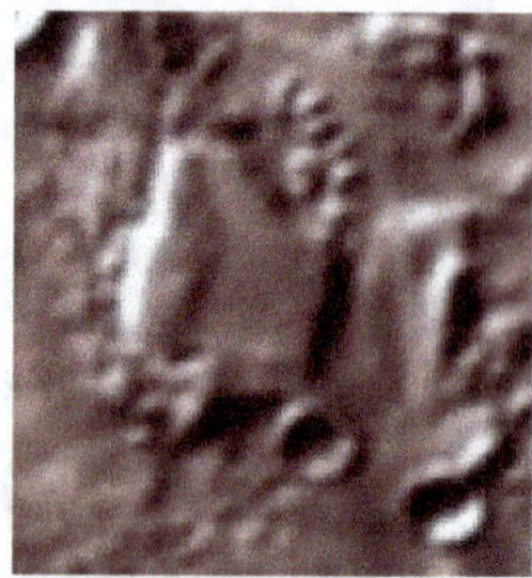

Random objects, Image 64.

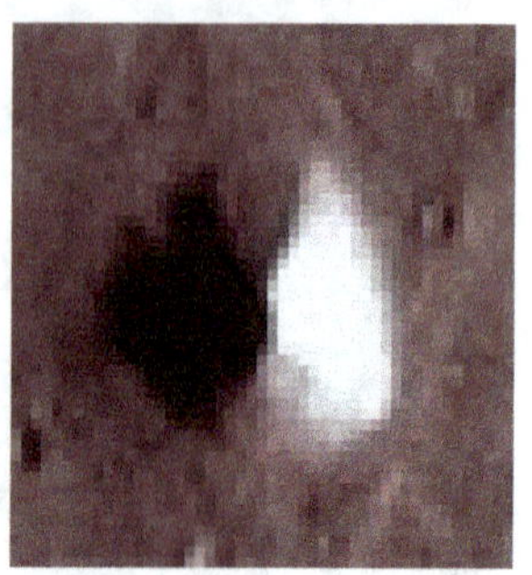

Random object, Image 65.

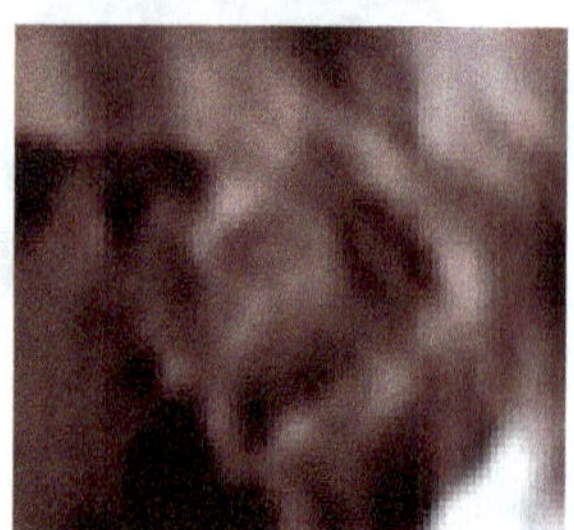

Random objects, Image 66.

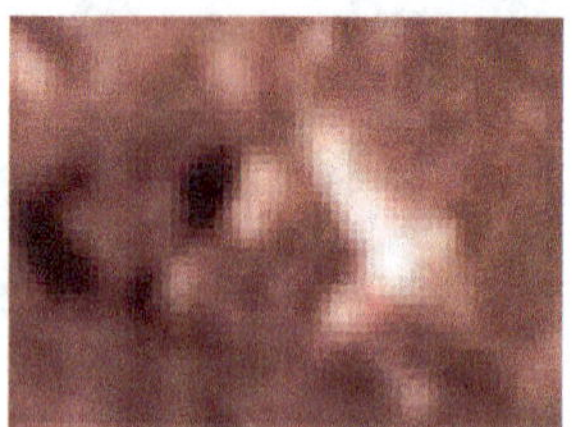

Random objects, Image 67.

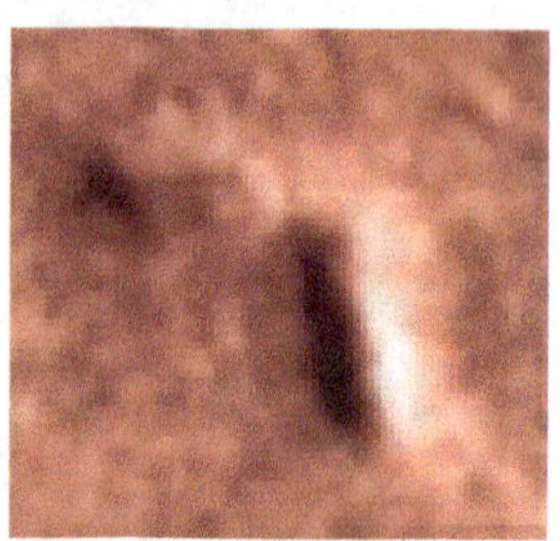

Random objects, Image 68.

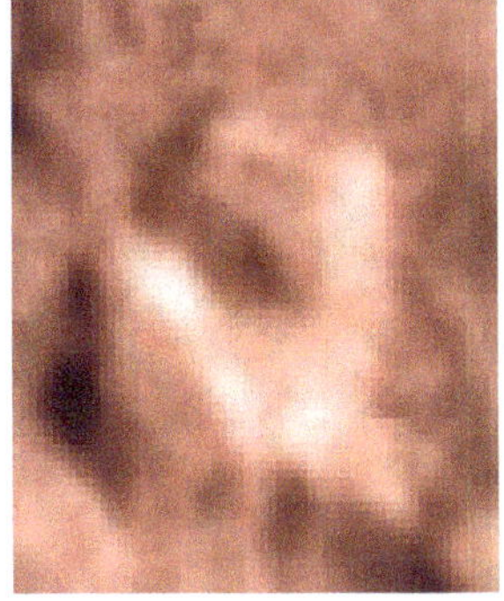

Random objects, Image 69.

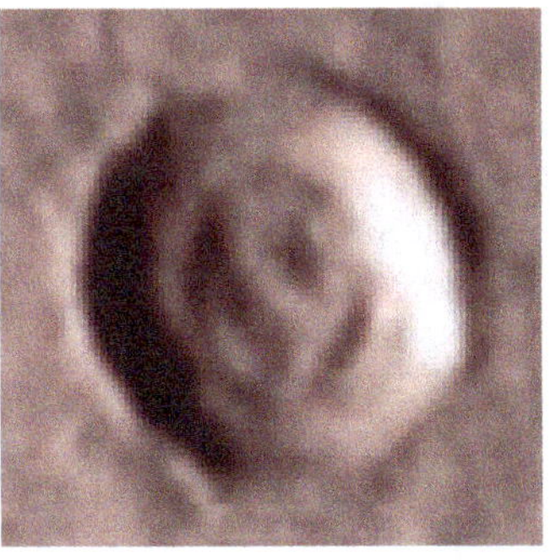

Random objects, Image 70.

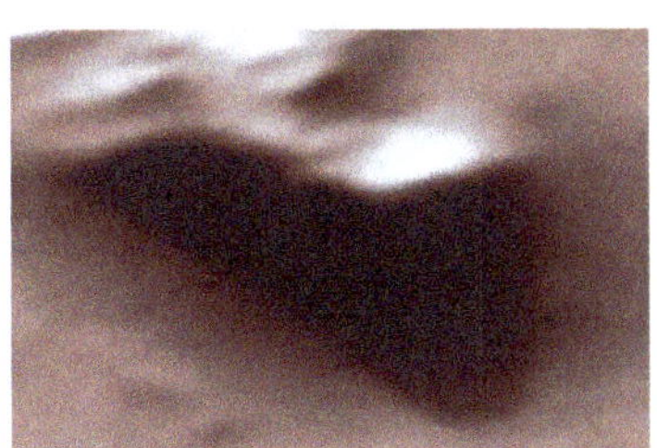

Random object, Image 71.

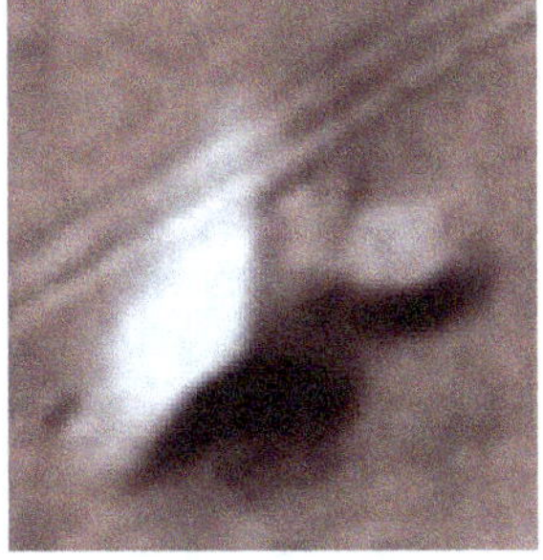

Random objects, Image 72.

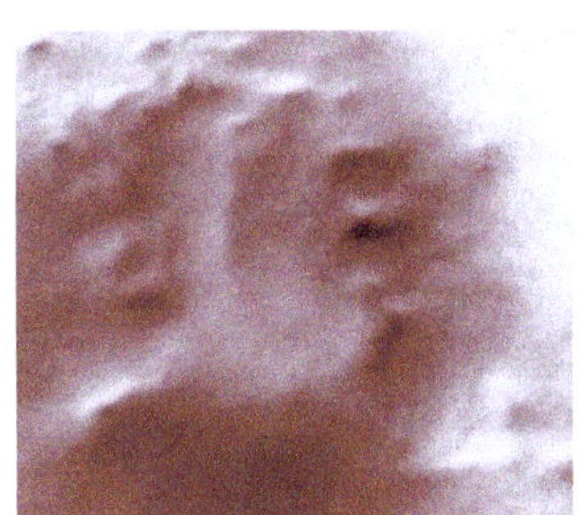

Random objects, Image 73.

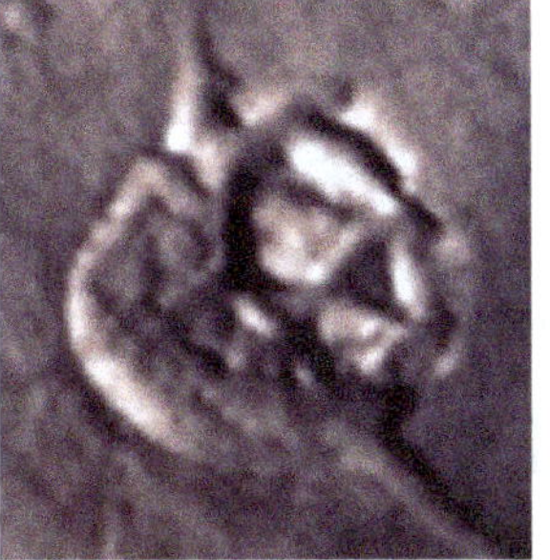

Random objects, Image 75.

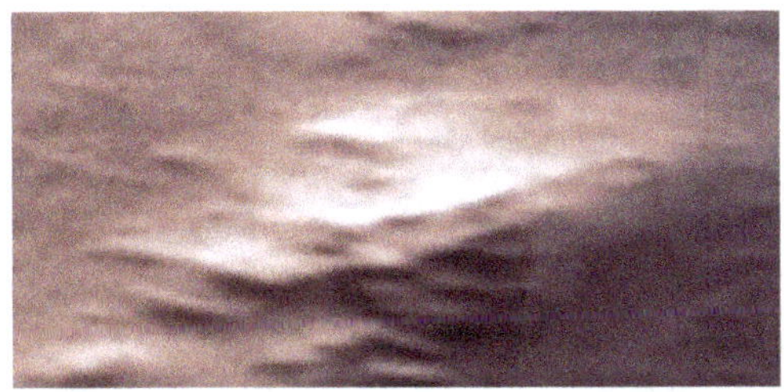

Random objects, Image 74.

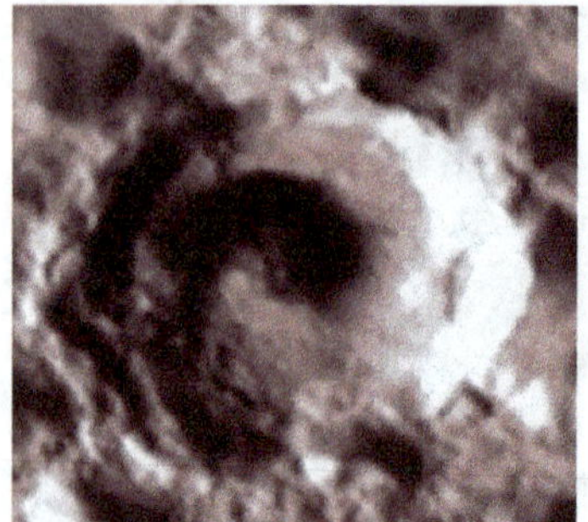

Random objects, Image 76.

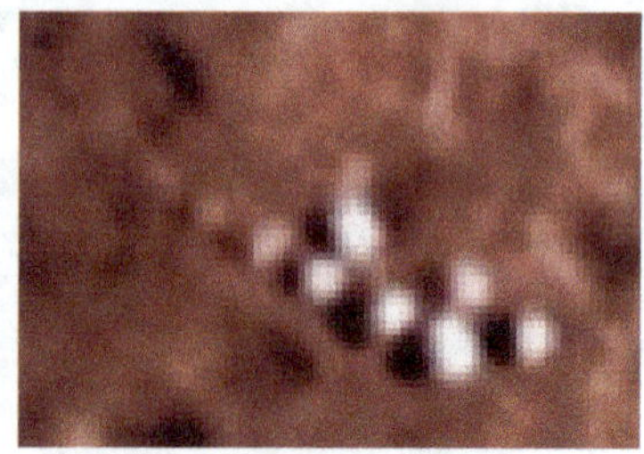

Random objects, Image 77.

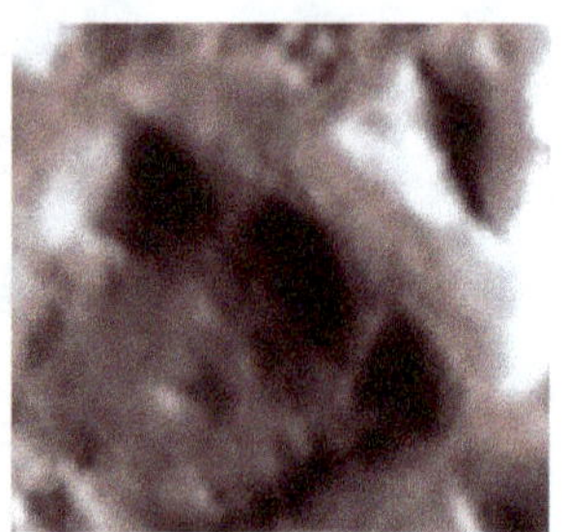

Random objects, Image 78.

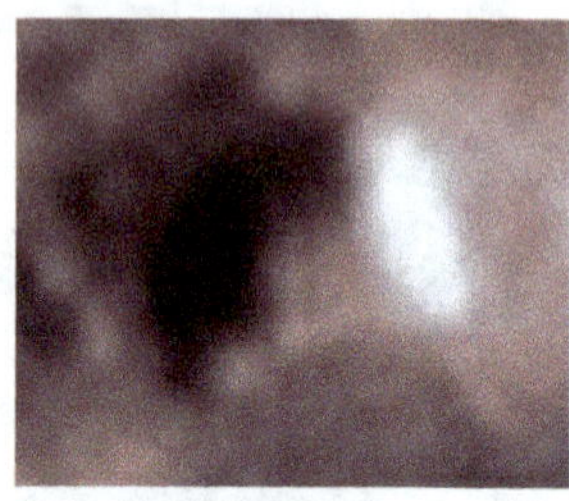

Random objects, Image 79.

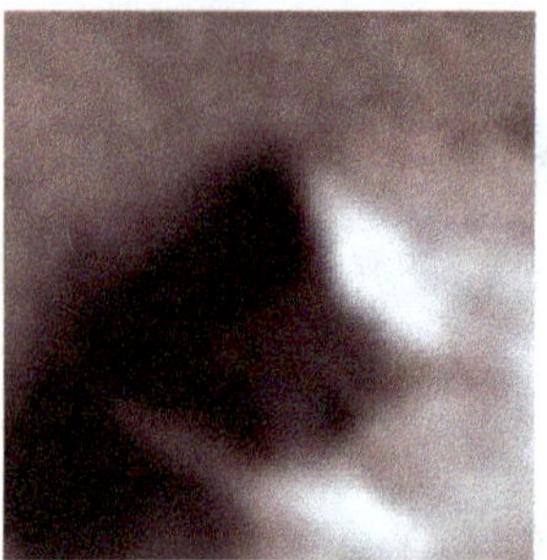

Random objects, Image 80.

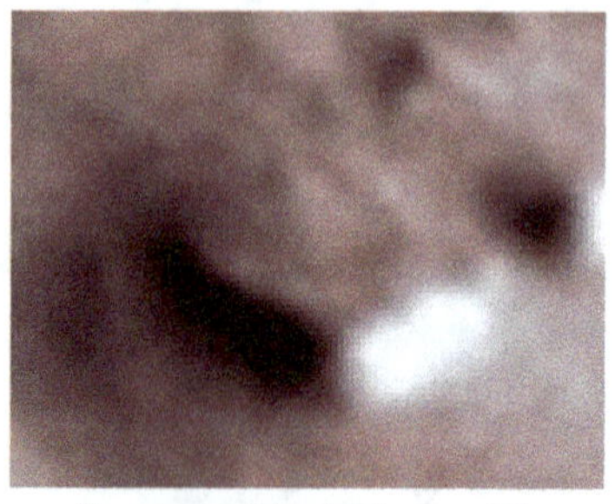

Random objects, Image 81.

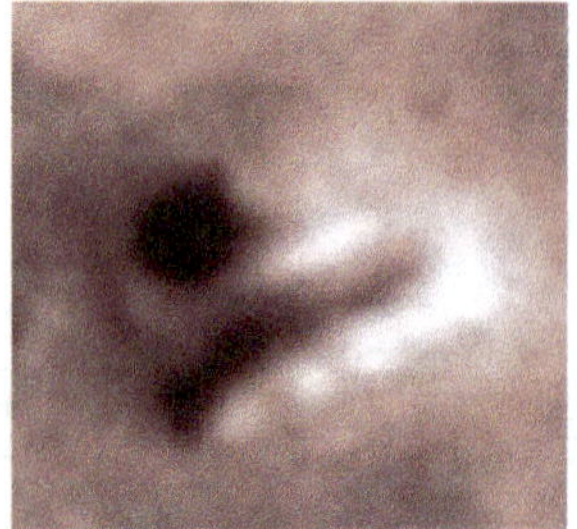

Random objects, Image 82.

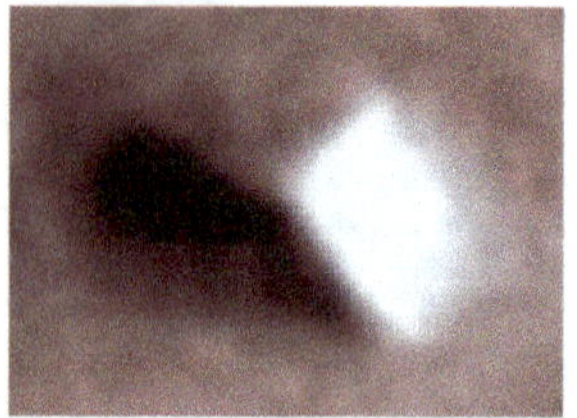

Random object, Image 83.

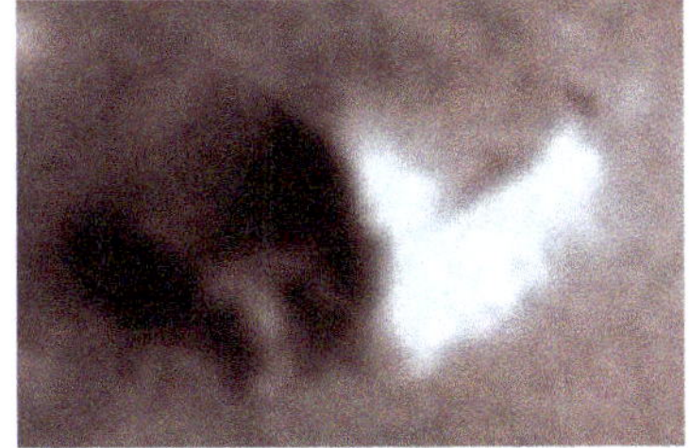
Random objects, Image 84.

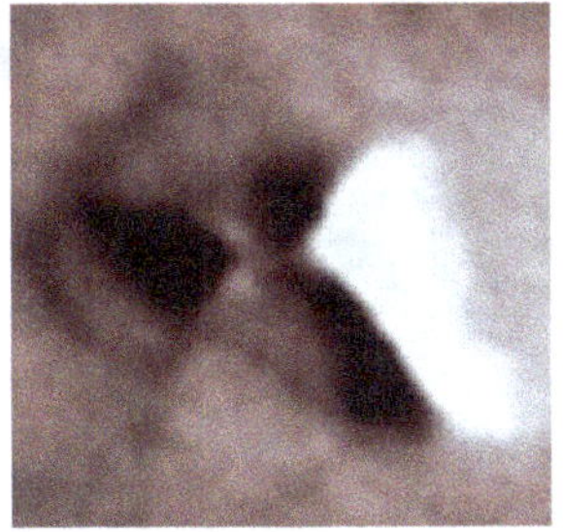
Random objects, Image 85.

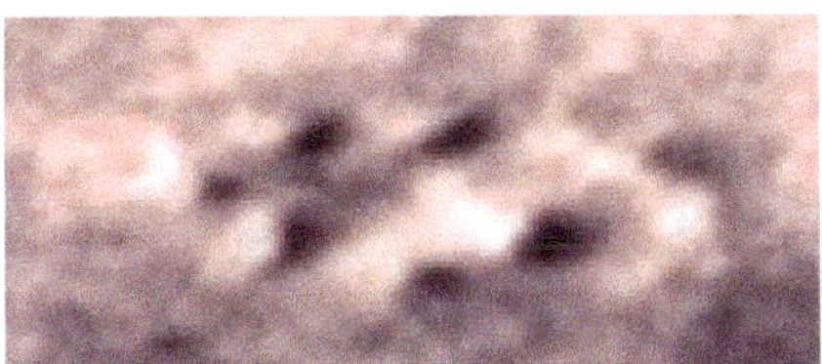
Random objects, Image 86.

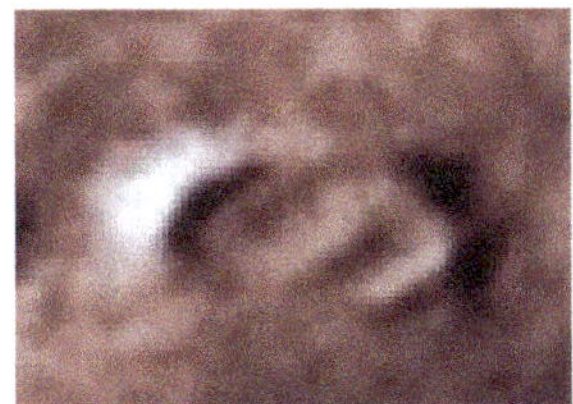
Random objects, Image 87.

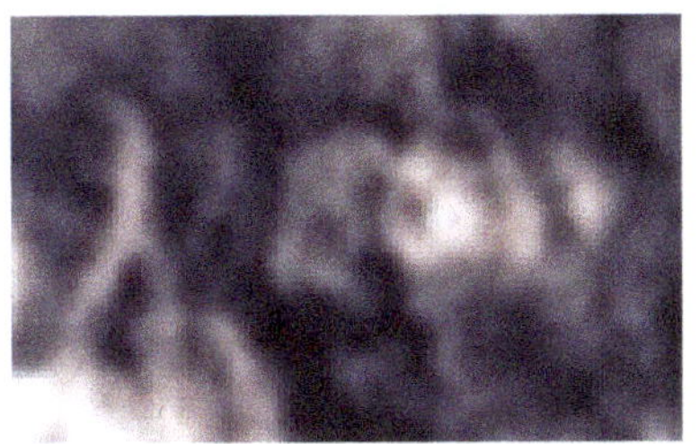
Random objects, Image 88.

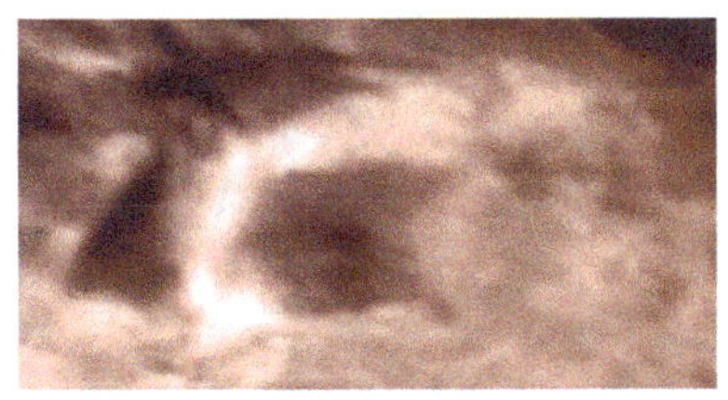
Random objects, Image 89.

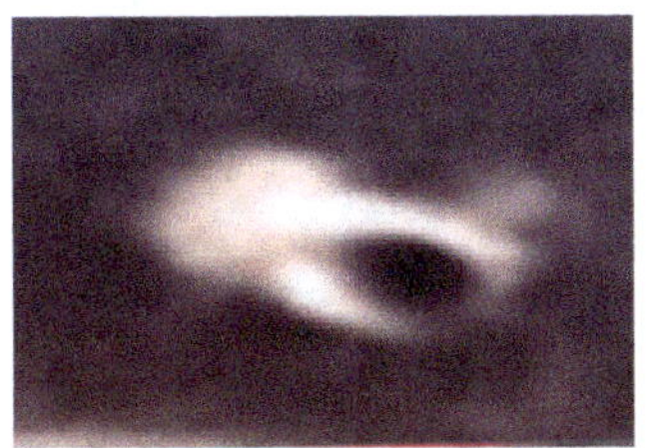
Random object, Image 90.

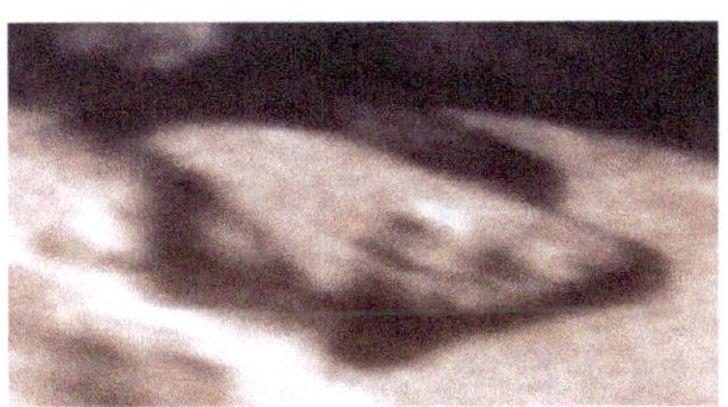
Random objects, Image 91.

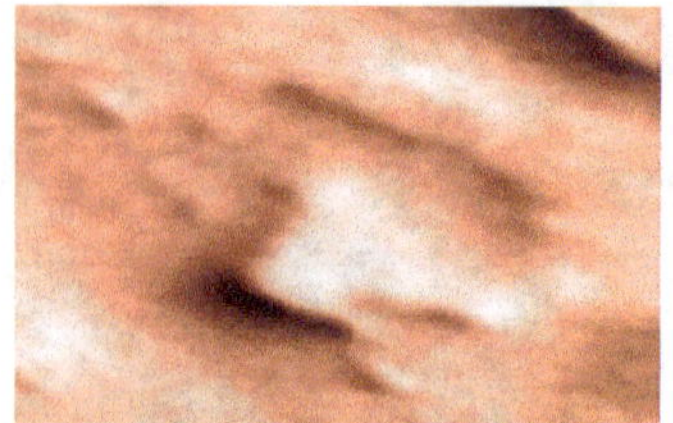
Random objects, Image 92.

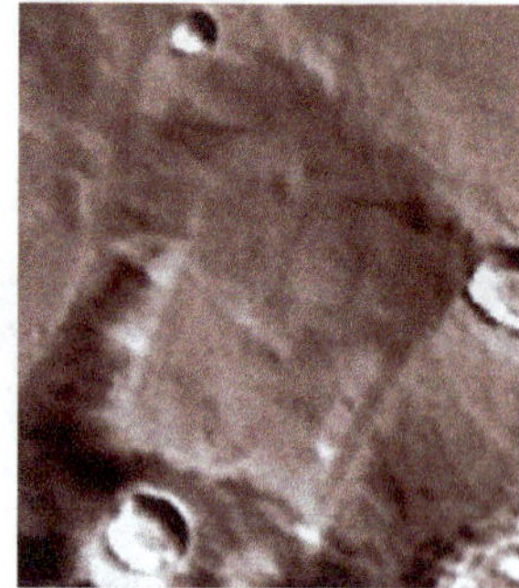
Random objects, Image 93.

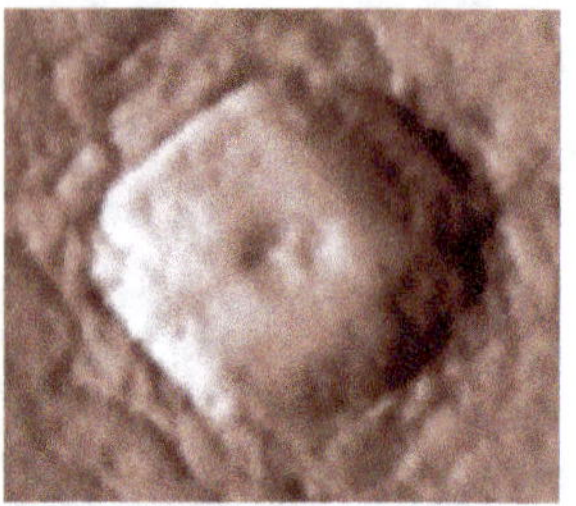
Random objects, Image 94.

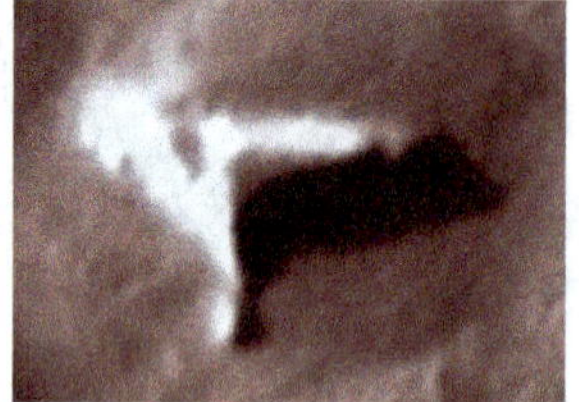
Random objects, Image 95.

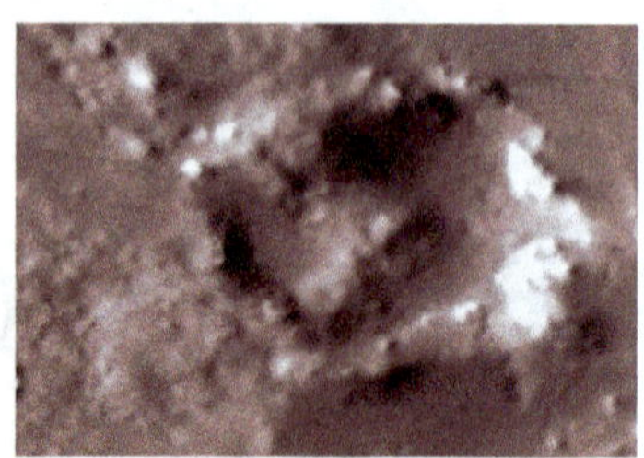
Random objects, Image 96.

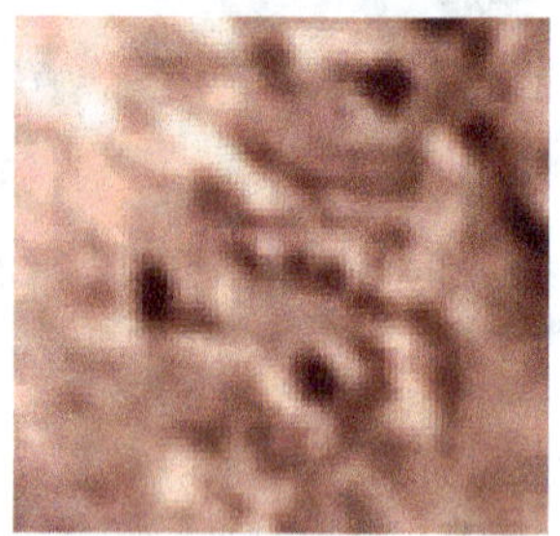
Random objects, Image 97.

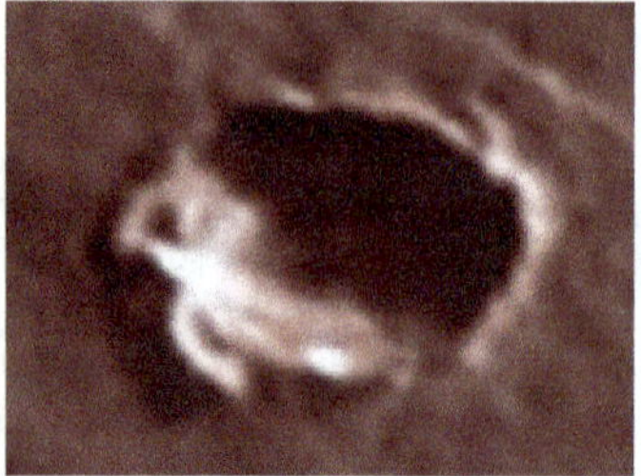
Random objects, Image 98.

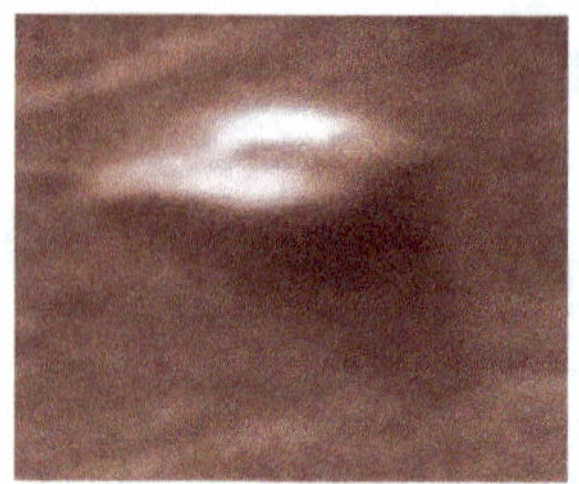
Random object, Image 99.

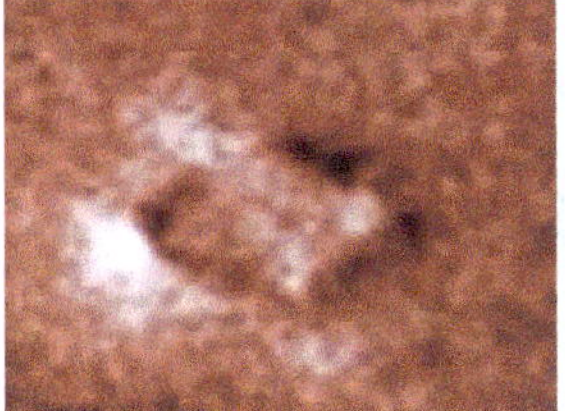
Random objects, Image 100.

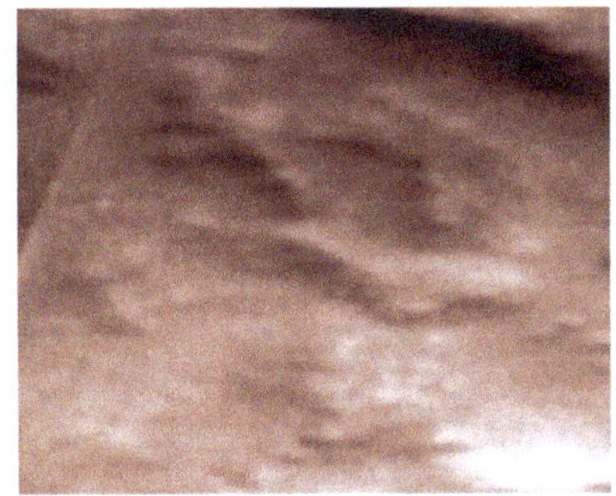
Random objects, Image 101.

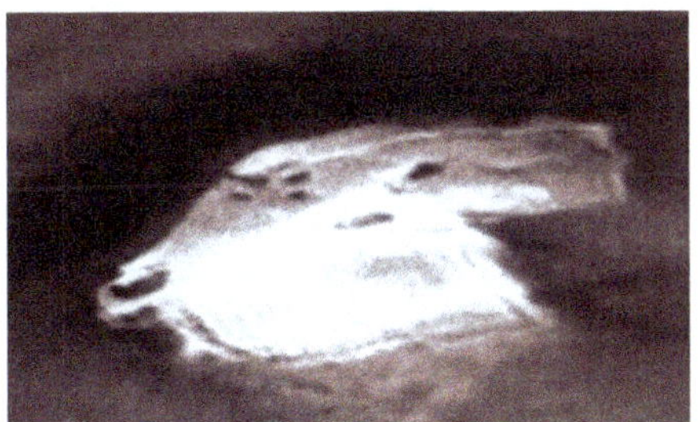
Random objects, Image 102.

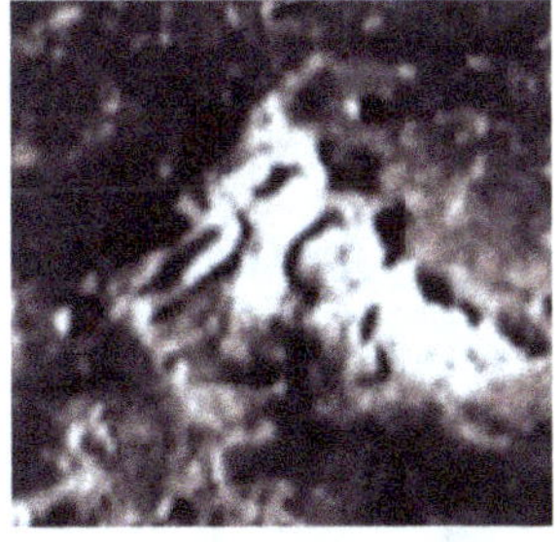
Random objects, Image 103.

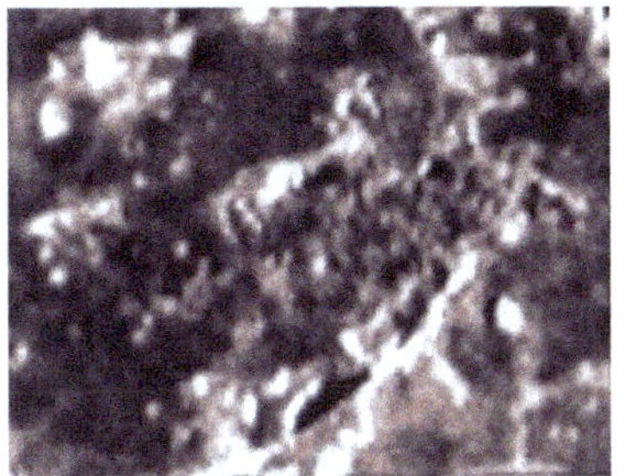
Random objects, Image 104.

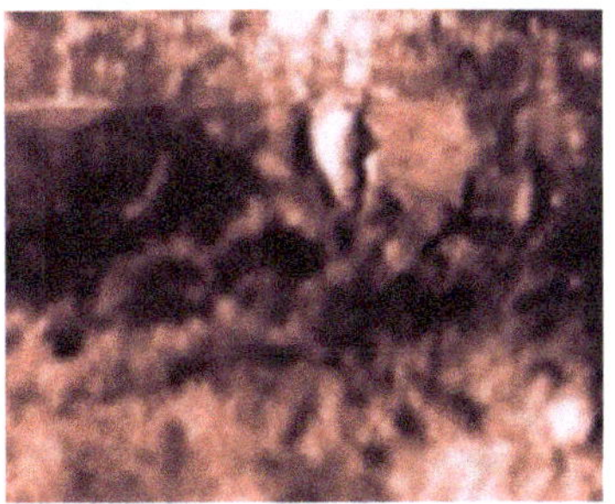
Random objects, Image 105.

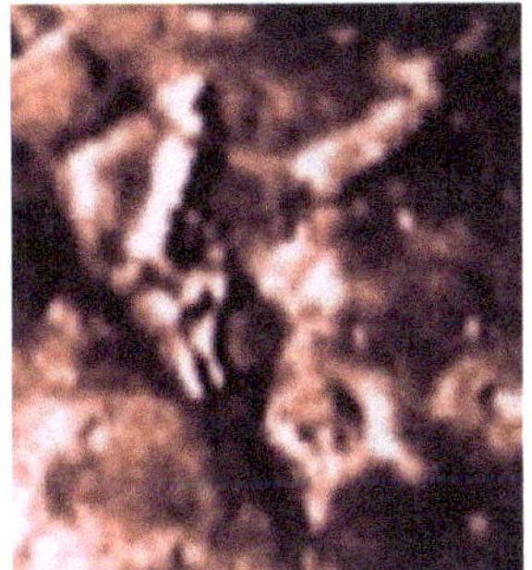
Random objects, Image 106.

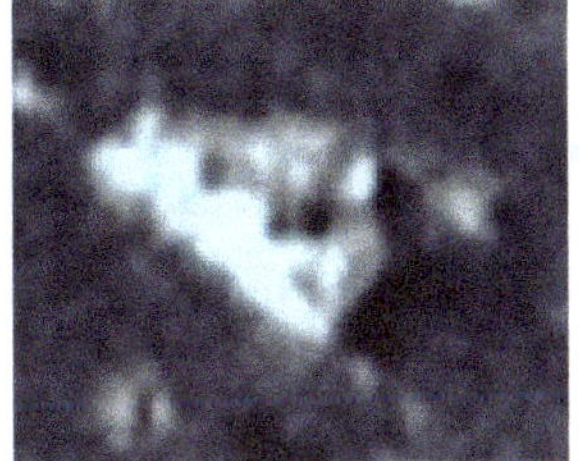
Random objects, Image 107.

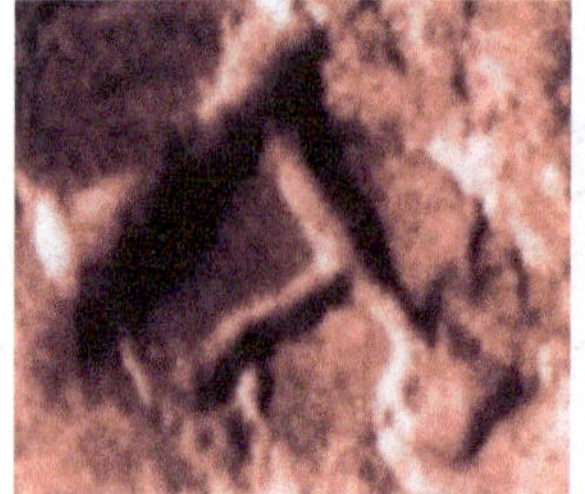

Random objects, Image 108.

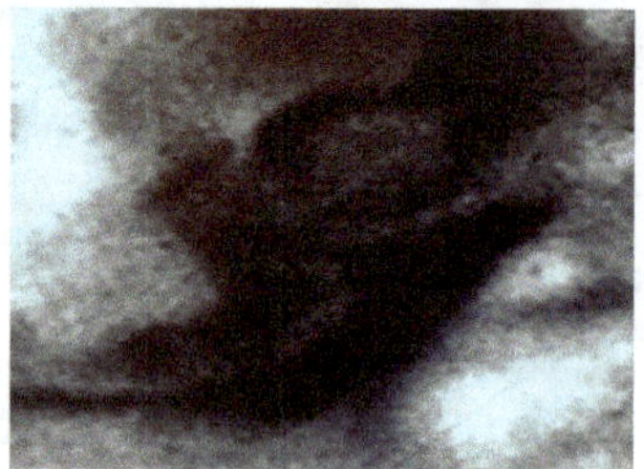

Random objects, Image 110.

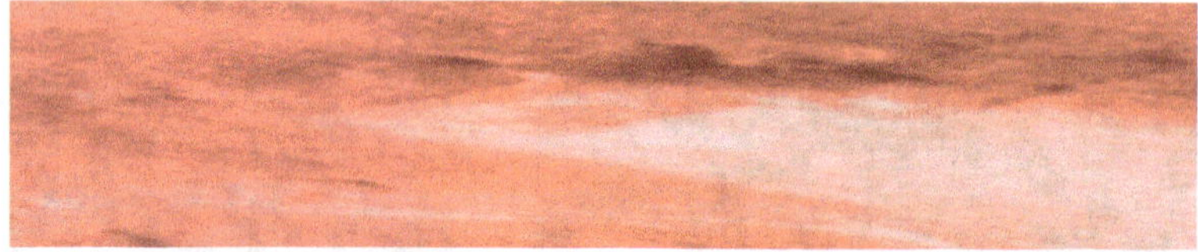

Random objects, Image 109 (taken from the Martian surface).

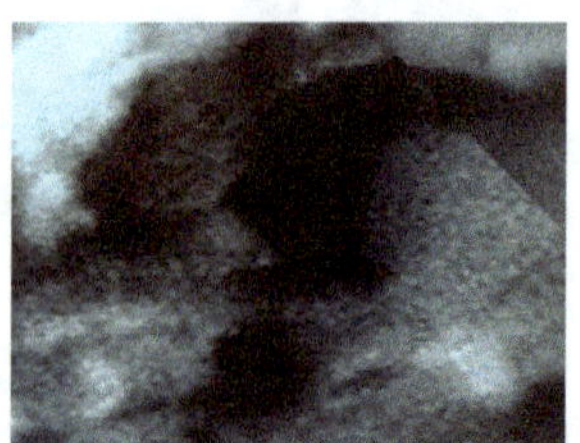

Random objects, Image 111.

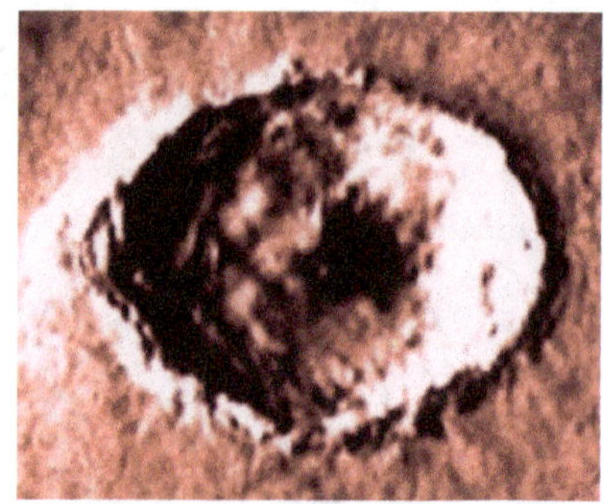

Random objects, Image 112.

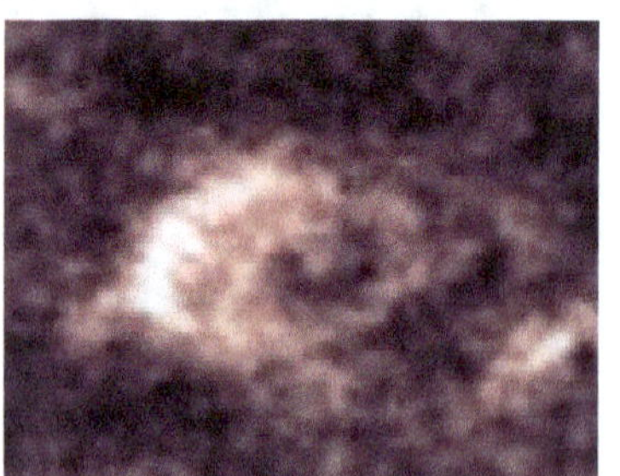

Random objects, Image 113.

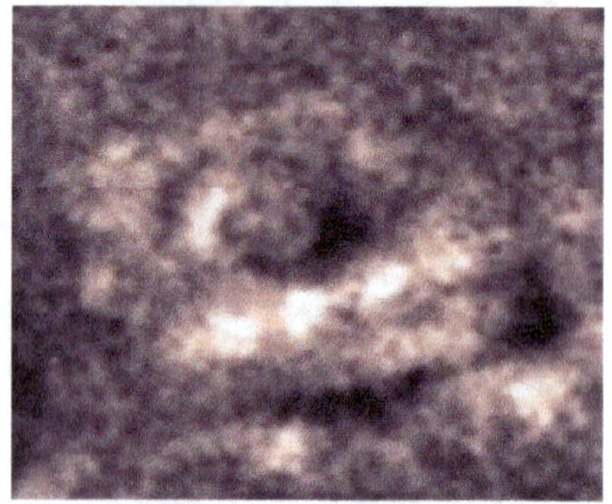

Random objects, Image 114.

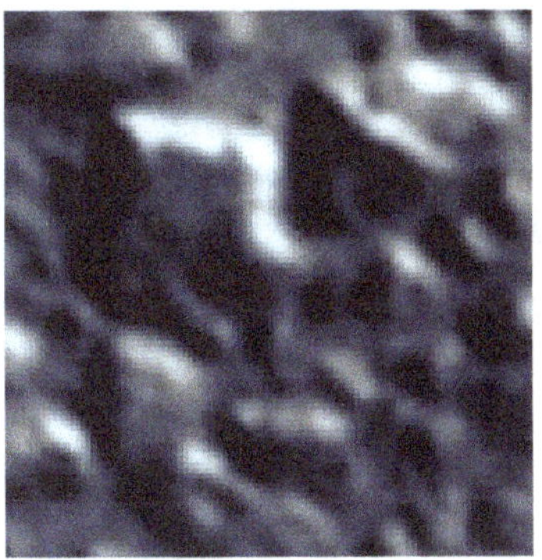

Random objects, Image 115.

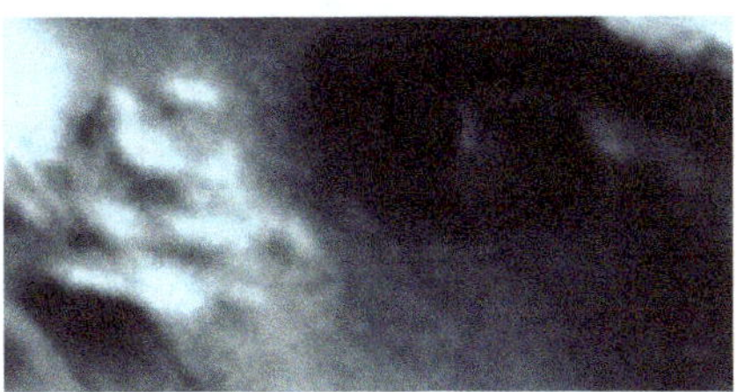

Random objects, Image 116.

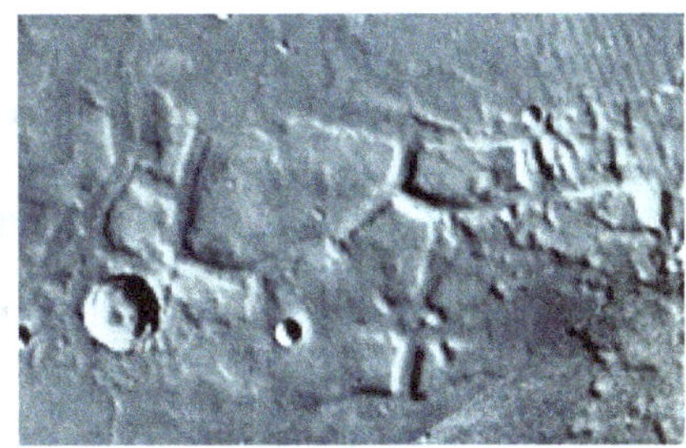

Random objects, Image 118.

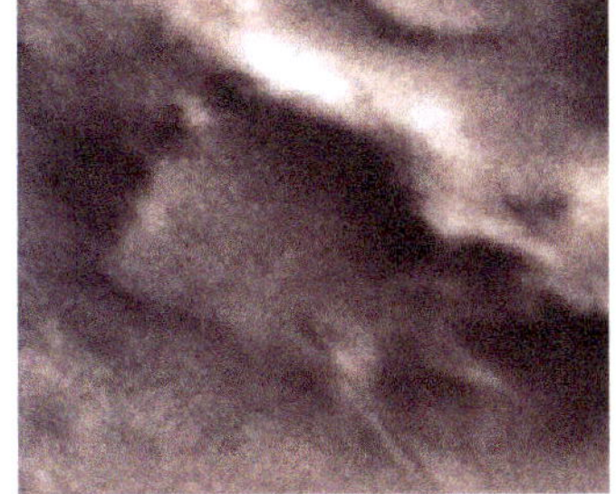

Random objects, Image 117.

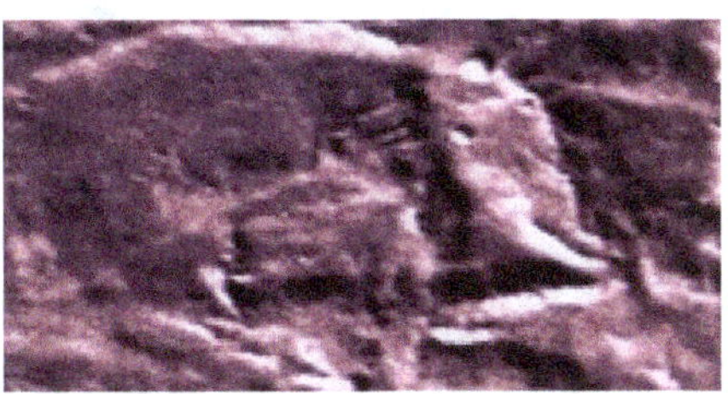

Random objects, Image 120.

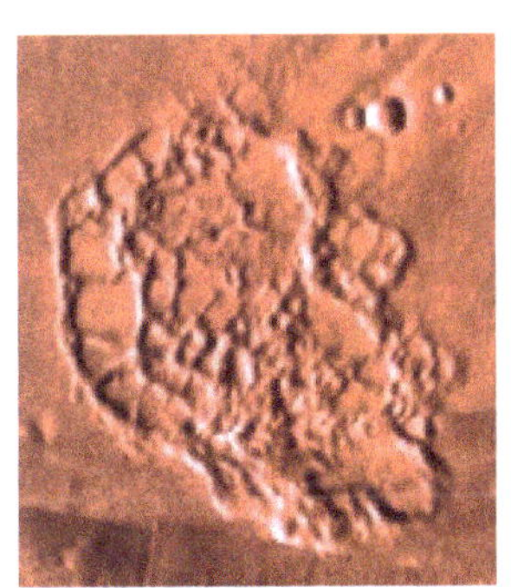

Random objects, Image 119.

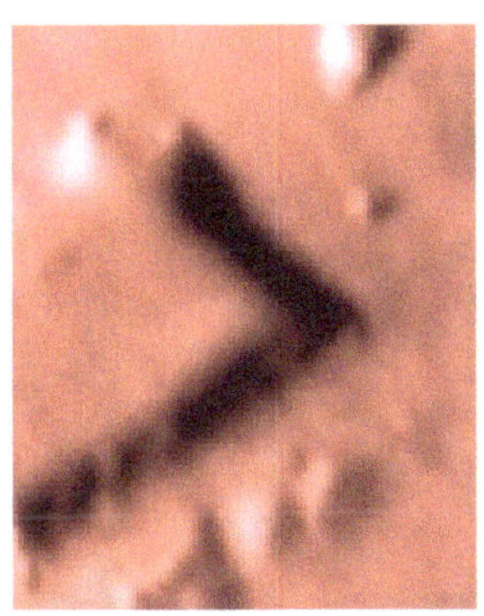

Random objects, Image 121.

Random objects, Image 122.

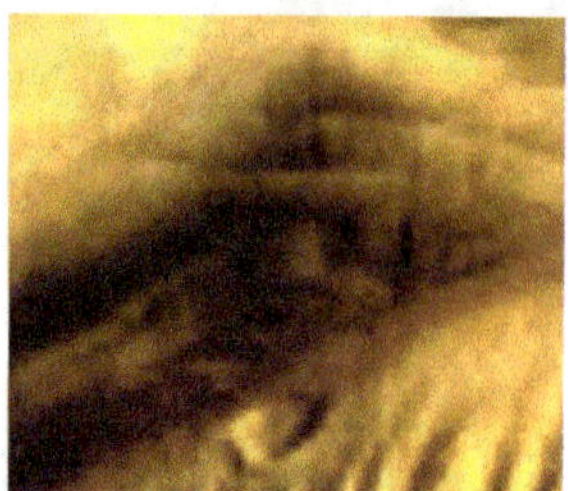

Random objects, Image 123.

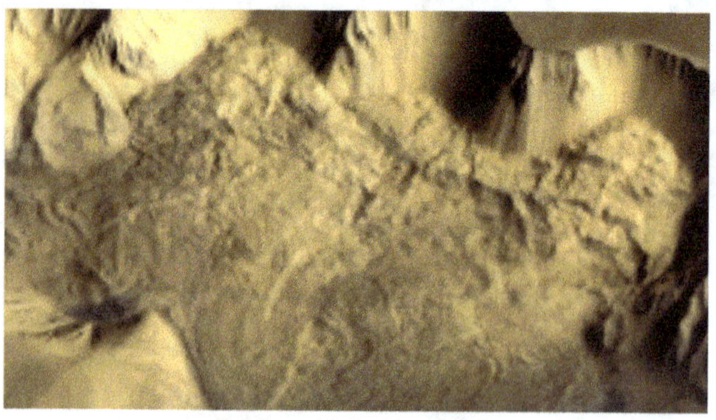

Random objects, Image 124.

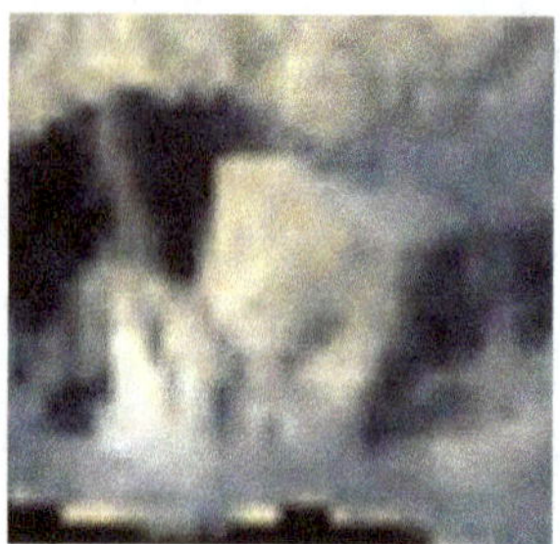

Random objects, Image 125.

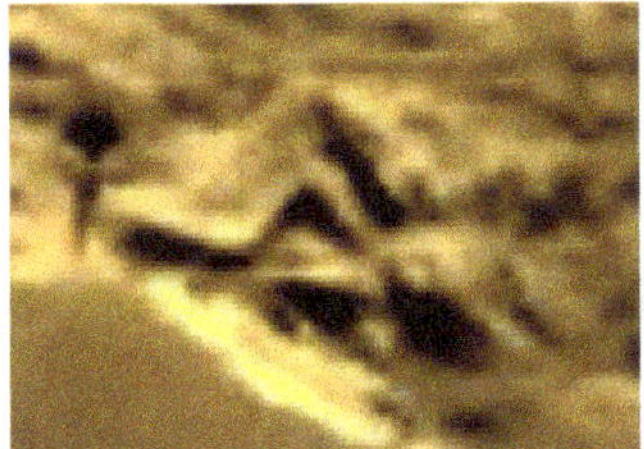

Random objects, Image 126.

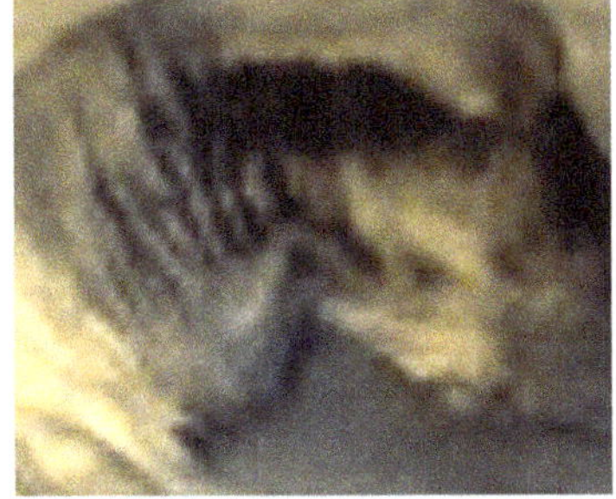

Random objects, Image 127.

Random objects, Image 128.

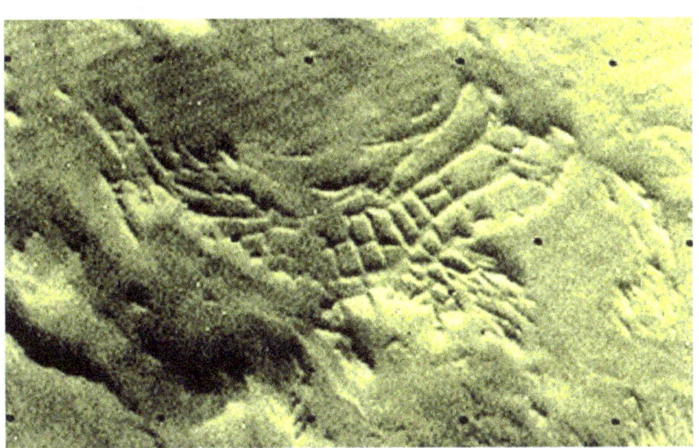

Random objects, Image 129, "Inca City" (named by the author).

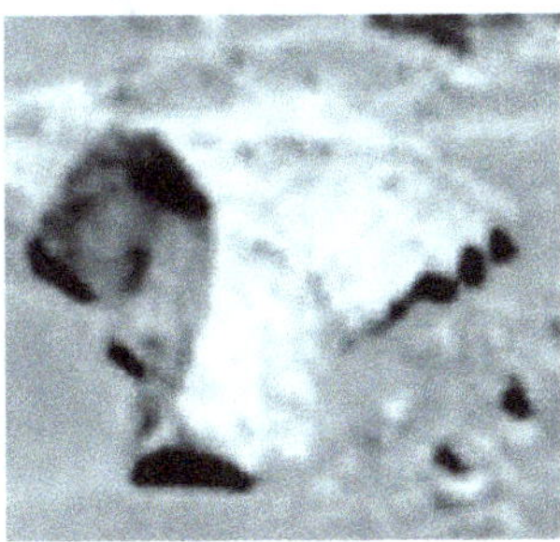

Random objects, Image 130 (taken from the Martian surface).

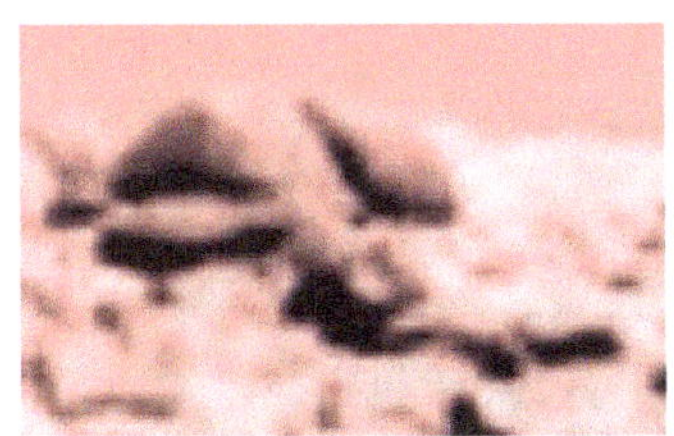

Random objects, Image 131.

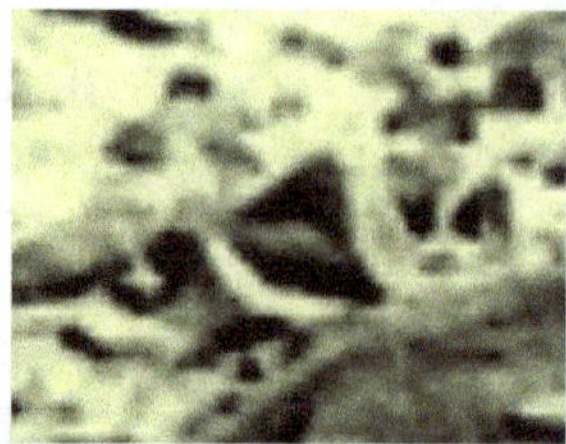

Random objects, Image 132.

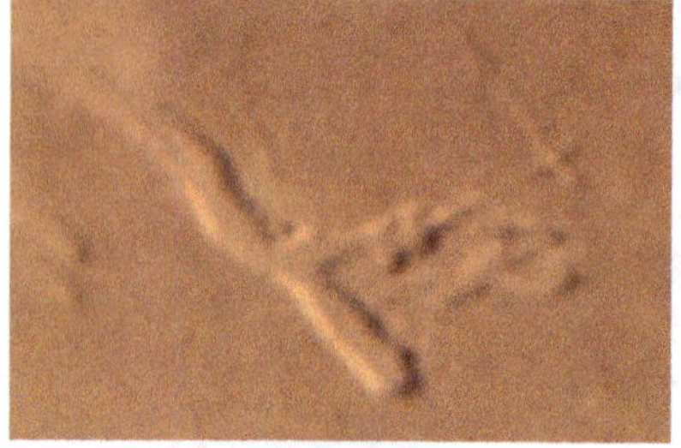

Random objects, Image 133.

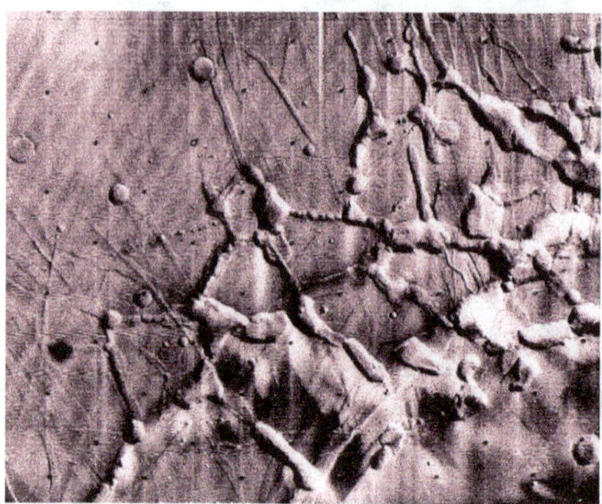

Random objects, Image 134.

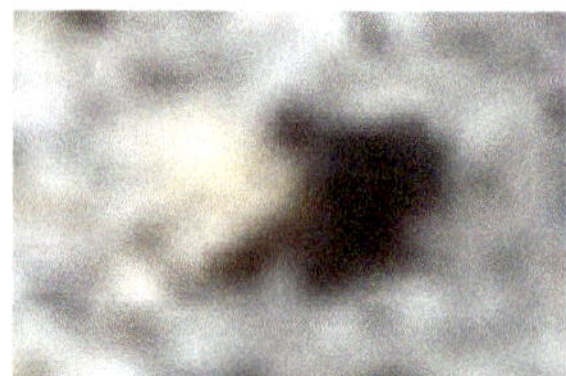

Random objects, Image 135.

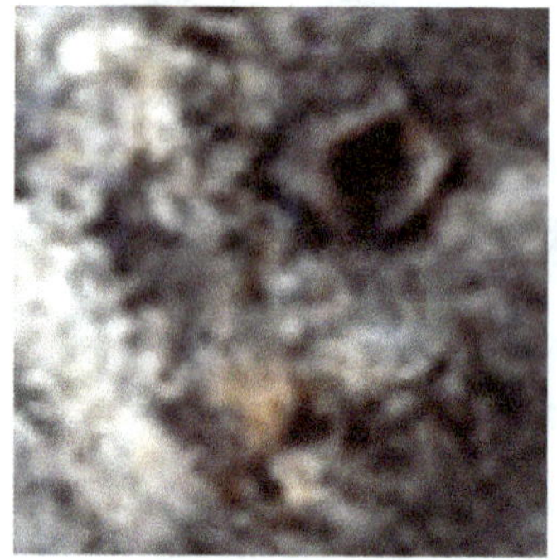

Random objects, Image 136.

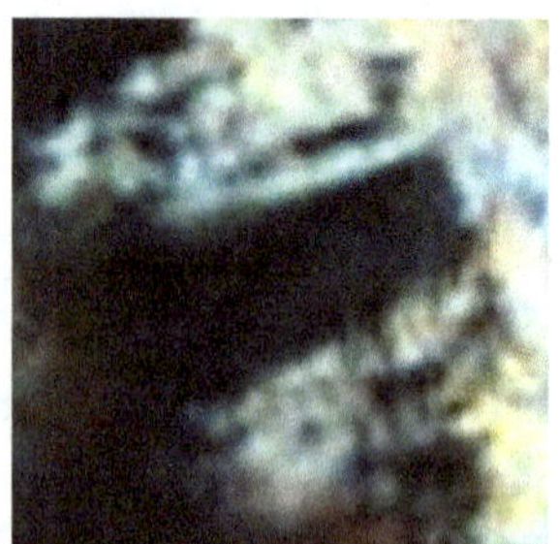

Random objects, Image 137.

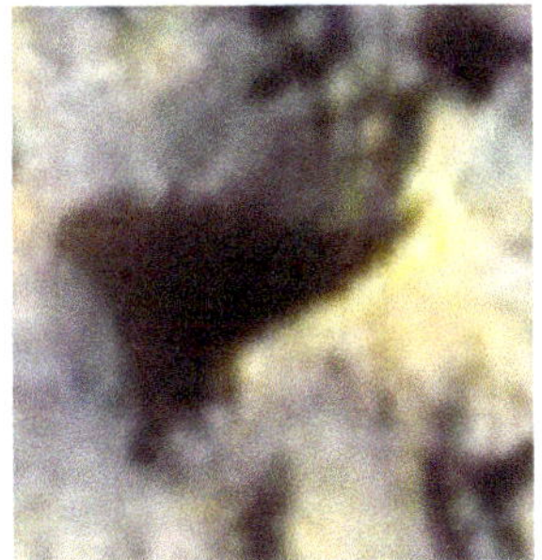

Random objects, Image 138.

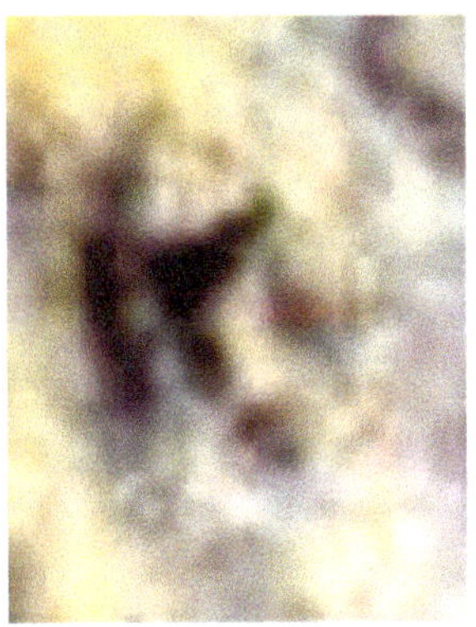

Random objects, Image 139.

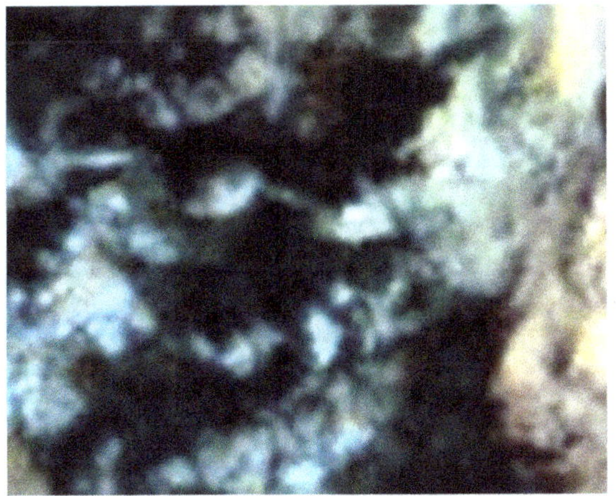

Random objects, Image 140.

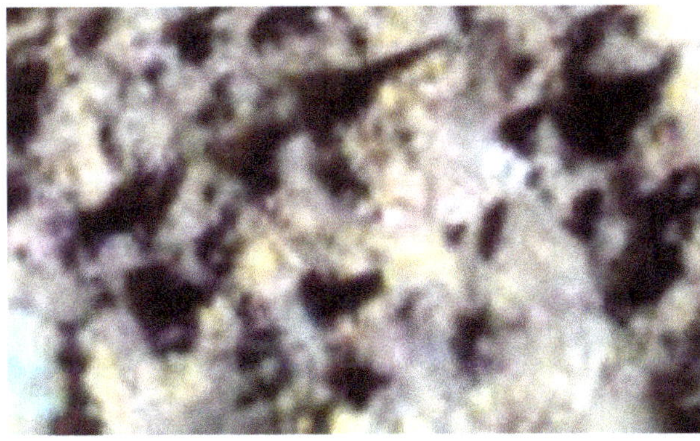

Random objects, Image 141.

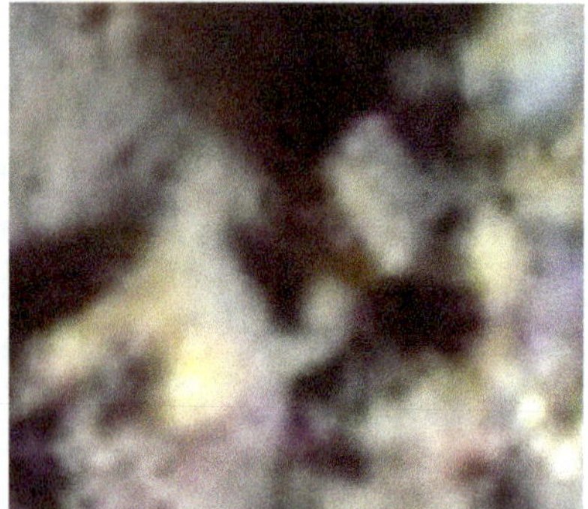

Random objects, Image 142.

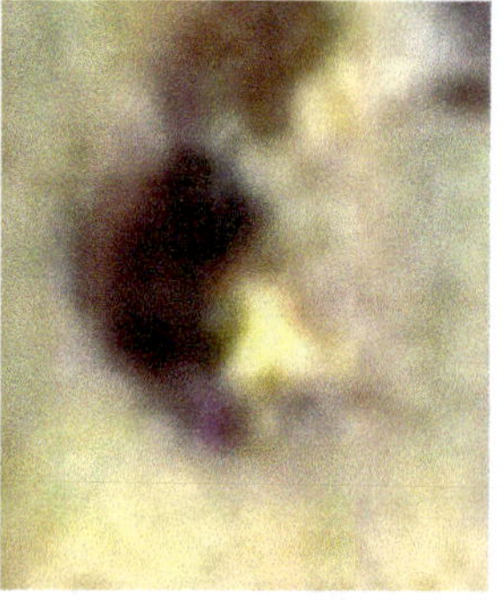

Random objects, Image 143.

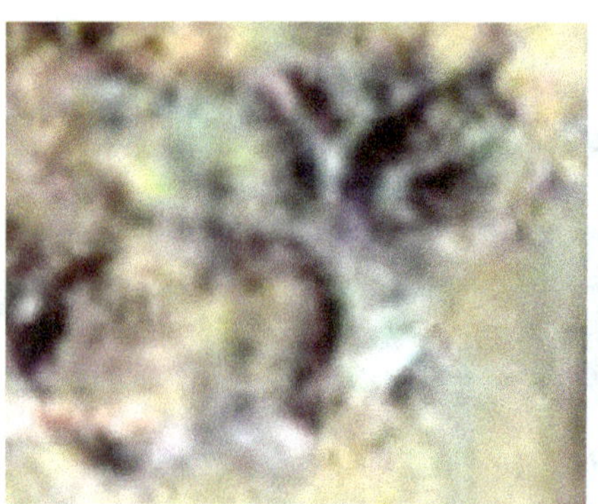

Random objects, Image 144.

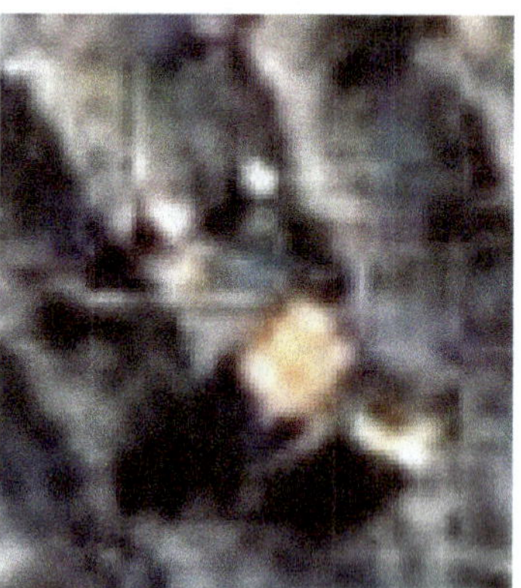

Random objects, Image 145.

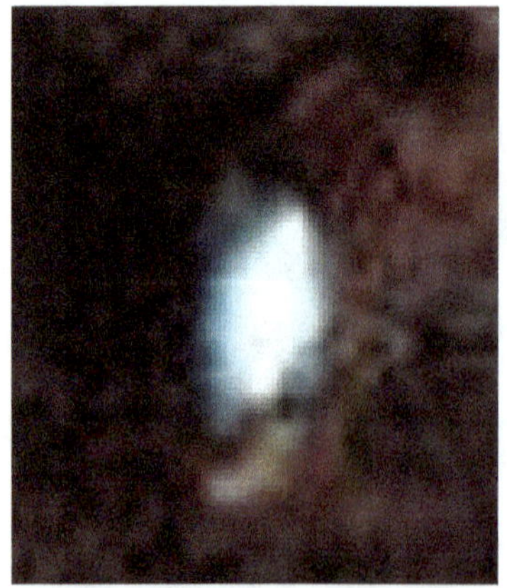

Random objects, Image 146.

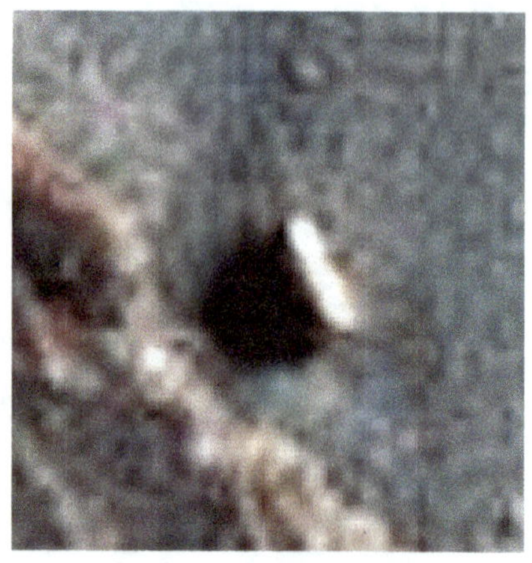

Random object, Image 147.

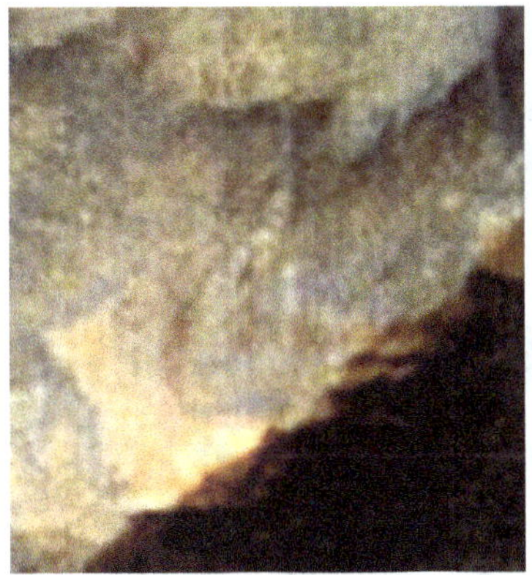

Random objects, Image 148.

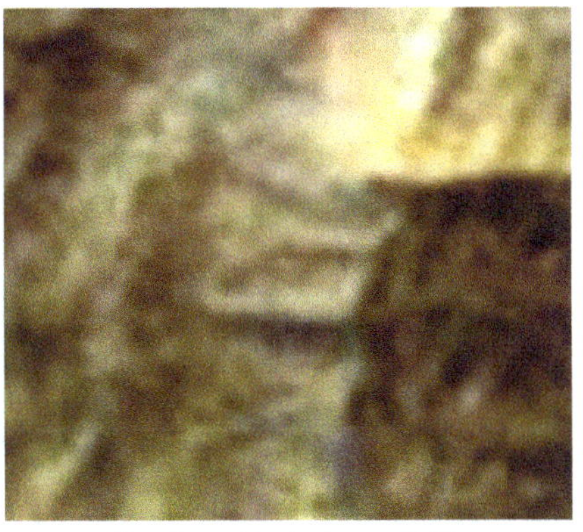

Random objects, Image 149.

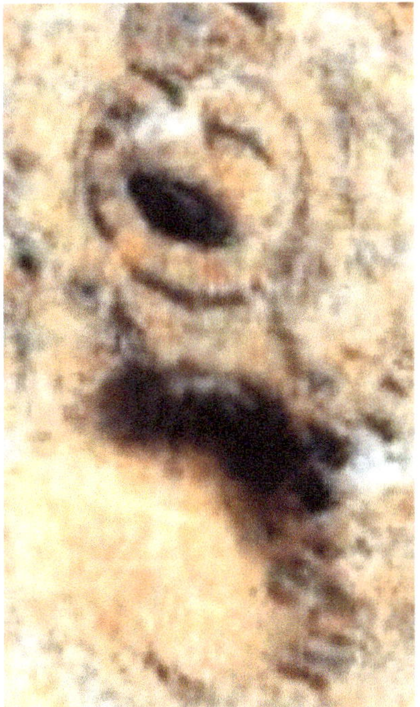

Random objects, Image 150.

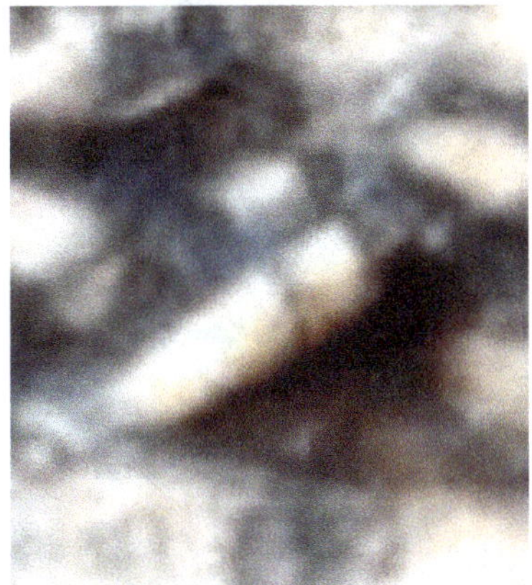

Random objects, Image 151.

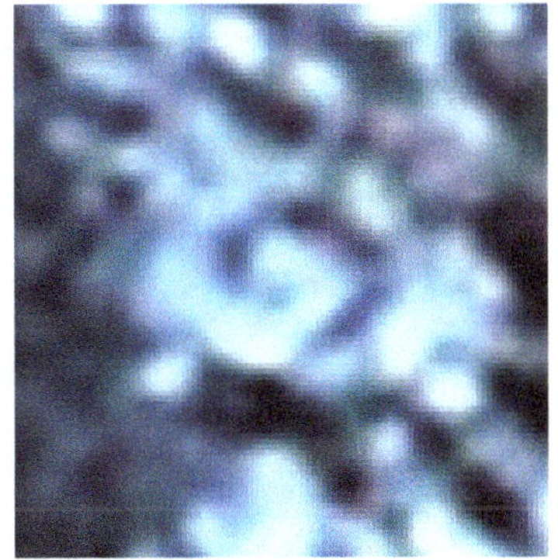

Random objects, Image 152.

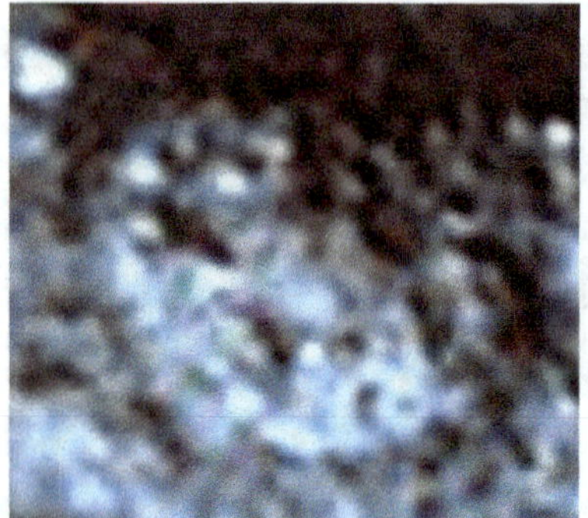

Random objects, Image 153.

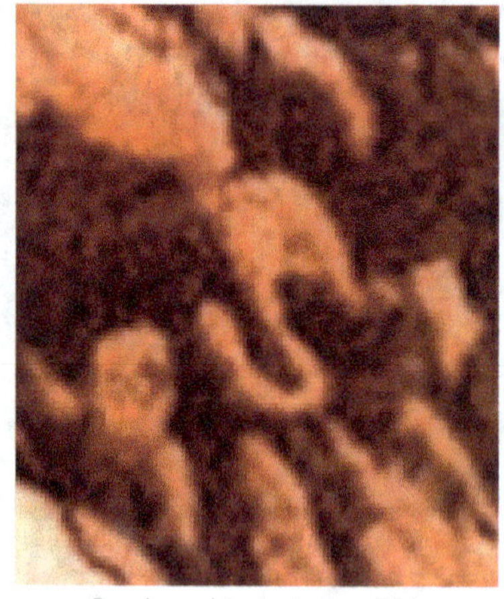

Random objects, Image 154.

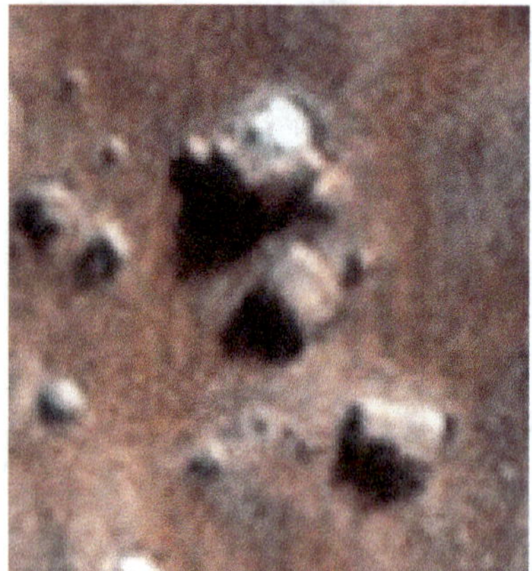

Random objects, Image 155.

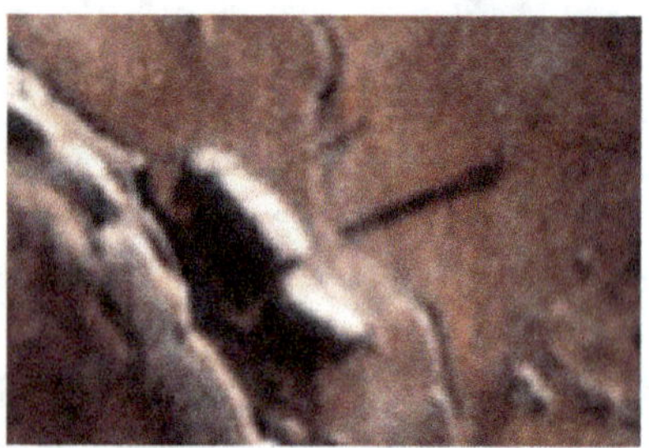

Random objects, Image 156.

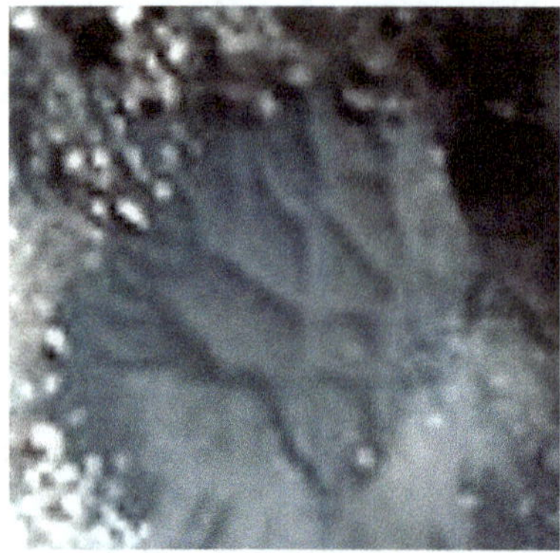

Random objects, Image 157.

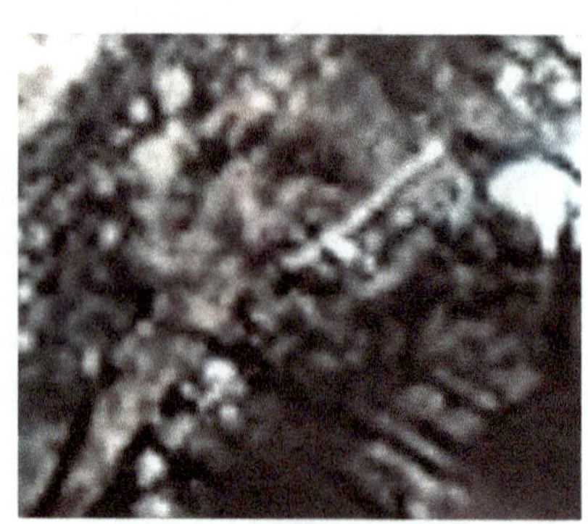

Random objects, Image 158.

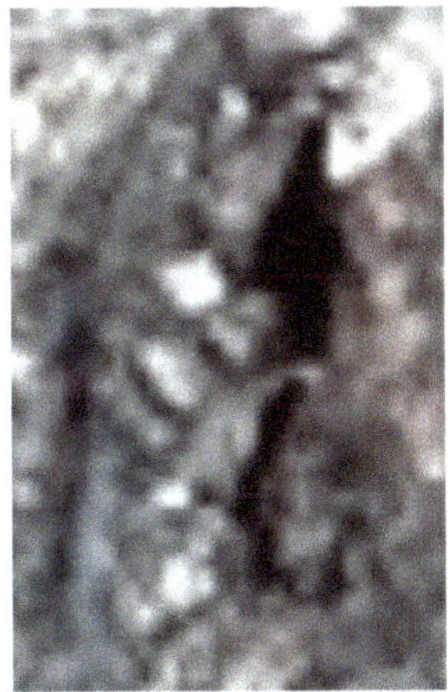

Random objects, Image 159.

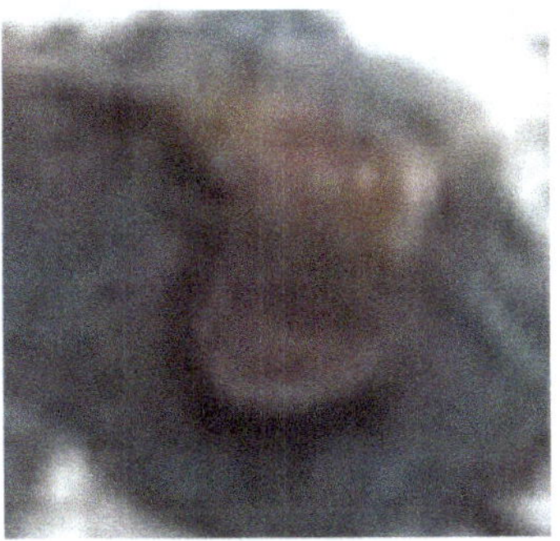

Random objects, Image 160.

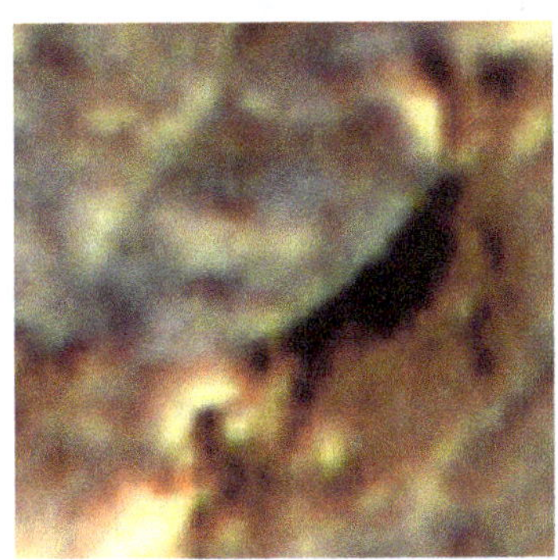

Random objects, Image 161.

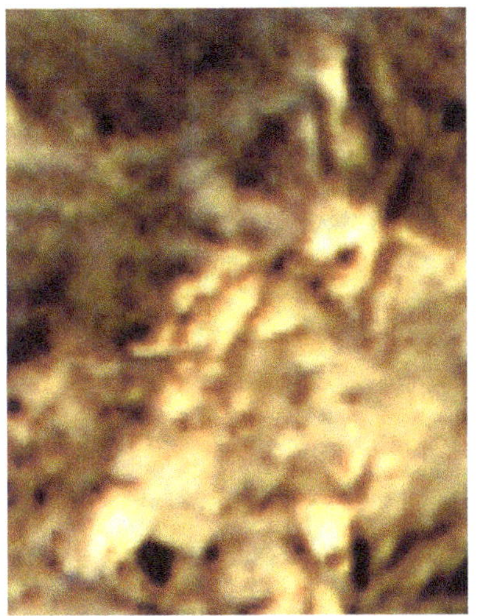

Random objects, Image 162.

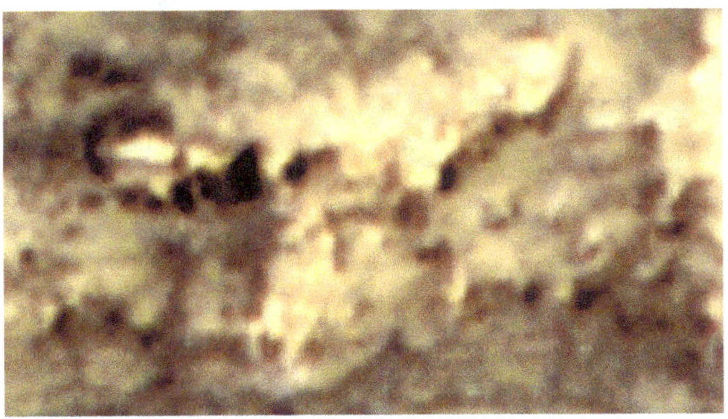

Random objects, Image 163.

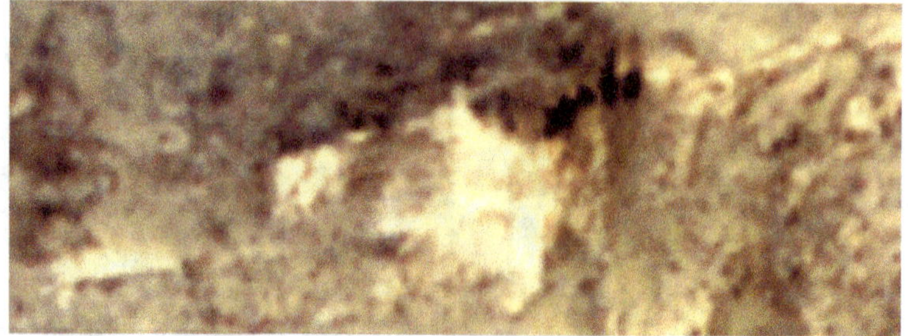

Random objects, Image 164.

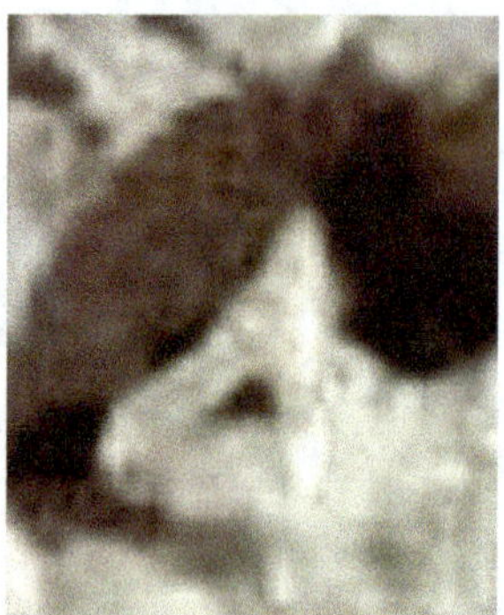

Random objects, Image 165.

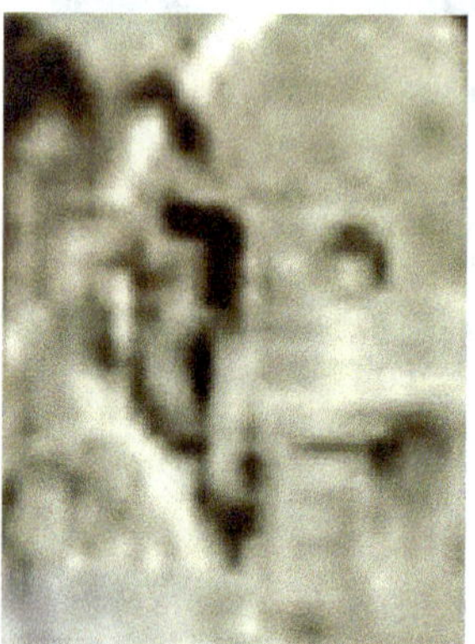

Random objects, Image 166.

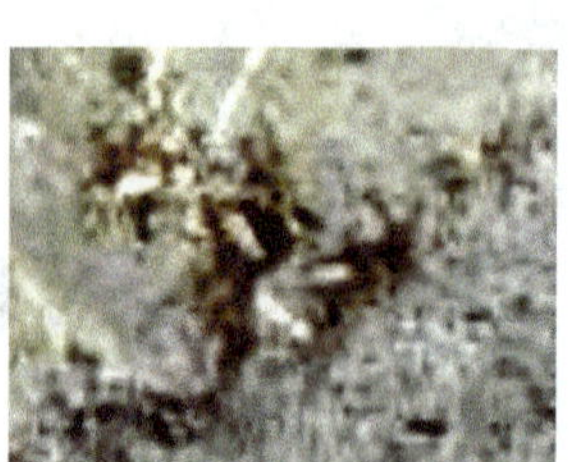

Random objects, Image 167.

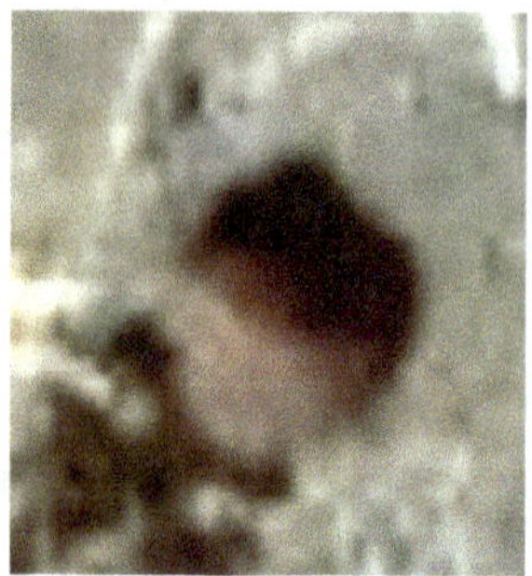

Random objects, Image 168.

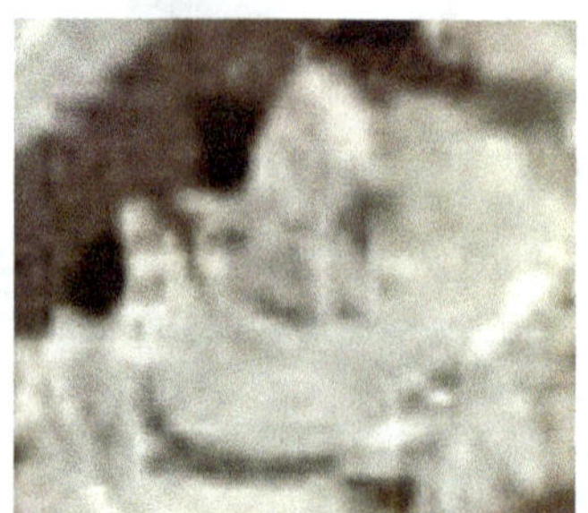

Random objects, Image 169.

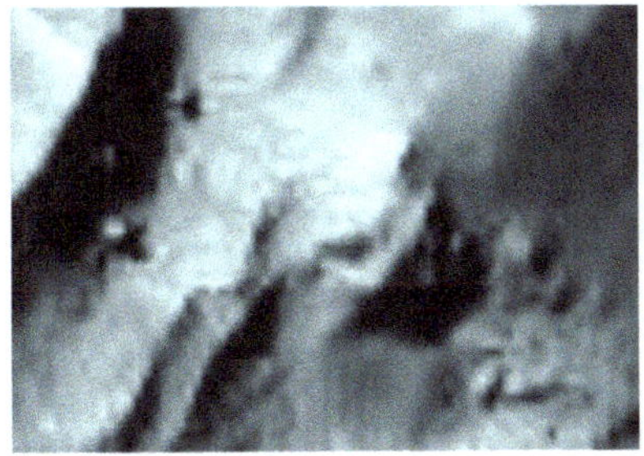
Random objects, Image 170.

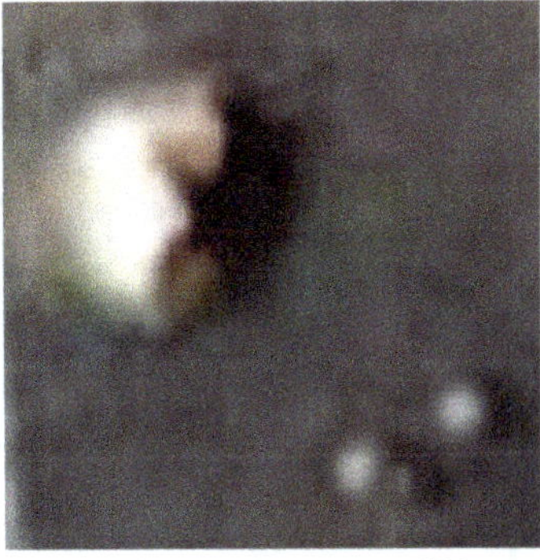
Random objects, Image 171.

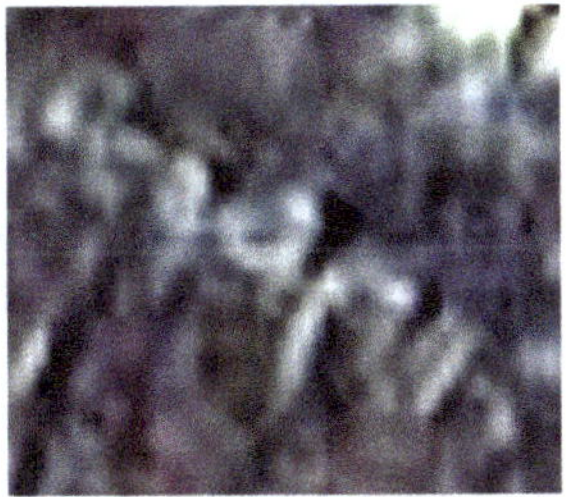
Random objects, Image 172.

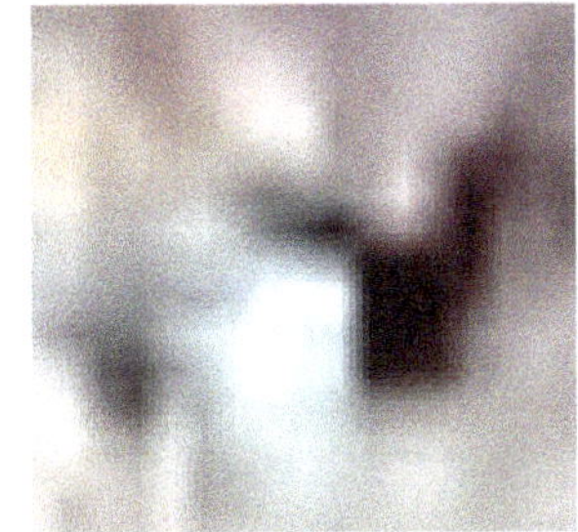
Random objects, Image 173.

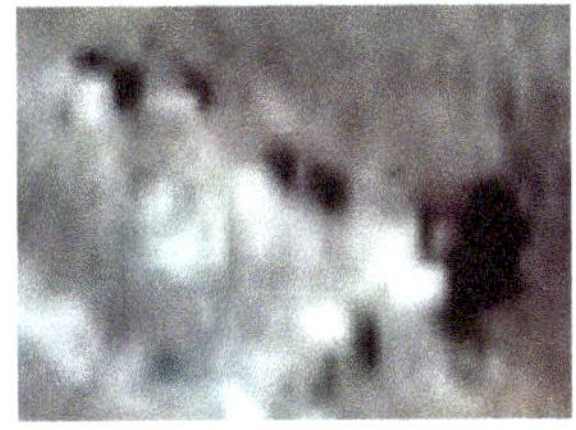
Random objects, Image 174.

Random objects, Image 175.

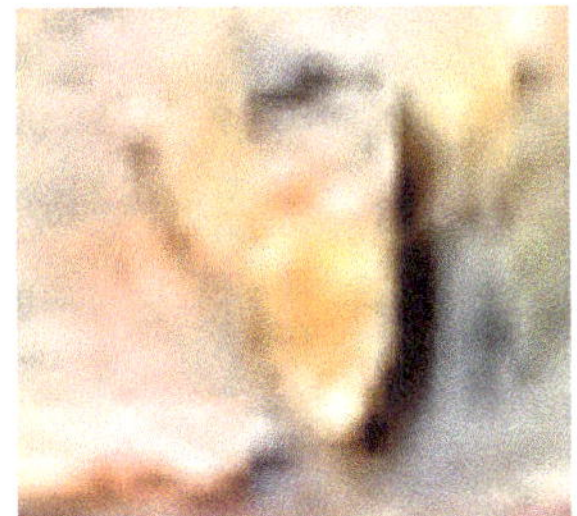
Random objects, Image 176.

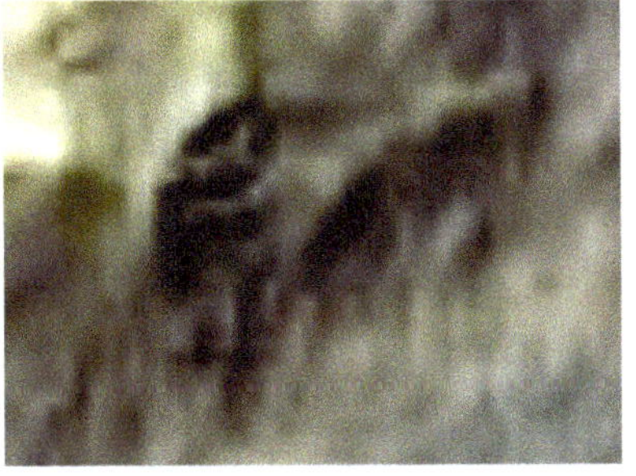
Random objects, Image 177.

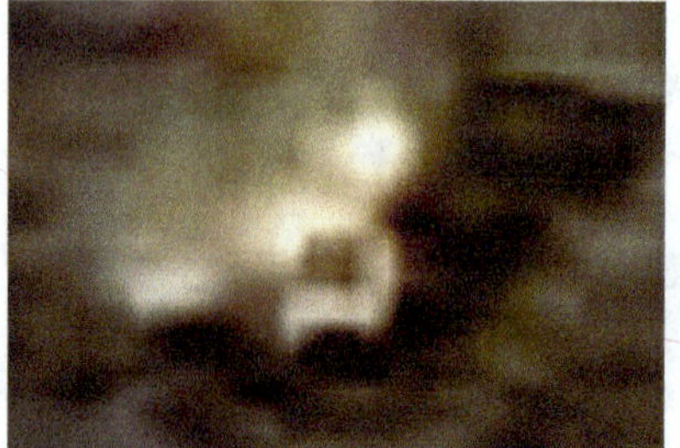
Random objects, Image 178.

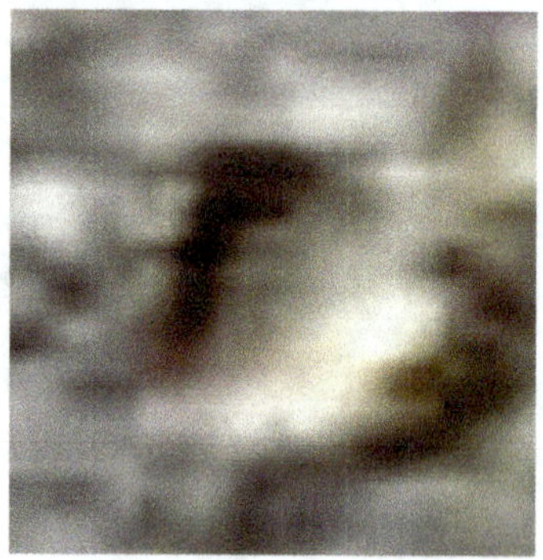
Random objects, Image 179.

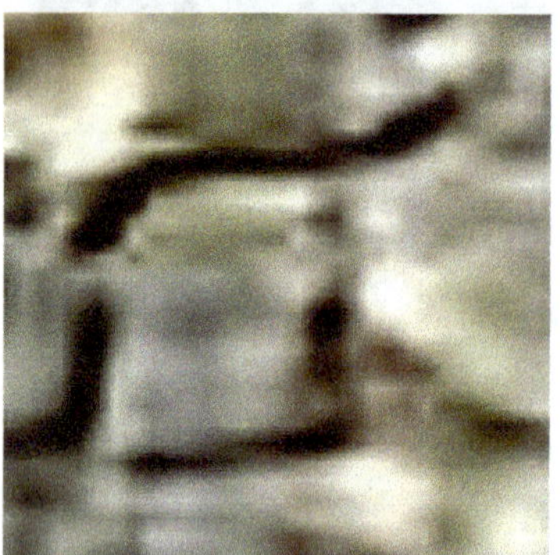
Random objects, Image 180.

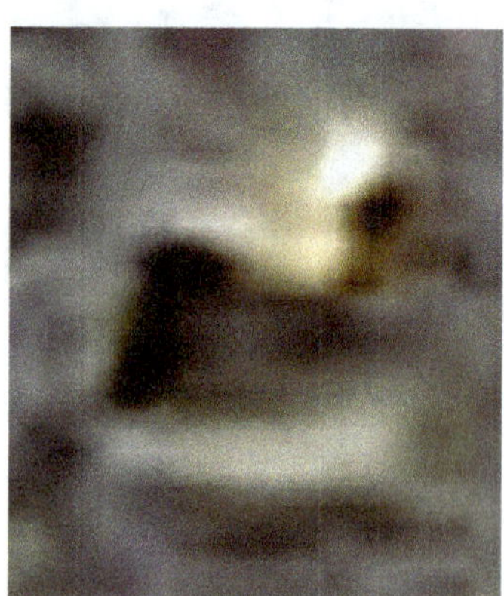
Random objects, Image 181.

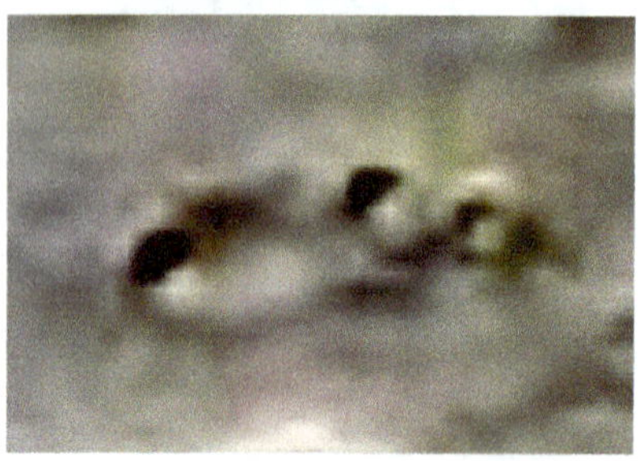
Random objects, Image 182.

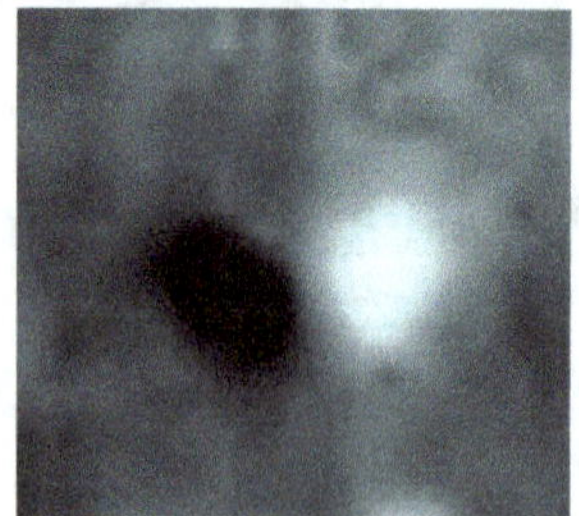
Random objects, Image 183.

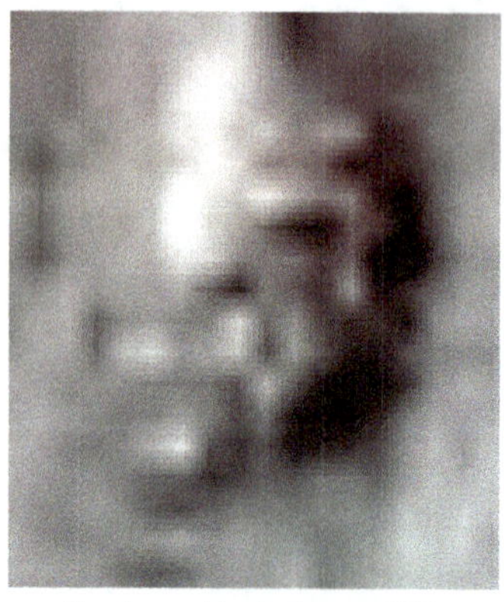
Random objects, Image 184.

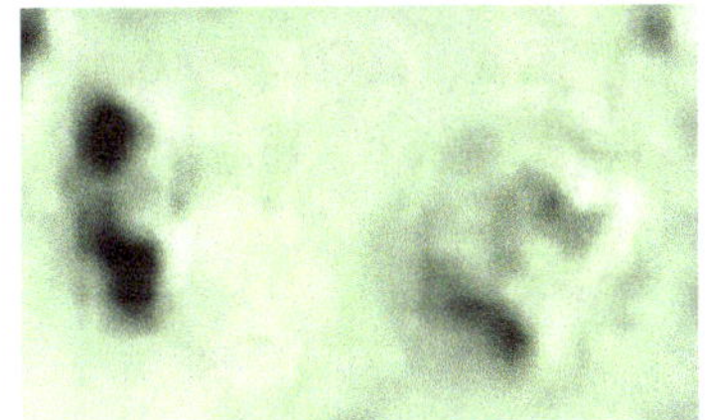

Random objects, Image 185.

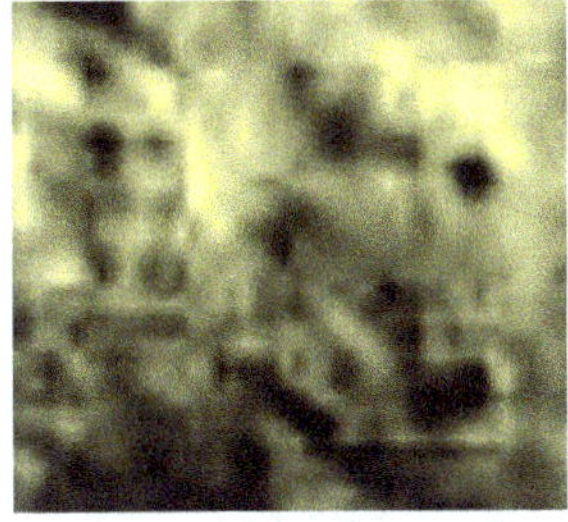

Random objects, Image 186.

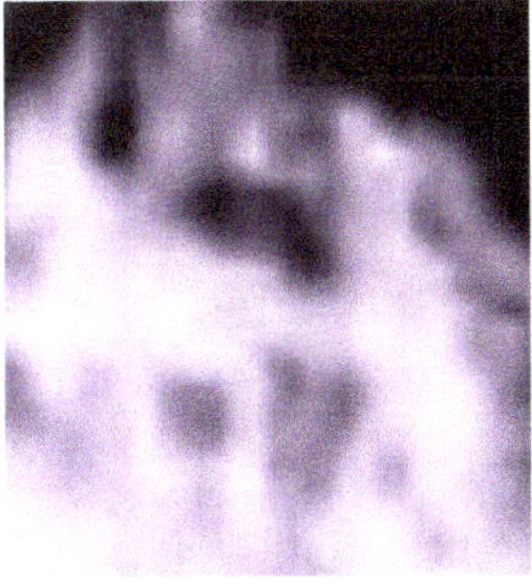

Random objects, Image 187.

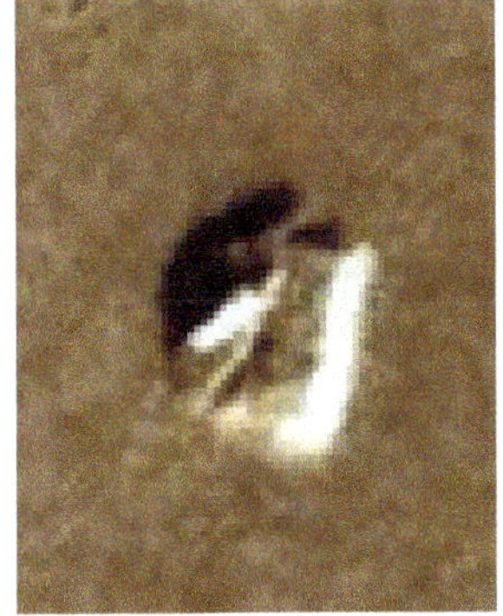

Random objects, Image 188.

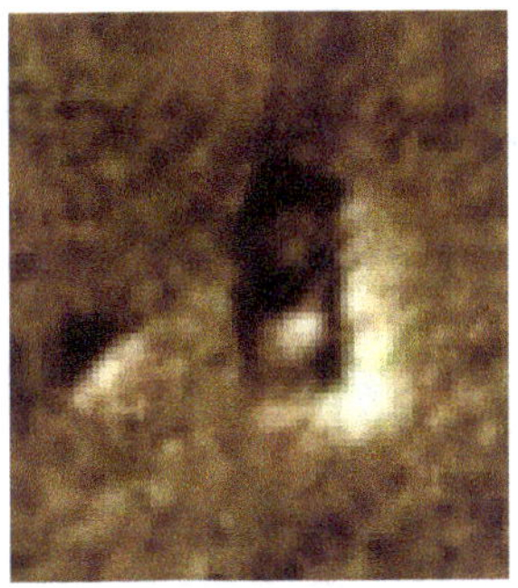

Random objects, Image 189.

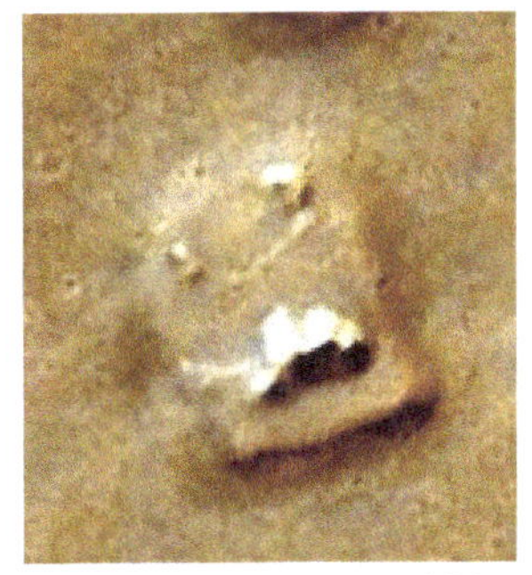

Random objects, Image 190.

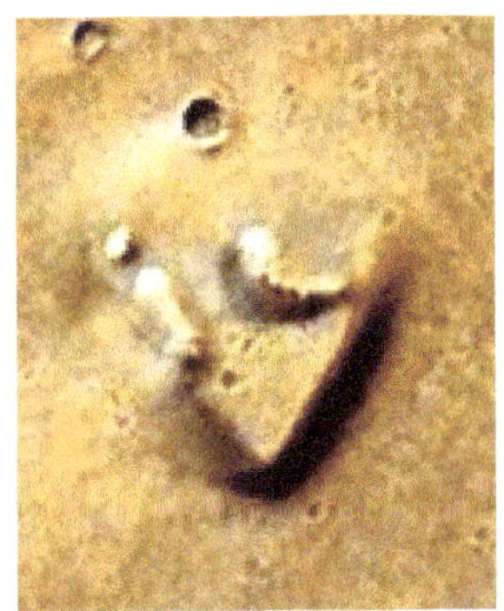

Random objects, Image 191.

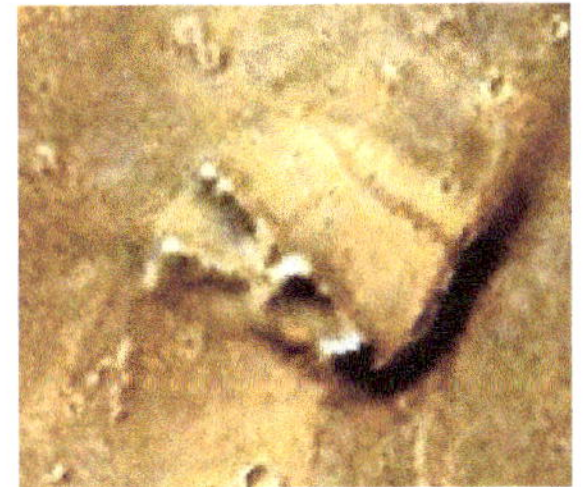

Random objects, Image 192.

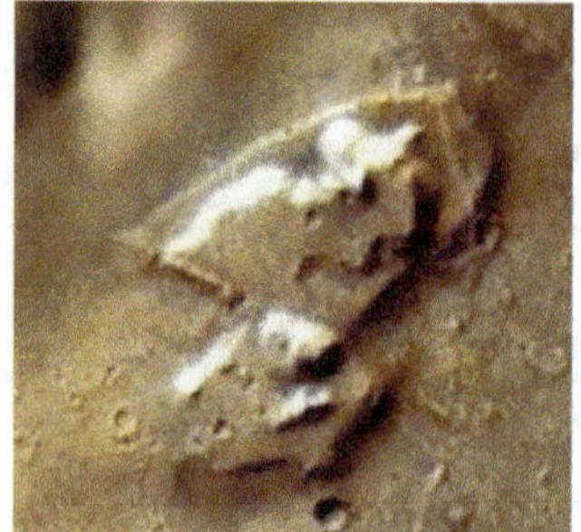
Random objects, Image 193.

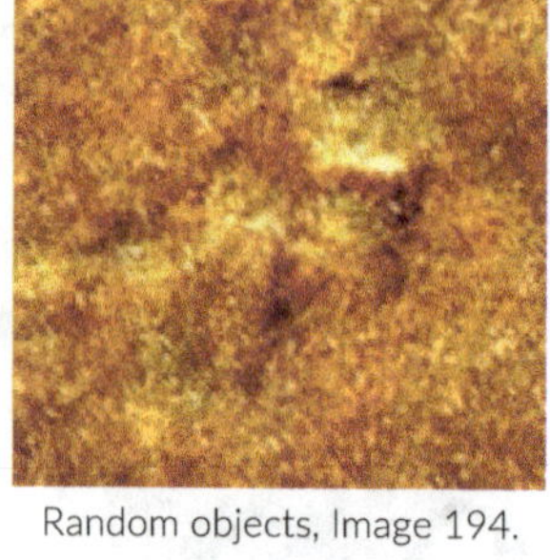
Random objects, Image 194.

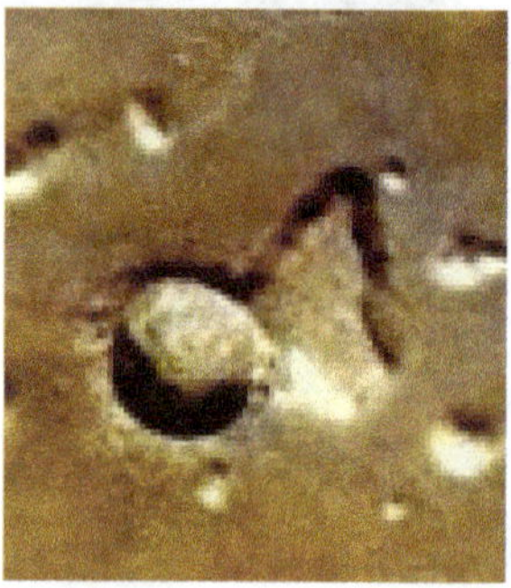
Random objects, Image 195.

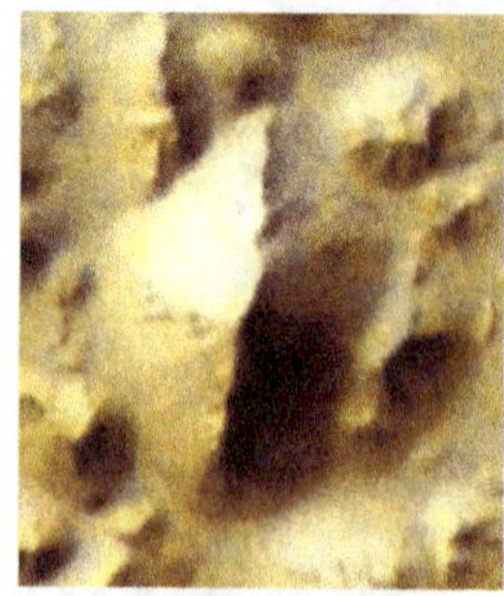
Random objects, Image 196.

Random objects, Image 197.

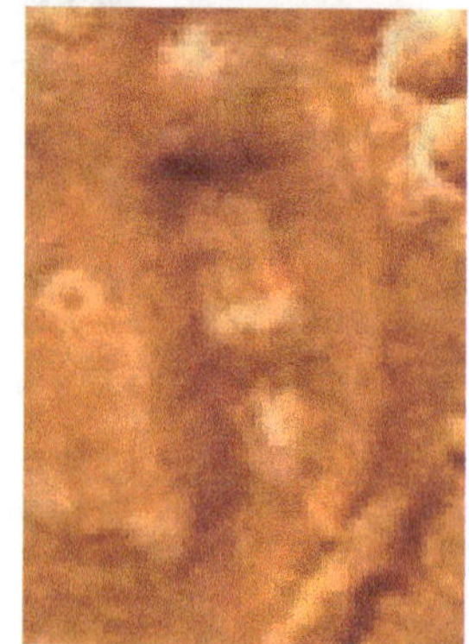
Random objects, Image 198.

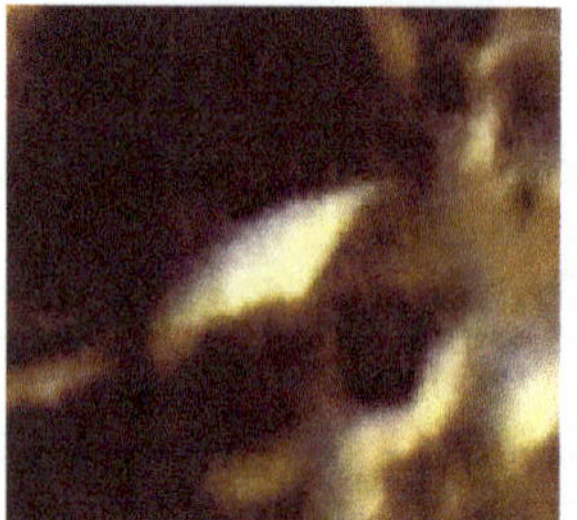
Random objects, Image 199.

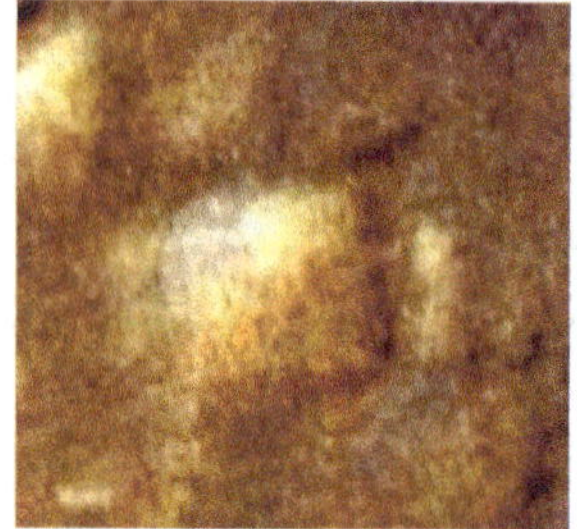
Random objects, Image 200.

Chapter 48

ANOMALIES ON PHOBOS

One of Mars' Two Moons

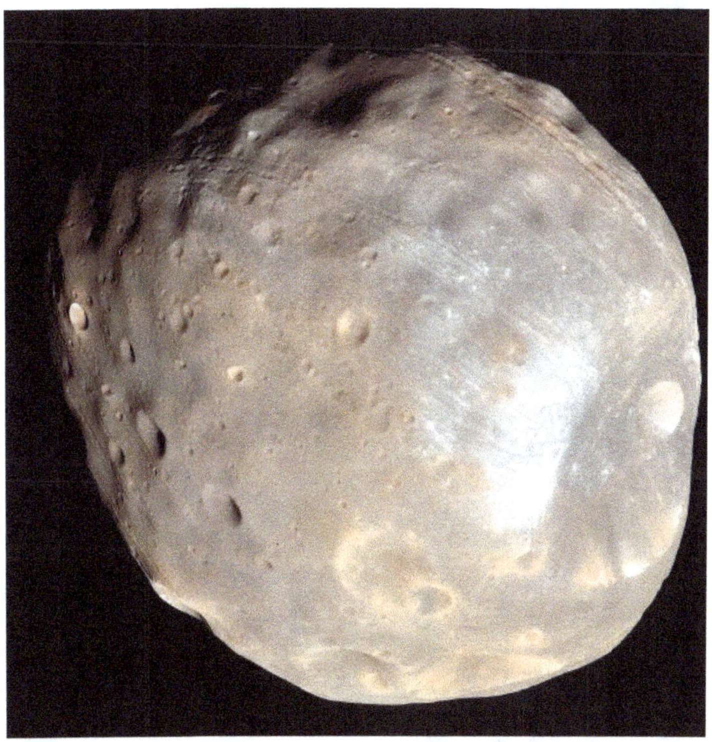

Original image. "The High Resolution Imaging Science Experiment (HiRISE) camera on NASA's Mars Reconnaissance Orbiter took two images of the larger of Mars' two moons, Phobos, within 10 minutes of each other on March 23, 2008. This is the first, taken from a distance of about 6,800 kilometers (about 4,200 miles)." Image credit: NASA/JPL-Caltech/University of Arizona.

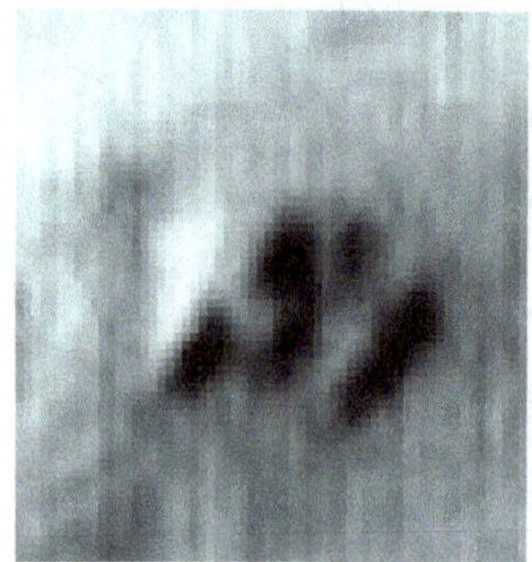

Random objects 1, Phobos.

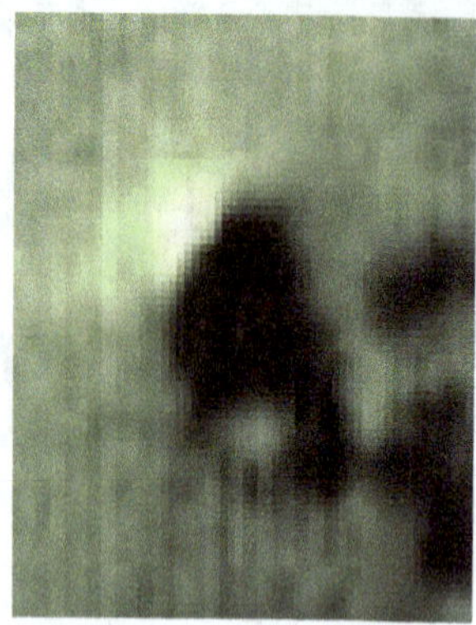

Random objects 2, Phobos.

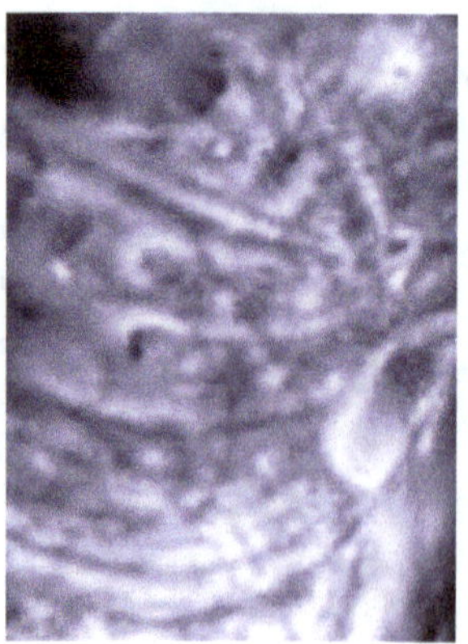

Random objects 3, Phobos.

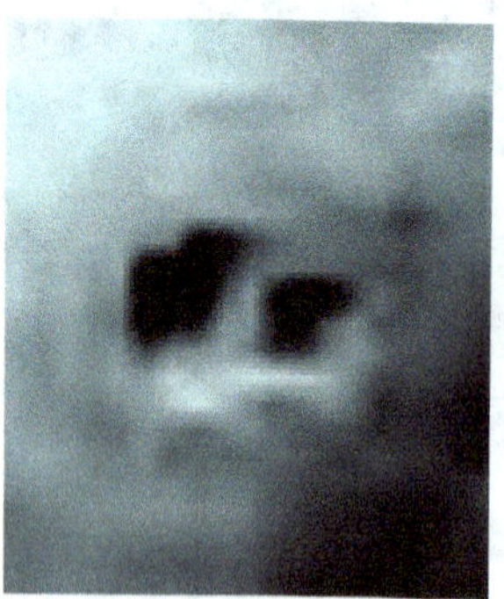

Random objects 4, Phobos.

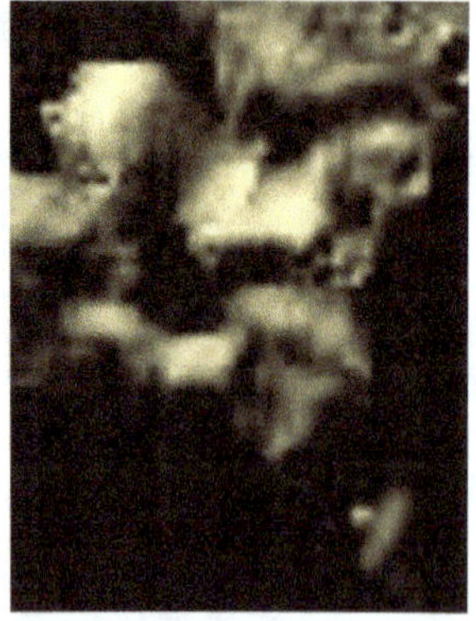

Random objects 5, Phobos.

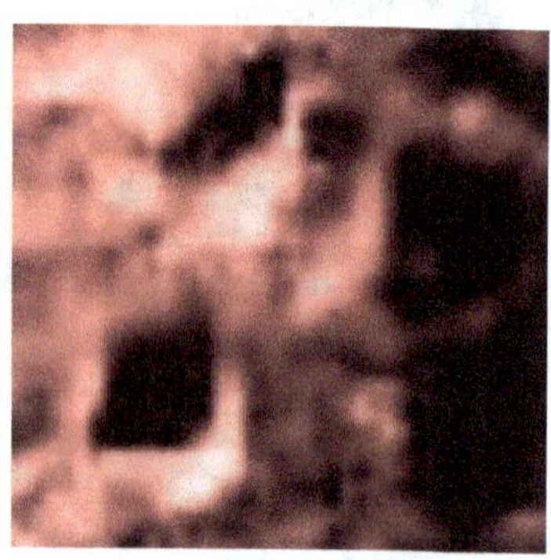

Random objects 6, Phobos.

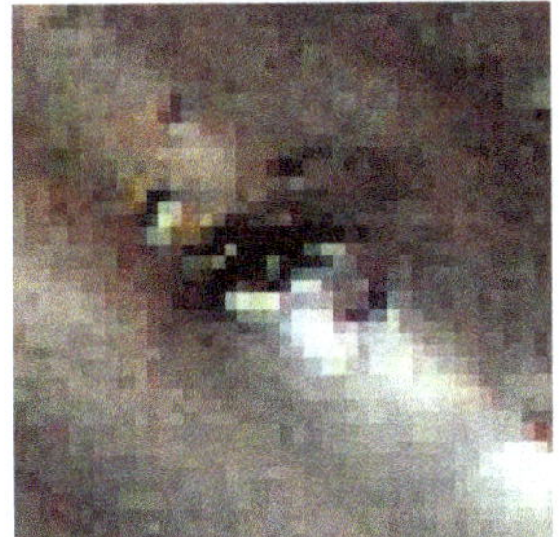

Random objects 7, Phobos.

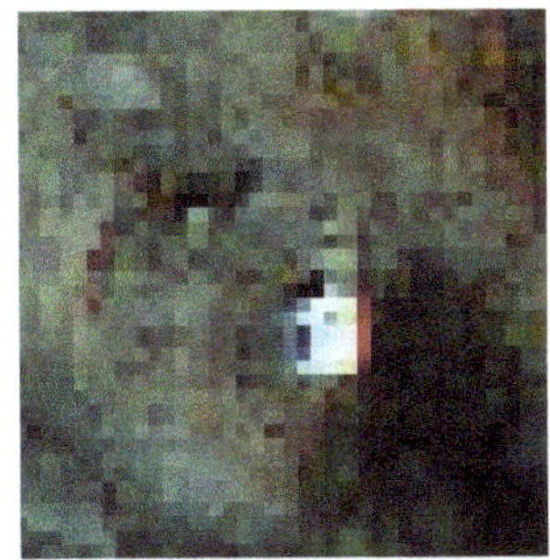

Random objects 8, Phobos.

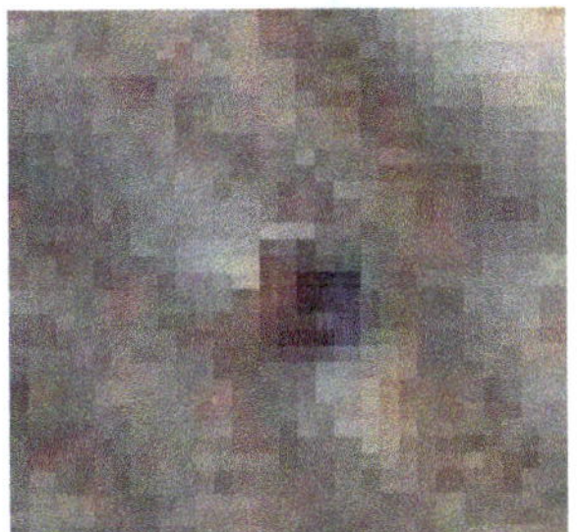

Random objects 9, Phobos.

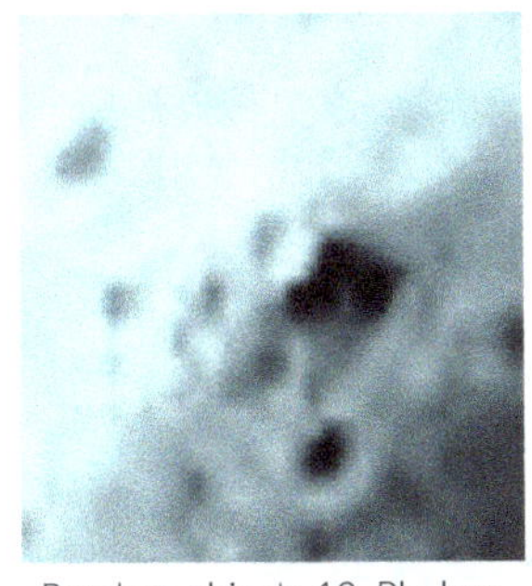

Random objects 10, Phobos.

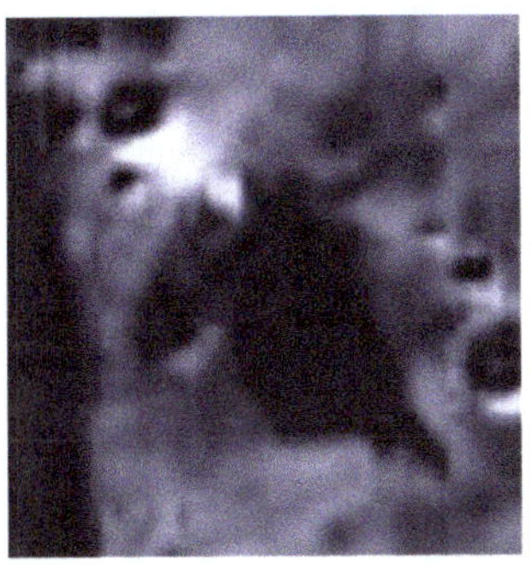

Random objects 11, Phobos.

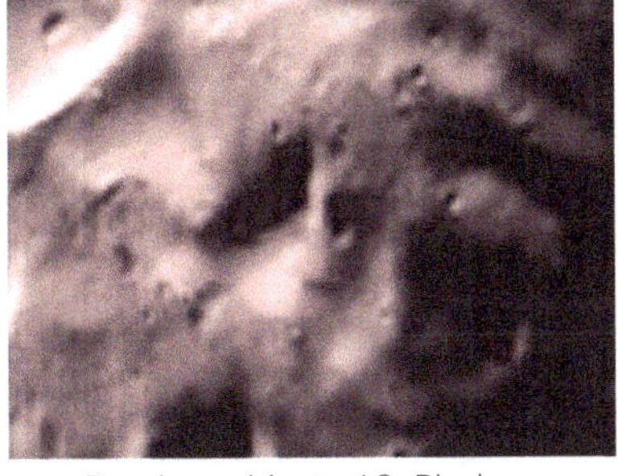

Random objects 12, Phobos.

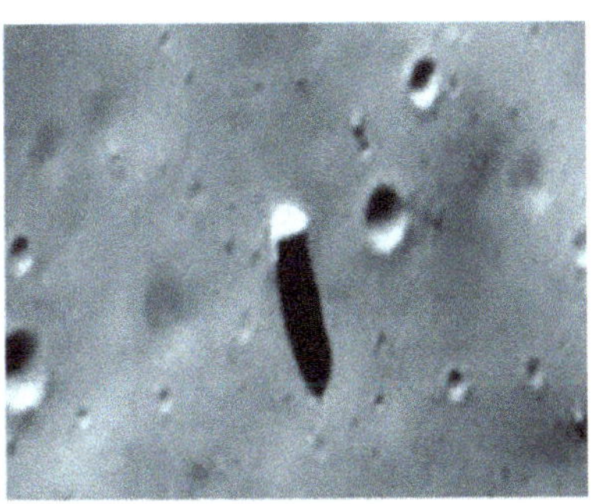

The roughly 300 foot tall Phobos "Monolith" (left): On July 19, 2009, Buzz Aldrin said of it: "There's a monolith [on Phobos]—a very unusual structure on this little potato shaped object that goes around Mars once every seven hours. When people find out about that they are going to say, 'Who put that there?'" Image credit: NASA, Mars Global Surveyor, 1998.

"It is far more absurd to think that there has never been intelligent life on Mars than to think there has." — LOCHLAINN SEABROOK

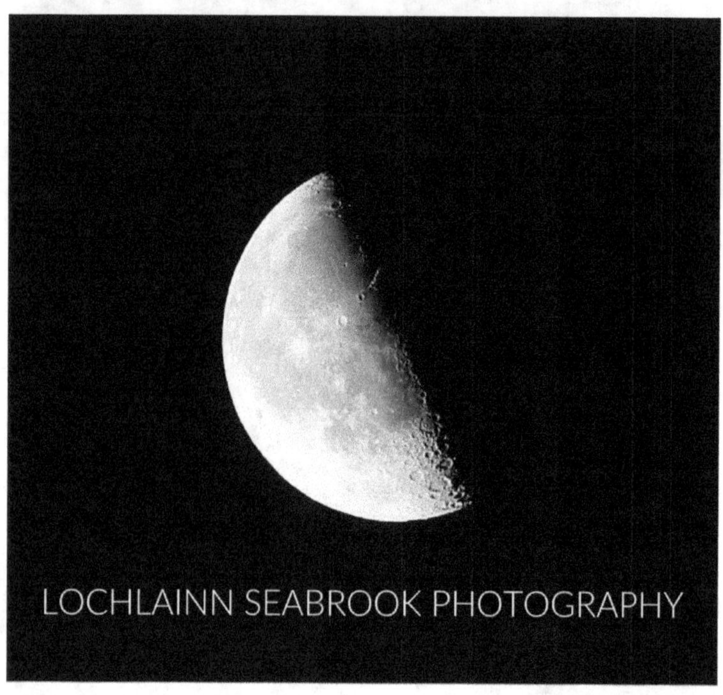

Above: Like Mars, our Moon, known to the ancient Romans as Luna and to the ancient Greeks as Selene, also bears overt evidence of probable extraterrestrial intelligence. As just one example, for centuries astronomers have been observing what are now called "transient lunar phenomenon" (TLP) or "lunar transient phenomenon" (LTP): static lights, moving lights, changing lights, mysterious traveling objects, colored lights, odd or moving shadows, and flashes of light (fire, explosions, weapons systems?) on and above the Moon's surface (the Aristarchus crater is one of many TLP "hot spots"). Not even NASA has been able to ignore the massive number of TLP reports that have been recorded—and continue to be generated. In response, in 1968 it published a 64 page document called "Chronological Catalog of Reported Lunar Events" (recorded between 1540 and 1967). One day the space agency may find itself publishing a similar document on the planet Mars. Image credit: Lochlainn Seabrook, taken January 25, 2022, Nashville, Tennessee, USA; Nikon, 100-400mm f/4.5-5.6 VR S lens. (NASA's TLP report can be found at: https://ntrs.nasa.gov/citations/19680018720.)

Appendices

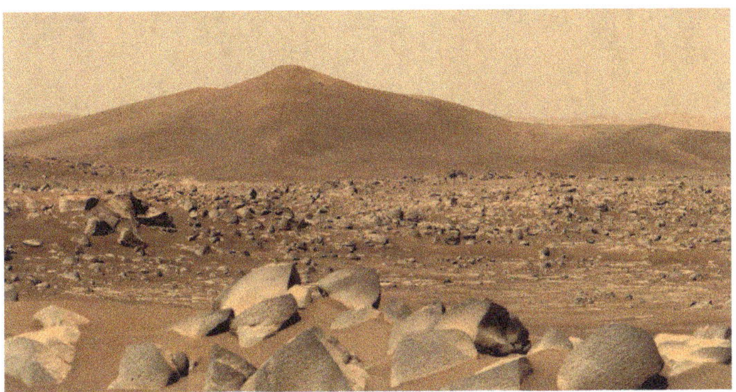

NASA's Mars rover Perseverance took this photo on April 29, 2021, from inside the Red Planet's Jezero crater. An interesting group of large anomalies is located at the 9:00 position. As of yet they remain positively unidentified. Image credit: NASA/JPL-Caltech/ASU/MSSS.

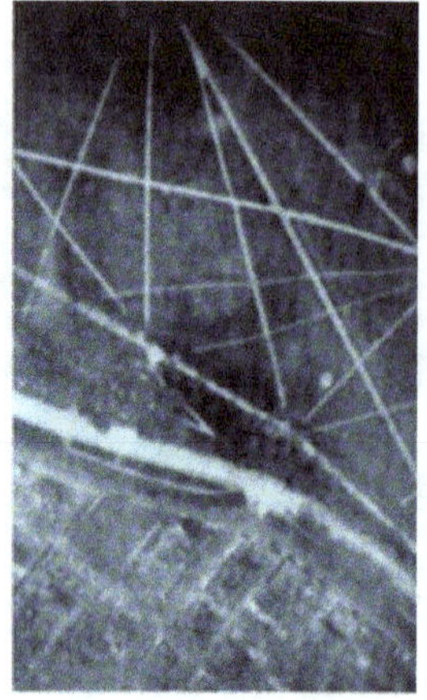

Left, an aerial photo of Hyde Park, London, UK, 1908: In the early 1900s astronomer Percival Lowell saw an analogy between the lines laid out at the famous English common and the lines on Mars. According to orthodox science, Lowell's "Martian geometry" is nothing more than Martian geology.

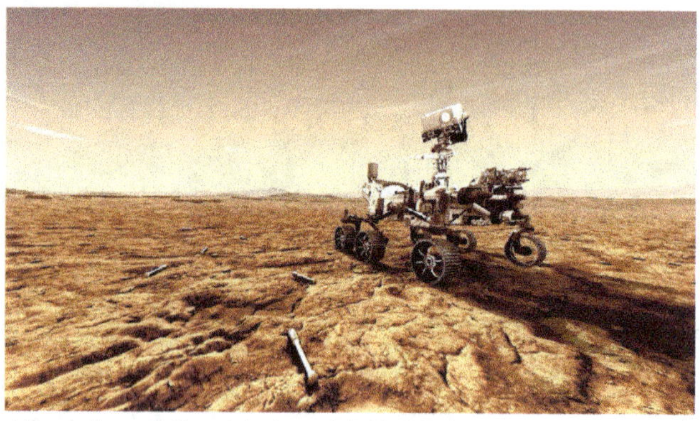

Above: An artist's rendering of the NASA rover Perseverance as it explores Mars. Note the rock and soil sample tubes it uses to collect geological specimens. Image credit: NASA/JPL-Caltech.

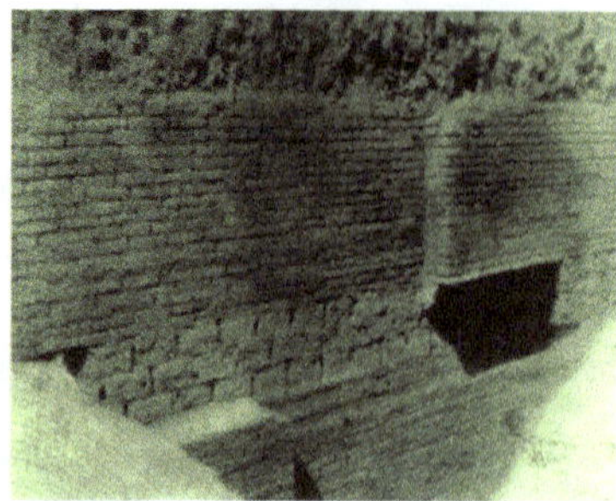

Above: This wall of an ancient Babylonian ziggurat bears numerous similar traits to the many brick wall like structures on Mars.

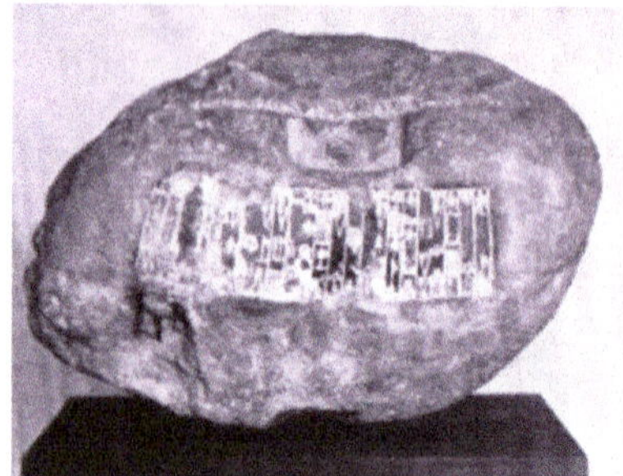

Above: An engraved, 4,500 year old rock found in the ruins of Ur, ancient Sumer. Mars hosts many similar objects. Are they connected?

APPENDIX A

SEABROOK'S BASIC MARS FACT SHEET

COMPILED BY AUTHOR/HISTORIAN LOCHLAINN SEABROOK

☛ Named After: Mars, the Roman god of war, who is often associated with red—the color of blood.

☛ Age of Mars: Around 4.5 billion years (roughly the same as Earth).

☛ Weight: 0.642 x 10^{24} kg; or 1,408,000,000,000,000,000,000,000 pounds (642,000,000 trillion metric tons).

☛ Diameter: 4,220 miles, making it the third largest and the second smallest planet in our solar system. For comparison Earth is 7,926 miles in diameter, our Moon is 2,159 miles in diameter. (The total dry land area of Earth is similar to the total amount of dry land on Mars.)

☛ Radius of Mars: 2,106 miles (3,390 km).

☛ Circumference of Mars: About 13,300 miles (21,342 km).

☛ Shape Type of Mars: Oblate spheroid.

☛ Volume of Mars: 1.6 x 10^{11} cubic km—15% of Earth's volume (meaning, six Mars could fit inside of one Earth).

☛ Mass of Mars: 642 sextillion kg, or about one-tenth the mass of Earth.

☛ Density of Mars: 3.9 q/cm^3, or around 71% as dense as the Earth.

☛ Average Distance from the Sun: 142 million miles (229 million km), or 1.5 AUs.

☛ Average Orbital Speed: 14.5 miles per second, or 52,200 miles an hour.

☛ Length of a Martian Day: 24 hours and 37 mins., or 24.6 hours, the length of time it takes Mars to completely rotate on its own axis.

☛ Name of a Martian Day: Sol, an abbreviation of "solar day."

☛ Length of a Martian Year: 669.6 sols, or 687 Earth days (the length of time it takes Mars to completely circle the Sun)—nearly twice as long as an Earth year.

☛ Tilt of Axis: 25 degrees (almost the same as Earth, 23.5 degrees).

☛ Gravity on Mars: Weak, 37.6% less than on Earth. (Thus if you weigh 100 lbs on Earth you would weigh about 38 lbs on Mars.)

☛ Type of Planet: Terrestrial, due to having a solid rocky surface.

- ☛ Type of Structure: Unknown, but believed to be similar to Earth; that is, with a crust, a rocky mantle, a liquid outer core, and a solid core.
- ☛ Main Soil Components: Iron, nickel, and sulfur, along with magnesium, calcium, potassium, and aluminum.
- ☛ Number of Mars' Moons: Two asymmetrical bodies that may actually be captured asteroids: Phobos (13.8 miles in diameter) and Deimos (7.8 miles in diameter); named after the two horses that pull the chariot of the god Mars.
- ☛ Time it Takes Light to Travel From the Sun to Mars: 12.6 mins.
- ☛ Position in Our Solar System: The fourth planet from the Sun. Earth is on one side of Mars (closer to the Sun) while Jupiter is on the other side of Mars (further away from the Sun).
- ☛ Weather on Mars: Has four seasons (spring, summer, fall, winter), strong winds, and massive global dust storms (known as "planet-encircling dust events") that can last for many months.
- ☛ Temperature: Winter average: -148°F (-100°C); summer average: 68°F (20°C). Overall average: -81°F (-62.7°C).
- ☛ Polar Caps?: Yes, with north and south poles covered in dry ice (frozen carbon dioxide).
- ☛ Rings Around Mars?: No.
- ☛ Water?: Yes, in the past; and probably today—mainly locked up in ice and clouds and some of it underground. Additionally, salty water has recently been discovered seeping out of hills and crater walls. Past water systems included streams, rivers, lakes, and seas. Early in Mars development large floods may have been common. (Note: Mars possesses certain types of surface rocks that could have only been created in liquid water.)
- ☛ Topography: Harsh, cold rocky plains and rugged desert, with channels, craters, canyons, ridges, rocky outcroppings, scree, hills, mountains, and dormant volcanoes. Inhospitable to humans.
- ☛ Atmosphere on Mars: Thin and comprised mostly of carbon dioxide (95.9%), as well as nitrogen (2%), argon (2%), and traces of water vapor and oxygen. (Note: While Mars seems to have had a magnetosphere, or global magnetic field, during its early development, it is absent today, for reasons unknown—but which

could be related to observations made by the U.S. government remote viewer discussed in my introduction.)

☛ Cause of Reddish Color: The rusting of iron minerals in the soil. (Note: Mars also sports other colors, such as varieties of gold, brown, tan, yellow, white, black, and perhaps blue and green.) Dust storms spread its iron rich soil, giving Mars its overall reddish hue and atmosphere.

☛ Luminosity of Mars: 1.523 AUs.

☛ Visible With the Naked Eye?: Yes, both at night and, under the right conditions, during the day—primarily when it is closest to Earth.

☛ Time It Takes Us to Reach Mars: About 8 months (currently).

☛ Unique Attributes:

- Mars is one of the most heavily explored bodies in our solar system. Despite this, there is still far more that we do not know about Mars than what we do know.
- Mars possesses the largest volcano and tallest mountain in our solar system: Olympus Mons (inactive). It is 16 miles in height (three times as tall as Mt. Everest) and about 372 miles wide at its base (the size of the state of New Mexico).
- Mars possesses the biggest canyon in our solar system: Valles Marineris. It runs for some 2,500 miles over the surface of Mars, nearly the span of the entire United States of America. At its widest it is 200 miles across while at its deepest it plunges to 4.3 miles in depth. This makes Valles Marineris nearly 10 times the size of the Grand Canyon.
- The largest crater on Mars is Borealis Basin. With a diameter of 5,300 miles, it extends over nearly 40% of the Martian surface.
- Besides Earth, Mars is the most likely to have life (past and/or present) of any planet in our solar system. This book provides highly compelling circumstantial evidence of this fact.

* For more information on some of these stats, see Website: https://mars.nasa.gov/all-about-mars/facts/

APPENDIX B

NASA'S SCIENTIFIC MARS FACT SHEET

by Author/Curator Dr. David R. Williams
NASA Goddard Space Flight Center
(Current as of December 23, 2021)

Mars/Earth Comparison

Bulk Parameters

	MARS	EARTH	RATIO (MARS/EARTH)
Mass (10^{24} kg)	0.64169	5.9722	0.107
Volume (10^{10} km^3)	16.318	108.321	0.151
Equatorial radius (km)	3396.2	6378.1	0.532
Polar radius (km)	3376.2	6356.8	0.531
Volumetric mean radius (km)	3389.5	6371.0	0.532
Core radius (km)	1700	3485	0.488
Ellipticity (Flattening)	0.00589	0.00335	1.76
Mean density (kg/m^3)	3934	5513	0.714
Surface gravity (m/s^2)	3.71	9.80	0.379
Surface acceleration (m/s^2)	3.69	9.78	0.377
Escape velocity (km/s)	5.03	11.19	0.450
GM (x 10^6 km^3/s^2)	0.042828	0.39860	0.107
Bond albedo	0.250	0.306	0.817
Geometric albedo	0.170	0.434	0.392
V-band magnitude V(1,0)	-1.60	-3.99	-
Solar irradiance (W/m^2)	586.2	1361.0	0.431
Black-body temperature (K)	209.8	254.0	0.826
Topographic range (km)	30	20	1.500
Moment of inertia (I/MR^2)	0.366	0.3308	1.106
J_2 (x 10^{-6})	1960.45	1082.63	1.811
Number of natural satellites	2	1	
Planetary ring system	No	No	

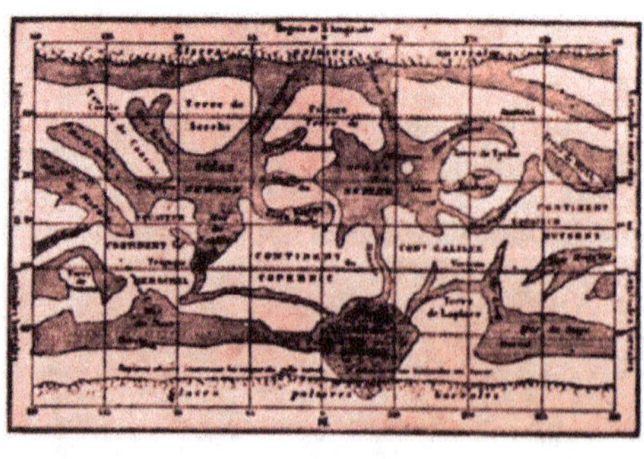

Left: In 1876, long before the dawn of high resolution telescopes, French astronomer Nicolas Camille Flammarion drew this now primitive map of Mars.

Orbital Parameters

	MARS	EARTH	RATIO (MARS/EARTH)
Semimajor axis (10^6 km)	227.956	149.598	1.524
Sidereal orbit period (days)	686.980	365.256	1.881
Tropical orbit period (days)	686.973	365.242	1.881
Perihelion (10^6 km)	206.650	147.095	1.405
Aphelion (10^6 km)	249.261	152.100	1.639
Synodic period (days)	779.94	-	-
Mean orbital velocity (km/s)	24.07	29.78	0.808
Max. orbital velocity (km/s)	26.50	30.29	0.875
Min. orbital velocity (km/s)	21.97	29.29	0.750
Orbit inclination (deg)	1.848	0.000	-
Orbit eccentricity	0.0935	0.0167	5.599
Sidereal rotation period (hrs)	24.6229	23.9345	1.029
Length of day (hrs)	24.6597	24.0000	1.027
Obliquity to orbit (deg)	25.19	23.44	1.075
Inclination of equator (deg)	25.19	23.44	1.075

Mars Observational Parameters

Discoverer: Unknown
Discovery Date: Prehistoric
Distance from Earth
    Minimum (10^6 km): 54.6
    Maximum (10^6 km): 401.4
Apparent diameter from Earth
    Maximum (seconds of arc): 25.6
    Minimum (seconds of arc): 3.5
Mean values at opposition from Earth
    Distance from Earth (10^6 km): 78.34
    Apparent diameter (seconds of arc): 17.8
    Apparent visual magnitude: -2.0
Maximum apparent visual magnitude: -2.94

Mars Mean Orbital Elements (J2000)

Semimajor axis (AU): 1.52366231
Orbital eccentricity: 0.09341233
Orbital inclination (deg): 1.85061
Longitude of ascending node (deg): 49.57854
Longitude of perihelion (deg): 336.04084
Mean Longitude (deg): 355.45332

North Pole of Rotation

Right Ascension: 317.681 - 0.106T
Declination: 52.887 - 0.061T
Reference Date: 12:00 UT 1 Jan 2000 (JD 2451545.0)
T = Julian centuries from reference date

Martian Atmosphere

Surface pressure: 6.36 mb at mean radius (variable from 4.0 to 8.7 mb depending on season) [6.9 mb to 9 mb (Viking 1 Lander site)]
Surface density: ~0.020 kg/m^3
Scale height: 11.1 km
Total mass of atmosphere: ~2.5 x 10^{16} kg
Average temperature: ~210 K (-63 C)
Diurnal temperature range: 184 K to 242 K (-89 to -31 C) (Viking 1 Lander site)
Wind speeds: 2-7 m/s (summer), 5-10 m/s (fall), 17-30 m/s (dust storm) (Viking Lander sites)
Mean molecular weight: 43.34
Atmospheric composition (by volume):
Major: Carbon Dioxide (CO_2) - 95.1%; Nitrogen (N_2) - 2.59%
Argon (Ar) - 1.94%; Oxygen (O_2) - 0.16%; Carbon Monoxide (CO) - 0.06%
Minor (ppm): Water (H_2O) - 210; Nitrogen Oxide (NO) - 100; Neon (Ne) - 2.5; Hydrogen-Deuterium-Oxygen (HDO) - 0.85; Krypton (Kr) - 0.3; Xenon (Xe) - 0.08

Satellites of Mars

	Phobos	Deimos
Semimajor axis* (km)	9378	23459
Sidereal orbit period (days)	0.31891	1.26244
Sidereal rotation period (days)	0.31891	1.26244
Orbital inclination (deg)	1.08	1.79
Orbital eccentricity	0.0151	0.0005
Subplanetary axis radius (km)	13.0	7.8
Along-orbit axis radius (km)	11.4	6.0
Polar axis radius (km)	9.1	5.1
Mass (10^{15} kg)	10.6	2.4
Mean density (kg/m^3)	1900	1750
Geometric albedo	0.07	0.08
Visual magnitude V(1,0)	+11.8	+12.89
Apparent visual magnitude (V_0)	11.3	12.40

*Mean orbital distance from the center of Mars.

Source: https://nssdc.gsfc.nasa.gov/planetary/factsheet/marsfact.html

APPENDIX C

Genuine Natural Phenomena Found on Mars

A FEW EXAMPLES

- Craters
- Lava flows
- Mudstone
- Sandstone
- Shale
- Ridges
- Highlands
- Lowlands
- Canyons
- Valleys
- Scarps
- Polar ice caps
- Clay minerals
- Wind-borne deposits
- Conglomerate rocks
- Sulfate minerals
- Sedimentary rock
- Graben
- Erosion
- Plains
- Basins
- Silicon dioxide dust
- Columnar basalt
- Meteorite (iron-nickel)
- Scoria
- Faults
- Mesas
- Tectonic deposits
- Volcanoes
- Sand dunes
- Ejecta rings (around impact craters)
- Crustal deformation

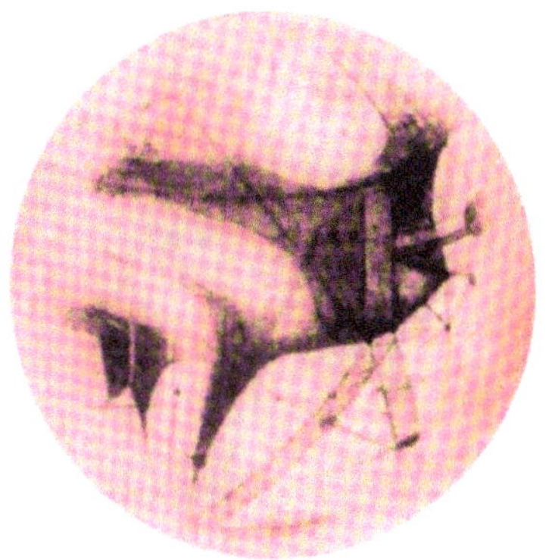

Another early drawing of Mars by Percival Lowell.

APPENDIX D

THE PENTAGON'S UFO INVESTIGATIVE GROUP

PENTAGON ANNOUNCES FORMATION OF GROUP TO INVESTIGATE UFOS

IMMEDIATE RELEASE
DoD Announces the Establishment of the Airborne Object Identification and Management Synchronization Group (AOIMSG)
NOV. 23, 2021

Today, Deputy Secretary of Defense Kathleen Hicks, in close collaboration with the Director of National Intelligence, directed the Under Secretary of Defense for Intelligence & Security to establish within the Office of the USD (I&S) the Airborne Object Identification and Management Synchronization Group (AOIMSG) as the successor to the U.S. Navy's Unidentified Aerial Phenomena Task Force. The AOIMSG will synchronize efforts across the Department and the broader U.S. government to detect, identify and attribute objects of interests in Special Use Airspace (SUA), and to assess and mitigate any associated threats to safety of flight and national security. To provide oversight of the AOIMSG, the Deputy Secretary also directed the USD (I&S) to lead an Airborne Object Identification and Management Executive Council (AOIMEXEC) to be comprised of DoD and Intelligence Community membership, and to offer a venue for U.S. government interagency representation.

Incursions by any airborne object into our SUA pose safety of flight and operations security concerns, and may pose national security challenges. DOD takes reports of incursions—by any airborne object, identified or unidentified—very seriously, and investigates each one. This decision is the result of planning efforts and collaboration conducted by OUSD (I&S) and other DoD elements at the direction of Deputy Secretary Hicks, to address the challenges associated with assessing UAP occurring on or near DOD training ranges and installations highlighted in the DNI preliminary assessment report submitted to Congress in June 2021. The report also identified the need to make improvements in processes, policies, technologies, and training to improve our ability to understand UAP.

In coming weeks, the Department will issue implementing guidance, which will contain further details on the AOIMSG Director, organizational structure, authorities, and resourcing.

Source: www.defense.gov/News/Releases/Release/Article/2853121
See also: www.dni.gov/files/ODNI/documents/assessments/Prelimary-Assessment-UAP-20210625.pdf

BIBLIOGRAPHY

And Suggested Reading

Belzoni, Giovanni. *Narrative of the Operations and Recent Discoveries Within the Pyramids, Temples, Tombs, and Excavations, in Egypt and Nubia*. London, UK: John Murray, 1829.

Besant, Annie. *Thought Power: Its Control and Culture*. London, UK: The Theosophical Publishing Society, 1901.

Bird, James Malcolm (ed.). *Einstein's Theories of Relativity and Gravitation*. New York: Scientific American Publishing Co., 1921.

Brinkley, John. *Brinkley's Astronomy*. Dublin, Ireland: Hodges, Foster, and Co., 1871.

Bristol, Claude M. *The Magic of Believing*. 1948. New York: Simon and Schuster, 1969 ed.

Capra, Fritjof. *The Tao of Physics: An Exploration of the Parallels Between Modern Physics and Eastern Mysticism*. New York: Bantam, 1977.

Dallmann, William. *Jesus: His Words and His Works According to the Four Gospels*. Milwaukee, WI: Northwestern Publishing House, 1914.

Department of the Interior. *Geological Survey Professional Paper: Shorter Contributions to General Geology*. (Issues 165-167.) Washington, D.C.: United States Government Printing Office, 1930.

Einstein, Albert. *Relativity: The Special and General Theory*. New York: Henry Holt and Co., 1920.

——. *The Meaning of Relativity: Four Lectures Delivered at Princeton University, May, 1921*. Princeton, NJ: Princeton University Press, 1923.

Hall, James. *Key to a Chart of the Successive Geological Formations*. Boston, MA: Gould and Lincoln, 1852.

Hartmann, W. K., and O. Raper (NASA). *The New Mars*. Washington, D.C.: United States Government Printing Office, 1974.

Hudson, Thomas Jay. *The Law of Psychic Phenomenon: A Working Hypothesis for the Systemic Study of Hypnotism, Spiritism, Mental Therapeutics, Etc.* 1893. Chicago, IL: A. C. McClurg and Co., 1905 ed.

James, George Wharton. *In and Around the Grand Canyon: The Grand Canyon of the Colorado River in Arizona*. Boston, MA: Little, Brown, and Co., 1900.

Kaku, Michio. *The Future of the Mind: The Scientific Quest to Understand, Enhance, and Empower the Mind*. New York: Doubleday, 2014.

——. *The Future of Humanity: Terraforming Mars, Interstellar Travel, Immortality, and Our Destiny Beyond Earth*. New York: Doubleday, 2018.

Kenrick, John. *Ancient Egypt Under the Pharaohs*. Clinton Hall, NY: Redfield, 1852.

Lawrence, Byrem. *A Concise Description of the Geological Formations and Mineral Localities of the Western States*. Boston, MA: Samuel N. Dickinson, 1843.

Leonard, Louise. *Percival Lowell: An Afterglow*. Boston, MA: The Gorham Press, 1921.

Love, John D. *Split Rock Formation (Miocene) and Moonstone Formation (Pliocene) in Central Wyoming*. Washington, D.C.: United States Government Printing Office, 1961.

Lowell, Percival. *Mars*. Boston, MA: Houghton, Mifflin and Co., 1895.

——. *Mars and Its Canals*. New York: Macmillan Co., 1906.

——. *The Evolution of Worlds*. New York: Macmillan Co., 1909.

——. *Mars as the Abode of Life*. New York: Macmillan Co., 1910.

MacPherson, Hector. *Astronomers of Today and Their Work*. Edinburgh, Scotland: Gall and Inglis, 1905.

Martin, Thomas Commerford. *The Inventions, Researches and Writings of Nikola Tesla*. New York: The Electrical Engineer, 1894.

National Aeronautics and Space Administration (NASA). *Mars as a Planet: One of a Series of NASA Facts About the Exploration of Mars*. Washington, D.C.: United States Government Printing Office, 1976.

Osborn, Henry Fairfield. *Men of the Stone Age: Their Environment, Life and Art*. New York: Charles Scribner's Sons, 1923.

Osburn, William. *The Monumental History of Egypt, as Recorded on the Ruins of Her Temples, Palaces, and Tombs*. London, UK: Trübner and Co., 1854.

Peters, John Punnett. *Nippur or Explorations and Adventures of the Euphrates: The Narrative of the University of Pennsylvania Expedition to Babylonia in the Years 1888-1890*. New York: G. P. Putnam's Sons, 1897.

Petrie, Willam M. F. *The Pyramids and Temples of Gizeh*. London, UK: Field and Tuer, 1883.

Rhine, Joseph Banks. *Extra-Sensory Perception After Sixty years: A Critical Appraisal of the Research in Extra-Sensory Perception*. 1940. Boston, MA: Bruce Humphries, 1966 ed.

Ruins of Sacred and Historic Lands. (No author.) London, UK: Thomas Nelson, 1852.

Seabrook, Lochlainn. *Britannia Rules: An Academic Look at the United Kingdom's Matricentric Spiritual Past*. 1999. Franklin, TN: Sea Raven Press, 2020 ed.

——. *UFOs and Aliens: The Complete Guidebook*. 2005. Franklin, TN: Sea Raven Press, 2015 ed.

——. *Christmas Before Christianity: How the Birthday of the "Sun" Became the Birthday of the "Son."* 2010. Franklin, TN: Sea Raven Press, 2018 ed.

——. *Jesus and the Law of Attraction: The Bible-Based Guide to Creating Perfect Health, Wealth, and Happiness Following Christ's Simple Formula*. Franklin, TN: Sea Raven Press, 2013.

——. *The Bible and the Law of Attraction: 99 Teachings of Jesus, the Apostles, and the Prophets*. Franklin, TN: Sea Raven Press, 2013.

——. *Christ Is All and In All: Rediscovering Your Divine Nature and the Kingdom Within*. Franklin, TN: Sea Raven Press, 2014.

——. *Seabrook's Bible Dictionary of Traditional and Mystical Christian Doctrines*. Spring Hill, TN: Sea Raven Press, 2016.

Seiss, Joseph A. *A Miracle in Stone: Or the Great Pyramid of Egypt*. Philadelphia, PA: Porter and Coates, 1877.

Stuart, Villiers. *Nile Gleanings Concerning the Ethnology, History and Art of Ancient Egypt as Revealed by Egyptian Paintings, and Bas-Reliefs*. London, UK: John Murray, 1879.

Talbot, Michael. *The Holographic Universe*. New York: Harper Perennial, 1991.

Vyse, Howard. *Operations Carried on at the Pyramids of Gizeh in 1837*. London, UK: James Fraser, 1840.

Wallace, Alfred Russel. *Is Mars Habitable?: A Critical Examination of Professor Percival Lowell's Book "Mars and Its Canals," With an Alternative Explanation*. London, UK: Macmillan and Co., 1907.

Weiss, Sara. *Journeys to the Planet Mars or Our Mission to Ento (Mars)*. Rochester, NY: Austin Publishing Co., 1905.

Wilkinson, Gardner. *The Manners and Customs of the Ancient Egyptians*. 5 vols. London, UK: John Murray, 1847.

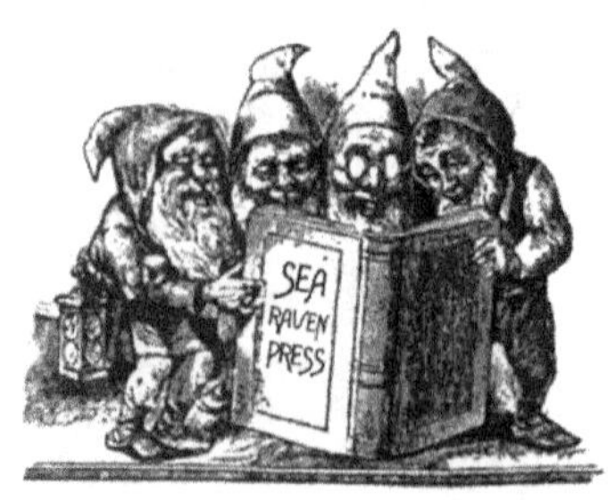

MEET THE AUTHOR

NEO-VICTORIAN SCHOLAR LOCHLAINN SEABROOK, a descendant of the families of Alexander Hamilton Stephens, John Singleton Mosby, Edmund Winchester Rucker, and William Giles Harding, is a 7th generation Kentuckian and one of the most prolific and widely read writers in the world today. Known by literary critics as the "new Shelby Foote" and the "American Robert Graves," and by his fans as the "Voice of the Traditional South," he is a recipient of the prestigious Jefferson Davis Historical Gold Medal. As a lifelong writer he has authored and edited books ranging in topics from history, politics, science, religion, astronomy, and biography, to nature, music, humor, gastronomy, genealogy, and the paranormal; books that his readers describe as "game changers," "transformative," and "life altering."

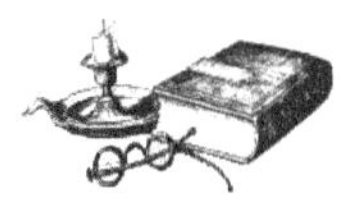

One of the world's most popular living historians, he is a 17th generation Southerner of Appalachian heritage who descends from dozens of patriotic Revolutionary War soldiers and Confederate soldiers from Kentucky, Tennessee, North Carolina, and Virginia. Also a history, wildlife, and nature preservationist, he began life as a child prodigy, later transforming into an archetypal Renaissance Man. Besides being an accomplished and well respected author-historian and Bible authority, he is also a Kentucky Colonel, eagle scout, screenwriter, nature, wildlife, and landscape photographer, artist, graphic designer, songwriter (3,000 songs), film composer, multi-instrument musician, vocalist, session player, music producer, genealogist, former history museum docent, and a former ranch hand, zookeeper, and wrangler.

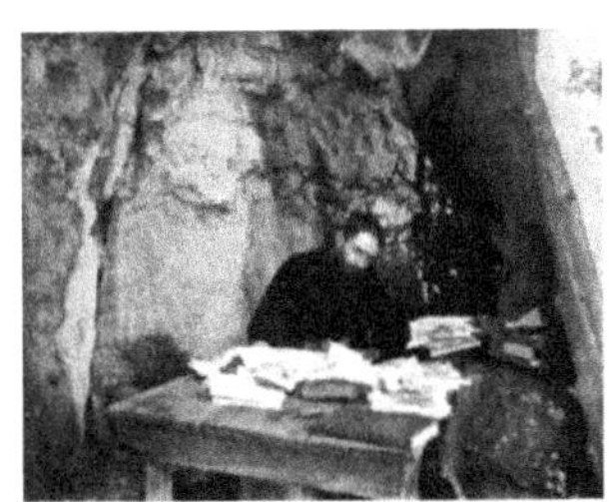

Currently Seabrook is the author and editor of 80 adult and children's books (a total of 12,562,500 words) that have earned him accolades from around the globe. His works, which have sold on every continent except Antarctica, have introduced hundreds of thousands to vital facts that have been left out of our mainstream books. He has been endorsed internationally by leading experts, museum curators, award-winning historians, bestselling authors, celebrities, filmmakers, noted scientists, well regarded educators, TV show hosts and producers, renowned military artists, esteemed heritage organizations, and distinguished academicians of all races, creeds, and colors.

Of northern, western, and central European ancestry, he is the 6th great-grandson of the Earl of Oxford and a descendant of European royalty. His modern day cousins include: Johnny Cash, Elvis Presley, Lisa Marie Presley, Billy Ray and Miley Cyrus, Patty Loveless, Tim McGraw, Lee Ann Womack, Dolly Parton, Pat Boone, Naomi, Wynonna, and Ashley Judd, Ricky Skaggs, the Sunshine Sisters, Martha Carson, Chet Atkins, Patrick J. Buchanan, Cindy Crawford, Bertram Thomas Combs (Kentucky's 50th governor), Edith Bolling (second wife of President Woodrow Wilson), Andy Griffith, Riley Keough, George C. Scott, Robert Duvall, Reese Witherspoon, Lee Marvin, Rebecca Gayheart, and Tom Cruise.

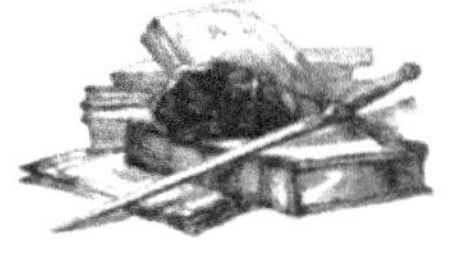

A constitutionalist and avid outdoorsman and gun advocate, Seabrook is the author of the international blockbuster, *Everything You Were Taught About the Civil War is Wrong, Ask a Southerner!* He lives with his wife and family in beautiful historic Middle Tennessee, the heart of the Old South.

For more information on author Mr. Seabrook visit

LOCHLAINNSEABROOK.COM

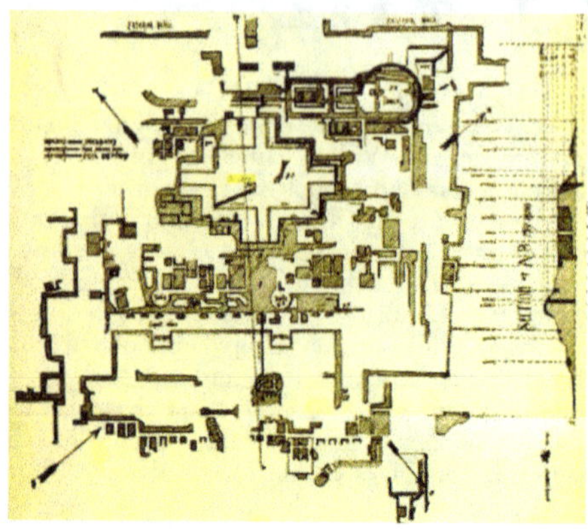

Above: Original caption—"Plan of Temple of Bel, at Nippur; after the Excavation of the Babylonian Expedition; End of Season, May 1890." There are countless similarities between this map and the "layout" of the alleged cityscapes on Mars.

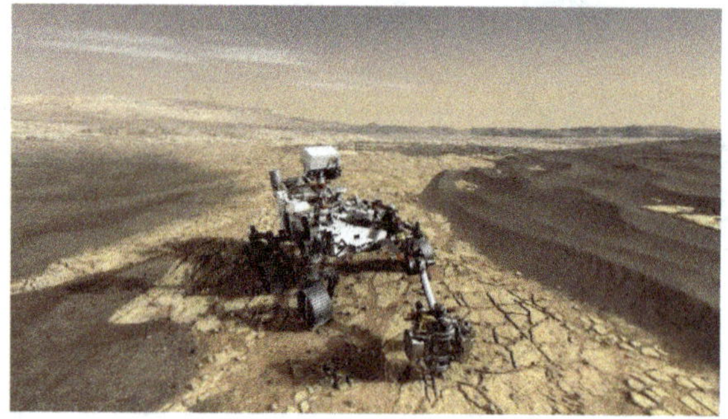

Above: Another artist's concept of the NASA rover Perseverance on Mars. Image credit: NASA/JPL-Caltech.

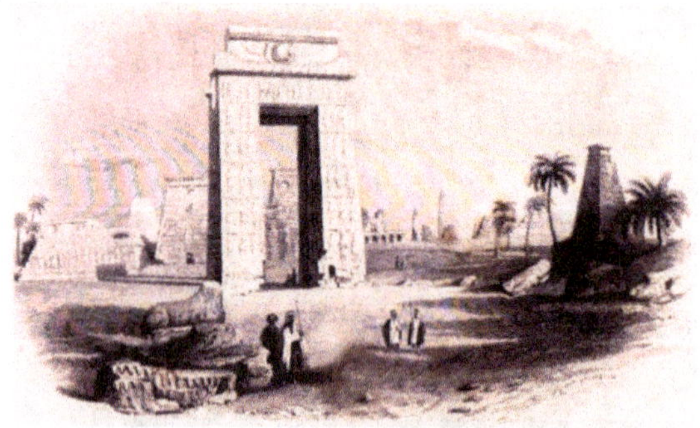

Above: The ruins of Karnak, near Luxor, Egypt. Mars may have its own Karnak.

www.ingramcontent.com/pod-product-compliance
Lightning Source LLC
LaVergne TN
LVHW020053110826
845155LV00022B/76

* 9 7 8 1 9 5 5 3 5 1 1 4 0 *